STRUCTURED SURFACES AS OPTICAL METAMATERIALS

Optical metamaterials are an exciting new field in optical science. A rapidly developing class of these metamaterials allow the manipulation of volume and surface electromagnetic waves in desirable ways by suitably structuring the surfaces they interact with. They have applications in a variety of fields, such as materials science, photovoltaic technology, imaging and lensing, beam shaping, and lasing.

Describing techniques and applications, this book is ideal for researchers and professionals working in metamaterials and plasmonics, as well as for those just entering this exciting new field. It surveys different types of structured surfaces, their design and fabrication, their unusual optical properties, recent experimental observations, and their applications. Each chapter is written by an expert in that area, giving the reader an up-to-date overview of the subject. Both the experimental and theoretical aspects of each topic are presented.

ALEXEI A. MARADUDIN is a Research Professor in the Department of Physics and Astronomy, at the University of California, Irvine. His research interests have included lattice dynamics of perfect and imperfect crystals; surface excitations on perfect and imperfect elastic, dielectric, and magnetic media; and the scattering of light from elementary excitations in solids.

STRUCTURED SURFACES AS OPTICAL METAMATERIALS

Edited by

ALEXEI A. MARADUDIN
University of California, Irvine

CAMBRIDGE
UNIVERSITY PRESS

CAMBRIDGE
UNIVERSITY PRESS

University Printing House, Cambridge CB2 8BS, United Kingdom

One Liberty Plaza, 20th Floor, New York, NY 10006, USA

477 Williamstown Road, Port Melbourne, VIC 3207, Australia

314-321, 3rd Floor, Plot 3, Splendor Forum, Jasola District Centre, New Delhi - 110025, India

103 Penang Road, #05-06/07, Visioncrest Commercial, Singapore 238467

Cambridge University Press is part of the University of Cambridge.

It furthers the University's mission by disseminating knowledge in the pursuit of education, learning and research at the highest international levels of excellence.

www.cambridge.org
Information on this title: www.cambridge.org/9780521119610

First published 2011

A catalogue record for this publication is available from the British Library

Library of Congress Cataloging in Publication data
Structured surfaces as optical metamaterials / [edited by] Alexei A. Maradudin.
p. cm.
Includes bibliographical references and index.
ISBN 978-0-521-11961-0 (hardback)
1. Optical materials. 2. Surfaces (Technology) 3. Shapes. I. Maradudin, A. A.
QC374.S77 2011
621.36 – dc22 2010054603

ISBN 978-0-521-11961-0 Hardback

Contents

Contributors

Alú, Andrea
Department of Electrical and Computer Engineering, The University of Texas at Austin, Austin, TX 78712, USA.

Bonod, Nicolas
Institut Fresnel, Aix-Marseille Université, CNRS, Unité Mixte de Recherche 6133, Domaine Universitaire de Saint Jerome, 13397 Marseille Cedex 20, France.

Catrysse, Peter B.
E. L. Ginzton Laboratory, Stanford University, Stanford, CA 94305, USA.

Davis, Christopher C.
Department of Electrical and Computer Engineering, University of Maryland, College Park, MD 20742, USA.

Engheta, Nader
Department of Electrical and Systems Engineering, University of Pennsylvania, Philadelphia, PA 19104, USA.

Fan, Shanhui
E. L. Ginzton Laboratory, Stanford University, Stanford, CA 94305, USA.

Fernández-Domínguez, A. I.
Departamento de Fisica Teorica de la Materia Condensada, Universidad Autonoma de Madrid, E-28049 Madrid, Spain.

García-Vidal, F.
Departamento de Fisica Teorica de la Materia Condensada, Universidad
Autonoma de Madrid, E-28049 Madrid, Spain.

Grebel, Haim
Electronic Imaging Center, and the ECE Department at the New Jersey Institute
of Technology, Newark, NJ 07102, USA.

Izrailev, F. M.
Instituto de Física, Universidad Autónoma de Puebla, Apdo. Post. J-48, Puebla
72570, México.

Leskova, T. A.
Department of Physics and Astronomy and Institute for Surface and Interface
Science, University of California, Irvine, CA 92697 USA.

Lu, Wentao Trent
Department of Physics and Electronic Materials Research Institute, Northeastern
University, Boston, MA 02115, USA.

Maradudin, A. A.
Department of Physics and Astronomy and Institute for Surface and Interface
Science, University of California, Irvine, CA 92697 USA.

Makarov, N. M.
Instituto de Ciencias, Universidad Autónoma de Puebla, Priv. 17 Norte No. 3417,
Col. San Miguel Hueyotlipan, Puebla 72050, México.

Martín-Moreno, L.
Departamento de Fisica de la Materia Condensada, ICMA-CSIC, Universidad de
Zaragoza, E-500009 Zaragoza, Spain.

Méndez, E. R.
División de Física Aplicada, Centro de Investigación Científica y de Educación
Superior de Ensenada, Carretera Ensenada-Tijuana No. 3918, Ensenada, B. C.,
22860, México.

Plum, Eric
Optoelectronics Research Centre, University of Southampton, Southampton
SO17 1BJ, UK.

Popov, Evgeny

Institut Fresnel, Aix-Marseille Université, CNRS, Unité Mixte de Recherche 6133, Domaine Universitaire de Saint Jerome, 13397 Marseille Cedex 20, France.

Shin, Hocheol

E. L. Ginzton Laboratory, Stanford University, Stanford, CA 94305, USA, and Intel Corporation, Santa Clara, CA 95054, USA.

Smolyaninov, Igor I.

Department of Electrical and Computer Engineering, University of Maryland, College Park, MD 20742, USA.

Sridhar, Srinivas

Department of Physics and Electronic Materials Research Institute, Northeastern University, Boston, MA 02115, USA.

Teperik, Tatiana V.

Instituto de Optica, CSIC, Serrano 121, 28006 Madrid, Spain.

Zheludev, Nikolay I.

Optoelectronics Research Centre, University of Southampton, Southampton SO17 1BJ, UK.

Preface

If a metamaterial can be defined as a deliberately structured material that possesses physical properties that are not possible in naturally occurring materials, then deliberately structured surfaces that possess desirable optical properties that planar surfaces do not posses can surely be considered to be optical metamaterials. The surface structures displaying these properties can be periodic, deterministic but not periodic, or random.

In recent years interest has arisen in optical science in the study of such surfaces and the optical phenomena to which they give rise. A wide variety of these phenomena have been predicted theoretically and observed experimentally. They can be divided roughly into those in which volume electromagnetic waves participate and those in which surface electromagnetic waves participate. Both types of optical phenomena and the surface structures that produce them are described in this volume.

The first several chapters are devoted to optical interactions of volume electromagnetic waves with structured surfaces. One of the earliest examples of a structured surface that acts as an optical metamaterial, and the one that today is perhaps the best known and most widely studied, is a metal film pierced by a two-dimensional periodic array of holes with subwavelength diameters. It was shown experimentally by Ebbesen *et al.* [1] that the transmission of *p*-polarized light through this structure can be extraordinarily high at the wavelengths of the surface plasmon polaritons supported by the film. "Extraordinarily high" in this context refers to the observation that more than twice as much light is transmitted as impinges on the holes. This paper stimulated a great deal of theoretical and experimental work directed at elucidating the mechanism(s) responsible for the extraordinarily high transmissivity, and at enhancing it even more. In the first chapter, E. Popov and N. Bonod describe the theoretical and experimental studies of this phenomenon, whose explanation at times has been the subject of some controversy.

Not all optical enhancement effects occur in transmission through structured surfaces. In Chapter 2, T. V. Teperik discusses recent theoretical and experimental work on the diffraction of light from a two-dimensional periodic lattice of sub-micron voids (nanopores) situated beneath the surface of a metal in contact with vacuum. This kind of structure supports both dispersive surface plasmon polaritons at the vacuum–metal interface, and nondispersive void plasmons associated with each void. One of the interesting and important consequences of the existence of the latter type of excitation is the possibility of achieving omnidirectional total absorption of *p*- and *s*-polarized light of a specified wavelength incident on the structure when the voids are filled with a dielectric medium. Moreover, as a consequence of Kirchhoff's law, such a structured surface can also exhibit omnidirectional blackbody emission at a resonant frequency that can be varied by varying the radius of the dielectric-filled voids. Other interesting optical properties of nanoporous metal surfaces are also discussed in this chapter.

The reflection of an optical plane wave from, and its transmission through, yet another type of two-dimensional periodic planar structure is discussed by A. Alú and N. Engheta in Chapter 3. The structure considered is a dense planar array of nanoparticles, primarily metallic nanospheres whose diameter and periods are smaller than the wavelength of the illuminating electromagnetic field, that are treated in the dipolar approximation. The reflection and transmission spectra display features arising from the plasmonic resonances of the individual nanoparticles, and from the two-dimensional periodicity of the structure as a whole. It is shown that structures of this type offer the possibility of basing highly reflective and/or frequency-selective surfaces at optical frequencies on them, which can be used for filtering, absorption, and radiation purposes.

A planar metamaterial is a planar two-dimensional surface of zero thickness that is periodically structured on the sub-wavelength scale. In practice such a material is represented by a single periodically patterned metal or dielectric layer that is very thin compared to the wavelength of the light incident on it, and is often supported by a transparent substrate. In a comprehensive review in Chapter 4, E. Plum and N. Zheludev analyze polarization and propagation properties of these metamaterials on the basis of such general principles as symmetry, Lorentz reciprocity, and energy conservation. They show that suitably structured planar metamaterials can display circular birefringence and circular dichroism, linear birefringence and linear dichroism, as well as asymmetric transmission of circularly polarized light incident on them from opposite directions.

The ability to control the propagation of light is important for a variety of applications. In recent years a great interest has arisen in the negative refraction of light as it passes through the interface between two media. This interest is due to the fundamental importance of this effect, as well as to possible applications of it. For example, a perfect lens can be created on the basis of a medium that produces negative refraction, and sub-wavelength imaging can also be achieved by the use of such a medium. Negative refraction has been achieved in two types of materials. The first type is a metamaterial that possesses simultaneously a negative dielectric permittivity and a negative magnetic permeability within some frequency range [2]. Such a medium has a negative-index of refraction, and hence is often referred to as a negative-index material. The first material with these properties was fabricated by embedding arrays of split-ring resonators in a lattice of metal wires [3]. The second type of material is a nonmagnetic metamaterial with a positive dielectric permittivity. Such a material has a positive index of refraction, and is often referred to as a positive-index material. Photonic crystals formed from dielectric components can serve as positive-index materials that display negative refraction. One of the mechanisms responsible for negative refraction in such media is the presence in their photonic band structure of a surface of constant frequency with a negative group velocity in some frequency range [4]. In this case the Poynting vector of a wave packet is directed opposite to its wave vector, which leads to negative refraction [5]. The negative group velocity of circularly polarized electromagnetic waves of one handedness propagating in a gyrotropic medium also leads to negative refraction in certain frequency ranges [6].

The types of metamaterials just described are bulk materials. However, negative refraction of volume electromagnetic waves can also be achieved by the use of suitably structured surfaces. In a recent study, Lu *et al.* [7] showed that negative refraction can be achieved when light is incident from a dielectric medium with a real positive refractive index $n > 1$ on a periodically corrugated interface with air, at an angle of incidence θ_0 that is greater than the critical angle for total internal reflection, $\theta_0 > \theta_c = \sin^{-1}(1/n)$. In this situation the zeroth and all positive orders of the light refracted into the air are suppressed, and by a suitable choice of the period of the corrugation of the interface only the (-1)-order refracted beam is nonzero. This mechanism for negative refraction has been confirmed experimentally. These authors also show that by introducing the periodic surface not on a homogeneous semi-infinite dielectric medium but on a planar multilayered medium, the restriction $\theta_0 > \theta_c$ can be lifted. This prediction has also been verified experimentally. W. T. Lu and S. Sridhar review this work in Chapter 5, and present descriptions of several optical devices based on this approach to negative refraction.

A more general type of refraction is described by A. A. Maradudin *et al.* in Chapter 6, where it is shown how to design and fabricate a two-dimensional randomly rough surface that transforms a beam with a specified transverse intensity distribution into a beam with a different specified intensity distribution on its transmission through that surface. Such beam shaping is used in a variety of applications from laser surgery to optical scanning. In this chapter it is also shown how to design and fabricate a circularly symmetric but radially random surface that transforms a plane wave incident on it into a transmitted beam that does not spread over a finite distance along the symmetry axis of the structure from the surface – a pseudo-nondiffracting beam. Such beams can be used in precision alignment and in laser machining, for example.

The preceding examples of structured surfaces that act as optical metamaterials have all consisted of surfaces that are illuminated by volume electromagnetic waves. However, the propagation of surface electromagnetic waves, and even their existence, can be modified in specified desirable ways by structuring in suitable ways the surfaces on which they propagate. Similarly, novel applications of these waves can be realized by a suitable structuring of the surfaces supporting them.

For example, it has been known for some time that the planar surface of a semi-infinite perfect conductor does not support a surface electromagnetic wave. However, if a perfectly conducting surface is periodically corrugated, as in a classical grating, or is doubly periodically corrugated, as in a bigrating, it can support a surface electromagnetic wave. These theoretical predictions have recently been confirmed experimentally. The interesting properties of these surface waves, which owe their existence to the structuring of the surfaces on which they propagate, are described by A. I. Fernández-Domínguez *et al.* in Chapter 7.

As we have noted above, the negative refraction of volume electromagnetic waves has been studied theoretically and experimentally by many investigators, and several mechanisms for accomplishing such refraction have been explored, including the use of a periodically corrugated dielectric surface [7]. Recently, attention has been directed at the negative refraction of surface plasmon polaritons. Shin and Fan [8] proposed a metal–dielectric–metal structure that produces all-angle negative refraction of a surface plasmon polariton incident on it. The negative refraction they predicted is not due to the structure producing it possessing simultaneously a negative dielectric permittivity and a negative magnetic permeability in some frequency range. Instead it arises because each structure supports a surface plasmon polariton whose dispersion curve possesses a branch with an isotropic negative group velocity. It has been known for some time that the existence in a medium of an elementary excitation that possesses a negative group velocity within some frequency range is a sufficient condition for that medium to display in that frequency range the negative refraction of light incident on it with a frequency in that range [4, 5]. The theoretical and experimental aspects of the negative refraction of a surface plasmon polariton are presented in Chapter 8 by P. B. Catrysse *et al.*

There exists a commonly held belief that any randomness in a long one-dimensional conductor leads to an exponentially small transmission due to the Anderson localization of all of its eigenstates. However, the actual situation is subtler than this. It has been shown [9] that specific long-range correlations in a scattering potential give rise to perfect electron wave transmission within any given energy/frequency window. This result, which is known as selective transparency, was confirmed in experiments on a single-mode waveguide possessing this type of disorder. The experimental results clearly showed the mobility edges that separate regions of perfect transparency from those with localized transport. As F. M. Izrailev and N. M. Makarov point out in Chapter 9, these results suggested to them that similar results should be observed in single-mode or multimode planar waveguides with one of their surfaces randomly rough, when the rough surface profile function has long-range correlations of a specific type. These authors present results confirming their expectation for both single-mode and multimode waveguides.

A recently introduced class of metamaterials is one consisting of materials designed in such a way that an object embedded in one of them is cloaked from observation by electromagnetic waves propagating through the material. Perhaps the most commonly employed approach to the design of such cloaks is transformation optics [10, 11]. It predicts materials with dielectric permittivities and magnetic permeabilities that possess coordinate dependencies that deform the path of electromagnetic waves propagating in them to avoid spatial regions occupied by the objects to be cloaked. This approach to the cloaking of two- and three-dimensional objects, and other approaches that have been proposed, are reviewed by C. C. Davis and I. I. Smolyaninov in Chapter 10. They then show how the approach to the cloaking of two-dimensional objects in metamaterials designed by transformation optics can be extended to the design of surface structures that cloak surface defects from detection by surface plasmon polaritons, and produce the "trapped rainbow" effect for guided waves, in which a suitably designed plasmonic waveguide slows down and stops light of different wavelengths at different spatial points along the waveguide. Experimental results demonstrating both effects are presented.

In a planar waveguide consisting of a thin oxide layer sandwiched between an air superstrate and a metallic substrate the electric field intensity of the surface electromagnetic wave guided by this structure becomes a maximum at the interface between air and the oxide layer as the waveguide thickness is made extremely thin but finite. If the oxide layer is patterned with a periodic structure, e.g. by an array of holes, a standing electromagnetic surface wave can be formed. Such a standing wave enhances the interaction between a molecule placed on the air–oxide interface and the electromagnetic field of the surface wave. This enhanced interaction can be useful in surface-enhanced Raman spectroscopy, in the detection of molecules on a surface, and as a source for coherent radiation (lasers). These applications, and the physics underlying them, are described by H. Grebel in Chapter 11.

The chapters constituting this book present an up-to-date survey of many aspects of optical effects produced by structured surfaces. Yet, the topics covered in it do not exhaust the optical phenomena to which suitably structured surfaces can give rise. Indeed, they are limited only by our imagination. Nevertheless they provide a good indication of the variety of these phenomena, and the kinds of surfaces required for their realization, and help to indicate why this emerging field in optical science will continue to generate more research activity and applications in the future.

The editorial staff at the Cambridge University Press have my thanks for their help in producing this book. Special thanks are due to Ms. Irene Pizzie for her excellent copyediting of each manuscript.

I owe an enormous debt of gratitude to my colleague Dr. Tamara A. Leskova for the many hours spent in ensuring the correct formatting of the chapters, in helping to prepare the subject index, and in checking and correcting the references.

Finally, I wish to express my appreciation to the authors for the thought and care they put into preparing their contributions.

Irvine, California   *Alexei A. Maradudin*

References

[1] T. W. Ebbesen, H. J Lezec, H. F. Ghaemi, T. Thio, and P. A. Wolff, "Extraordinary optical transmission through sub-wavelength hole arrays," *Nature* **391**, 667–669 (1998).

[2] V. G. Veselago, "The electrodynamics of substances with simultaneously negative values of ϵ and μ," *Soviet Physics-Uspekhi* **10**, 509–514 (1968).

[3] R. A. Shelby, D. R. Smith, and S. Schultz, "Experimental verification of a negative index of refraction," *Science* **292**, 77–79 (2001).

[4] N. Notomi, "Theory of light propagation in strongly modulated photonic crystals: refractionlike behavior in the vicinity of the photonic band gap," *Phys. Rev. B* **62**, 10696–10705 (2000).

[5] L. I. Mandel'shtam, "Group velocity in crystalline arrays," *Zh. Eksp. Teor. Fiz.* **15**, 475–478 (1945).

[6] V. M. Agranovich, Yu. N. Gartstein, and A. A. Zakhidov, "Negative refraction in gyrotropic media," *Phys. Rev. B* **73**, 045114 (1-12) (2006).

[7] W. T. Lu, Y. J. Huang, P. Vodo, R. K. Banyal, V. H. Perry, and S. Sridhar, "A new mechanism for negative refraction and focusing using selective diffraction from surface corrugation," *Opt. Express* **15**, 9166–9175 (2007).

[8] H. Shin and S. Fan, "All-angle negative refraction for surface plasmon waves using a metal-dielectric-metal structure," *Phys. Rev. Lett.* **96**, 073907 (1-4) (2006).

[9] F. A. B. F. de Moura and M. L. Lyra, "Delocalization in the 1D Anderson model with long-range correlated disorder," *Phys. Rev. Lett.* **81**, 3735–3738 (1998).

[10] U. Leonhardt, "Optical conformal mapping," *Science* **312**, 1777–1780 (2006).

[11] J. B. Pendry, D. Schurig, and D. R. Smith, "Controlling electromagnetic fields," *Science* **312**, 1780–1782 (2006).

1

Physics of extraordinary transmission through subwavelength hole arrays

EVGENY POPOV AND NICOLAS BONOD

1.1 A brief reminder of the history of grating anomalies and plasmon surface waves

The recent history of the research and development around plasmon surface waves that was initiated by the work published in *Nature* in 1998 by Ebbesen *et al.* [1] looks like a ten-fold compressed version of studies initiated more than a century ago by Robert Wood with his discovery of anomalies in the efficiency of metallic diffraction gratings, now known as Wood's anomalies [2]. In 1902, R. Wood wrote: "I was astounded to find that under certain conditions, the drop from maximum illumination to minimum, a drop certainly from 10 to 1, occurred within a range of wavelengths not greater than the distance between the sodium lines," an observation that marked the discovery of grating anomalies.

The first period of the search for their explanation is marked by the attempt of Lord Rayleigh [3, 4] to link Wood's anomalies to the redistribution of the energy due to the passing-off (cut-off) of higher diffraction orders of the grating (transfer from propagating into evanescent type). As pointed out by Maystre [5], his prediction was all the more remarkable as the author first ignored the groove frequency of the grating used by Wood, and thus could not verify this assumption with experimental data.

It took more than 30 years for the second period of experimental and theoretical studies to establish another explanation of Wood's anomalies. In 1941, Fano [6] was the first to distinguish between two types of anomaly: (i) an edge anomaly, with a sharp behavior connected with the passing-off of a higher diffraction order, and (ii) an anomaly, generally consisting of a minimum and a maximum in the efficiency, which appears in a much broader interval. The second type of anomaly was described by Fano as a resonance one, linked with the excitation of a guided (leaky) wave along the grating surface. Hessel and Oliner [7] published a

Structured Surfaces as Optical Metamaterials, ed. A. A. Maradudin. Published by Cambridge University Press.
© Cambridge University Press 2011.

pioneering paper that shows for the first time, using a theory based on an analysis of electromagnetic scattering from a generic model of a periodic structure yielding a simple closed form solution, that Wood's anomaly resonances are of two types: one due to branch point singularities that correspond physically to the onset of a new propagating spectral order (first indicated by Lord Rayleigh), and the other due to pole singularities that correspond to the condition of resonance for leaky surface waves guided by the structure. In addition, Hessel and Oliner developed a so-called phenomenological approach to the resonant anomalies that permitted describing the anomaly by a very small number of physical parameters: a pole of the scattering matrix, a zero of the diffracted amplitudes, and smoothly varying coefficients.

The third period that continues even today contains studies in three different directions. First are the grating manufacturers and instrumental optics users for whom it is strongly advisable to avoid anomalies because of their devastating effects for spectral instrument performance. Second, the excitation of surface plasmons can lead to a total absorption of the incident light, a phenomenon predicted and observed by Hutley and Maystre [8] in a single polarization for classical gratings with one-dimensional (1D) periodicity, with the magnetic field vector parallel to the groove direction (TM, transverse magnetic, polarization). By the use of crossed gratings with two-dimensional periodicity, it is possible to obtain total light absorption in unpolarized light [9, 10]. In both cases, it was necessary to use gratings with subwavelength periods that can support only the specular (zeroth) reflected order. The total light absorption by metallic gratings evidenced that the coupling between the incident light and metals can be strongly enhanced by the excitation of surface plasmons, and this effect opened the way to many applications based on the strongly enhanced light–matter interaction. Third, as the electromagnetic field is localized in the vicinity of the metallic surface, light absorption leads to very strong optical intensities at the surface. As any surface presents natural roughness that can excite the surface wave, the field enhancement obtained under specific conditions was sufficient to provide a proper physical explanation of the surface enhanced Raman scattering (SERS) effect [11]. The same effect is also used to enhance otherwise weak nonlinear phenomena [12]. Biosensors based on surface plasmons are highly dependent on the refractive index of the surrounding media. Binding or adsorption of molecules on the metallic surface induces a change of the local refractive index of the dielectric medium, so that such biosensors can be called refractometric sensors [13–17].

1.2　Generalities of the surface waves on a single interface

Before discussing in detail the historical development of the studies of enhanced light transmission through arrays of holes in a metallic screen, let us introduce several notations and basic principles. Let us consider a plane metal–dielectric interface in the $x0z$ plane that separates two nonmagnetic media with relative dielectric permittivities ε_1 and ε_2. In TM (transverse magnetic) polarization and incidence in the $x0y$ plane, the two components of the electric and magnetic field that are parallel to the interface, and thus continuous across it, are:

$$\omega\mu_0 H_z = \exp(ik_x x)\left[\exp(-ik_{1y}y) + r\exp(ik_{1y}y)\right],$$
$$E_x = \frac{k_{1y}}{k_0^2 \varepsilon_1}\exp(ik_x x)\left[\exp(-ik_{1y}y) - r\exp(ik_{1y}y)\right] \tag{1.1}$$

in the cladding and

$$\omega\mu_0 H_z = t\exp(ik_x x)\exp(-ik_{2y}y)$$
$$E_x = \frac{k_{2y}}{k_0^2 \varepsilon_2}t\exp(ik_x x)\exp(-ik_{2y}y) \tag{1.2}$$

in the substrate. The first terms in the brackets in Eqs. (1.1) correspond to the incident wave, and the second terms correspond to the reflected wave, with r the reflection coefficient for the magnetic field amplitude. The transmission coefficient is denoted by t. Note that k_x is the x-component of the incident wavevector, and that k_{1y} and k_{2y} are the y-components of the wavevectors in the cladding and in the substrate:

$$k_{jy} = \sqrt{k_0^2\varepsilon_j - k_x^2}, \qquad j = 1, 2, \tag{1.3}$$

with k_0 the free-space wavenumber. The Fresnel reflection coefficients depend on the polarization and have the following form for transverse electric (TE) polarization:

$$r_{TE} = \frac{k_{1y} - k_{2y}}{k_{1y} + k_{2y}}, \tag{1.4}$$

and for TM polarization:

$$r_{TM} = \frac{k_{1y}/\varepsilon_1 - k_{2y}/\varepsilon_2}{k_{1y}/\varepsilon_1 + k_{2y}/\varepsilon_2}. \tag{1.5}$$

It is well-known that r_{TE} has neither a pole nor a zero. In contrast, when both ε_1 and ε_2 are real and positive, there is a zero of r_{TM} called the Brewster effect. There also exists a pole (a zero of the denominator) if one of the media is a dielectric and the other a metal, a pole that corresponds to a surface wave that can propagate along the interface. When expressed in terms of the wavevector component parallel to the interface, the solution has the same form for the Brewster effect and the pole,

$$k_x = k_0\sqrt{\frac{\varepsilon_1\varepsilon_2}{\varepsilon_1 + \varepsilon_2}}, \tag{1.6}$$

due to the ambiguity of the choice in Eq. (1.3) of the sign of the square root for complex arguments. Indeed, combining Eqs. (1.3) and (1.6), we obtain the classical form of the Brewster angle in the incident medium: $\tan \theta_1 = k_x/k_{1y} = \sqrt{\varepsilon_2/\varepsilon_1}$. When the second medium is a metal with the real part of ε_2 negative and smaller than $-\varepsilon_1$, the real part of k_x in Eq. (1.6) is greater than the wavenumber $k_0\sqrt{\varepsilon_1}$ in the upper medium; i.e., the wave is evanescent in the cladding (and inside the metal), with increasing distance from the interface, representing a surface wave with a propagation constant equal to k_x, a solution that we shall note as k_g, the index g standing for "guided," and its normalized propagation constant will be denoted as $\alpha_g = k_g/k_0$. As ε_2 always has a non-zero imaginary part due to the absorption losses in the metal, the surface wave decays as it propagates. Quite often the negative permittivity is due to the collective oscillations of the free electron plasma in the metal, which gives the names surface plasmon or plasmon surface wave (PSW) to these surface waves. As an incident electric field creates polarization states of the plasma, some authors call this wave a surface plasmon polariton (SPP). In the case of a polar crystal/vacuum interface, the corresponding surface waves represent surface phonon polaritons. They all have common properties from an electromagnetic point of view, although the background solid state physics can be quite different. As they represent a zero of the denominator of r_{TM}, they are solutions of the homogeneous problem – a scattered field with zero incident field – and thus represent proper (eigen) modes of the system.

When considering an idealized presentation of a perfectly conducting metal with $\varepsilon_2 \to -\infty$, the propagation constant in Eq. (1.6) becomes equal to the vacuum wavenumber, and thus the solution represents a plane wave propagating parallel to the interface inside the cladding, with its electric field vector perpendicular to the surface; i.e., the solution is not localized to the surface.

When α_g is greater than n_1, such a wave cannot be excited with an incident plane propagating wave. The excitation of the surface wave is possible through the Kretschmann configuration [13]: the surface plasmon is excited on the lower surface of a metallic layer having on its upper surface a prism with refractive index higher than the index of the substrate in order to match the horizontal component of the incident wavevector to the real part of the PSW wavenumber on the lower interface. The surface plasmon is then coupled to the incident light by tunneling through the metallic layer. A surface plasmon can also be excited in a prism coupler in the Otto configuration. In that case, the metallic film is coated on a glass substrate. The strength of excitation depends on the distance between the prism and the metallic layer. In both cases, the reflection of light is strongly attenuated when the surface plasmon is coupled to the incident wave, and the angle of incidence where the absorption is maximum depends on the refractive index of the dielectric medium surrounding the metallic layer.

Much more efficient coupling occurs when gratings are used with a periodicity that serves as a generator of wavevectors parallel to the surface. An incident plane wave generates an infinite number of diffraction orders. In the case of 1D periodicity in the x-direction with period d, the wavevector component of each diffraction order m is given by the grating equation:

$$k_{1x,m} = k_{1x,0} + mK, \qquad K = \frac{2\pi}{d}, \tag{1.7}$$

where $k_{1x,0}$ is equal to the x-component of the incident wavevector and m is an integer. If, for a certain value of m, $k_{1x,m}$ is close to k_g, a surface wave can be excited. A simplified notation leads to the condition of excitation:

$$Re(\alpha_g) = \sin \theta_i + m\frac{\lambda}{d} \tag{1.8}$$

if the upper interface is air (more precisely, a vacuum).

The coupling of the incident wave to the surface wave (mode) is reciprocal; i.e., the surface wave can be radiated into propagating diffraction orders in the cladding following Eq. (1.7). This phenomenon is called leakage and the surface wave becomes a leaky one, which leads to an increase of the imaginary part of α_g. Another important feature that is not obvious from Eq. (1.8) is that the real part of the propagation constant, as well as the electromagnetic field distribution of the surface wave characteristics, are modified by the presence of surface corrugation. Another possibility to excite a PSW realizes itself in SERS, where the surface roughness scatters the incident plane wave into waves with different k_x, and, in particular, with $k_x > k_0$, with part of the incident energy coupled to the PSW.

1.3 Extraordinary transmission and its first explanations

Just as Robert Wood 95 years earlier was astounded by his experimental observation, Thomas Ebbesen and his collaborators found it quite surprising to observe that when light tries to pass through an array of holes of subwavelength dimensions in an optically thick (opaque) metallic sheet (Fig. 1.1(a)), and whose entire area is much smaller than the total illuminated surface, there are spectral regions with anomalously high transmission (Fig. 1.1(b)) compared with the predictions of classical diffraction theory. The surprise was so great that it prevented the publication of the results for almost ten years from their first observation [18] of the effect in the NEC laboratories.

As with Wood's anomalies, it is possible to separate the studies on this extraordinary transmission into three much shorter and more dynamic periods. In 1998, in contrast to the situation at the start of the twentieth century, electromagnetic theories

Evgeny Popov and Nicolas Bonod

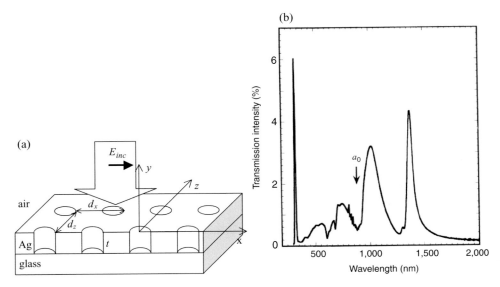

Figure 1.1. (a) Schematic representation and notations of a two-dimensional hole array perforated in a metallic screen deposited on a glass substrate and illuminated from above with a linearly polarized incident wave. (b) Spectral dependence of the transmission of the structure presented in (a), with $d_x = d_z = 0.9 \, \mu m$, $t = 0.2 \, \mu m$, and a hole diameter of $0.2 \, \mu m$ [1]. Reprinted with permission from Macmillan Publishers Ltd. © 1998.

of gratings (1D or 2D) were largely developed. The understanding of the role of the PSW (and surface waves in general) in grating anomalies, field enhancement, etc., was much deeper, and the number of scientists working in the field incomparably larger. The end of the Cold War moved large human resources from defense micro-electronics, solid state and high energy physics into optics, creating neologisms like photonic crystals, photonics, metamaterials, etc., causing, for instance, the change of *Optics News* into *Optics and Photonic News*. In addition, production resources such as optical photolithography, focused ion-beam and laser-beam writing and etching, became more available in optics laboratories, which permitted structuring metals and dielectric media at the nanometer scale and developing photonic devices able to control light at the subwavelength scale. Already in the original paper [1], the authors clearly indicated that the transmission enhancement appears at spectral positions closely given by PSW excitation by a bi-periodic structure, but they were not satisfied by this qualitative explication. The publication of these results by Ebbesen and coworkers in *Nature* immediately strongly impacted the newly enlarged optical community that was closely interested in photonics, to find a new but similar interest in plasmon surface waves, giving birth to another neologism, plasmonics.

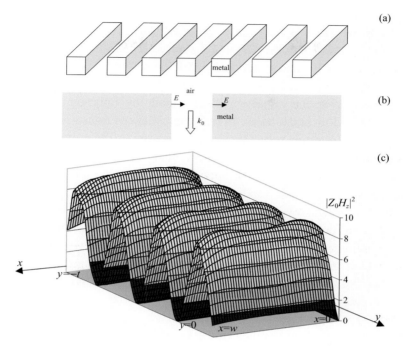

Figure 1.2. (a) Slit grating having a one-dimensional periodicity and characterized by vertical straight channels. (b) Propagation of a TM electromagnetic field inside the slit with perfectly conducting walls. The electric field vector perpendicular to the slit walls satisfies the boundary conditions there and the vertical wave propagates as in free space. (c) Propagating character of the electromagnetic field inside a narrow slit ($w = 40$ nm). Reprinted with permission from ref. [22]. © 2000, American Physical Society.

The first numerical results were so close to the experimental enhancement that doubts were generated whether the transmission increase was so extraordinary. The main characteristic of this first period (see, for example, refs. [19] and [20]) was the use by theoreticians of gratings with 1D periodicity made of periodic slits in a metallic screen (Fig. 1.2(a)). The results were quite nice, with a clearly visible flow of the electromagnetic field inside the slits, so that some authors started to indicate the decisive role of another wave – a vertical plasmon wave that propagates inside the slits of the metal–dielectric interface – that is responsible for the enhanced transmission, acting simultaneously with the grating-induced resonances of the horizontally propagating PSW, excited at spectral positions given by Eq. (1.8). Figure 1.2(b) represents such a vertical wave which propagates inside the slits as in free space for perfectly conducting walls. For metals with a finite conductivity, this wave represents a hybrid wave that can propagate in the vertical direction formed by two coupled plasmons on the slit walls [21].

The finite conductivity of the metal changes the boundary conditions on the vertical walls, but even for very narrow slits and a silver grating, the TM electromagnetic field preserves its propagating nature, as seen in Fig. 1.2(c).

These idyllic conclusions were common for the first period of about two to three years following 1998. The main problem was that they were not applicable to the geometry involved in the initial experiment made by Ebbesen *et al.*, where the slits were replaced by small holes. In fact, the enhanced transmission through slit metallic gratings (or gratings having similar grooves) in TM polarization had been known for quite a long time and resulted in commercially available wire-grating polarizers (see, for example, [10]). Such gratings (as represented in Fig. 1.2(a)) with subwavelength periods small enough to support just the specular reflected and transmitted orders, reflect incident light of TE polarization almost completely, while light of TM polarization can be transmitted almost totally for a proper choice of grating parameters. The reason lies in the existence of a waveguide mode inside each slit, which in TM polarization has no cut-off wavelength. Let us consider a slit, neglecting the absorption losses inside the metal walls. As in the case of the plane horizontal interface between a lossless metal and a dielectric in TM polarization, a vertical interface also supports a wave of plane-wave type inside the dielectric propagating parallel to the interface with a magnetic field vector parallel to the interface (Fig. 1.2(b)). The same wave can propagate inside slits with lossless walls, as it satisfies the boundary conditions on both walls, whatever the width of the slit. The mode is characterized by a real propagation constant in the vertical direction (neglecting absorption losses, as assumed). On the upper and lower interfaces (Fig. 1.2(a)), the mode is reflected backwards in the slit, and is partially transferred into propagating waves in the cladding and in the substrate, representing a Fabry–Perot resonator. If the system is symmetrical (the same optical index of the substrate and the cladding, as in wire polarizers), the Fabry–Perot resonance maxima can reach 100% in transmission. In contrast, in TE polarization the corresponding slit mode has a cut-off, because the electric field vector is parallel to the slit walls. The electric field vanishes on both walls and satisfies the following relation (Fig. 1.3(a)):

$$E_z \sim \sin\left(\frac{\pi}{w}x\right), \tag{1.9}$$

so that the y-component of its wavevector, given by

$$q_y = \sqrt{k_0^2 - \frac{\pi^2}{w^2}} = \pi\sqrt{\frac{4}{\lambda^2} - \frac{1}{w^2}}, \tag{1.10}$$

becomes imaginary for small widths w; i.e., the mode is evanescent in the vertical y-direction if the slit width is smaller than $\lambda/2$. Thesmaller the width, the faster the

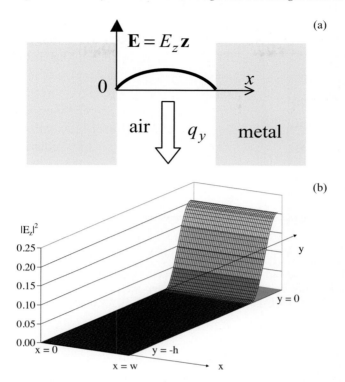

Figure 1.3. (a) TE mode inside a slit with perfectly conducting walls. The x-dependence of E_z is given by the thick line, and it has to vanish on the walls. (b) A map of the evanescent TE mode inside a subwavelength slit ($w = 40$ nm) with silver walls. The slit is so small compared to the wavelength that the sinusoidal dependence that appears for perfectly conducting walls (a) cannot be distinguished in the x-dependence. Reprinted with permission from ref. [22]. © 2000, American Physical Society.

exponential decrease of the mode amplitude inside the slit depth. The narrower the slit and the thicker the metal layer, the smaller the amount of transmitted energy, and thus the polarizing properties of the device.

When finite conductivity is taken into account, the cut-off width is slightly smaller for finitely conducting walls than the $\lambda/2$ value obtained from Eq. (1.10). In addition, very narrow slits absorb an electromagnetic field, and the transverse variation of the field becomes very weak, as can be observed in Fig. 1.3(b) for silver walls and a 40 nm slit width.

In contrast to what happens with slits, holes with a finite width in both directions of their cross-section do not support modes without cut-off; i.e., below a given width of the hole, the field of the modes inside is always evanescently decreasing. Although obvious, these considerations were not taken into account in the first modelizations, when the hole array was replaced by periodic slits. However, these

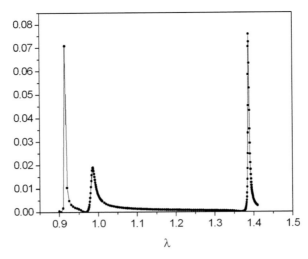

Figure 1.4. Computed spectral dependence of the transmission intensity of a square-hole array in a 200 nm thick silver screen deposited on a glass substrate. Reprinted with permission from ref. [22]. © 2000, American Physical Society.

first works were relatively easily carried out from theoretical and computational points of view, and the large number of studies based on this assumption attracted substantial interest to metallic gratings and surface plasmons.

1.4 The role of the evanescent mode

The second period of the studies of extraordinary transmission started with the first rigorous electromagnetic modeling of the array of holes with finite subwavelength cross-section dimensions [22]. The numerical results are similar to the experimental observations (Fig. 1.4).

The grating period in both directions is the same and equal to 0.9 µm, the cladding is air, and the substrate is glass. The metal is silver 0.2 µm thick, and the holes have a square cross-section with a width of 0.25 µm, much below the cut-off dimensions for the spectral interval under study. Two peaks are clearly distinguished, the shorter-wavelength one, lying around 1 µm, corresponds to the excitation of PSW on the upper air–silver interface. The long-wavelength peak is due to the excitation of PSW on the lower glass–silver interface.

For an infinitely conducting metal, the fundamental TE mode of the hollow square waveguide formed inside each hole has a propagation constant of the order of $q_y \approx i11 \, \mu m^{-1}$, which corresponds to a decay constant in the y-direction, $\gamma = Im(q_y)/k_0 \approx 2.5$. When the finite conductivity of the metal is taken into account,

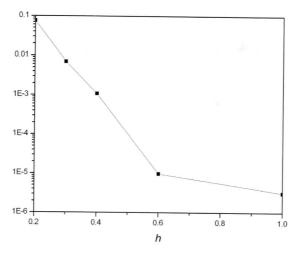

Figure 1.5. Zero-order transmittivity as a function of the silver layer thickness. Reprinted with permission from ref. [23]. © 2002, IOP publishing.

the electromagnetic field of the mode penetrates inside the hole walls, so that its propagation constant changes, and the decay constant becomes smaller and equal to $\gamma \simeq 2.1$ at a wavelength of 1.4 µm, a value more than four times smaller than the imaginary part of the refractive index of the silver film, $n_{Ag} = 0.1 + i8.94$, in the IR. When compared to the direct tunneling through the metal layer, with a decay constant given by $Im(n_{Ag})$, the decay of the fundamental mode, even though it has an exponentially decreasing amplitude, within the film thickness of 0.2 µm is 20 000 times smaller than without the perforations.

Numerical simulations can easily confirm these conclusions, because one can smoothly vary the layer thickness quite easily numerically. Figure 1.5 presents the transmission under the same conditions ($\lambda = 1.38$ µm) as a function of the metal layer thickness. As can be observed, the dependence for $h < 0.6$ µm is linear on a semi-logarithmic scale, and the slope close to 20 µm^{-1} corresponds quite well to $2q_y$ because the intensity decreases as the square of the amplitude.

In parallel with the rigorous electromagnetic study of the enhanced transmission through hole arrays, several authors have presented approximate models that not only provide an easier physical understanding of the various interactions, but also in some cases predicted new phenomena. In the early 2000s, a simple model that we describe below was developed due to the understanding of the role of the evanescent waveguide mode inside the hole [24, 25]. Let us consider an interface between air and a perforated metal with infinite thickness, in order to eliminate the role of the lower interface. When the array period(s) is small enough to avoid propagation of higher diffracted orders in the cladding, and when the hole cross-section of width w is small in comparison with the period and thus with the wavelength, the predominant fields in TM incident polarization are the incident and specularly reflected fields, as given in Eqs. (1.1). Inside the rectangular hole with perfectly conducting walls, the transverse electric fundamental mode is parallel to the x-axis, it does not depend on z, and it must vanish at $x = 0$ and w, as in Eq. (1.9), but permuting the z- and x-axes, because the TE mode with electric field oriented in the z-direction cannot be excited with an incident electric field oriented in the $x0y$ plane. If the amplitude of H_z is denoted by τ, the x- and z-components are given by

$$\omega\mu_0 H_z = \tau e^{-iq_y y} \sin\left(\frac{\pi z}{w}\right),$$

$$E_x = \begin{cases} \frac{\tau}{q_y} e^{-iq_y y} \sin\left(\frac{\pi z}{w}\right), & 0 \le x, z \le w \\ 0, & x, z \notin [0, w], \end{cases} \tag{1.11}$$

so that the boundary conditions require that the tangential electric field components vanish on the walls at $x = 0$ and w, and at $z = 0$ and w.

The requirement of the continuity of the tangential (x and z) components of the electromagnetic field at $y = 0$ links the system of Eqs. (1.1) and (1.11) and yields two equations:

$$e^{ik_x x}(1 + r) = \tau \sin\frac{\pi z}{w}, \tag{1.12}$$

$$\frac{k_y}{k_0^2} e^{ik_x x}(1 - r) = \begin{cases} \frac{\tau}{q_y} \sin\frac{\pi z}{w}, & x, z \in [0, w] \\ 0, & x, z \notin [0, w]. \end{cases} \tag{1.13}$$

The first equation must be satisfied within the aperture opening, because the magnetic field inside the metal is not known. The basic modal functions of the modes in the apertures are $\sin(m\pi z/w)$ and $\cos(p\pi x/w)$, where m and p are integers. If we multiply Eq. (1.12) by $\sin(\pi z/w)$ and integrate over the aperture with respect to z and x, the result is as follows

$$\frac{2}{\pi}(1 + r)\left(\sin\left(\frac{k_x w}{2}\right) \Big/ \left(\frac{k_x w}{2}\right)\right) = \frac{\tau}{2}. \tag{1.14}$$

The second boundary condition must be valid within the entire cell, with basic modal functions in x and y being the exponential functions. By multiplying the equation by $\exp(-ik_x x)$ and integrating from 0 to d in x and z, we obtain a similar relation:

$$\frac{k_y}{k_0^2} d^2(1 - r) = \frac{\tau}{q_y} \frac{2a^2}{\pi}\left(\sin\left(\frac{k_x w}{2}\right) \Big/ \left(\frac{k_x w}{2}\right)\right). \tag{1.15}$$

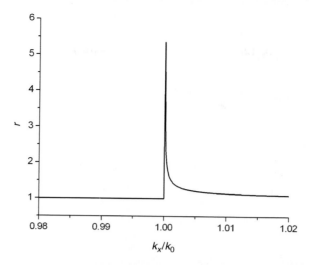

Figure 1.6. Reflection coefficient r from Eq. (1.16) as a function of the x-component of the incident wave wavevector k_x/k_0.

It is straightforward to obtain the reflection coefficient, which has a form similar to Eq. (1.5):

$$r = \frac{k_y q_y/k_0^2 - I^2}{k_y q_y/k_0^2 + I^2},$$

(1.16)

where

$$I = \frac{w}{d}\frac{2\sqrt{2}}{\pi}\left(\sin\left(\frac{k_x w}{2}\right)\Big/\left(\frac{k_x w}{2}\right)\right)$$

represents the coupling integral between the electromagnetic field in the cladding and in the apertures. Assuming that the aperture size is much smaller than the wavelength, the dependence of I on the angle of incidence can be neglected. In the lossless case and for evanescent modes (imaginary q_y), the modulus of r is equal to unity, as for a planar metal–air interface without losses.

The situation can change if we consider incidence with $k_x > k_0$, as is the case with the PSW. With k_y becoming imaginary, it is possible to find a zero of the denominator of r in Eq. (1.16) representing also a resonance of the mode amplitude inside the aperture, because the pole of the reflection coefficient is a pole in transmission too. It must be stressed that this resonance can be excited only with evanescent incident waves, as observed in Fig. 1.6, where the reflection coefficient r is plotted as a function of k_x/k_0, with $d = 1\,\mu\text{m}$, $w = 0.2\,\mu\text{m}$, wavelength $\lambda = 1.3\,\mu\text{m}$, and the refractive index inside the aperture $n = 1$. A sharp peak can be observed for an evanescent incident wave. The periodicity of the array can provide such waves through the grating equation, and a PSW excited by the grating will

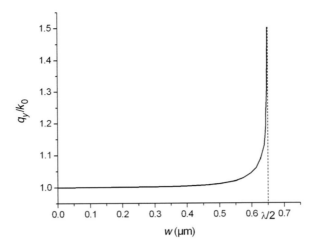

Figure 1.7. Pole of the reflection coefficient given by Eq. (1.16) as a function of the aperture width w. Period $d = 1\,\mu$m, wavelength $\lambda = 1.3\,\mu$m.

serve as a strong source to excite the evanescent mode resonantly through the zero of the denominator of Eq. (1.16).

Let us see, from the example provided by the original paper [1], how this could happen. If the ratio between the wavelength and the hole side dimensions is approximately equal to $1.4/0.2 = 7$, the value q_y/k_0 is equal to $i3.35$, as given by Eq. (1.10). The normalized propagation constant of a PSW on a highly conducting surface is slightly greater than unity, $k_x/k_0 \simeq 1.01$, whereas $k_y/k_0 \simeq 0.01$. The product $k_y q_y/k_0^2 \simeq -0.034$. With $a/d = 0.2$, the value of I^2 in Eq. (1.16) is close to 0.032, so that $r = 33$, indicating the existence of a resonance.

In the case of a single aperture, light can be coupled to a localized surface plasmon, as discussed in Section 1.6, that is able to excite further a waveguide mode. However, this effect is much weaker than when using the perodicity of the hole array.

Of course, the model discussed here is quite simplified. The interaction between the PSW and the evanescent mode changes the PSW propagation constant, as can be observed in Fig. 1.7 for the case of the perfectly conducting model represented by Eq. (1.16). A larger aperture width leads to a stronger interaction between the mode and the incident wave, and a shift of the resonance to longer wavevectors is observed, with a cut-off width equal to $\lambda/2$, above which q_y becomes real.

In practice, the mode properties are also modified by the finite conductivity of the metal walls, and its cut-off wavelength is increased whereas its cut-off hole dimension is reduced. In addition, higher diffraction orders will play a role in the grating scattering. However, such a model serves well to unveil the physics of the process. Another, more sophisticated, model is proposed in ref. [26] based on the assumption of small aperture diameter. It shows that the amplitude of the electric

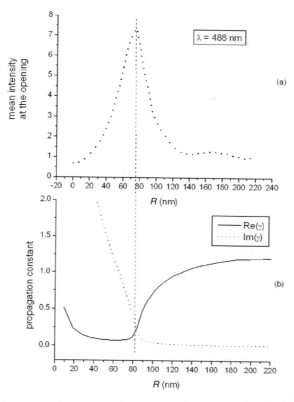

Figure 1.8. Field enhancement inside a circular aperture made in an aluminum film with thickness 220 nm deposited on a glass substrate and filled with a solvent with refractive index of 1.4, with incidence from the substrate side and wavelength 488 nm. (a) Mean electric field intensity I_S at a depth of $z = -5$ nm inside the aperture as a function of aperture radius. (b) Real and imaginary parts of the normalized propagation constant of the mode inside the hollow metallic waveguide inside the aperture. Reprinted with permission from ref. [27]. © 2006, Optical Society of America.

field inside small apertures grows linearly with the aperture diameter. For larger apertures this approximation is no longer valid, but rigorous theoretical calculations and fluorescence measurements show that a maximum of the electromagnetic field intensity is obtained when the hole dimensions are just below the cut-off of the fundamental mode. This can be understood by taking into account the fact that the real part of the propagation constant q_y of the waveguide mode at its cut-off is almost zero, as if the field were accumulated at the entrance of the aperture. As a consequence, the field inside the single aperture can be enhanced several times when compared to its value in free space, which leads to enhanced transmission through the waveguide mode inside the hole. An example well confirmed by fluorescent measurements [27] is given in Fig. 1.8(a), which presents the mean electric field intensity I_S as measured just below the aperture entrance and defined as

$$I_S = \frac{1}{\pi R^2} \int_S |\mathbf{E}|^2 \, dS,$$ (1.17)

where the integral is taken over the cross-section of the aperture. For small radii its dependence on the radius is quadratic, and a seven-fold enhancement of it when compared to the field inside the solvent without apertures is observed close to the cut-off of the fundamental mode. Figure 1.8(b) presents the real and imaginary parts of the normalized mode constant $\gamma = q_y/k_0$.

1.5 Enhanced Fabry–Perot resonances through evanescent modes

A direct consequence of the fact that a plasmon surface wave propagating on the horizontal surface can resonantly interact with the vertical evanescent mode inside a hole is discussed in ref. [24]. If the model presented in the preceding section is extended to a layer with two metallic surfaces, a perforated layer, there will be additional reflection of the mode when it reaches the lower surface. Multiple reflection leads to Fabry–Perot resonances, which, however, are quite weak when evanescent waves are used instead of propagating waves. Indeed, a textbook formula is applicable to both propagating and evanescent modes. The transmission of a Fabry–Perot resonator is proportional to a denominator that contains the mode propagation constant q_y, the layer thickness h (waveguide length), and the product of the reflection coefficients of the mode on the upper (r_+) and lower (r_-) surfaces:

$$t \sim \frac{1}{1 - r_+ r_- \exp(2i q_y h)}.$$ (1.18)

The usual case of evanescent waves with imaginary q_y cannot ensure strong resonances because the reflection coefficients are smaller than unity in modulus. However, this is not the case discussed in the preceding section. We have observed the possibility of having a ten-fold or more increase in the coefficient of reflection r, although we have considered the reflection from the cladding into the cladding, rather than from inside the hole backwards to the hole. A similar analysis shows that r_+ and r_- are enhanced in a similar manner to compensate the exponential decay due to $\exp(2i q_y h)$, leading to enhanced transmission through evanescent-wave Fabry–Perot resonances.

1.6 What resonance predominates?

As shown in Sections 1.4 and 1.5, the evanescent fundamental TE mode plays a crucial role in the transmission mechanism. However, this role alone cannot explain the sharp spectral and angular variations of the transmission, because its wavevector component k_x/k_0 is much larger than unity. Indeed, in the example considered at the end of Section 1.4, the propagation constant in the vertical direction, $q_y/k_0 = i3.35$, corresponds to a value of $k_x/k_0 = 3.5$ in air or 3.67 in glass. Such a mode is quite difficult to excite by an incident plane wave, even with the help of the grating periodicity.

On the other hand, we have observed in Section 1.4 that the interaction between the incident plane wave and the fundamental TE_{11} mode inside each aperture can have a resonance close to k_x/k_0 slightly greater than unity (in air), provided that both the incident wave and the mode are evanescent. On the other hand, the excitation of the PSW on a plane surface also requires evanescent incident fields with $k_x/k_0 > 1$. A natural question that arises is whether there is a difference between these resonances, or whether they represent two physical interpretations of the same resonance. If there is a real difference obtained through a complete diffraction analysis (and not by a simplified model), this means that the excitation of the evanescent mode on the aperture opening would create a new surface resonance on the interface between the perforated metal and the dielectric cladding, in addition to the PSW. A resonance of the reflection coefficient means the existence of a scattered field without the existence of an incident one. In addition, as far as $k_x/k_0 > 1$, this diffracted field will be evanescent in the cladding, i.e. localized close to the interface, in the same manner as for the PSW. As a consequence of this hypothesis, it would be possible to excite both resonances with the help of the grating periodicity, and the excitation would occur, in general, at different wavelength and angular conditions. However, both experimental and numerical results do not show such a double resonance. Another possibility is that the two resonances exist, and that they represent different phenomena, but the values of k_x at which they appear coincide, as for the case of phase-assisted second-harmonic generation in periodically corrugated (or poled) dielectric waveguides with waveguide modes excited at both the fundamental and the harmonic frequencies [28]. However, if such a case appears, the resonant response will be observed as a double Lorentzian (with two identical poles). Moreover, a rigorous numerical analysis carried out in ref. [23] shows that the transmission intensity T of the structure analyzed experimentally in ref. [1] can be extremely well approximated by a single pole of the transmission amplitude:

$$T = \left| \frac{k_x - k_x^z}{k_x - k_x^p} \right|^2 \sim \left| \frac{\lambda - \lambda^z}{\lambda - \lambda^p} \right|^2, \tag{1.19}$$

where p stands for a pole, and z for zero.

Although, in general, there are four surface waves excited in normal incidence, those propagating in the $+x$-, $-x$-, $+z$-, and $-z$-directions, they are all coupled

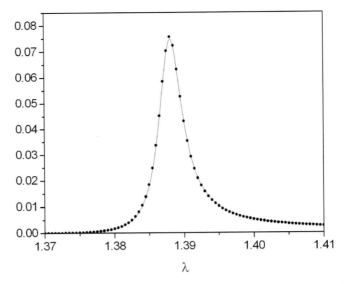

Figure 1.9. Comparison between the result of the rigorous electromagnetic analysis (dots) and the approximation given by Eq. (1.19) with $\lambda^p = 1.3878 + i0.002$
and $\lambda^z = 1.3689 + i0.00048$. Values of the transmission in normal incidence
through a perforated silver screen of 200 nm thickness deposited on a glass substrate. Hole size is 250×250 nm. Reprinted with permission from ref. [23].
© 2002, IOP publishing.

by the grating to form four different standing waves, symmetric or antisymmetric
with respect to the origin. There is only one solution symmetric with respect to the
change of sign of the z-axis and antisymmetric with respect to the sign of the x-axis,
having the same symmetry as the incident wave is polarized along the x-axis. The
other three solutions cannot be excited in normal incidence by an incident plane
wave.

The Lorentzian resonance response is deformed by the existence of a zero, a
fact that is well known in grating theories, and is necessary in order to limit the
value of T and to compensate the pole when the modulation due to the grating
(the hole dimensions in our case) tends to zero, because no anomaly is observed
without the grating. As demonstrated for classical one-dimensional gratings [29],
the zero λ^z can become real for a certain value of the grating depth, which leads to
a total light absorption in TM polarization. The comparison between the numerical
values obtained by rigorous electromagnetic theory and the results obtained using
Eq. (1.19) are given in Fig. 1.9, which clearly demonstrates the existence of only a
single pole (the square arises because it is the intensity that is being calculated).

An additional argument (presented in ref. [23]) that the resonance responsible
for the enhanced transmission is due to the PSW is its trajectory in the complex
λ-plane when the aperture is shrunk to zero. The resonant wavelength gradually

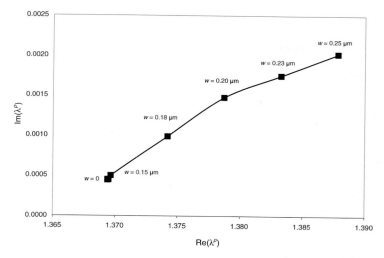

Figure 1.10. The real and imaginary parts of the pole in the transmission amplitude for different hole widths w. The structure is described in Fig. 1.9. Reprinted with permission from ref. [23]. © 2002, IOP publishing.

tends to the resonant wavelength of the PSW on the unperturbed plane metal–glass interface, as observed in Fig. 1.10, in the same manner as the pole of Eq.(1.16) for a perfectly conducting metal tends towards the wavenumber of the cladding; see Fig. 1.7.

The presence of a hole modifies the constant of propagation of the PSW due to scattering, to the radiation losses through the grating periodicity, and to the interaction with the waveguide evanescent mode. That is the reason why the analytical formula given by Eq. (1.6) for the complex value of λ^p in the case of an unperturbed flat surface is no longer valid in the case of a periodic array of holes. The red-shift of the pole explains why the transmission maximum appears at longer wavelengths than those predicted by Eq. (1.6). The zero λ^z (see the caption of Fig. 1.9) is almost real, leading to the almost zero transmission when $\lambda = \lambda^z$ observed in Fig. 1.4. In addition, the real part of λ^z almost coincides with the initial position of the pole, when $w = 0$ (Fig. 1.10), which explains why the transmission drops to zero at a wavelength corresponding to PSW excitation at a flat metallic surface without holes, a fact that has put in question the role of the PSW in the enhanced transmission. The correct explanation comes from a proper understanding of the red-shift of the pole observed in Fig. 1.10, presenting Fano-type anomalies in transmission [30, 31].

A direct proof that the resonance of the mode reflection coefficient, discussed in Section 1.4 as a consequence of Eq. (1.16), is the same as the PSW on the perforated structure, can be found using a more sophisticated model, which analyzes the scattering by a single aperture in a real metal in the approximation of a very small aperture diameter. The excitation of the waveguide mode in such small

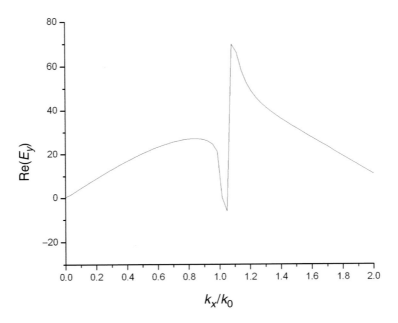

Figure 1.11. Spatial spectrum of the vertical component of the electric field on the entrance surface of a single aperture. It is zero in the vertical direction, and presents a resonance anomaly corresponding to a PSW. Reprinted with permission from ref. [33]. © 2005, Optical Society of America.

cross-section waveguides can be represented by a single magnetic dipole, as in Bethe's paper [32], but buried inside the metallic layer [26]. Its coupling to the outside radiation is achieved through the classical Fresnel coefficients of transmission and reflection at the dielectric–metal interface, which presents a pole, known as the PSW. When the hole diameter is larger, the coupling is stronger, as expected, and one always finds in the spatial spectrum of the scattered light the signature of the PSW pole, as observed in Fig. 1.11 for the k_x decomposition of E_y on the entrance surface of a circular aperture (250 nm diameter) in a silver sheet (thickness 200 nm) illuminated normally with TM polarized light (wavelength 500 nm) from air. The position of the resonance anomaly coincides with the normalized constant of propagation of the PSW on the planar surface. When a hole array is considered, this spatial component can be enhanced by the periodicity under the conditions given by the grating equation, Eq. (1.8).

1.7 Nonplasmonic contributions

The role of PSW in producing the minima and maxima observed in the spectral features of the transmission through subwavelength holes is now clearly established.

However, the question remains as to whether the PSW is the only propagating wave along the metallic surface launched by subwavelength holes. This question has opened the third period of research since the discovery of Ebbesen and co-workers. Let us recall that the first period (1998–2000) was dedicated to the study of enhanced transmission though subwavelength slits. The second period (2000–2004) was marked by the development of rigorous numerical methods able to tackle arrays of subwavelength apertures. Numerical and experimental studies permitted understanding the role of the shape and size of a hole. This period corresponds also to the emergence of the complementary thematic of light diffraction by single sub-wavelength apertures and its applications in biosensing, single-molecule analysis, membrane analysis, etc.

During the entire decade, however, there have always been studies that contested the role of the surface plasmon in the enhanced transmission (see, for example, the discussion after Fig. 1.10). In 2004, another model, called the composite diffracted evanescent wave (CDEW) model, was introduced into the theory of light diffraction by subwavelength indentations [34]. The CDEW model predicts that the summation of all diffracted inhomogeneous waves by a subwavelength indentation results in a propagating wave along the metallic surface, with a propagation constant k_0 and an amplitude decaying with increasing distance from an indentation as $1/r$. This paper attempted to explain the minima and maxima observed in the transmission of light by an array of subwavelength holes with the use of the CDEW model only and by fully neglecting the role of the PSW. It is now established that PSWs are a key component of the mechanism of enhanced transmission. However, this study attracted the attention of researchers to other kinds of electromagnetic waves propagating close to the surface. The fundamental problem that arose was the determination of the properties of waves launched by a subwavelength indentation when illuminated by an incident plane wave. It was then necessary to simplify considerably the device under study to determine quantitatively the properties of this wave. Gay *et al.* experimentally studied the interference between light transmitted through a slit and waves launched by a neighboring nanoslit as a function of the distance between the two neighboring slits [35]. They explained their result with the use of the CDEW model, but rapidly this formalism has been strongly debated [36, 37]. This model has been therefore replaced by the so-called quasi-cylindrical wave (QCW) model.

The existence of this wave is better understood on the basis of the experiment made by Aigouy *et al.* [38]. In 2007, they studied experimentally the near field distribution created by a slit-doublet on a metallic sheet. When the slits were illuminated in TM polarization, they observed the presence of an interference pattern between the two slits generated by two counter-propagating waves (Fig. 1.12). The interference pattern has been thoroughly studied at different distances from

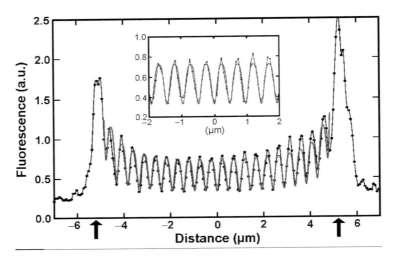

Figure 1.12. Near-field interference observed by the fluorescence of erbium molecules. Thin line with dots: fluorescence signal proportional to $|E|^4$. The two slits represented by the arrows are separated by a distance of 10.44 μm and are illuminated in TM polarization at the wavelength $\lambda = 974.32$ nm. Solid thick line: fitted curve obtained with PSW and QCW models. The insets shows the fitted curve with the PSW model only in the central zone where the QCW can be neglected. Reprinted with the permission from ref. [38]. © 2007, American Physical Society.

the slit, and Aigouy *et al.* confirmed in the near field the observations made by Gay *et al.* in the far field: a rapid decrease of the amplitude of the waves close to the slits followed by a persistent wave farther from the slit, i.e. in the central part of the doublet. The fringes observed in the central part are well fitted by the interference of two counter-propagating PSWs only. The persistent surface wave was then identified with PSWs and the rapidly decaying wave was attributed to QCWs (see the inset in Fig. 1.12). Both waves propagate on the metallic surface, the PSW with a propagation constant k_{SP} slightly higher than k_0, and the QCW with a propagation constant equal to k_0. Thorough experimental, numerical, and theoretical studies revealed that QCWs present two decay rates with respect to the distance ρ from the line source along the surface, different from the decay rate obtained with the CDEW model. It has been established that the QCW decays as $\rho^{-1/2}$ and $\rho^{-3/2}$. An analytical expression of the QCW can be obtained in the case of a line source located at a metal–dielectric interface. It is the product of an Erf-like envelope function of ρ, and a term proportional to $\rho^{-3/2}$ with a wavevector parallel to the interface. The two decay rates observed experimentally are the two asymptotes of the Erf-like envelope: very close to the slit the decay rate of the QCW is $\rho^{-1/2}$; farther away, the envelope vanishes and the rate tends towards

$\rho^{-3/2}$ [39]. A numerical study [33] made in 2005 of light diffracted by a single circular aperture on a metallic sheet in a direction parallel to the surface had already shown that the electric field decay as a function of the distance ρ from the hole is determined mainly by two contributions: the PSW and a spherical wave decaying as r^{-1}, which represents the field radiated by a single dipole, as predicted by Bethe's model. Along the surface, the propagation constant of this spherical wave is equal to the free space wavevector k_0 (in vacuum or air). If we consider a slit, or a chain of holes, as discussed in the next paragraph, they radiate like a line antenna, with an electric field decreasing as $\rho^{-1/2}$, a fact that provides a simple explanation of the QCW.

In 2008, Liu and Lalanne published a microscopic analysis of the enhanced light transmission based on the scattering of PSW on an array of subwavelength holes [40]. This analysis permitted tackling separately the contribution of the PSW and quantifying its role in the energy transmitted. Microscopic here means that the holes are considered individually as single scatterers. More precisely, the authors considered a one-dimensional linear chain of holes, and they listed the elementary scattering coefficients when the PSWs interact with the chain: PSWs can be reflected (with reflection coefficient P), transmitted (transmission coefficient T) on the interface plane, or they can be scattered into outgoing plane waves and modes in the holes (with coefficient α). The coefficient of reflection R_A of the Bloch mode supported by the array on the front surface of the holes can be written in the case of normal incidence as

$$R_A = R + \frac{2\alpha^2}{u^{-1} - (P + T)}, \qquad (1.20)$$

where R is the coefficient of reflection of the Bloch mode of a single hole chain, $u = \exp(ik_{sp}d)$, and d is the distance between the hole chains. This analytical expression takes into account the excitation of a PSW by a linear chain of holes and the scattering of this PSW by the infinity of other linear chains of holes. In addition, the authors of ref. [40] were able to account for the contribution of the QCW wave. Not surprisingly, in conditions close to those of the initial experiment, this model reveals all the spectral features of light transmission by a 2D array of subwavelength holes. Maxima and minima are well represented, and a comparison with a rigorous numerical study confirms that their frequencies are well predicted by the pure PSW model. However, this comparison also reveals that this model does not predict the exact portion of light transmitted. More precisely, the pure PSW model predicts about half of the total light transmitted. This fact demonstrates that PSWs are fully involved in the transmission of light through 2D arrays of subwavelength holes but that QCWs are responsible for the other half of the light energy transmitted. The amount of light energy transmitted by PSWs diminishes for lower frequencies, where QCWs are predominant. Liu and Lalanne extracted the PSW contribution from the total field and they represented the real part of the y-component of the magnetic field along the x-axis. They confirmed the fact that the field scattered by subwavelength holes is not a pure PSW mode but that it also contains a QCW contribution. The two waves are excited with a very moderate phase difference, their phase velocities k_{sp} and k_0 are very close, and their contributions interfere constructively. The radiative decay of the QCW scales as $\rho^{-1/2}$, so that it is much faster than the PSW (as observed on the slit model in Fig. 1.12). However, close to the linear chain, the QCW contributes equally with the PSW to the total light transmitted.

1.8 Conclusions

We are now able to understand quite well the phenomenon of enhanced light transmission through an array of subwavelength apertures in a metallic screen. The scattering of the incident field on the aperture boundaries creates waves with a variety of wavevector directions, in particular the surface plasmon along the upper boundary. When the array periodicity corresponds to the resonant conditions of PSW excitation, as given by the grating equation, the PSW wave is enhanced significantly, which creates a local field enhancement on the metal surface. This enhancement leads to a resonant excitation of the fundamental waveguide mode inside each aperture, due to the fact that the transmission coefficient for its excitation has a resonance corresponding to the PSW on the interface. Of course, the propagation constant of the PSW is modified by the interaction, which can be used to create similar surface waves in the microwave domain. At normal incidence, the PSW forms a standing wave on the entrance interface, so that, at both sides of each aperture, the electromagnetic energy flows towards that opening, somehow as if the incident intensity is gathered inside the aperture, as seen in Fig. 1.13. In addition to the PSW, each hole generates a radiated field, a part of which propagates in grazing directions to the surface, according to refs. [39] and [40], and can reach the neighboring holes to enhance the field there.

The fact that the fundamental mode is resonantly enhanced in this process explains why the field transmitted by the mode is not negligible at the exit side, even if the mode is evanescent for a small hole cross-section. At the exit side, the mode excites both the PSW and a radiated field, and the radiation from each hole interferes to form the transmitted zeroth diffracted order. If the system is symmetrical with identical substrate and cladding, the resonant conditions for excitation of the PSW on the entry interface are the same as for a resonant emission of the PSW on the exit interface into the zeroth transmission order, which enhances the transmission even more.

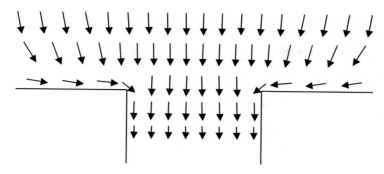

Figure 1.13. Schematic view of the Poynting vector flow on the upper surface and in the vicinity of the hole entrance.

In the case when the periodicity is not suitable to excite the PSW on the entry interface, another resonance can be observed when a PSW is excited on the lower interface. The waveguide mode inside the holes acts to enhance the tunneling of light from the entry to the exit surface. When the grating period is suitably chosen, the periodicity can add in phase the PSW generated on each exit aperture, enhancing the PSW. In addition, the same periodicity also enhances the radiation of the PSW into the substrate, which explains the existence of the transmission peak close to 1.4 μm in the experiment of Ebbesen *et al.* Another way of explaining this case is to use the reciprocity theorem, which implies in particular that the transmission into the substrate when light is incident from the cladding is the same in the reciprocal case of transmission into the cladding when light is incident from the substrate side. The latter provides suitable conditions for the excitation of the PSW on the entry side with all the resulting transmission enhancement, as explained above.

Other factors can modify the system response and contribute to its quantitative understanding. For example, the hole form modifies the polarization response. If the dimensions are close to the cut-off, the field inside the aperture is enhanced as the mode propagation constant in the vertical direction approaches zero.

References

[1] T. W. Ebbesen, H. J. Lezec, H. F. Ghaemi, T. Thio, and P. A. Wolff, "Extraordinary optical transmission through sub-wavelength hole arrays," *Nature* **391**, 667–669 (1998).

[2] R. W. Wood, "On a remarkable case of uneven distribution of light in a diffraction grating spectrum," *Phil. Mag.* **4**, 396–402 (1902).

[3] Lord Rayleigh, "On the dynamical theory of gratings," *Proc. Roy. Soc. (London) Ser. A* **79**, 399–416 (1907).

[4] Lord Rayleigh, "Note on the remarkable case of diffraction spectra described by Prof. Wood," *Phil. Mag.* **14**, 60–65 (1907).

[5] D. Maystre, "General study of grating anomalies from electromagnetic surface modes," in *Electromagnetic Surface Modes*, ed. A. D. Boardman (New York: Wiley, 1982), chap. 7.

[6] U. Fano, "The theory of anomalous diffraction gratings and of quasi-stationary waves on metallic surfaces (Sommerfeld's waves)," *J. Opt. Soc. Am.* **31**, 213–222 (1941).

[7] A. Hessel and A. A. Oliner, "A new theory of Wood's anomalies on optical gratings," *Appl. Opt.* **4**, 1275–1297 (1965).

[8] M. C. Hutley and D. Maystre, "Total absorption of light by a diffraction grating," *Opt. Commun.* **19**, 431–436 (1976).

[9] M. Neviére, D. Maystre, R. C. McPhedran, G. H. Derrick, and M. C. Hutley, "On the total absorption of unpolarized monochromatic light," *Proceedings of the ICO-11 Conference*, Madrid, Spain, pp. 609–612 (1978).

[10] R. C. McPhedran, G. H. Derrick, and L. C. Botten, "Theory of crossed gratings," in *Electromagnetic Theory of Gratings*, ed. R. Petit (Berlin: Springer, 1980), chap. 7.

[11] D. A. Weitz, T. J. Gramila, A. Z. Genack, and J. I. Gersten, "Anomalous low-frequency Raman scattering from rough metal surfaces and the origin of the surface-enhanced Raman scattering," *Phys. Rev. Lett.* **45**, 355–358 (1980).

[12] R. Reinisch and M. Neviére, "Electromagnetic theory of diffraction in nonlinear optics and surface enhanced nonlinear optical effects," *Phys. Rev. B* **28**, 1870–1885 (1983).

[13] E. Kretschmann, "Determination of optical constants of metals by excitation of surface plasmons," *Z. Phys.* **241**, 313–324 (1971).

[14] H. Raether, *Surface Plasmons on Smooth and Rough Surfaces and on Gratings* (Berlin: Springer-Verlag, 1988).

[15] M. J. Jory, P. S. Vukusic, and J. R. Sambles, "Development of a prototype gas sensor using surface plasmon resonance on gratings," *Sens. Actuators B* **17**, 203–209 (1994).

[16] U. Schroter and D. Heitmann, "Grating couplers for surface plasmons excited on thin metal films in the Kretschmann-Raether configuration," *Phys. Rev. B* **60**, 4992–4999 (1999).

[17] J. Homola, S. S. Yee, and G. Gauglitz, "Surface plasmon resonance sensors: review," *Sens. Actuators. B* **54**, 3–15 (1999).

[18] T. W. Ebbesen, "Des photons passe-muraille," *La Recherche* **329**, 50–52 (2000).

[19] J. A. Porto, F. T. García-Vidal, and J. B. Pendry, "Transmission resonances on metallic gratings with very narrow slits," *Phys. Rev. Lett.* **83**, 2845–2848 (1999).

[20] P. Lalanne, J. P. Hugonin, S. Astilean, M. Palamaru, and K. D. Möller, "One-mode model and Airy-like formulae for one-dimensional metallic gratings," *J. Opt. A: Pure Appl. Opt.* **2**, 48–51 (2000).

[21] T. Lopez-Rios, D. Mendoza, F. J. García-Vidal, J. Sanchez-Dehesa, and B. Pannetier, "Surface shape resonances in lamellar metallic gratings," *Phys. Rev. Lett.* **81**, 665–668 (1998).

[22] E. Popov, M. Neviére, S. Enoch, and R. Reinisch, "Theory of light transmission through subwavelength periodic hole arrays," *Phys. Rev. B* **62**, 16100–16108 (2000).

[23] S. Enoch, E. Popov, M. Neviére, and R. Reinisch, "Enhanced light transmission by hole arrays," *J. Opt. A: Pure Appl. Opt.* **4**, S83–S87 (2002).

[24] L. Martín-Moreno, F. J. García-Vidal, H. J. Lezec, K. M. Pellerin, T. Thio, J. B. Pendry, and T. W. Ebbesen, "Theory of extraordinary optical transmission through subwavelength hole arrays," *Phys. Rev. Lett.* **86**, 1114–1117 (2001).

[25] F. J. García-Vidal, L. Martín-Moreno, and J. B. Pendry, "Surfaces with holes in them: new plasmonic metamaterials," *J. Opt. A: Pure Appl. Opt.* **7**, S97–S101 (2005).

[26] E. Popov, M. Neviére, A. Sentenac, N. Bonod, A.-L. Ferrenbach, J. Wenger, P.-F. Lenne, and H. Rigneault, "Single-scattering theory of light diffraction by a

circular subwavelength aperture in a finitely conducting screen," *J. Opt. Soc. Am. A* **24**, 339–358 (2007).

[27] E. Popov, M. Neviére, J. Wenger *et al.*, "Field enhancement in single subwavelength apertures," *J. Opt. Soc. Am. A* **23**, 2342–2348 (2006).

[28] M. Neviére, E. Popov, and R. Reinisch, "Electromagnetic resonances in linear and nonlinear optics: phenomenological study of grating behavior through the poles and zeros of the scattering operator," *J. Opt. Soc. Am. A* **12**, 513–523 (1995).

[29] D. Maystre and R. Petit, "Brewster incidence for metallic gratings," *Opt. Commun.* **17**, 196–200 (1976).

[30] M. Sarrazin, J. P. Vigneron, and J. M. Vigoureux, "Role of Wood anomalies in optical properties of thin metallic films with a bidimensional array of subwavelength holes," *Phys. Rev. B* **67**, 085415(1-8) (2003).

[31] C. Genet, M. P. Van Exter, and J. P. Woerdman, "Fano-type interpretation of red-shifts and red-tails in hole array transmission spectra," *Opt. Commun.* **225**, 331–336 (2003).

[32] H. A. Bethe, "Theory of diffraction by small holes," *Phys. Rev.* **66**, 163–182 (1944).

[33] E. Popov, N. Bonod, M. Neviére, H. Rigneault, P.-F. Lenne, and P. Chaumet, "Surface plasmon excitation on a single subwavelength hole in a metallic sheet," *Appl. Opt.* **44**, 2332–2337 (2005).

[34] H. J. Lezec and T. Thio, "Diffracted evanescent wave model for enhanced and suppressed optical transmission through subwavelength hole arrays," *Opt. Express* **12**, 3629–3651 (2004).

[35] G. Gay, O. Alloschery, J. Weiner, H.J. Lezec, C. O'Dwyer, M. Sukharev, and T. Seindeman, "The response of nanostructured surfaces in the near field," *Nature Phys.* **2**, 792 (2006).

[36] F. J. García-Vidal, S. G. Rodrigo, and L. Martín-Moreno, "Foundations of the composite diffracted evanescent wave model," *Nature Phys.* **2**, 790 (2006).

[37] J. Weiner and H. J. Lezec, "Reply: Foundations of the composite diffracted evanescent wave model," *Nature Phys.* **2**, 791 (2006).

[38] L. Aigouy, P. Lalanne, J. P. Hugonin, G. Juié, V. Mathet, and M. Mortier, "Near-field analysis of surface waves launched at nano-slit apertures," *Phys. Rev. Lett.* **98** 153902(1-4) (2007).

[39] P. Lalanne, J. P. Hugonin, H. T. Liu, and B. Wang, "A microscopic view of the electromagnetic properties of sub-wavelength metallic surfaces," *Surface Sci. Rep.* **64**, 453–469 (2009).

[40] H. T. Liu and P. Lalanne, "Microscopic theory of the extraordinary optical transmission," *Nature* **452**, 728–731 (2008).

2

Resonant optical properties of nanoporous metal surfaces

TATIANA V. TEPERIK

2.1 Introduction

The structuring of a metal at nanoscale dimensions results in novel optical properties that are not present for bulk metals. Metallic photonic crystals, metal-based structures with periodicities on the scale of the wavelength of light, have attracted particular attention due to their unique optical properties. Among the approaches taken to prepare a three-dimensional photonic crystal is to take advantage of the self-assembly of spheres from a colloidal solution. Spherical colloidal particles of polymers or silica with diameters ranging from 20 nm up to 1 μm and larger, with low coefficients of variation in their diameter, are readily available. The methods of producing monodispersive colloids are well discussed in ref. [1]. The importance and interest of these particles lies in the fact that it is possible to induce them into a close-packed structure analogous to an ordinary close-packed crystal. There are several methods for self-assembly of colloidal spheres, in particular, sedimentation, evaporation, and electrophoresis. These close-packed arrays of uniform particles offer an attractive and, in principle, simple means to template the three-dimensional structure of a variety of materials.

Generally, self-assembly is restricted to the formation of close-packed two-dimensional or three-dimensional assemblies of colloidal particles. However, the low cost and availability of a relatively easy protocol to obtain this type of photonic crystals, artificial opals, make the self-assembly technique very attractive and widely used. The next step in the development of this technique to prepare metallic photonic crystal is to infiltrate the sample with some appropriate material, removing the original structure, and obtaining in this way inverted opals. The first results obtained concerning metallic inverse opal structures were reported by Jiang et al. [2] and by Velev et al. [3]. Jiang et al. used nanocrystal catalyzed electroless

Structured Surfaces as Optical Metamaterials, ed. A. A. Maradudin. Published by Cambridge University Press.
© Cambridge University Press 2011.

Figure 2.1. Typical scanning electron microscope images of a gold sample for different thicknesses; taken from ref. [11].

deposition to infiltrate the structures with metal [2, 4]. Velev *et al.* [3, 5] infiltrated the structure with colloidal gold particles. Subsequent work has improved the quality of metal porous structures by the use of electrochemical deposition through a colloidal template [6–8], while the first example of the use of electrochemical deposition was demonstrated by Braun and Wiltzius by infiltrating semiconductors such as CdSe and CdS through a colloidal template [9]. It has been shown that electrochemical deposition has a number of significant advantages for fabricating high-quality metal inverse opals with complete control over sample thickness, surface topography, and the structural openness (metal filling fraction) [6–8]. Metal structures made from gold, silver, copper, platinum, palladium, and nickel have been produced. Zhou and Zhao provided an outstanding overview of the self-assembly approaches to the fabrication of three-dimensional porous photonic materials [10], and supplied this work with a large list of references.

In this chapter we discuss the optical properties of nanoporous metal structures. Among the metals, gold is a very promising material due to its strong plasmon properties and chemical inactivity. As a pictorial example of a metal nanostructure we consider a gold nanoporous surface. Nevertheless, the results discussed here can be generalized to other noble metals. The experimental samples of nanoporous gold we will consider were prepared using a nanoscale casting technique with electrochemical deposition of metal through a self-assembled latex template. The templates were produced using a capillary force method, by which the latex spheres were initially deposited on a gold-coated glass slide from a colloidal solution, allowing a monolayer of well ordered spheres to be produced. Electrodeposition, while measuring the total charge passed, allows the accurate growth of a metal to a required thickness t. The resulting metallic mesh reflects the order of the self-assembled close-packed template, allowing convenient control of the pore diameters and regularity of the array. After deposition the template is dissolved, leaving a free-standing structure. This allows the production of shallow, well-spaced dishes as well as nearly encapsulated spherical voids on a single sample. Figure 2.1 shows scanning electron microscope images taken from [11] for a gold sample with a void diameter of 600 nm at three normalized thicknesses $t = 0.2d$, $0.5d$, and $0.9d$, where d is the void diameter. Optical and electron microscopy shows that the resulting surfaces are smooth on the sub-10 nm scale. The sub-3% size dispersion of the latex spheres results in an identical size dispersion of voids, while detailed diffraction measurements confirm the single-domain nature of the samples. A detailed description of the technique can be found in refs. [7] and [12].

2.2 Resonant optical properties of metal surfaces with spherical pores

The optical properties of metallic gratings (in other words one-dimensional metallic photonic crystals) have been the subject of extensive research since the early 1900s. One of the early achievements in the field of optics of metal gratings was the discovery and understanding of the Wood's anomalies [13] in the reflection spectra, which were later assigned into two groups. One type of anomaly is caused by excitation of surface plasmons, the density waves of electrons that propagate along the surface. Another type is the diffractive anomaly, which is associated with the opening of new orders of diffraction into the surrounding media (also called Rayleigh anomalies).

Considerable advances in the assembly of microporous and nanoporous metal structures have prompted the investigation of their optical properties and renewed the interest in these problems. A lattice of voids beneath a metal surface can act as a coupling element, which couples incoming light to the surface plasmons and diffracted beams. Furthermore, in this specific type of resonant grating coupler, the inherent confined resonances in the voids (void plasmon resonances) can be excited. Therefore, one can expect a complex optical response associated with different types of plasmon excitations in periodic porous metal structures.

In this chapter we describe the role of surface plasmons, void plasmons, and diffractive anomalies in molding reflection and absorption spectra of porous metal surfaces. We discuss the interaction between surface plasmons and plasmons localized in buried voids. We also focus on the interaction between diffracted beams and void plasmons.

Before embarking on a treatment of the optical problems discussed above with the use of rigorous electromagnetic methods, we make several estimates with the help of simplified approaches. Let us consider light incident at an angle θ to the surface normal, on a planar surface of a metal that contains a two-dimensional hexagonal lattice of voids with primitive vectors $\mathbf{a}$ and $\mathbf{b}$ ($|\mathbf{a}| = |\mathbf{b}|$) just beneath the surface. The plane of incidence of the incident light is defined by the azimuthal angle ϕ measured with respect to the x-axis (Fig. 2.2).

The diffracted beams emerge at the frequencies of the grazing photons, which are given by

$$\omega = c|\mathbf{q}_{pq}|, \tag{2.1}$$

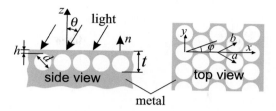

Figure 2.2. Theoretical model of a nanoporous metal surface with a two-dimensional hexagonal lattice of spherical voids.

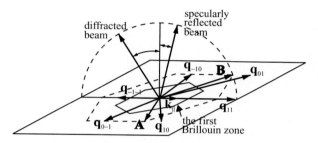

Figure 2.3. The first Brillouin zone and the photons' wavevectors. The incidence here is along the Γ–M direction.

where c is the speed of light, $\mathbf{q}_{pq} = \mathbf{k}_\parallel + \mathbf{g}$, $\mathbf{g} = p\mathbf{A} + q\mathbf{B}$ are reciprocal void-lattice vectors, $\mathbf{A} = 2\pi(\mathbf{b} \times \mathbf{n})/|\mathbf{a} \times \mathbf{b}|$ and $\mathbf{B} = 2\pi(\mathbf{n} \times \mathbf{a})/|\mathbf{a} \times \mathbf{b}|$ are the primitive vectors of the reciprocal two-dimensional void lattice (Fig. 2.3), $\mathbf{n}$ is the normal to the porous layer, p and q are integers, $k_\parallel = k_0 \sin\theta$ is the in-plane component of the wavevector of the incident light, which is equal to zero in the case of normal incidence, and k_0 is the wavenumber of light in the surrounding vacuum.

The dispersion relation for a surface plasmon propagating along a planar vacuum–metal interface can be written in the following form [14]:

$$k_{sp}^2 = \left(\frac{\omega}{c}\right)^2 \frac{\varepsilon(\omega)}{1 + \varepsilon(\omega)}, \tag{2.2}$$

where k_{sp} is the surface plasmon wavenumber and $\varepsilon(\omega)$ is the frequency-dependent dielectric function of the metal. Let us use the local Drude model to describe the dielectric response of the metal to an electric field $\mathbf{E}\exp(-i\omega t)$,

$$\varepsilon(\omega) = 1 - \frac{\omega_p^2}{\omega(\omega + i\nu_e)}, \tag{2.3}$$

where ω_p is the bulk plasma frequency and ν_e is a phenomenological bulk electron relaxation rate. (The case of gold corresponds to $\hbar\omega_p = 7.9$ eV and $\hbar\nu_e = 90$ meV.) Then, for high-conductivity metals (for which $\omega_p \gg \omega$), from Eq. (2.2) one can obtain

$$k_{sp}^2 = \left(\frac{\omega}{c}\right)^2 \frac{\omega^2 - \omega_p^2}{2\omega^2 - \omega_p^2}. \tag{2.4}$$

For a periodic structure the dispersion relation of surface plasmons propagating along the nanostructured surface can be estimated in the "empty lattice approximation." In this case in Eq. (2.2) and Eq. (2.4) the surface plasmon wavenumber k_{sp} should be replaced by $|\mathbf{q}_{pq}| = |\mathbf{k}_\| + \mathbf{g}|$. Thus, for a given wavevector $\mathbf{k}_\|$ the frequencies of surface plasmons are slightly below the frequencies of grazing photons.

The frequencies of void plasmons can be estimated in the framework of a simple model of plasmon modes supported by a spherical void in an infinite metallic medium. The modes of a void are given by the zeros of the denominator in the corresponding scattering-matrix element familiar from Mie scattering theory [15]. (We give the explicit expressions for the scattering matrices of spherical voids, which are used in rigorous electromagnetic calculations, in Section 2.3.) More precisely, for electrical *plasmon* modes supported by a void in a metal (those with zero radial component of the magnetic field)

$$\varepsilon(\omega)h_l^+(\rho_1)[\rho_0 j_l(\rho_0)]' = [\rho_1 h_l^+(\rho_1)]' j_l(\rho_0), \tag{2.5}$$

where $\rho_0 = k_0 d/2$, $\rho_1 = k_0 d\sqrt{\varepsilon(\omega)}/2$, $k_0 = \omega/c$, d is the diameter of the void, $j_l(x)$ is a spherical Bessel function of the first kind of lth order (l is the orbital momentum quantum number), $h_l^+(x)$ is a spherical Hankel function, and the prime denotes differentiation with respect to argument. Note that the magnetic modes, i.e. those with zero radial component of the electric field, do not couple to plasma oscillations in a spherical geometry and therefore are not considered here.

For a small void ($d \ll \lambda$, where λ is the wavelength of light) one can use the asymptotic form of the spherical Bessel and Hankel functions. In this case Eq. (2.5) can be greatly simplified, and with the use of the Drude model (Eq. (2.3)) for the description of the dielectric function of the metal one can obtain the frequencies of plasmon modes in a small void. For $\omega \gg v_e$ one has

$$\omega = \omega_p\sqrt{\frac{l+1}{2l+1}}. \tag{2.6}$$

The fundamental ($l = 1$) plasmon mode of a void with a frequency $\omega = \omega_p\sqrt{2/3}$ is known as the Fröhlich mode [15]. For an electric field with a short spatial distribution we can take $l \to \infty$ and thus obtain the asymptotic nonretarded surface plasmon solution [14], $\omega = \omega_p/\sqrt{2}$. This result shows the common origin of plasmons excited on a flat surface and plasmons localized in voids. Note that for nanoporous metal surfaces, which we consider in this chapter, the pore size is comparable with the wavelength of light. Thus, Eqs. (2.2) and (2.5) should be used to estimate the frequencies of the plasmon modes of a nanostructured surface.

2.3 Self-consistent electromagnetic model: scattering-matrix layer-KKR approach

Theoreticians increasingly come up against the problem of the choice between the use of universal well developed, but still time consuming, methods and methods that are specially elaborated for solving a particular optical problem. We have had the chance to use a method particularly adapted to the problem at hand. In this section we consider the framework of the self-consistent electromagnetic multiple-scattering layer-Korringa–Kohn–Rostoker (KKR) approach first developed for studying the properties of photonic crystals [16, 17] and later adapted for studying the optical properties of metal surfaces with pores [18, 19]. The purpose of this section is to acquaint the reader with the main ideas of the method; a detailed description can be found in refs. [16] and [17].

Let us again address Fig. 2.2, where the theoretical model of a periodic two-dimensional hexagonal lattice of spherical voids inside a metal is presented. The multiple-scattering layer-KKR approach is based on a rigorous solution of Maxwell's equations that makes use of a re-expansion of the plane-wave representation of the electromagnetic field in terms of spherical harmonics [16, 17, 20–23]. This approach involves the following steps. First, the whole structure is divided into parts separated by parallel planes that form two homogeneous semi-infinite media and a planar layer in between that contains the periodic lattice of voids. The homogeneous semi-infinite space below the periodic layer is filled with a homogeneous metal medium. The periodic layer is treated within the multiple-scattering KKR method [22, 23] in a spherical-wave representation, whereas the interaction between the fields in the periodic layer and the field in the homogeneous half-spaces is treated separately in a plane-wave representation.

One of the advantages of this scattering-matrix approach is that it employs explicitly the decomposition of the total field into a sum of waves propagating (or decaying) along and opposite to the **n**-direction (see Fig. 2.2). It allows one to avoid difficulties with convergence when describing the evanescent waves. The total fields in the homogeneous media surrounding the periodic metal layer result from the superposition of plane waves

$$\mathbf{E}_{\text{tot}}^{\pm} = \sum_g \mathbf{E}_g^{\pm} \exp(i\mathbf{K}_g^{\pm}\mathbf{r}) \tag{2.7}$$

with transverse wavevectors

$$\mathbf{K}_g^\pm = \left[\mathbf{k}_{||} + \mathbf{g}, \pm\sqrt{k^2 - (\mathbf{k}_{||} + \mathbf{g})^2} \right],$$

where the sum in Eq. (2.7) runs only over propagating and evanescent waves, $k = k_0\sqrt{\varepsilon(\omega)}$ inside the metal, k_0 is the wavenumber of light in the surrounding vacuum, $\mathbf{g} = p\mathbf{A} + q\mathbf{B}$ are the in-plane reciprocal lattice vectors, $\mathbf{A}$ and $\mathbf{B}$ are the primitive vectors of the two-dimensional reciprocal lattice, p and q are integers, and $\mathbf{r}$ is the radius vector. The superscripts "+" and "−" label waves that propagate (or decay) along and opposite to the $\mathbf{n}$-direction, respectively. The square root of the frequency-dependent dielectric function $\varepsilon(\omega)$ is chosen here to have a non-negative imaginary part. It should be noted that every plane wave in the metal substrate is evanescent at frequencies below the bulk plasma frequency.

The total field inside the layer with a periodic lattice of voids is represented as a superposition of the incoming plane waves (both propagating and evanescent) and the field scattered from every void:

$$\mathbf{E}_{sc}(\mathbf{r}) = \sum_l \sum_{m=-l}^{l} \left(\frac{i}{k} b_{lm}^E \nabla \times \sum_{\mathbf{R}_n} \exp\left(i\mathbf{k}_{||}\mathbf{R}_n\right) h_l^+(kr_n)\mathbf{X}_{lm}(\hat{\mathbf{r}}_n) \right.$$

$$\left. + b_{lm}^H \sum_{\mathbf{R}_n} \exp\left(i\mathbf{k}_{||}\mathbf{R}_n\right) h_l^+(kr_n)\mathbf{X}_{lm}(\hat{\mathbf{r}}_n) \right), \quad (2.8)$$

where $h_l^+(kr)$ is the spherical Hankel function of lth order, which has the asymptotic form describing an outgoing spherical wave $h_l^+(kr) \approx (-i)^l \exp(ikr)/ikr$ at $r \to \infty$, $\mathbf{r}_n = \mathbf{r} - \mathbf{R}_n$, $\mathbf{R}_n$, is the radius vector of the center of the nth void, $\mathbf{X}_{lm}(\hat{\mathbf{r}})$ is a vector spherical harmonic defined by

$$\sqrt{l(l+1)}\mathbf{X}_{lm}(\hat{\mathbf{r}}) = -i\hat{\mathbf{r}} \times \nabla Y_{lm}(\hat{\mathbf{r}}),$$

where the unit vector $\hat{\mathbf{r}}$ denotes the angular variables (θ, ϕ) of the radius vector $\mathbf{r}$ in spherical coordinates, and Y_{lm} are spherical harmonics. The amplitude coefficients $b_{lm}^{E,H}$ of the scattered spherical waves with E and H polarizations in Eq. (2.8) are determined using the scattering matrix of a void that relates the total electromagnetic field incident on a given void to the electromagnetic field scattered from this void:

$$\begin{pmatrix} \mathbf{b}^E \\ \mathbf{b}^H \end{pmatrix} = \begin{pmatrix} \mathbf{T}^E & 0 \\ 0 & \mathbf{T}^H \end{pmatrix} \begin{pmatrix} \mathbf{a}^E \\ \mathbf{a}^H \end{pmatrix}, \quad (2.9)$$

where $\mathbf{b}^{E,H} \equiv \{b_{lm}^{E,H}\}$ are column matrices of $l_{max}(l_{max} + 2)$ elements, $\mathbf{a}^{E,H} \equiv \{a_{lm}^{E,H}\}$ are the column matrices with the amplitude coefficients of the combined electromagnetic field incident on a given void as their elements, and l_{max} is the cutoff value of the angular momentum in the spherical-wave expansion to reach a desired level of convergence. The elements of the scattering matrices of spherical voids depend only on l as

$$T_{lm;l'm'}^{E,H} = T_l^{E,H} \delta_{ll'} \delta_{mm'},$$

where [15, 20]

$$T_l^E = \frac{-\varepsilon(\omega) j_l(\rho_1)[\rho_0 j_l(\rho_0)]' + [\rho_1 j_l(\rho_1)]' j_l(\rho_0)}{\varepsilon(\omega) h_l^+(\rho_1)[\rho_0 j_l(\rho_0)]' - [\rho_1 h_l^+(\rho_1)]' j_l(\rho_0)},$$
$$T_l^H = \frac{-j_l(\rho_1)\rho_0 j_l'(\rho_0) + \rho_1 j_l'(\rho_1) j_l(\rho_0)}{h_l^+(\rho_1)\rho_0 j_l'(\rho_0) - \rho_1 [h_l^+(\rho_1)]' j_l(\rho_0)},$$

(2.10)

$\rho_0 = k_0 d/2$, $\rho_1 = k_0 d\sqrt{\varepsilon(\omega)}/2$, $k_0 = \omega/c$, and d is the diameter of the void. The prime denotes differentiation with respect to argument. The zeros of the denominator in the scattering-matrix elements T_l^E and T_l^H give the frequencies of the electrical and magnetic modes of a single void, respectively. Note that only electrical modes are associated with plasmon oscillations. The magnetic modes have a zero radial component of the electric field, and thus do not couple to plasma oscillations in a spherical geometry.

The field scattered from each void reaches out to other voids and contributes to their scattered fields. The layer-KKR method incorporates this effect by translating outgoing spherical waves from each void to other voids, where they are expressed as spherical components of plane waves. This translation procedure is detailed in refs. [16] and [17]. Then a summation over all of these in-plane scattering contributions is performed directly in real space. Indeed, in our case the metal provides a natural space cutoff distance beyond which the voids cannot *see* each other through the inter-void metal portions.

Therefore, the combined field incident on a given single void is decomposed into spherical waves, each of which is scattered according to Eqs. (2.9) and (2.10). Then the combined field scattered from all voids is back-transformed into the plane-wave representation that is expressed as a sum over the in-plane reciprocal vectors $\mathbf{g}$,

$$\mathbf{E}_{sc}^{\pm} = \sum_g [\mathbf{E}_{sc}^{\pm}]_g \exp(i\mathbf{K}_g^{\pm}\mathbf{r}).$$

Finally, the boundary conditions are applied at the interfaces of the layer containing the lattice of voids with the surrounding media. As a result, the total scattering matrix is constructed, yielding the reflectivity, R, transmittivity, T, and absorption, $A = 1 - R - T$, of the entire structure. This procedure is explained in great detail in refs. [16] and [17], together with the extension to an arbitrary number of layers.

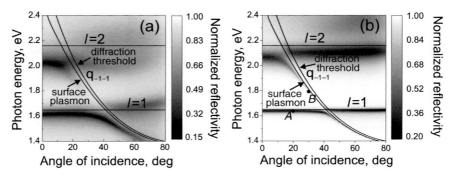

Figure 2.4. Reflectivity spectra measured (a) and calculated (b) for a nanoporous metal surface formed by periodic arrangements of close-packed spherical voids.

The electromagnetic approach described here allows us to obtain the reflection and absorption spectra of nanoporous metal surfaces and to elucidate the interaction of plasmons localized in voids with delocalized surface plasmons, and diffracted beams.

2.4 Optical spectra of nanoporous metal surfaces

In this section we discuss the optical spectra of nanoporous metal surfaces. We present theoretical and experimental results for gold samples. The measurements and rigorous electromagnetic calculations are analyzed on the basis of simplified approaches (see Eqs. (2.1), (2.2), and (2.5)) that allow us to unravel the main physical phenomena.

Figure 2.4(a) shows the measured intensity of the zero-order reflected beam (specular reflection spectrum) as a function of photon energy $\hbar\omega$ and angle of incidence θ for a nanoporous gold surface formed by the periodic arrangement of close-packed spherical nanovoids (with $d = 500$ nm) grown up to the void diameter $t = d$. The angle θ is measured with respect to the direction perpendicular to the surface (see Fig. 2.2). The incident light has p polarization with its plane of incidence along the $\Gamma-M$ direction of the first Brillouin zone (see Fig. 2.3), which corresponds to a zero azimuthal angle ($\phi = 0$).

Calculations (Fig. 2.4(b)) are performed in the framework of the self-consistent electromagnetic multiple-scattering layer-KKR approach described in Section 2.3. In the theoretical model a single two-dimensional hexagonal lattice ($|\mathbf{a}| = 515$ nm) of close-packed voids of diameter $d = 500$ nm is buried just beneath a planar gold surface. The distance from the planar metal surface to the top of the voids, h, is chosen to be 5 nm, which is much smaller than the skin depth (25 nm for gold), in order to model a strong coupling of the incoming light to the nanoporous metal surface. Note that there are residual void openings and interconnections of voids in the sample, which are not described by our theoretical model, although we expect that their effect lies more in the details than in the qualitative optical response. We also assume that there is a residual dielectric material inside the nanopores, which we account for by an effective dielectric constant different from unity ($\varepsilon_{void} = 1.3$).

The reflectivity spectra exhibit dips that are associated with the excitation of plasmons in voids and surface plasmons (see Fig. 2.4). The stronger and almost dispersionless resonances are associated with the excitation of plasmons in the voids. The frequencies of these void-plasmon resonances are close to the frequencies of the void-plasmon modes with orbital quantum numbers $l = 1$ and $l = 2$ of a single void in bulk gold given by Eq. (2.5) (marked by horizontal lines in Fig. 2.4). The slight red shifts are due to the effects of coupling between plasmons in adjacent voids and the disturbance of the void-plasmon mode by the proximity of the planar surface of the metal. The other (dispersive) resonance in the reflectivity spectra originates from the excitation of the surface-plasmon mode with wavevector $\mathbf{q}_{-1-1}$ (see Fig. 2.3) on the planar metal surface. The frequency of this mode is close to that of the corresponding surface-plasmon mode estimated in the "empty lattice approximation" given by Eq. (2.2) for $(p, q) = (-1, -1)$.

The void-plasmon resonances in the calculated specular reflection spectra become stronger near grazing incidence (see Fig. 2.4(b) for $\theta \sim 70° - 80°$). This behavior can be explained for the $l = 1$ dipolar mode, where the dipole orientation should follow approximately the incident electric field polarization (p polarization in our case). Therefore, at normal incidence, the dipole is buried beneath the metal surface at a depth roughly equal to half of the void diameter, whereas increasing the angle of incidence produces rotation of the dipole in such a way that its pole becomes closer to the surface of the metal, which in turn leads to stronger coupling of the void-plasmon mode with the incident light. Similar arguments can be applied to explain the $l = 2$ mode behavior. As to the experiment, such an effect is smeared out in the spectra because the light incident at grazing angles is more sensitive to imperfections of the metal surface that lead to additional scattering, and this, in turn, results in an increased divergence of the reflected beam (see Fig. 2.4(a)).

From Fig. 2.4 it is clear that plasmons excited in the voids produce stronger resonance dips compared with those from surface-plasmon resonances. The reason for this is that the void plasmons are radiative excitations [24, 25], and for the case of a void lattice slightly buried in a metal (void to surface distance h much smaller than the skin depth, which is about 25 nm for gold) these plasmons couple to light efficiently. The surface plasmons are nonradiative excitations and can couple to light only via a coupling element [26] (which is the lattice of voids itself in the structure here). The surface-plasmon resonances on a planar surface of nanoporous metals with nearly encapsulated close-packed voids are rather weak since the effective grating aspect ratio of such a lattice is too small (i.e. the fractional volume occupied by metal in the porous layer is too small) to couple light effectively to nonradiative surface plasmons. Note that the coupling of localized void plasmons and delocalized surface plasmons with light can be tuned in the process of sample growth. Thus, for samples with $t < 0.3d$ the nanostructured surfaces take the form of well-spaced dishes (see Fig. 2.1(a)) that appear to be a very efficient two-dimensional grating for surface-plasmon excitation [11]. In contrast, the localized plasmons are not exhibited in the optical spectra since the voids are not properly molded. Once $t \geq d$, the sample with nearly encapsulated voids supports the localized excitations. The efficiency of their coupling with light is controlled by the void to surface distance h (Fig. 2.2). If the lattice of voids is buried beneath the planar surface of metal at a distance h deeper than the skin depth (about 25 nm for gold), the regime of weak coupling takes place.

One can see in Fig. 2.4 that at a certain angle of incidence the localized plasmons in voids and surface plasmons come into a resonant interaction. Thus, the surface-plasmon mode exhibits an avoided crossing with the first and second void-plasmon modes at angles of incidence $\approx 20°$ and $\approx 45°$, respectively [27]. In the anticrossing regime the interaction between void plasmons and surface plasmons produces two mixed plasmon modes comparable in absorption efficiency.

Additional information about the resonant structures in the reflectivity spectra of nanoporous metal surfaces, and the proof of the localized and delocalized origin of plasmon modes, come from the analysis of the calculated near fields. In Fig. 2.5 the calculated distributions of the normal-to-surface electric-field component induced at the planar surface of a nanoporous metal by the incident light, are plotted over several unit cells of the void lattice. One can perceive that the normal-to-surface electric-field component is the most representative of the main characteristic features of the different plasmon modes in the structure under investigation. Indeed, for example, for normal incidence of the incoming light, this component of the electric field originates entirely from the excitation of plasmons (either void plasmons or surface plasmons) in the structure.

Let us consider electric-field distributions in the uncoupled fundamental ($l = 1$) void-plasmon mode and in the surface-plasmon mode with $(p, q) = (-1, -1)$ excited at points A and B of the dispersion plane in Fig. 2.4, respectively. Note that at points A and B the void plasmon and surface plasmon do not interact with each other. As expected, the void plasmon oscillates with a dumbbell-like field distribution localized in the voids (Fig. 2.5A), which is oriented along the external electric field direction (i.e. the Γ–M direction). At oblique incidence, the orbitals are slanted with respect to the planar metal surface and the void-plasmon oscillations are displaced in phase along the in-plane component of the wavevector of the incident light wave, $k_{\parallel}$ (along the Γ–M direction). The uncoupled

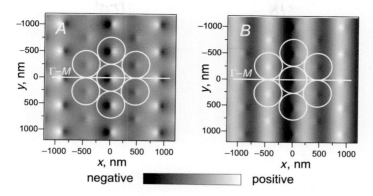

Figure 2.5. Snapshots of the normal-to-surface electric-field component in plasmon modes excited at points *A* and *B* of the dispersion plane in Fig. 2.4.

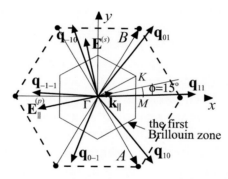

Figure 2.6. The first Brillouin zone and the wavevectors $\mathbf{q}_{pq}$ of surface plasmons in the lowest-frequency subband. For further details see Section 2.2.

surface-plasmon mode is a delocalized plane wave with wavevector $\mathbf{q}_{-1-1}$ propagating along the $\Gamma-M$ direction (Fig. 2.5*B*).

In general, at an arbitrary azimuthal angle and oblique incidence on the hexagonal lattice of voids there are six surface-plasmon resonances observed in the reflectivity [28]. This can be understood from the first Brillouin zone and wavevectors of surface plasmons belonging to the lowest-energy subband ($|q| \leq 1, |p| \leq 1$) shown schematically in Fig. 2.6. However, due to symmetry reasons, some of the six surface-plasmon resonances are degenerate along the $\Gamma-K$ and $\Gamma-M$ directions, and only one surface-plasmon mode is seen in the frequency range of the spectra presented in Fig. 2.4 with $\mathbf{k}_{\parallel}$ directed along $\Gamma-M$.

In Fig. 2.7 calculated reflection spectra of *s*- and *p*-polarized incident light are shown as functions of the photon energy $\hbar\omega$ and the in-plane light wavevector $k_{\parallel}$. The results correspond to a planar gold surface with a two-dimensional hexagonal

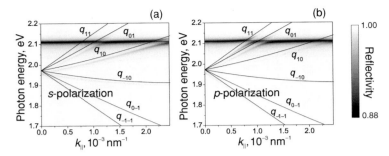

Figure 2.7. Reflectivity spectra for (a) *s*-polarized and (b) *p*-polarized light incident with an azimuthal angle $\phi = 15°$ onto a nanoporous metal surface.

lattice ($|a| = 705$ nm) of voids of diameter $d = 630$ nm. The plane of incidence of the incoming light is defined by the azimuthal angle $\phi = 15°$. The lattice of voids is buried beneath the planar surface of the metal at a distance h of the order of the skin depth (about 25 nm for gold). In such a way the regime of weak coupling of light with plasmon modes of the porous layer can be realized. In this case the resonant dips in the reflectivity spectra are very close to those estimated with the help of the simplified models (see Eqs. (2.2) and (2.5)). In the weak coupling regime the nanoporous metal surface is a poor absorber, $A = 1 - R \leq 0.12$. One can see the series of dips in the reflectivity spectra associated with excitation of surface plasmons and plasmons localized in voids. Again, the stronger and almost dispersionless resonance is associated with excitation of plasmons in the voids. The other (dispersive) resonances in the reflectivity spectra are related to the excitation of surface plasmons on the planar metal surface.

Note that different surface plasmons are excited by *p*-polarized or *s*-polarized light with different efficiencies, even at a small angle of incidence, when the in-plane components of the electric field of the *p*-polarized and *s*-polarized light have nearly equal amplitudes. In the case of a small angle of incidence θ, measured with respect to the surface normal, one has $k_{||} \ll g$. The directions of propagation of different surface-plasmon modes are then nearly parallel to the reciprocal lattice vectors g (Fig. 2.6). Therefore, for an azimuthal angle $\phi = 15°$ as in Fig. 2.7, surface plasmons with wavevectors $(p, q) = (1, 1)$ and $(p, q) = (-1, -1)$ (which nearly coincide with vectors $\mathbf{A} + \mathbf{B}$ and $-(\mathbf{A} + \mathbf{B})$, respectively) are efficiently excited by *p*-polarized light (see Fig. 2.7). This is because the in-plane electric-field component of *p*-polarized light $\mathbf{E}_{||}^{(p)}$ is nearly parallel to the wavevectors of these particular surface plasmons. For the same reason *s*-polarized light efficiently excites surface plasmons with $(p, q) = (1, 0)$ and $(p, q) = (-1, 0)$, which have wavevectors nearly parallel to the vectors $\mathbf{A}$ and $-\mathbf{A}$, respectively. Surface plasmons with $(p, q) = (0, -1)$ and $(p, q) = (0, 1)$ are excited by *p*-polarized and

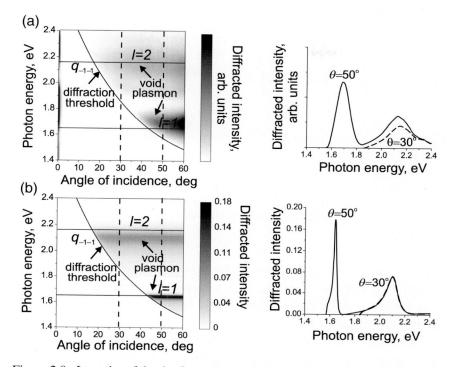

Figure 2.8. Intensity of the the first-order diffracted beam as a function of photon energy $\hbar\omega$ and angle of incidence θ (a) measured from and (b) calculated for a surface of nanoporous gold. The right-hand panels show the measured and calculated spectra corresponding to two angles of incidence, 30° (dashed) and 50° (solid), marked by dashed vertical lines in the contour maps.

s-polarized light with nearly the same moderate efficiency (see Fig. 2.7), since the in-plane electric-field components of either *p*-polarized or *s*-polarized light are directed at an angle of about 45° with respect to the wavevectors $\mathbf{q}_{0-1}$ and $\mathbf{q}_{01}$.

A lattice of voids beneath the metal surface can serve as a diffraction grating [29]. In what follows we will discuss the diffractive anomalies that are associated with the opening of new diffraction beams into the surrounding media (also called Rayleigh anomalies), as well as the interaction between diffracted beams and plasmons localized in voids.

One can notice from Fig. 2.4 that the void-plasmon resonances in the specularly reflected beam quench abruptly (a Rayleigh anomaly occurs) starting with the onset of the diffracted beam with in-plane wavevector $\mathbf{q}_{-1-1}$. The reason for this anomaly is that the diffracted beam takes a certain amount of energy out of the specularly reflected beam. The diffraction threshold in Fig. 2.4 is given by Eq. (2.1).

Figure 2.8 shows the measured and calculated intensities of the first-order diffracted beam with in-plane wavevector $\mathbf{q}_{-1-1}$ as a function of photon energy $\hbar\omega$

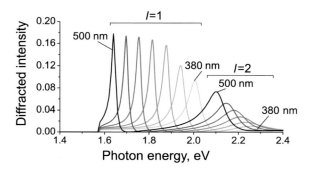

Figure 2.9. Calculated intensity of the first-order diffracted beam versus photon energy $\hbar\omega$ for a surface of nanoporous gold formed by a hexagonal arrangement of spherical voids of different diameters d from 380 to 500 nm (from right to left, in steps of 20 nm).

and angle of incidence θ. The p-polarized light is incident along the Γ–M direction of the first Brillouin zone (see Fig. 2.3), which corresponds to a zero azimuthal angle ($\phi = 0$). The structure parameters are the same as for Fig. 2.4. The calculated intensity of the diffracted beam is normalized to the intensity of the incident light. Naturally, the diffracted beam appears only in the frequency/angle-of-incidence region to the right of the dispersion curve for grazing photons with wavevector $\mathbf{q}_{-1-1}$. However, its intensity becomes strong only at the frequency of the dispersionless resonances associated with excitation of plasmons in the voids. The theoretical results agree qualitatively with experimental data, demonstrating that the diffracted beam takes significant intensity *only* at the void-plasmon resonances: the intensity of the diffracted beam at a void-plasmon resonance is two orders of magnitude stronger than that off resonance.

The frequencies of void-plasmon resonances can be tuned by varying the diameter of the voids. Figure 2.9 shows the calculated intensity of the first-order diffracted beam with in-plane wavevector $\mathbf{q}_{-1-1}$ as a function of photon energy $\hbar\omega$ at a given angle of incidence, $\theta = 50°$, for a surface of nanoporous gold with a hexagonal arrangement of spherical voids of different diameters. Resonances of the diffraction beam intensity follow the blue shift of the void-plasmon resonances ($l = 1$ and $l = 2$) with decreasing void diameter. Observe that the diffracted beam intensity decreases with decreasing void diameter.

In the case under consideration, the diffracted beam with the in-plane wavevector $\mathbf{q}_{-1-1}$ preserves the polarization of the incident wave (p-polarization) with respect to the plane of diffraction (which is the plane defined by the wavevector of the diffracted beam and the normal to the metal surface). The plane of diffraction coincides with the plane of incidence, the Γ–M plane, which is a plane of mirror symmetry of the structure (see Fig. 2.3). The polar angle of diffraction τ can be deduced from the expression $\sin \tau = c|\mathbf{k}_\parallel + \mathbf{g}|/\omega$. It can be easily perceived from Fig. 2.3 that for the diffracted beam with the in-plane wavevector $\mathbf{q}_{-1-1}$ the polar angle of diffraction τ decreases with increasing angle of incidence θ.

Note that the width of plasmon resonances in the experimental plot is broader than predicted by the theory (see Figs. 2.4(a, b) and Figs. 2.8(a, b)). One possible reason for such broadening of the plasmon resonances may be the formation of absorption bands caused by tunneling of light through void interconnections [30] (not accounted for by the theory here), although the role of additional inhomogeneous plasmon broadening caused by imperfections in the experimental sample (e.g. residual void openings) also cannot be ruled out. The experimental data confirm that the void-plasmon resonances are more pronounced in samples whose thickness t approaches the void diameter d, because the localized void-plasmon modes are well molded in almost fully enclosed voids [31].

In closing this section, we have discussed the different types of anomalies in reflectivity spectra of nanoporous metal surfaces caused by (i) plasmons localized in voids, (ii) surface plasmons propagating on the planar metal surface, and (iii) diffractive anomalies, which are associated with the opening of new diffraction beams into the surrounding media (Rayleigh anomalies). Particular attention has been given to the interaction between surface plasmons and diffractive beams with plasmons localized in buried voids.

2.5 Total light absorption in nanostructured metal surfaces

In this section we focus on the absorption properties of a nanoporous metal surface. Let us look again at the reflection spectra of a nanoporous metal surface presented in Fig. 2.4. Since there is no transmission through the structure (the nanoporous layer is located on a metal substrate), the resonant minimum in the reflectivity spectrum below the first diffraction threshold corresponds to the resonant maximum in the absorption spectrum ($A = 1 - R$). It is common knowledge that planar metal surfaces absorb light very poorly. However, if light gets trapped in contact with a nanostructured metal surface and interacts with it for a sufficiently long time, the electromagnetic energy can be absorbed considerably and even totally. To understand the physics behind the effect of total absorbtion by nanostructured metals we consider several approaches. In the following subsection we refer to electrical engineering to develop a simple physical model describing the optical properties of metallic nanoporous structures in terms of equivalent oscillating-current resonant circuits. We explain the total light absorption phenomenon by relating it to the impedance matching condition at the plasmon resonance. This model can be easily adapted to describe the scattering of electromagnetic waves from different nanostructures possessing plasmon resonances. In the framework of another approach we apply the general principles of formal scattering theory developed in the context of quantum mechanics to analyze the interaction of light with resonant nanostructures. The latter allows a description of the total light absorption effect on general physical grounds applicable for an arbitrary resonant layer.

2.5.1 Equivalent resonant RLC circuit model

All structures that transmit transverse electromagnetic waves, including free space, can be described within the transmission-line model [32]. In this model, a nanostructured metal surface can be considered as a load impedance at the end of the transmission line modeling free space. Introducing the effective surface impedance Z_{eff} of the metallic nanostructure, the reflection of light can be described on a common basis independent of the particular kind of resonant nanostructure at hand. In this subsection we present a simple physical model describing the total light absorption phenomenon in nanoporous metallic structures in terms of an equivalent oscillating-current resonant circuit model. This allows us to look at the problem from the electrical engineering point of view, and can be helpful in understanding the optical phenomenon.

Beyond the specific resonant conditions, planar metal surfaces absorb light very poorly. The reason for this is their high free-electron density, which reacts to the incident light by sustaining strong oscillating currents that, in turn, re-radiate light efficiently back into the surrounding medium. However, if one could match the free-space (purely resistive) impedance, $Z_0 = 377$ Ω, to the impedance of the metal surface, the incident-light energy would be completely transformed into ohmic losses inside the metal and the total absorption effect would be achieved.

The bulk conductivity of a metal can be described using the local Drude model as

$$\sigma_e(\omega) = \frac{e^2 N_e}{m(\nu_e - i\omega)},$$

where N_e is the bulk free-electron density, ν_e is a phenomenological bulk electron relaxation rate, ω is the angular frequency of an external electric field $\mathbf{E} \exp(-i\omega t)$, and e and m are the electron charge and mass, respectively. For a planar surface of a homogeneous bulk metal, its effective surface impedance can be written as

$$Z_e = \frac{1}{\sigma_e \delta} = R_e - i\omega L_e,$$

where δ is the skin depth, an intrinsic property of each metal. The effective areal electronic resistance, $R_e = m\nu_e/(e^2\delta N_e)$, determines the amount of power

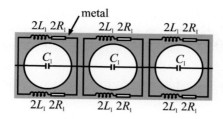

Figure 2.10. Lattice of voids in a metal, along with its equivalent circuit.

absorbed on the metal surface, whereas the areal reactance, $-\omega L_e$, determined by a kinetic electronic inductance, $L_e = m/(e^2 \delta N_e)$, accounts for the phase shift between the electric field and the surface current induced in the skin layer. For homogeneous bulk metals, the strong inequalities $R_e \ll Z_0$ and $\omega L_e \gg R_e$ hold, and the matching condition cannot be satisfied at optical frequencies.

The problem of impedance matching may be solved by using an intrinsic plasmon resonance, for which the effective electron resistance can be drastically enhanced due to the fact that only a small fraction of the total number of electrons participate in the plasmon mode, while the inductive reactance may be canceled by a capacitive reactance inherent in the resonance, leaving the impedance purely resistive. Therefore, the nanoporous metal surfaces exhibiting plasmon resonances in the visible part of the spectrum are candidates for achieving the total light absorption effect in that regime.

Every resonance can be described in terms of its equivalent resonant RLC circuit. We illustrate this equivalent model by considering a two-dimensional lattice of nanovoids in a metal, which we describe by a parallel-resonant circuit. The equivalent RLC circuits of the lattice shown in Fig. 2.10 describe the main physical features of the resonant structures: the effective electronic resistance R_l determines the amount of power absorbed due to ohmic losses associated to oscillating plasmonic electron currents; the kinetic electronic inductance L_l accounts for the phase shift between the oscillating electric field and the current flowing within the skin depth; and the capacitance C_l describes charge accumulation induced by the oscillating current. Note that here l refers to the lth plasmon mode.

The equivalent impedance of this circuit can be written as

$$Z_l = \frac{R_l - i\omega L_l}{1 - \omega^2 L_l C_l - i\omega R_l C_l}. \tag{2.11}$$

Here $R_l = 2m\nu_l/(e^2 \Delta_l \delta N_e)$ is the equivalent areal electronic resistance and $L_l = m/(e^2 \Delta_l \delta N_e)$ is the areal kinetic electronic inductance characteristic of the lth plasmon mode, where ν_l is the damping of the lth plasmon mode due to all dissipative processes, except radiative damping, and Δ_l is the fraction of free electrons participating in the plasma oscillations in the lth mode. We estimate the equivalent areal capacitance as $C_l = |f_l|^2 d\varepsilon_0$, where d is the void diameter, ε_0 is the electrical constant, and $|f_l|^2$ is the dimensionless phenomenological form-factor characteristic of a given lth multipole plasmon mode of a void, which is a free parameter in this simple model. Substituting the expressions for R_l, L_l, and C_l given above into Eq. (2.11), one finds

$$Z_l = \frac{m}{e^2 \Delta_l \delta N_e} \frac{\omega_l^2 (2\nu_l - i\omega)}{(\omega_l^2 - \omega^2 - 2i\omega\nu_l)},$$

where

$$\omega_l = \frac{1}{\sqrt{L_l C_l}} = \sqrt{\frac{e^2 \Delta_l \delta N_e}{|f_l|^2 d\varepsilon_0 m}} \tag{2.12}$$

is the frequency of the lth plasmon mode. In the vicinity of the resonance, $\omega \simeq \omega_l$, assuming that $2\nu_l \ll \omega_l$, we finally obtain

$$Z_l \simeq -i \frac{m}{2e^2 \Delta_l \delta N_e} \frac{\omega_l^2}{(\omega_l - \omega - i\nu_l)}.$$

With these considerations, we can easily obtain the total frequency-dependent equivalent areal impedance of the two-dimensional lattice of voids in the following form:

$$Z_{\text{eff}} \simeq -i \frac{m}{2e^2} \sum_{l=1}^{\infty} \frac{|\beta_l|^2}{\Delta_l \delta N_e} \frac{\omega_l^2}{(\omega_l - \omega - i\nu_l)}, \tag{2.13}$$

where $|\beta_l|^2 < 1$ is the phenomenological coefficient of coupling between the external oscillating electric field $\mathbf{E} \exp(-i\omega t)$ and the lth plasmon mode, which depends on the geometry of the particular structure under consideration.

In the vicinity of the lth plasma resonance, $\omega \simeq \omega_l$, the lth term of the summation dominates the right-hand side of Eq. (2.13) and we have

$$Z_{\text{eff}} \approx -i \frac{m|\beta_l|^2}{2e^2 \Delta_l \delta N_e} \frac{\omega_l^2}{(\omega_l - \omega - i\nu_l)}. \tag{2.14}$$

Now let us consider an electromagnetic plane wave incident normally from a vacuum on a planar surface of a bulk metal that contains a two-dimensional lattice of voids just beneath the surface. In this case of zero transmission through the structure we apply the impedance boundary condition in the form [33]

$$\mathbf{E}_{\text{in}} + \mathbf{E}_{\text{rf}} = Z_{\text{eff}} [\mathbf{n} \times (\mathbf{H}_{\text{in}} + \mathbf{H}_{\text{rf}})]. \tag{2.15}$$

By solving Maxwell's equations in the ambient medium (vacuum) together with the boundary condition Eq. (2.15), it is easy to obtain the complex amplitude reflection coefficient given by

$$r = \frac{Z_{\text{eff}} - Z_0}{Z_{\text{eff}} + Z_0}.$$ (2.16)

The surface impedance given by Eq. (2.14) leads to the following expression for the reflectance and absorptance of light in the neighborhood of the lth plasma resonance:

$$R = rr^* \approx \frac{(\omega_l - \omega)^2 + (\gamma_l - \nu_l)^2}{(\omega_l - \omega)^2 + (\gamma_l + \nu_l)^2},$$ (2.17a)

$$A = 1 - R \approx \frac{4\nu_l \gamma_l}{(\omega_l - \omega)^2 + (\gamma_l + \nu_l)^2}.$$ (2.17b)

Here

$$\gamma_l = |\beta_l|^2 \frac{m\omega_l^2}{2 Z_0 e^2 \Delta_l \delta N_e}$$ (2.18)

is the radiative damping of the lth plasmon mode on the nanoporous metal surface. The absorption resonance described by Eq. (2.17b) has a Lorentzian lineshape with the full width at half maximum $2(\nu_l + \gamma_l)$. The free parameters $|f_l|^2/\Delta_l$ and $|\beta_l|^2/\Delta_l$ can be obtained by fitting the resonance frequency and the full width at half maximum yielded by this simple model of a resonant porous metal surface to their values obtained by the rigorous self-consistent electromagnetic model described in Section 2.3.

Finally, at resonance, $\omega = \omega_l$, one finds

$$R_{\text{res}} \approx \frac{(\gamma_l - \nu_l)^2}{(\gamma_l + \nu_l)^2},$$

$$A_{\text{res}} \approx \frac{4\nu_l \gamma_l}{(\gamma_l + \nu_l)^2},$$

and it is readily seen that nearly total light absorption by the lth plasmon mode ($A_{\text{res}} \approx 1$) occurs when $\nu_l = \gamma_l$, while the reflectivity, R_{res}, drops to zero.

Therefore, using a simple physical model based on equivalent oscillating-current resonant circuits we explain the total light absorption phenomenon by a porous metallic nanostructure. This model can be easily adapted to describe different resonant nanostructures exhibiting total absorption of light at plasmon resonances. In particular, a similar approach has been applied in ref. [34] to describe the total light absorption effect by a two-dimensional square lattice of metallic nanospheres above a metal surface. This structure is inverted with respect to the nanoporous surface.

2.5.2 *General conditions for total light absorption*

It is worth noting that the total absorption effect is of quite general nature. In 1973 the effect of large anomalous absorption was registered in partially disordered silver films [35]. In 1976 the total absorption of light was experimentally and theoretically shown [36, 37] for a metal diffraction grating. These works were later extended both theoretically and experimentally to a variety of metallic nanostructures including metal gratings of different configurations [38–41], doubly periodic metal gratings [42–44], diffraction gratings consisting of cylindrical cavities in a metallic substrate [45], metal–semiconductor–metal nanostructures [46], multilayers of ordered metallic nanoparticles [34], as well as partially disordered metallic nanoparticle arrays [47]. It is also known that electromagnetic energy can be completely absorbed by an overdense plasma slab at the surface-plasmon resonance in the microwave frequency range [48]. Furthermore, according to Kirchhoff's law, the effect of enhanced absorption is relevant for resonant thermal emission, which has been experimentally observed and theoretically explained for metal gratings [49, 50] and doped silicon gratings [51–53]. More precisely, Kirchhoff's law states that the emissivity and absorptance of an object are equal for systems in thermal equilibrium.

Although the nanostructures that exhibit total light absorption might differ considerably, they possess some common characteristics. First, the transmission of light through the entire structure is forbidden. This can be achieved by making the nanostructure thick enough [47, 48] or by locating the nanostructured absorbing layer over a metal slab [34, 36, 37] or over a multilayer Bragg reflector [46] that effectively refocuses the light into the absorbing layer. Second, the total light absorption effect relies on the excitation of intrinsic resonances in the structure, and requires specific conditions of the effective coupling of light with resonant excitations in the system. Therefore, it is relatively common to observe total absorption associated with optical resonances (e.g. in gratings) [54, 55], although this effect has so far been realized only for specific directions of incidence. Only recently has the omnidirectional polarization independent total light absorption by a nanoporous metal surface been demonstrated [56].

Let us define the properties of the nanostructured surface that would lead to the effect of total light absorption sought. We find it appropriate to use a Hamiltonian formalism to recast Maxwell's equations in the form of a Schrödinger equation [57]. The general principles based on the formal scattering theory developed in quantum mechanics can then be applied [58, 59]. Thus, if ω is the frequency of the incident wave, then, for a given wavevector of the incident light parallel to the structure $\mathbf{k}_{||}$, the solution of the scattering problem can be written in the form of the Lippmann–Schwinger equation [60],

$$\Psi = \Psi_0 + G^+(\omega)(H - H_0)\Psi \equiv \Psi_0 + \Psi^+, \qquad (2.19)$$

where H and H_0 are Hamiltonians representing structured and planar metal surfaces, respectively, Ψ^+ is an outgoing wave function containing all components of the electric and magnetic fields for the structured surface, Ψ_0 is the wave function of the unperturbed planar-surface system that satisfies $(H_0 - \omega)\Psi_0 = 0$ and corresponds to a specific angle of incidence and polarization, and $G^+(\omega)$ is the retarded Green function of H.

The entire energy flux in the system is distributed in different scattering channels, among which we single out the *elastic* one (channel 1) describing the incident light and its specular reflection without polarization conversion. All other scattering channels represent reflection with converted polarization, higher-order diffractive beams, and absorption in the metal. Note that transmission through the structure is prevented.

The asymptotic magnetic field of channel 1 can be written as follows:

$$\mathbf{B_1} = \frac{1}{\sqrt{k_\perp}} \left[\hat{\mathbf{e}}_i \exp(-ik_\perp z) + \hat{\mathbf{e}}_r S_{11} \exp(ik_\perp z) \right], \qquad (2.20)$$

where $\hat{\mathbf{e}}_{i,r}$ defines the polarization direction for the incident (i) and reflected (r) waves, and $k_\perp$ is the wavevector component perpendicular to the structure (along the z direction). The first term in Eq. (2.20) represents the incident light, whereas S_{11} is the scattering amplitude. In the vicinity of a pole associated with a localized resonance, the Breit–Wigner multichannel scattering theory permits approximating S_{11} by [61]

$$S_{11} = \exp(2i\delta_{11}) - \frac{iM_{11}\Gamma}{\omega - \omega_0 + i\Gamma/2} \exp(2i\delta_{11}), \qquad (2.21)$$

where the second term describes resonant scattering, and the absorption of light by the nonstructured (flat) metal surface is neglected. Here, ω_0 and Γ are the frequency and width of the resonance, respectively, δ_{11} is a non-resonant scattering phase, and M_{11} is a constant. Flux conservation in the system implies

$$|S_{11}|^2 + J = 1, \qquad (2.22)$$

where J is the flux associated to channels other than channel 1. In what follows we assume that the partial decay rate into channel 1 (i.e. the relevant *radiative* decay rate) equals the sum of the decay rates into all other channels:

$$J = \left| \frac{M_{11}\Gamma}{\omega - \omega_0 + i\Gamma/2} \right|^2. \tag{2.23}$$

Now, using Eqs. (2.21), (2.22), and (2.23) we obtain

$$|M_{11}|^2\Gamma = Re\{i M_{11}(\omega - \omega_0) + M_{11}\Gamma/2\}. \tag{2.24}$$

Equation (2.24) should hold for all frequencies. Therefore $M_{11} = 1/2$, so that the partial decay rate into the first channel is equal to one-half the total decay rate of the resonance. The specular reflection coefficient (reflection into channel 1) is then given by

$$R = |S_{11}|^2 = \frac{(\omega - \omega_0)^2}{(\omega - \omega_0)^2 + \Gamma^2/4}. \tag{2.25}$$

Clearly, R drops to zero at the resonance frequency $\omega = \omega_0$, at which the incident flux is completely transferred into polarization conversion, diffraction, and absorption.

This allows us to assess the conditions for total resonant light absorption assisted by a localized resonance as follows: (i) there is only specular reflection with no diffracted beams; (ii) there is no polarization conversion; and (iii) the radiative decay rate of the resonance equals its dissipative decay rate (rate equipartition condition). Under these conditions the incident light is fully transformed into losses in the metal, such that the absorption $A = 1 - R$ is 100% at the resonance frequency $\omega = \omega_0$.

Note that because of the dispersion of the intrinsic resonant mode, $\omega_0 = \omega_0(\mathbf{k}_{||})$, the total light absorption can be a priori achieved only at a specific value of the incident wavevector $\mathbf{k}_{||}$. That is why only the directional total light absorption effect has been reported for the metallic nanostructures based on gratings that support dispersive surface plasmons. In order to obtain the omnidirectional (for arbitrary incident angles) total light absorption effect, a nondispersive intrinsic resonance in the system should be used. Thus, the nanoporous metal surface appears very promising.

2.5.3 Omnidirectional absorption by a nanoporous metal surface

We have already shown in Section 2.4 that a nanoporous metal surface can simultaneously support both delocalized (surface) plasmons and localized (void) plasmons. The surface plasmons exhibit a strong dispersion. They are highly sensitive to the

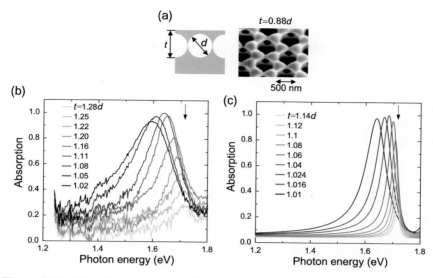

Figure 2.11. Total light absorption in nanostructured metal surfaces. (a) Sketch and scanning electron microscope (SEM) image of nanoporous gold surfaces. (b) Measured and (c) calculated absorption spectra under normal incidence. Results are presented for samples with different thicknesses t as indicated.

angle of incidence, thus they cannot enable omnidirectional total light absorption. We now turn to the properties of plasmons localized in voids. This type of plasmon excitations has no dispersion (see Fig. 2.4). Thus, one can expect that these properties allow us to obtain omnidirectional total light absorption.

We first investigate the absorption spectra of nanoporous metal surfaces for normal light incidence. Figure 2.11(a) presents a sketch and a scanning electron microscope (SEM) image of nanoporous gold surfaces, consisting of a layer of close-packed voids of diameter $d = 500$ nm covered with gold to a thickness t. The SEM image is taken at $t < d$ to help visualize the inner part of the nanovoids. The absorption measurements have been performed for different thicknesses of the nanorporous layer. Since the gold coating was thick enough to prevent transmission through the sample, the absorption is $A = 1 - R$, where R is the measured reflectivity of the specularly reflected beam.

We have simulated absorption spectra using the layer-KKR approach described in Section 2.3 for fully buried voids. The samples have openings at the top of the voids (Fig. 2.11(a)) and possibly also in the touching region between adjacent voids. The effect of partial penetration of the electromagnetic field through these openings has been phenomenologically accounted for by describing the metal using a Maxwell–Garnett dielectric function [33] formed by 55% empty pores embedded in gold, with the gold represented by measured optical data [62].

The calculated results (Fig. 2.11(c)) show good qualitative agreement with the measurements (Fig. 2.11(b)). One can observe the red shift of the resonance from the energy of the dipole plasmon of a single void in gold marked in Figs. 2.11(b) and 2.11(c) by arrows. The red shift as well as the width of the resonant absorption features are larger in the experiment for the reasons explained in Section 2.4. Nonetheless, the same evolution of the maximum absorption with t is observed in experiment and theory, with an optimum metal thickness at which absorption reaches 100%. In other words, the effect occurs at a specific coupling of light with plasmons in voids so that condition (iii) for the total light absorption is satisfied (see the preceding subsection). Note that with a hexagonal lattice of period 500 nm, the wavelength of the onset of diffraction (433 nm) (see Eq. (2.1)) is well below the minimum wavelength explored in Fig. 2.11 (689 nm). Thus, condition (i) is satisfied. As to the polarization conversion upon reflection: while generally it can occur for a metal surface with a hexagonal lattice of voids, it appears to be negligibly small in the present case. This is due to the dipole symmetry of the void-plasmon mode (that follows the polarization of the incident light), as well as to the small coupling between individual voids through the metal membranes. Therefore, upon specular reflection the polarization is preserved, so that condition (ii) is satisfied as well. Consequently, all three conditions are met to enable total absorption.

Now let us turn to the oblique incidence case. As we already discussed in Section 2.4 the nanoporous metal surface can exhibit different types of anomalies. Apart from plasmons localized in voids, at certain angles of incidence one can observe the anomalies in reflectivity (absorption) spectra caused by the excitation of delocalized surface plasmons and Rayleigh anomalies associated with the opening of new diffraction beams into the surrounding media. Thus, at oblique incidence the omnidirectionality in absorption might be strongly affected in some incident angle/frequency intervals, e.g. at the avoided crossing between void and surface plasmons (see Fig. 2.4). However, omnidirectional absorption can be restored using dielectric filling of the voids. Indeed, the void plasmon is very sensitive to the dielectric filling [63, 64]. One can reduce the frequency of void plasmons significantly by filling the pores with a dielectric material. In such a situation the dispersion of surface plasmons and grazing photons remains undisturbed. Indeed, the dispersion of grazing photons depends only on the lattice configuration. As for the surface plasmons propagating along the air–metal interface, they are weakly sensitive to the dielectric filling of voids. As a result, the anticrossing between two plasmon modes is removed.

Figure 2.12(a) shows the calculated incidence-angle dependence of absorption by a layer of 500 nm close-packed silica-filled inclusions buried in gold for p-polarized light incident along the $\Gamma-M$ direction of the void lattice. The metal

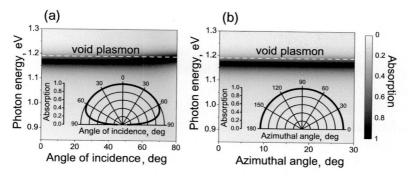

Figure 2.12. Omnidirectional light absorption by the structure formed by a layer of close-packed silica-filled inclusions buried in gold. Calculated (a) incidence-angle and (b) azimuthal-angle dependence of absorption for *p*-polarized light incident along the Γ–M direction of the void lattice.

extends 5 nm above the top of the inclusions to maximize absorption over a wide angular range. Under the same conditions as Fig. 2.12(a), Fig. 2.12(b) shows the azimuthal dependence of absorption for 20° off-normal incidence. The polar plot insets show the angular behavior of absorption at the frequency of the dipolar void plasmon, $\omega = 1.17$ eV. The frequencies of the plasmon resonances of a single silica-filled void are marked by horizontal dashed lines in Fig. 2.12. Comparing results presented in Fig. 2.12 with those in Fig. 2.4 (without dielectric filling of voids), it appears that by filling the pores with silica the energy of plasmons localized in voids is red shifted. There is no anticrossing between the void- and surface-plasmon modes. Note that we simultaneously avoid the scattering of light into nonzero diffraction channels since their energies are close to the energies of surface plasmons for noble metals with a small effective grating aspect ratio (i.e. the fractional volume occupied by the metal in the periodic layer is small), which is the case for a metal surface with a close-packed lattice of voids. Moreover, there is no polarization conversion due to the reasons given above, and conditions (i) and (ii) are well satisfied, while condition (iii) is achieved by tuning the overlayer thickness. Therefore, provided the proper choice of nanostructure parameters is made, all conditions are met for total light absorption, where the entire energy of the incident light is transformed into losses in the metal.

As soon as total light absorption is achieved for normal incidence, it is intuitively clear that for *s*-polarized light with the electric field of the incident light in the plane of the metal surface total light absorption can be achieved at any incident and azimuthal angles. We check the validity of this assumption by rigorous electromagnetic calculations and conclude that the total light absorption effect for a nanoporous metal surface is omnidirectional and polarization-independent as well.

Finally, as a consequence of Kirchhoff's law [65], the nanostructured metal surfaces should exhibit omnidirectional black-body emission at a resonant frequency that can be tuned by varying the size of the dielectric inclusions, thus resulting in efficient, spectrally narrow, wide-angle, thermal emitters.

Summarizing this chapter: we have discussed the optical properties of nanoporous metal surfaces prepared using a nanoscale casting technique with electrochemical deposition of metal through a self-assembled latex template. We have demonstrated that a nanoporous metal surface can support two types of plasmon oscillations: surface plasmons propagating at the planar metal surface and void-localized plasmons. The surface plasmons lead to the dispersive features in the light scattering. The features due to the void plasmons are nearly dispersionless owing to the spherical symmetry of the voids and the deep localization of the voids beneath a metal surface. By analyzing the near field calculations we have demonstrated the localized and delocalized origin of plasmon modes. Furthermore, we have discussed the interaction of localized and delocalized plasmon modes in the anticrossing regime. We have also addressed the diffractive properties of nanoporous metal surfaces, Rayleigh anomalies, that are strongly pronounced in the spectra when resonances are coupled with plasmons localized in voids.

Particular attention has been given to the total light absorption effect. While planar metal surfaces are poor absorbers, when nanostructured they can efficiently absorb electromagnetic energy. We have demonstrated that the physics underlying the total light absorption can be understood from general principles based on the resonance scattering theory developed in quantum mechanics. An additional understanding of the phenomenon can come from the use of the equivalent oscillating-current resonant circuit model. In the latter case the particular configuration of the periodic cell of a plasmonc nanostructure should be taken into account. We have also shown that omnidirectional polarization-independent total light absorption can be achieved using the dispersionless void-plasmon resonance in a nanoporous metal surface.

Acknowledgments

On the theoretical side, this work would not have been possible without strong contributions from F. J. García de Abajo, V. V. Popov, and A. G. Borisov. The experimental work reviewed here was performed by J. J. Baumberg, Y. Sugawara, T. A. Kelf, M. Abdelsalam, and P. N. Bartlett.

References

[1] M. Bardosova and R. H. Tredgold, "Ordered layers of monodispersive colloids," *J. Mater. Chem.* **12**, 2835–2842 (2002).

[2] P. Jiang, J. Cizeron, J. F. Bertone, and V. L. Colvin, "Preparation of macroporous metal films from colloidal crystals," *J. Am. Chem. Soc.* **121**, 7957–7958 (1999).

[3] O. D. Velev, P. M. Tessier, A. M. Lenhoff, and E. W. Kaler, "A class of porous metallic nanostructures," *Nature* **401**, 548 (1999).

[4] K. M. Kulinowski, P. Jiang, H. Vaswani, and V. L. Colvin, "Porous metals from colloidal templates," *Adv. Mater.* **12**, 833–838 (2000).

[5] P. M. Tessier, O. D. Velev, A. T. Kalambur, J. F. Rabolt, A. M. Lenhoff, and E. W. Kaler, "Assembly of gold nanostructured films templated by colloidal crystals and used in surface-enhanced Raman spectroscopy," *J. Am. Chem. Soc.* **122**, 9554–9555 (2000).

[6] P. N. Bartlett, J. J. Baumberg, S. Coyle, and M. E. Abdelsalam, "Optical properties of nanostructured metal films," *Faraday Discuss.* **125**, 117–132 (2004).

[7] M. E. Abdelsalam, P. N. Bartlett, J. J. Baumberg, and S. Coyle, "Preparation of arrays of isolated spherical cavities by self-assembly of polystyrene spheres on self-assembled pre-patterned macroporous films," *Adv. Mater.* **16**, 90–93 (2004).

[8] X.Yu, Y.-J. Lee, R. Furstenberg, J. O. White, and P. V. Braun, "Filling fraction dependent properties of inverse opal metallic photonic crystals," *Adv. Mater.* **19**, 1689–1692 (2007).

[9] P. V. Braun and P. Wiltzius, "Microporous materials: electrochemically grown photonic crystals," *Nature* **402**, 603–604 (1999).

[10] Z. C. Zhou and X. S. Zhao, "3D macroporous photonic materials templated by self assembled colloidal spheres," in *Nanoporous Materials: Science and Engineering, vol. 4*, eds. G. Q. Lu and X. S. Zhao (London: Imperial College Press, 2004), chap. 8.

[11] T. A. Kelf, Y. Sugawara, R. M. Cole *et al.* "Localized and delocalized plasmons in metallic nanovoids," *Phys. Rev. B* **74**, 245415(1–12) (2006).

[12] M. E. Abdelsalam, P. N. Bartlett, J. J. Baumberg, S. Cintra, T. A. Kelf, and A. E. Russell, "Electrochemical SERS at a structured gold surface," *Electrochem. Commun.* **7**, 740–744 (2005).

[13] R. W. Wood, "Anamolous diffraction grating," *Phys. Rev.* **48**, 928–936 (1935).

[14] H. Raether, *Surface Plasmons on Smooth and Rough Surfaces and on Gratings*, Springer Tracts in Modern Physics 111 (Berlin: Springer, 1988).

[15] C. Bohren and D. Hufmann, *Absorption and Scattering of Light by Small Particles* (New York: J. Wiley, 1998).

[16] N. Stefanou, V. Yannopapas, and A. Modinos, "Heterostructures of photonic crystals: frequency bands and transmission coefficients," *Comput. Phys. Commun.* **113**, 49–77 (1998).

[17] N. Stefanou, V. Yannopapas, and A. Modinos, "MULTEM 2: a new version of the program for transmission and band-structure calculations of photonic crystals," *Comput. Phys. Commun.* **132**, 189–196 (2000).

[18] N. Stefanou, A. Modinos, and V. Yannopapas, "Optical transparency of mesoporous metals," *Solid State Commun.* **118**, 69–73 (2001).

[19] T. V. Teperik, V. V. Popov, and F. J. García de Abajo, "Void plasmons and total absorption of light in nanoporous metallic films," *Phys. Rev. B* **71**, 085408(1-9) (2005).

[20] F. J. García de Abajo, "Multiple scattering of radiation in clusters of dielectrics," *Phys. Rev. B* **60** 6086–6102 (1999).

[21] F. J. García de Abajo, "Interaction of radiation and fast electrons with clusters of dielectrics: a multiple scattering approach," *Phys. Rev. Lett.* **82**, 2776–2779 (1999).

[22] J. Korringa, "On the calculation of the energy of a Bloch wave in a metal," *Physica* **13**, 392–400 (1947).

[23] W. Kohn and N. Rostoker, "Solution of the Schrödinger equation in periodic lattices with an application to metallic lithium," *Phys. Rev.* **94**, 1111–1120 (1954).

[24] L. A. Weinstein, *Open Resonators and Open Waveguides* (New York: Golem, 1969).

[25] T. V. Teperik, V. V. Popov, and F. J. García de Abajo, "Radiative decay of plasmons in a metallic nanoshell," *Phys. Rev. B* **69**, 155402(1-7) (2004).

[26] V. M. Agranovich and D. L. Mills, eds. *Surface Polaritons. Electromagnetic Waves at Surface and Interfaces* (Amsterdam: North-Holland Publishing Company, 1982).

[27] T. V. Teperik, V. V. Popov, F. J. García de Abajo, M. Abdelsalam, P. N. Bartlett, T. A. Kelf, Y. Sugawara, and J. J. Baumberg, "Strong coupling of light to flat metals via a buried nanovoid lattice: the interplay of localized and free plasmons," *Opt. Express* **14**, 1965–1972 (2006).

[28] T. V. Teperik, V. V. Popov, F. J. García de Abajo, and J. J. Baumberg, "Tuneable coupling of surface plasmon-polaritons and Mie plasmons on a planar surface of nanoporous metal," *phys. stat. sol.* (c) **2**, 3912–3915 (2005).

[29] T. V. Teperik, V. V. Popov, F. J. García de Abajo, T. A. Kelf, Y. Sugawara, J. J. Baumberg, M. Abdelsalam, and P. N. Bartlett, "Mie plasmon enhanced diffraction of light from nanoporous metal surfaces," *Opt. Express* **14**, 11964–11971 (2006).

[30] I. Romero and F. J. García de Abajo, "Plasmonics in buried structures," *Opt. Express* **17**, 18866–18877 (2009).

[31] T. A. Kelf, Y. Sugawara, J. J. Baumberg, M. Abdelsalam, and P. N. Bartlett, "Plasmonic band gaps and trapped plasmons on nanostructured metal surfaces," *Phys. Rev. Lett.* **95**, 116802(1-4) (2005).

[32] M. Dressel and G. Grüner, *Electrodynamics of Solids* (Cambridge: Cambridge University Press, 2002).

[33] J. D. Jackson, *Classical Electrodynamics* (New York: Wiley, 1999).

[34] T. V. Teperik, V. V. Popov, and F. J. García de Abajo, "Total light absorption in plasmonic nanostructures," *J. Opt. A: Pure Appl. Opt.* **9**, S458–S462 (2007).

[35] O. Hunderi and H. P. Myers, "The optical absorption in partially disordered silver films," *J. Phys. F: Metal Phys.* **3**, 683–690 (1973).

[36] M. C. Hutley and D. Maystre, "The total absorption of light by a diffraction grating," *Opt. Commun.* **19**, 431–436 (1976).

[37] D. Maystre and R. Petit, "Brewster incidence for metallic gratings," *Opt. Commun.* **17**, 196–200 (1976).

[38] L. B. Mashev, E. K. Popov, and E. G. Loewen, "Total absorption of light by a sinusoidal grating near grazing incidence," *Appl. Opt.* **27**, 152–154 (1988).

[39] E. K. Popov, L. B. Mashev, and E. G. Loewen, "Total absorption of light by gratings in grazing incidence: a connection in the complex plane with other types of anomaly," *Appl. Opt.* **28**, 970–975 (1989).

[40] E. Popov, L. Tsonev, and D. Maystre, "Lamellar metallic grating anomalies," *Appl. Opt.* **33**, 5214–5219 (1994).

[41] N. Bonod, G. Tayeb, D. Maystre, S. Enoch, and E. Popov, "Total absorption of light by lamellar metallic gratings," *Opt. Express* **16**, 15431–15438 (2008).

[42] G. H. Derrick, R. C. McPhedran, D. Maystre, and M. Nevière, "Crossed gratings: a theory and its applications," *Appl. Phys.* **18**, 39–52 (1979).

[43] W.-C. Tan, J. R. Sambles, and T. W. Preist, "Double-period zero-order metal gratings as effective selective absorbers," *Phys. Rev. B* **61**, 13177–13182 (1999).

[44] E. Popov, D. Maystre, R. C. McPhedran, M. Nevière, M. C. Hutley, and G. H. Derrick, "Total absorption of unpolarized light by crossed gratings," *Opt. Express* **16**, 6146–6155 (2008).

[45] N. Bonod and E. Popov, "Total light absorption in a wide range of incidence by nanostructured metals without plasmons," *Opt. Lett.* **33**, 2398–2400 (2008).

[46] S. Collin, F. Pardo, R. Teissier, and J.-L. Pelouard, "Efficient light absorption in metal–semiconductor–metal nanostructures," *Appl. Phys. Lett.* **85**, 194–196 (2004).

[47] S. Kachan, O. Stenzel, and A. Ponyavina, "High-absorbing gradient multilayer coatings with silver nanoparticles," *Appl. Phys. B* **84**, 281–287 (2006).

[48] Y. P. Bliokh, J. Felsteiner, and Y. Z. Slutsker, "Total absorption of an electromagnetic wave by an overdense plasma," *Phys. Rev. Lett.* **95**, 165003(1-4) (2005).

[49] M. Kreiter, J. Oster, R. Sambles, S. Herminghaus, S. Mittler-Neher, and W. Knoll, "Thermally induced emission of light from a metallic diffraction grating mediated by surface plasmons," *Opt. Commun.* **200**, 117–122 (1999).

[50] M. Laroche, C. Arnold, F. Marquier, R. Carminati, J. J. Greffet, S. Collin, N. Bardou, and J.-L. Pelouard, "Highly directional radiation generated by tungsten thermal source," *Opt. Lett.* **30**, 2623–2625 (2005).

[51] J.-J. Greffet, R. Carminati, K. Joulain, J.-P. Mulet, S. Mainguy, and Y. Chen, "Coherent emission of light by thermal sources," *Nature* **416**, 61–64 (2002).

[52] F. Marquier, K. Joulain, J.-P. Mulet, R. Carminati, J. J. Greffet, and Y. Chen, "Coherent spontaneous emission of light by thermal sources," *Phys. Rev. B* **69**, 155412(1-11) (2004).

[53] F. Marquier, M. Laroche, R. Carminati, and J.-J. Greffet, "Anisotropic polarized emission of a doped silicon lamellar grating," *J. Heat Trans.* **129**, 11–16 (2007).

[54] R. Petit, *Electromagnetic Theory Of Gratings* (Berlin: Springer Verlag, 1980).

[55] E. G. Loewen and E. Popov, *Diffraction Gratings And Applications* (New York: Marcel Dekker, Inc., 1997) chap. 8, p. 298.

[56] T. V. Teperik, F. J. García de Abajo, A. G. Borisov, M. Abdelsalam, P. N. Bartlett, Y. Sugawara, and J. J. Baumberg, "Omnidirectional absorption in nanostructured metal surfaces," *Nature Photon.* **2**, 299–301 (2008).

[57] A. G. Borisov and S. V. Shabanov, "Lanczos pseudospectral method for initial-value problems in electrodynamics and its applications to ionic crystal gratings," *J. Comput. Phys.* **209**, 643–664 (2004).

[58] A. G. Borisov, F. J. García de Abajo, and S. V. Shabanov, "Role of electromagnetic trapped modes in extraordinary transmission in nanostructured materials," *Phys. Rev. B* **71**, 075408(1-7) (2005).

[59] N. A. Gippius, S. G. Tikhodeev, and T. Ishihara, "Optical properties of photonic crystal slabs with an asymmetrical unit cell," *Phys. Rev. B* **72**, 045138(1–7) (2005).

[60] B. A. Lippmann and J. Schwinger, "Variational principles for scattering processes. I," *Phys. Rev.* **79**, 469–480 (1950).

[61] L. D. Landau and E. M. Lifshitz, *Quantum Mechanics: Non-Relativistic Theory* (Oxford: Pergamon Press, 1981).

[62] P. B. Johnson and R. W. Christy, "Optical constants of the noble metals," *Phys. Rev. B* **6**, 4370–4379 (1972).

[63] E. Prodan, P. Nordlander, and N. J. Halas, "Effects of dielectric screening on the optical properties of metallic nanoshells," *Chem. Phys. Lett.* **368**, 94–101 (2003).

[64] S. Lal, S. Link, and N. J. Halas, "Nano-optics from sensing to waveguiding," *Nature Photon.* **1**, 641–648 (2007).

[65] F. Reif, *Fundamentals of Statistical and Thermal Physics* (New York: McGraw-Hill, 1965).

3

Optical wave interaction with two-dimensional arrays of plasmonic nanoparticles

ANDREA ALÚ AND NADER ENGHETA

3.1 Introduction

Nanotechnology has seen enormous progress in recent years, and various techniques are now available for the realization of ordered periodic arrays of particles with nanoscale dimensions. Electron-beam [1] and interference lithography [2], polymer-based nanofabrication [3], and self-assembly techniques [4] indeed enable producing ordered one-dimensional (1-D), two-dimensional (2-D), and even three-dimensional (3-D) arrays of metallic or dielectric nanoparticles with sizes much smaller than the wavelength of operation. As is well established in the field of optical metamaterials, such arrays may interact with light in anomalous and exotic ways, provided that their unit cells are sufficiently close to the individual or collective resonance of these arrays.

The electromagnetic response of optical metamaterials and metasurfaces is very distinct from that of gratings and photonic crystals. In photonic crystals, for which lattice periods are comparable to the wavelength of operation, it is possible to tailor the optical interaction operating near the Bragg collective resonances and Wood's anomalies associated with their period, whereas in optical metamaterials and metasurfaces, we operate near the plasmonic resonances of the individual inclusions, leading to the advantage of a much broader response in terms of the angle of incidence, and the absence of grating lobes in the visible angular spectrum. On the other hand, unlike photonic crystals, optical metamaterials and metasurfaces require a much smaller scale for their unit cells. Moreover, plasmonic materials, required to support the required resonances at the nanoscale, are usually characterized by intrinsic non-negligible loss and absorption.

Plasmonic materials have indeed experienced a dramatic growth of interest in recent years. In particular, at optical frequencies where they are available in

Structured Surfaces as Optical Metamaterials, ed. A. A. Maradudin. Published by Cambridge University Press. © Cambridge University Press 2011.

nature [5, 6], various potential applications of plasmonic phenomena have been proposed, including transport of optical energy with tight lateral confinement [7–14], left-handed wave propagation [15–18], and optical radiation patterning [19–22]. In the recent past, we have proposed and analyzed different periodic geometries that may realize optical nanotransmission lines in the form of 1-D linear chains of plasmonic nanoparticles [11], planar (2-D) plasmonic nanolayers [14], and plasmonic 3-D nanoarrays of particles, to form optical metamaterials [12]. One relevant aspect of these plasmonic waveguides at IR and optical frequencies is based on their sub-diffractive properties, i.e. the possibility of supporting a guided beam with a sub-wavelength cross-section that may travel over reasonably long propagation distances with a sufficiently large bandwidth of operation. When these eigen-modes couple with propagating modes in free-space, anomalous reflection and transmission properties arise, which may be used for a variety of applications.

In the following, we analyze in detail the interaction of an optical plane wave with 2-D periodic arrays of sub-wavelength plasmonic nanoparticles, and present an analytical solution with fast convergence in the limit of the dipolar approximation for the particles in the periodic array excited by a plane wave with generic polarization and angle of incidence. These results generalize recent analytical approaches for the electromagnetic modeling of planar arrays of nanoparticles [23–27]. We discuss several features of this theoretical solution that highlight the main features and power balance of this anomalous wave interaction in the case of small plasmonic nanoparticles near resonance. Moreover, we apply these theoretical results to specific designs of such arrays, to suggest highly reflective frequency-selective response with relatively flat angular response. These geometries may offer novel possibilities for thin patterned surfaces for filtering, absorbing, and/or radiation patterning applications. In the following, we assume an $\exp(-i\omega t)$ time convention.

3.2 Plane wave excitation of two-dimensional arrays of nanoparticles: theoretical analysis

Consider the geometry of Fig. 3.1, namely, a periodic ordered array of nanoparticles with a rectangular lattice with periods d_x, d_y in a suitable Cartesian reference system. Without loss of generality, the array is positioned in the plane $z = 0$ and is excited by an external plane wave of arbitrary polarization and angle of incidence, which may model with a good degree of approximation a laser beam, or an analogous optical excitation of the array. As we have done extensively in our previous works on wave interactions with small nanoparticles [11, 12], consistent with analogous approaches in the literature [23–25], we approach this problem by

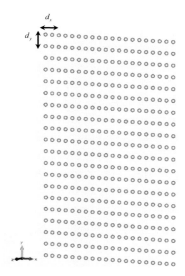

Figure 3.1. Periodic planar array of plasmonic nanoparticles.

assuming that the plasmonic nanoparticles may be described by a polarizability tensor α that relates the induced dipole moment **P** to the local electric field $\mathbf{E}_{loc}$ at the particle position as

$$\mathbf{P} = \alpha \cdot \mathbf{E}_{loc}. \tag{3.1}$$

This assumption implies that the nanoparticles are dominated by their dipole moment, and that they are not too closely packed (which implies that the effect of higher-order multipole moments is negligible when considering the coupling between neighboring elements). Moreover, for simplicity we assume that the orientation and geometry of the particles is chosen in such a way as to avoid polarization mixing, i.e. that the polarizability tensor α is diagonal in the chosen reference system. This assumption includes the special cases of isotropic particles, with a scalar polarizability, and strongly anisotropic particles, like nanodiscs or nanorods, which may become polarized only along one or two axes. Extension of this analysis to a full polarizability tensor is also possible, but beyond the interest of the present work.

Since at optical frequencies materials interact mostly with the impinging electric field (natural magnetism is very weak above the infrared range), the assumption of a dominant electric dipole is justified, and the transverse-electric (TE) polarization excitation is easier to analyze, since it implies an electric field always polarized in the same direction and with constant magnitude, independent of the angle of incidence. In the following subsection we focus on this polarization, and will analyze the transverse-magnetic (TM) case later in the chapter.

3.2.1 TE excitation

For this polarization we can assume, without losing generality, that the impinging plane wave has an electric field of the form

$$\mathbf{E}_{inc} = E_0 \exp(ik_y y) \exp\left[i\sqrt{k_0^2 - k_y^2}\, z\right] \hat{\mathbf{x}}. \tag{3.2}$$

From symmetry, the local field acting on each particle is expected also to have the same polarization $\mathbf{E}_{loc} = E_{loc}\hat{\mathbf{x}}$, ensuring that Eq. (3.1) responds solely to the α_{xx} component of the polarizability tensor. Moreover, due to phase matching considerations, we expect the generic nanoparticle at position $(x = N_x d_x,\ y = N_y d_y)$, where N_x and N_y may be positive or negative integers, to be characterized by a dipole moment

$$\mathbf{P}_{N_x N_y} = \exp(ik_y N_y d_y)\mathbf{P}_{00}, \tag{3.3}$$

where $\mathbf{P}_{00}$ is the dipole moment induced on the nanoparticle placed at the origin, which is yet to be determined.

As is commonly done in analogous problems (see refs. [23]–[25] and references therein), the value of $\mathbf{P}_{00}$ may be evaluated by considering the full coupling among the whole array of interacting nanoparticles. In general, for each nanoparticle we may write

$$\mathbf{P}_{N_x N_y} = \alpha \mathbf{E}_{loc}$$

$$= \alpha \left[\mathbf{E}_{inc} + \sum_{N_x' \neq N_x}^{\infty} \sum_{N_y' \neq N_y}^{\infty} \mathbf{G}\left(\mathbf{r}_{N_x N_y} - \mathbf{r}_{N_x' N_y'}\right) \cdot \mathbf{P}_{N_x' N_y'} \right], \tag{3.4}$$

where $\mathbf{G}$ is the dyadic Green's function in free-space given by

$$\mathbf{G}(\mathbf{r}) = \left(\nabla\nabla + k_0^2 \mathbf{I}\right) \frac{e^{ik_0 r}}{r}, \tag{3.5}$$

where $\mathbf{r}_{N_x N_y}$ is the position of the $\mathbf{P}_{N_x N_y}$ nanoparticle, k_0 is the free-space wave number, and $r = |\mathbf{r}|$. Equation (3.4), combined with Eqs. (3.2) and (3.3), implies that

$$\mathbf{P}_{00} = P_{00}\hat{\mathbf{x}} = \frac{\mathbf{E}_{inc}}{\alpha^{-1} - \sum_{N_x' \neq 0}^{\infty} \sum_{N_y' \neq 0}^{\infty} \mathbf{x} \cdot \mathbf{G}\left(\mathbf{r}_{N_x' N_y'}\right) \cdot \mathbf{x}\, e^{ik_y N_y' d_y}}, \tag{3.6}$$

which contains the full dynamic coupling among the infinite array of nanoparticles [23]. Although the electromagnetic problem is formally solved with Eq. (3.6), the convergence of the summation in the denominator of Eq. (3.6) is usually very slow from the numerical point of view in this direct expression. In order to solve this convergence issue, we can calculate Eq. (3.6) using an approach consistent with the one used in ref. [28] and in ref. [12], but here applied to 2-D arrays of nanoparticles. In this sense, Eq. (3.6) may be rewritten as follows:

$$P_{00} = \frac{E_0}{\alpha^{-1} - \beta_{TE}}, \tag{3.7}$$

where the interaction constant β_{TE}, corresponding to the summations in Eq. (3.6), may be split into two contributions: one that takes into account the coupling from parallel linear arrays of nanoparticles oriented along x, which we have calculated in closed-form in ref. [11], and a second that takes into account the interaction among the infinite set of these parallel arrays, which may be calculated in terms of cylindrical wave functions, similar to what was suggested in refs. [28] and [29], as a Floquet wave expansion. By combining these contributions, a rapidly convergent series equivalent to the summation in Eq. (3.6) for the interaction constant is given by

$$\beta_{TE} = \sum_{N_x' \neq N_x}^{\infty} \sum_{N_y' \neq N_y}^{\infty} \mathbf{G}\left(\mathbf{r}_{N_x' N_y'}\right) \exp(ik_y N_y' d_y) \cdot \hat{\mathbf{x}} \cdot \hat{\mathbf{x}}$$

$$= \frac{k_0^3}{\epsilon_0} \left[\frac{Li_3\left(\exp(i\overline{d}_x)\right) - i\overline{d}_x Li_2\left(\exp(i\overline{d}_x)\right)}{\pi \overline{d}_x^3} \right]$$

$$+ \frac{k_0^3}{\epsilon_0} \sum_{N_y=1}^{\infty} \sum_{N_x=-\infty}^{\infty} \frac{\left(\overline{d}_x^2 - 4\pi^2 N_x^2\right) U\left(4\pi^2 N_x^2 - \overline{d}_x^2\right)}{\pi \overline{d}_x^3}$$

$$\times K_0 \left(N_y \overline{d}_y \sqrt{\frac{4\pi^2 N_x^2}{\overline{d}_x^2} - 1}\right) \cos(N_y \overline{d}_y \overline{k}_y)$$

$$+ \frac{k_0}{2\epsilon_0} \sum_{N_x=-\infty}^{\infty} \frac{\left(\overline{d}_x^2 - 4\pi^2 N_x^2\right) U\left(\overline{d}_x^2 - 4\pi^2 N_x^2\right)}{\overline{d}_x^3 \overline{d}_y} \left\{ \frac{1}{\sqrt{\overline{k}_y^2 - 1 + 4\pi^2 N_x^2/\overline{d}_x^2}} \right.$$

$$+ \frac{\overline{d}_x^2 - 4\pi^2 N_x^2 + 2\overline{k}_y^2 \overline{d}_x^2}{8\pi^3 \overline{d}_y^{-3} \overline{d}_x^2 \xi(3)^{-1}} + \frac{\overline{d}_y}{\pi} \left[\ln \frac{\sqrt{\left|\overline{d}_x^2 - 4\pi^2 N_x^2\right|}}{4\pi \overline{d}_y^{-1} \overline{d}_x} + \gamma \right] \right]

$$-\frac{i\overline{d}_y}{2} - \sum_{N_y=1}^{\infty} \left[\frac{\overline{d}_y}{N_y\pi} + \frac{\overline{d}_y^3 \left(\overline{d}_x^2 - 4\pi^2 N_x^2 + 2\overline{k}_y^2\overline{d}_x^2 \right)}{8(N_y\pi)^3 \overline{d}_x^2} \right.$$

$$-\frac{1}{\sqrt{\left(\overline{k}_y - 2\pi N_y/\overline{d}_y\right)^2 - 1 + 4\pi^2 N_x^2/\overline{d}_x^2}}$$

$$\left. -\frac{1}{\sqrt{\left(\overline{k}_y + 2\pi N_y/\overline{d}_y\right)^2 - 1 + 4\pi^2 N_x^2/\overline{d}_x^2}} \right] \Bigg\}. \tag{3.8}$$

In this expression, $Li_N(z) = \sum_{k=1}^{\infty} z^k/k^N$ is the polylogarithm function [30], $K_0(\ldots)$ is the modified Bessel function of the second kind and zeroth order, $U(\ldots)$ is the step function, which is unity for a positive argument and zero for a negative argument, ϵ_0 is the free-space permittivity, $\overline{d}_i = k_0 d_i$, $\overline{k}_y = k_y/k_0$, $\xi(\ldots)$ is the Riemann zeta function, and γ is the Euler–Mascheroni gamma constant [31]. The first term in Eq. (3.8) takes into account the coupling among particles lying along the x axis, and it coincides with the closed-form expression derived in ref. [11] for a linear array with longitudinally polarized dipoles, calculated here for the case in which $k_x = 0$. The second term corresponds to the infinite summation of evanescent Floquet modes below cut-off, coupling the parallel arrays of nanoparticles forming the planar array. This summation has a very rapid convergence, since the cylindrical waves considered here are all below cut-off, and the arguments of the $K_0(\ldots)$ functions are all real. Finally, the third addend takes into account the propagating (above cut-off) Floquet modes coupling the parallel chains, which are finite in number. In the case of sufficiently small periods ($\overline{d}_x < 2\pi$), only the dominant Floquet mode propagates and carries power, and this summation is limited to the $N_x = 0$ term. Although cumbersome looking, this series in general has a very rapid convergence, in particular when $\overline{d}_x$ and $\overline{d}_y$ are sufficiently small to ensure that only one Floquet cylindrical wave dominates the coupling among the parallel linear arrays, and that only the dominant reflected and transmitted plane waves radiate away from the surface.

Averaged description of the periodic surface

Once P_{00} is obtained from Eq. (3.7), the electromagnetic problem is solved, and it is possible to derive various other electromagnetic quantities of interest. In particular, it is possible to define averaged quantities that may describe in an accurate, but simplified, way the interaction of a plane wave with the array. For example, the averaged surface current density is given by

$$\mathbf{J}_{av} = J_{av}\hat{\mathbf{x}} = \frac{-i\omega \mathbf{P}_{00}\exp(ik_y y)}{d_x d_y} = \frac{-i\omega E_0 \exp(ik_y y)\hat{\mathbf{x}}}{d_x d_y(\alpha^{-1} - \beta_{TE})}. \tag{3.9}$$

Such an averaged current sheet sustains the radiation from the dominant reflected and transmitted plane waves from the array, associated with the dominant Floquet plane wave. This is an accurate description of the electromagnetic properties of the array, as long as the periods are small enough to ensure that only one Floquet harmonic affects the far-field radiation from the surface. We will derive a formal condition for this to happen later in this section. The current expression presented in Eq. (3.9) produces a reflected wave with electric field given by

$$\mathbf{E}_r = -\frac{\omega\mu_0}{\sqrt{k_0^2 - k_y^2}}\frac{J_{av}}{2}\exp\left[-i\sqrt{k_0^2 - k_y^2}\,z\right]\hat{\mathbf{x}} \tag{3.10}$$

and a transmitted wave formed by the superposition of the impinging wave in the half-space behind the array and a scattered wave with the same magnitude as Eq. (3.10):

$$\mathbf{E}_t = \mathbf{E}_{inc} - \frac{\omega\mu_0}{\sqrt{k_0^2 - k_y^2}}\frac{J_{av}}{2}\exp\left[i\sqrt{k_0^2 - k_y^2}\,z\right]\hat{\mathbf{x}}. \tag{3.11}$$

In this limit of one propagating Floquet plane wave on each side of the array, the reflection and transmission coefficients are easily calculated as

$$R = \frac{i\omega^2\mu_0}{2\sqrt{k_0^2 - k_y^2}d_x d_y(\alpha^{-1} - \beta_{TE})}, \tag{3.12a}$$

$$T = 1 + R. \tag{3.12b}$$

The averaged surface impedance, namely the ratio of the local electric field to the averaged current density, is given by

$$Z_s = \frac{E_0}{J_{av}} - \frac{\omega\mu_0}{2\sqrt{k_0^2 - k_y^2}}. \tag{3.13}$$

These averaged relations are valid as long as only one plane wave is radiated out from each side of the array. In general, as discussed in the following, when the periods become larger, side lobes and higher-order Floquet harmonics may propagate out of the surface, and these averaged descriptions may not be sufficient for an accurate description of interaction of the array with the impinging plane wave.

Power balance

It is instructive to analyze in more detail the analytical expressions obtained and, in particular, their energy relations, which are associated with the imaginary part of the denominator in Eq. (3.7). The overall power extracted from the impinging plane wave by each nanoparticle is given by

$$P_{ext} = \frac{\omega |E_0|^2}{2} Im \left[\frac{1}{\alpha^{-1} - \beta_{TE}} \right] = \frac{\omega |E_0|^2}{2} \frac{Im[\beta_{TE} - \alpha^{-1}]}{|\alpha^{-1} - \beta_{TE}|^2}, \quad (3.14)$$

which is also equal to the energy extracted from the impinging electric field by the averaged current density, Eq. (3.9), as expected.

Equation (3.14) may be further manipulated by considering the physical requirements on the imaginary parts of α. In particular, the imaginary part of α^{-1} is associated with the radiation and absorption loss from each individual nanoparticle [11, 32], implying that

$$Im[\alpha^{-1}] = -\frac{k_0^3}{6\pi \epsilon_0} - \alpha_{loss}^{-1} \leq -\frac{k_0^3}{6\pi \epsilon_0}, \quad (3.15)$$

where the equal sign holds when $\alpha_{loss}^{-1} = 0$, i.e. when the particles are lossless. In Eq. (3.15) the first addend gives the radiation loss from an individual particle and the second gives the possible Ohmic loss in it. Note that Eq. (3.15) is totally independent of the specific nature of the nanoparticles, and it generally applies as long as the dipolar approximation holds. The geometry of the nanoparticles, and in particular its resonant features, are directly related to $Re[\alpha^{-1}]$ (which is zero at the resonance of the individual isolated nanoparticle), whereas $Im[\alpha^{-1}]$ is directly associated with the power balance, as described in Eq. (3.15). Moreover, by applying an analytical derivation analogous to the one in ref. [28] it is possible to express in closed-form the imaginary part of β_{TE}. In general, by inspecting Eq. (3.8) and using the results in refs. [11] and [28], we can write

$$Im[\beta_{TE}] = -\frac{k_0^3}{6\pi \epsilon_0} + k_0 \sum_{N_x, N_y = -\infty}^{\infty}$$

$$\times \frac{\left(1 - 4\pi^2 N_x^2 \overline{d}_x^{-2}\right) U \left(1 - 4\pi^2 N_x^2 \overline{d}_x^{-2} - \left[\overline{k}_y - 2\pi N_y \overline{d}_y^{-1}\right]^2\right)}{2 d_x d_y \epsilon_0 \sqrt{1 - 4\pi^2 N_x^2 \overline{d}_x^{-2} - \left(\overline{k}_y - 2\pi N_y \overline{d}_y^{-1}\right)^2}}.$$

$$(3.16)$$

The first term compensates the radiation loss addend in Eq. (3.15), due to the coupling of the whole periodic array, whereas the second term takes into account the power radiated by each of the propagating Floquet modes. It is evident, therefore, that the condition that ensures that the array supports only one dominant Floquet harmonic (and therefore the averaged quantities derived above may accurately describe the far-field interaction with the array), is given by the following condition on the periods:

$$
\frac{4\pi^2 N_x^2}{\overline{d}_x^2} + \left(\overline{k}_y - \frac{2\pi N_y}{\overline{d}_y}\right)^2 > 1 \qquad \forall N_x, N_y \neq 0. \tag{3.17}
$$

Evidently, this condition is satisfied for periods sufficiently smaller than the wavelength of operation. When this condition is satisfied, the summation in Eq. (3.16) has only one nonzero element for $N_x = N_y = 0$, and

$$
Im[\beta_{TE}] = -\frac{k_0^3}{6\pi\epsilon_0} + \frac{1}{2d_x d_y} \frac{\omega^2\mu_0}{\sqrt{k_0^2 - k_y^2}}. \tag{3.18}
$$

The power extracted by each particle may therefore be written in the general case as

$$
P_{ext} = \frac{\omega|E_0|^2}{2\left|\alpha^{-1} - \beta_{TE}\right|^2}\left\{\frac{k_0}{2d_x d_y\epsilon_0} \sum_{N_x, N_y=-\infty}^{\infty} \left[1 - \frac{4\pi^2 N_x^2}{\overline{d}_x^2}\right]\right.
$$
$$
\times \frac{U\left(1 - 4\pi^2 N_x^2\overline{d}_x^{-2} - \left[\overline{k}_y - 2\pi N_y\overline{d}_y^{-1}\right]^2\right)}{\sqrt{1 - 4\pi^2 N_x^2\overline{d}_x^{-2} - \left(\overline{k}_y - 2\pi N_y\overline{d}_y^{-1}\right)^2}} + \alpha_{loss}^{-1}\right\}, \tag{3.19}
$$

which for one single propagating Floquet plane wave becomes

$$
P_{ext} = \frac{\omega|E_0|^2}{2\left|\alpha^{-1} - \beta_{TE}\right|^2}\left[\frac{\omega^2\mu_0}{2d_x d_y\sqrt{k_0^2 - k_y^2}} + \alpha_{loss}^{-1}\right]. \tag{3.20}
$$

For lossless particles in particular, Eq. (3.20) simplifies to

$$
P_{ext} = \frac{\omega^3\mu_0|E_0|^2}{4d_x d_y\sqrt{k_0^2 - k_y^2}\left|\alpha^{-1} - \beta_{TE}\right|^2}. \tag{3.21}
$$

The power radiated from the array is given by the sum of the power flows carried by the radiated Floquet plane waves above cut-off on the two sides of the surface, which from symmetry considerations have the same magnitude. In particular, when condition (3.17) holds and the averaged current density, Eq. (3.9), is sufficient to describe the array, the reflected and transmitted waves in Eqs. (3.10)–(3.11) each radiate an overall power per unit cell given by

$$P_{rad} = \frac{\omega \mu_0 |J_{av}|^2}{8\sqrt{k_0^2 - k_y^2}} d_x d_y = \frac{\omega^3 \mu_0 |E_0|^2}{8 d_x d_y \sqrt{k_0^2 - k_y^2} |\alpha^{-1} - \beta_{TE}|^2}. \tag{3.22}$$

As expected from energy considerations, this quantity is exactly one-half of the extracted power in the lossless scenario, Eq. (3.21). Equation (3.19) indeed represents a generalized power balance relation for the array, and each term corresponds to a specific contribution to the total energy extracted from the impinging plane wave: when Ohmic losses are present, the absorbed power is given by

$$P_{abs} = \frac{\omega |E_0|^2}{2 |\alpha^{-1} - \beta_{TE}|^2} \alpha_{loss}^{-1}, \tag{3.23}$$

and if higher-order Floquet modes of order (N_x, N_y) are propagating away from the surface, each of them contributes an additional term to the summation (3.19) with magnitude:

$$P_{rad} = \frac{\omega |E_0|^2 k_0}{4 d_x d_y \epsilon_0 |\alpha^{-1} - \beta_{TE}|^2} \frac{\left(1 - 4\pi^2 N_x^2 \overline{d}_x^{-2}\right)}{\sqrt{1 - 4\pi^2 N_x^2 \overline{d}_x^{-2} - \left(\overline{k}_y - 2\pi N_y \overline{d}_y^{-1}\right)^2}}. \tag{3.24}$$

Of course, this quantity is real only when Eq. (3.17) is not satisfied for the specific (N_x, N_y) pair.

Finally, it is noted that, in the lossless scenario, power balance also requires

$$\frac{1}{2} Re[E_0 J_{av}^*] = \frac{\omega \mu_0}{4\sqrt{k_0^2 - k_y^2}} |J_{av}|^2, \tag{3.25}$$

which implies

$$Re\left[\frac{E_0}{J_{av}}\right] = \frac{\omega \mu_0}{2\sqrt{k_0^2 - k_y^2}}, \tag{3.26}$$

a relation that is valid as long as condition (3.17) holds. This ensures, using Eq. (3.13), that $Re[Z_s] = 0$ for any lossless array when only the dominant Floquet wave propagates, as expected.

Resonance condition

The largest induced dipole moments in Eq. (3.7) are obtained when $Re[\alpha^{-1}] = Re[\beta_{TE}]$, which corresponds to the surface resonance. It is observed that there are two possibilities of achieving a resonance, which correspond to two distinct phenomena supported by the array. The first possibility is that $Re[\alpha^{-1}] \simeq 0$ for sufficiently small periods, such that also $Re[\beta_{TE}] \simeq 0$, which implies nonresonant coupling in the array. This case implies that the nanoparticles are relatively close to their individual resonance frequency, with a slight shift associated with the coupling with the other nanoparticles in the array. In order to achieve this situation, special nanoparticles should be employed that may support a strong electric resonance in a small volume. At optical frequencies, plasmonic materials serve this purpose, since plasmonic nanoparticles may resonate even for a very small electrical size [6]. As a second possibility, nanoparticles far from their electric dipole resonance (such as dielectric nanoparticles), characterized by a large $Re[\alpha^{-1}]$, may still support a collective resonance as long as $Re[\beta_{TE}]$ is sufficiently large that the difference $Re[\alpha^{-1}] - Re[\beta_{TE}]$ in the denominator of Eq. (3.7) vanishes. This implies that the periods of the array are not small and a lattice resonance is in place. As we describe in the following, these two physical mechanisms have very different resonant properties for the array.

At the surface resonance, the average surface current density assumes the value

$$
\begin{aligned}
\mathbf{J}_{av} &= \frac{-\omega E_0 \exp(ik_y y)\hat{\mathbf{x}}}{d_x d_y Im\left[\alpha^{-1} - \beta_{TE}\right]} \\
&= \frac{\omega E_0 \exp(ik_y y)\hat{\mathbf{x}}}{d_x d_y} \left[\frac{1}{2d_x d_y} \frac{\omega^2 \mu_0}{\sqrt{k_0^2 - k_y^2}} + \alpha_{loss}^{-1}\right]^{-1},
\end{aligned} \tag{3.27}
$$

where we have assumed condition (3.17), which is necessary for a proper meaning of the averaged surface current density. This description is particularly relevant when plasmonic resonances are considered, for which the array periods are smaller than the wavelength of operation. It may be seen that $\mathbf{J}_{av}$ at resonance is purely in phase with the impinging plane wave, as expected (this implies a maximized power interaction with the impinging wave and zero net stored energy). In the absence of losses, this expression simplifies to

$$
\mathbf{J}_{av} = \frac{2\sqrt{k_0^2 - k_y^2} E_0 \exp(ik_y y)\hat{\mathbf{x}}}{\omega \mu_0}. \tag{3.28}
$$

The reflection and transmission coefficients at resonance are given by

$$R = - \left(1 + \alpha_{loss}^{-1} \frac{2\sqrt{k_0^2 - k_y^2}d_x d_y}{\omega^2 \mu_0} \right)^{-1}, \qquad (3.29a)$$

$$T = \frac{2\alpha_{loss}^{-1}\sqrt{k_0^2 - k_y^2}d_x d_y}{\omega^2 \mu_0 + 2\alpha_{loss}^{-1}\sqrt{k_0^2 - k_y^2}d_x d_y}, \qquad (3.29b)$$

which tend to

$$R = -1 \qquad (3.30a)$$

and

$$T = 0 \qquad (3.30b)$$

in the lossless case, realizing a perfect electric conducting surface at the ideal lossless resonance of such a planar array. Indeed, at resonance, for a lossless surface with periods satisfying Eq. (3.17), we obtain from Eqs. (3.28), (3.26), and (3.13)

$$Z_s = 0. \qquad (3.31)$$

Note that these limiting results, Eqs. (3.30)–(3.31), are independent of the specific choice of the array periods; i.e., in the ideal limit of zero loss, even a very sparse array may support total reflection and zero transmission.

3.2.2 TM excitation

For TM plane wave incidence, the situation is complicated by the fact that the electric field also has a nonzero normal component of polarization. In this case, without loss of generality, the impinging plane wave may be assumed to have a magnetic field,

$$\mathbf{H}_{inc} = \frac{E_0}{\eta_0} \exp(ik_x x) \exp\left[i\sqrt{k_0^2 - k_x^2}\, z \right] \hat{\mathbf{y}}, \qquad (3.32)$$

which implies

$$\mathbf{E}_{inc} = E_0 \left\{ \sqrt{1 - \frac{k_x^2}{k_0^2}} \exp(ik_x x) \exp\left[i\sqrt{k_0^2 - k_x^2}\, z\right] \hat{\mathbf{x}} \right.$$
$$\left. - \frac{k_x}{k_0} \exp(ik_x x) \exp\left[i\sqrt{k_0^2 - k_x^2}\, z\right] \hat{\mathbf{z}} \right\}. \tag{3.33}$$

Since the magnetic field does not interact directly with plasmonic nanoparticles (we are assuming here nonmagnetic materials), each of the two components of the electric field is responsible for the polarization of the array. In particular, for symmetry reasons the two components of $\mathbf{E}_{inc}$ separately induce a polarization vector on the whole array parallel to $\hat{\mathbf{x}}$ and $\hat{\mathbf{z}}$, respectively, and due to the linearity of the problem we can write, in analogy with Eq. (3.7),

$$P_{00x} = \sqrt{1 - \frac{k_x^2}{k_0^2}} \frac{E_0}{\alpha_{xx}^{-1} - \beta_{xx}}, \tag{3.34a}$$

$$P_{00z} = -\frac{k_x}{k_0} \frac{E_0}{\alpha_{zz}^{-1} - \beta_{zz}}, \tag{3.34b}$$

where α_{xx} and α_{zz} are the polarizability components of the nanoparticles in the two directions, and β_{xx}, β_{zz} are the corresponding interaction factors. As an example, the accelerated form of β_{xx} in this case is as follows:

$$\beta_{xx} = \frac{k_0^3}{\epsilon_0} \left\{ \frac{Li_3\left(\exp[i\overline{d}_x(1 + \overline{k}_x)]\right) + Li_3\left(\exp[i\overline{d}_x(-1 + \overline{k}_x)]\right)}{\pi \overline{d}_x^3} \right.$$
$$\left. -i \frac{Li_2\left(\exp[i\overline{d}_x(1 + \overline{k}_x)]\right) + Li_2\left(\exp[i\overline{d}_x(-1 + \overline{k}_x)]\right)}{\pi \overline{d}_x^2} \right\}$$
$$+ \frac{k_0^3}{\epsilon_0} \sum_{N_y=1}^{\infty} \sum_{N_x=-\infty}^{\infty} \frac{\left(\frac{2\pi N_x}{\overline{d}_x} + \overline{k}_x\right)^2 - 1}{\pi \overline{d}_x} U\left(\left(\frac{2\pi N_x}{\overline{d}_x} + \overline{k}_x\right)^2 - 1\right)$$
$$\times K_0 \left(N_y \overline{d}_y \sqrt{\left(\frac{2\pi N_x}{\overline{d}_x} + \overline{k}_x\right)^2 - 1}\right)$$
$$- \frac{k_0}{2\epsilon_0} \sum_{N_x=-\infty}^{\infty} \left[\overline{d}_x^{-1}\overline{d}_y^{-1}\left(\frac{2\pi N_x}{\overline{d}_x} + \overline{k}_x\right)^2 - 1\right] U\left([2\pi N_x + \overline{k}_x \overline{d}_x]^2 - \overline{d}_x^2\right)$$

$$\times \left\{ \frac{\overline{d}_x}{\sqrt{\left(2\pi N_x + \overline{k}_x \overline{d}_x\right)^2 - \overline{d}_x^2}} + \frac{\overline{d}_x^2 - \left(2\pi N_x + \overline{k}_x \overline{d}_x\right)^2}{8\pi^3 \overline{d}_x^2 \overline{d}_y^{-3} \xi(3)^{-1}} \right.$$

$$+ \frac{\overline{d}_y}{\pi} \left[\ln \frac{\sqrt{\left| \left(2\pi N_x + \overline{k}_x \overline{d}_x\right)^2 - \overline{d}_x^2 \right|}}{4\pi \overline{d}_y^{-1} \overline{d}_x} + \gamma \right] - \frac{i \overline{d}_y}{2}$$

$$+ \sum_{N_y=1}^{\infty} \left[\frac{2}{\sqrt{4\pi^2 N_y^2 \overline{d}_y^{-2} + (2\pi N_x \overline{d}_x^{-1} + \overline{k}_x)^2 - 1}} \right.$$

$$\left. \left. - \frac{\overline{d}_y}{N_y \pi} - \frac{\overline{d}_x^2 - (2\pi N_x + \overline{k}_x \overline{d}_x)^2}{8(N_y \pi \overline{d}_y^{-1})^3 \overline{d}_x^2} \right] \right\}, \tag{3.35}$$

where the first term corresponds to the closed-form interaction in a linear array of nanoparticles aligned along x with phase shift k_x, consistent with ref. [11], and the following summations refer once again to the Floquet expansion of the cylindrical waves radiated by each array, divided into cylindrical waves below cut-off (second addend in Eq. (3.35)) and propagating cylindrical waves (third addend).

Averaged description of the periodic array

For small enough periods, an averaged description of the array granularity is still in place, since only the dominant radiated plane waves will constitute the far-field of the array. For this polarization, the corresponding averaged surface current density also has two components:

$$\mathbf{J}_{av} = J_{avX} \hat{\mathbf{x}} + J_{avZ} \hat{\mathbf{z}}$$

$$= \frac{-i\omega E_0 \exp(ik_x x)}{d_x d_y} \left[\sqrt{1 - \frac{k_x^2}{k_0^2}} \frac{\hat{\mathbf{x}}}{\alpha_{xx}^{-1} - \beta_{xx}} - \frac{k_x}{k_0} \frac{\hat{\mathbf{z}}}{\alpha_{zz}^{-1} - \beta_{zz}} \right], \tag{3.36}$$

which in turn produces, under this assumption of one single propagating Floquet harmonic, reflected and transmitted plane waves with magnetic fields given by

$$\mathbf{H}_r = \left[\frac{J_{avX}}{2} + \frac{k_x J_{avZ}}{2\sqrt{k_0^2 - k_x^2}} \right] \exp\left[-i\sqrt{k_0^2 - k_x^2} \, z \right] \hat{\mathbf{y}}, \tag{3.37a}$$

$$\mathbf{H}_t = \mathbf{H}_{inc} - \left[\frac{J_{avX}}{2} + \frac{k_x J_{avZ}}{2\sqrt{k_0^2 - k_x^2}} \right] \exp\left[i\sqrt{k_0^2 - k_x^2} \, z \right] \hat{\mathbf{y}}. \tag{3.37b}$$

The corresponding reflection and transmission coefficients for magnetic fields may be written as follows:

$$R = -\frac{i}{2\epsilon_0 \sqrt{k_0^2 - k_x^2} d_x d_y} \left\{ \frac{k_0^2 - k_x^2}{\alpha_{xx}^{-1} - \beta_{xx}} + \frac{k_x^2}{\alpha_{zz}^{-1} - \beta_{zz}} \right\}, \qquad (3.38a)$$

$$T = 1 + \frac{i}{2\epsilon_0 \sqrt{k_0^2 - k_x^2} d_x d_y} \left\{ \frac{k_0^2 - k_x^2}{\alpha_{xx}^{-1} - \beta_{xx}} - \frac{k_x^2}{\alpha_{zz}^{-1} - \beta_{zz}} \right\}. \qquad (3.38b)$$

It is easily verified that for normal incidence this reflection coefficient collapses to the opposite sign of Eq. (3.12) for TE polarization, as expected (these are reflection and transmission coefficients for the magnetic fields). Moreover, when the particles cannot be polarized longitudinally ($\alpha_{zz} = 0$, as in the case of flat nanodiscs composing the surface), the coefficients have the dual form of Eq. (3.12), as expected:

$$R = \frac{-i\omega}{2d_x d_y (\alpha_{xx}^{-1} - \beta_{xx})} \frac{\sqrt{k_0^2 - k_x^2}}{\omega \epsilon_0}, \qquad (3.39a)$$

$$T = 1 - R. \qquad (3.39b)$$

In general, however, both polarizability factors come into play for TM oblique incidence. For this reason, the equivalent surface impedance now has a shunt component, the dual of Eq. (3.13), and a series component related to the normal polarization of the surface, which is always present unless $\alpha_{zz} = 0$ or $k_x = 0$. In this latter case, the shunt surface impedance has the following expression:

$$Z_s = \frac{E_0}{J_{av}} \sqrt{1 - \frac{k_x^2}{k_0^2}} \frac{\sqrt{k_0^2 - k_x^2}}{2\omega \epsilon_0}. \qquad (3.40)$$

Power balance

The overall power extracted by each nanoparticle from a TM impinging plane wave may be written in this case as

$$P_{ext} = \frac{\omega |E_0|^2}{2k_0^2} Im \left[\frac{k_0^2 - k_x^2}{\alpha_{xx}^{-1} - \beta_{xx}} + \frac{k_x^2}{\alpha_{zz}^{-1} - \beta_{zz}} \right]. \qquad (3.41)$$

The radiated power from the array has a more challenging derivation here, since on each side of the array the radiated wave is composed of the contribution from the two coexisting polarizations, which, in general, may interfere constructively or destructively. Limiting ourselves to the case of one dominant Floquet mode for simplicity, for which Eq. (3.38) holds, the radiated power per nanoparticle in the backward direction, associated with the reflected wave, reads as

$$
P_{rad}^{(R)} = \frac{\left| \sqrt{k_0^2 - k_x^2} J_{avX} + k_x J_{avZ} \right|^2 d_x d_y}{8\omega \sqrt{k_0^2 - k_x^2} \epsilon_0}
$$

$$
= \frac{\omega |E_0|^2}{k_0^2} \frac{1}{8\sqrt{k_0^2 - k_x^2} \epsilon_0 d_x d_y} \left| \frac{k_0^2 - k_x^2}{\alpha_{xx}^{-1} - \beta_{xx}} - \frac{k_x^2}{\alpha_{zz}^{-1} - \beta_{zz}} \right|^2. \tag{3.42}
$$

Similarly, the plane wave radiated along the positive z axis, in the transmission region, carries an overall power per unit cell given by

$$
P_{rad}^{(T)} = \frac{\left| \sqrt{k_0^2 - k_x^2} J_{avX} - k_x J_{avZ} \right|^2 d_x d_y}{8\omega \sqrt{k_0^2 - k_x^2} \epsilon_0}
$$

$$
= \frac{\omega |E_0|^2}{k_0^2} \frac{1}{8\sqrt{k_0^2 - k_x^2} \epsilon_0 d_x d_y} \left| \frac{k_0^2 - k_x^2}{\alpha_{xx}^{-1} - \beta_{xx}} + \frac{k_x^2}{\alpha_{zz}^{-1} - \beta_{zz}} \right|^2. \tag{3.43}
$$

The overall radiated power per unit cell is readily obtained as

$$
P_{rad} = P_{rad}^{(R)} + P_{rad}^{(T)}
$$

$$
= \frac{\omega |E_0|^2}{k_0^2} \frac{1}{4\sqrt{k_0^2 - k_x^2} \epsilon_0 d_x d_y} \left\{ \left| \frac{k_0^2 - k_x^2}{\alpha_{xx}^{-1} - \beta_{xx}} \right|^2 + \left| \frac{k_x^2}{\alpha_{zz}^{-1} - \beta_{zz}} \right|^2 \right\}. \tag{3.44}
$$

It is seen that, due to the polarization along z, the radiated power on the two sides of the array is not the same, due to the break in the symmetry. However, interestingly enough, the *total* radiated power is simply given by the sum of the power flows independently associated with the two current polarizations. As expected, Eqs. (3.42) and (3.43) coincide for normal incidence (equaling Eq. (3.22)), and in the case of $\alpha_{zz} = 0$, for which the symmetry is restored, due to the absence of normal polarization in the array.

The analogous expression to Eq. (3.16) for β_{xx} may be derived from Eq. (3.35):

$$Im[\beta_{xx}] = -\frac{k_0^3}{6\pi\epsilon_0}$$

$$+k_0 \sum_{N_x,N_y=-\infty}^{\infty} \frac{\left[1 - \left(\overline{k}_x - 2\pi N_x \overline{d}_x^{-1}\right)^2\right]}{2d_x d_y \epsilon_0 \sqrt{1 - \left(\overline{k}_x - 2\pi N_x \overline{d}_x^{-1}\right)^2 - 4\pi^2 N_y^2 \overline{d}_y^{-2}}}$$

$$\times U\left(1 - \left(\overline{k}_x - 2\pi N_x \overline{d}_x^{-1}\right)^2 - 4\pi^2 N_y^2 \overline{d}_y^{-2}\right). \tag{3.45}$$

Consistent with the TE case, for small enough periods such that

$$\left(\overline{k}_x - \frac{2\pi N_x}{\overline{d}_x}\right)^2 + \frac{4\pi^2 N_y^2}{\overline{d}_y^2} > 1, \qquad \forall N_x, N_y \neq 0, \tag{3.46}$$

only one dominant TM plane wave is radiated on each side of the array, simplifying Eq. (3.45) to

$$Im[\beta_{xx}] = -\frac{k_0^3}{6\pi\epsilon_0} + \frac{1}{2d_x d_y}\frac{\sqrt{k_0^2 - k_x^2}}{\epsilon_0}, \tag{3.47}$$

which is the dual of Eq. (3.18). Following similar arguments, the equivalent of Eq. (3.47) may be derived for $Im[\beta_{zz}]$ as

$$Im[\beta_{zz}] = -\frac{k_0^3}{6\pi\epsilon_0} + \frac{1}{2d_x d_y}\frac{k_x^2}{\sqrt{k_0^2 - k_x^2}\epsilon_0}. \tag{3.48}$$

In the absence of Ohmic absorption, and when condition (3.46) is satisfied, the extracted power from each nanoparticles may be written as

$$P_{ext} = \frac{\omega|E_0|^2}{4k_0^2\epsilon_0 d_x d_y}\left[\frac{(k_0^2 - k_x^2)^{3/2}}{|\alpha_{xx}^{-1} - \beta_{xx}|^2} + \frac{k_x^4}{\sqrt{k_0^2 - k_x^2}|\alpha_{zz}^{-1} - \beta_{zz}|^2}\right]$$

$$= P_{rad}, \tag{3.49}$$

which confirms the power conservation requirements. As noticed in the TE case, the presence of Ohmic loss in the array, or excitation of higher-order Floquet modes, increases the extracted power accordingly.

Finally, it is noticed in this polarization that when $\alpha_{zz} = 0$ and a purely shunt surface reactance is available, power balance requires

$$\frac{1}{2} Re \left[E_0 \sqrt{1 - \frac{k_x^2}{k_0^2}} J_{av}^* \right] = \frac{\sqrt{k_0^2 - k_x^2}}{4\omega\epsilon_0} |J_{av}|^2, \tag{3.50}$$

which implies

$$Re \left[\frac{E_0}{J_{av}} \right] = \frac{\sqrt{\mu_0/\epsilon_0}}{2}, \tag{3.51}$$

as long as condition (3.46) holds. In parallel with the case of TE polarization, this ensures, using Eq. (3.40), that $Re[Z_s] = 0$ for any lossless array when only the dominant Floquet wave propagates, as expected. This result may be generalized to arrays with a nonzero normal component of polarizability α_{zz} by also considering a series component of the surface reactance.

All the theoretical results derived in this section hold for a generic planar periodic array, independent of its nature, material, and geometry of the nanoparticles, as long as the dipolar approximation holds. In the following section, we describe some specific examples of interest to validate the preceding theory, and provide further insights into the design of these arrays for practical applications of interest.

3.3 Numerical results and design principles

After having established the main electromagnetic properties of a general 2-D array of nanoparticles excited by an external plane wave, we report in the following some numerical examples and verification of the preceding results that address the potentials of plasmonic arrays of nanoparticles in a variety of practical problems. We focus for simplicity on the TE polarization for most of this section, since we have highlighted the differences and analogies with the TM case in Section 3.2.

3.3.1 TE polarization: lossless nanoparticles

Consider a planar array of nanospheres with permittivity ϵ and radius $a \ll \lambda_0$, where λ_0 is the free-space wavelength, with an isotropic polarizability tensor

$$\boldsymbol{\alpha} = \alpha \mathbf{I} = -\frac{6\pi i \epsilon_0 c_1^{TM}}{k_0^3} \mathbf{I} \simeq 4\pi \epsilon_0 a^3 \frac{\epsilon - \epsilon_0}{\epsilon + 2\epsilon_0} \mathbf{I}, \tag{3.52}$$

where c_1^{TM} is the first TM Mie scattering coefficient as defined in ref. [32] and $\mathbf{I}$ is the identity tensor. The approximate expression in the last term of Eq. (3.52) holds for small spherical nanoparticles in the "quasi-static" small-radius limit. In

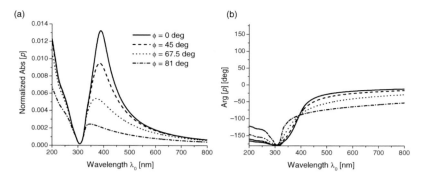

Figure 3.2. Dispersion of the magnitude and phase of the induced dipole moments on a periodic array with $a = 10$ nm, $d_x = d_y = 22$ nm.

our calculations, we use the exact expression for the Mie scattering coefficient c_1^{TM} in Eq. (3.52), which ensures complete power balance [33].

Let us assume first the case of lossless spheres, with permittivity $\epsilon = Re[\epsilon_{Ag}]$, where ϵ_{Ag} is the experimentally retrieved permittivity of silver in the visible, as found in ref. [5]. This allows us to consider a realistic dispersion model for the permittivity of the nanosphere, neglecting for the moment the presence of realistic losses. This is particularly instructive in order to analyze the effects of radiation loss and higher-order Floquet modes in the wave interaction with the array. Later in this section we will also consider realistic loss in the materials of the nanospheres.

Figure 3.2 reports the normalized induced dipole moment $p = (k_0^3/6\pi\epsilon_0)P_{00}$ for TE plane wave incidence for a uniform periodic array of nanospheres with radius $a = 10$ nm and equal periods $d_x = d_y = 22$ nm. The angle of incidence is varied from the normal $\phi = 0$ to the grazing angle, and the figure reports the amplitude and phase of the normalized dipole moment.

It is seen that the induced dipole moment on each nanoparticle experiences a significant peak at the frequency (around $\lambda_0 = 400$ nm) for which $Re[\alpha^{-1}] = \beta_{TE}$. Due to the different coupling among particles for different angles of incidence, the peak shifts slightly in frequency. It is also seen that a sharp minimum in the induced dipole moment is experienced, independent of the angle of incidence, near $\lambda_0 = 300$ nm. This is simply associated with the fact that around this wavelength $Re[\epsilon_{AG}] \simeq \epsilon_0$, and we are neglecting absorption by the particles.

Figure 3.3 reports the corresponding reflection coefficient for the same planar array, showing a sharp reflection peak, always centered around the same resonance frequency, independent of the incidence angle. This stability of the frequency of maximum reflection is associated with the sub-wavelength features of this array (this is drastically different from regular Wood's anomalies associated with lattice resonances, for which the angle of incidence significantly affects the resonance

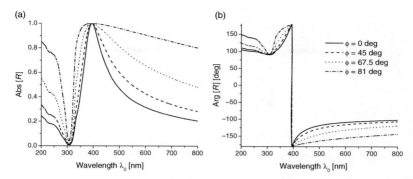

Figure 3.3. Dispersion of the reflection coefficient from the periodic array of Fig. 3.2 ($a = 10$ nm, $d_x = d_y = 22$ nm).

frequency, as we show in the following for larger periods). Note that, despite the presence of gaps between neighboring particles, the reflection may become total at the array resonance in this lossless scenario. In fact, as shown in the preceding section, total reflection is achieved at resonance independent of the period, and even sparse arrays would produce total reflection at the resonance frequency (in this lossless scenario). The array density affects the bandwidth and Q-factor of the resonance. As anticipated in Section 3.2, the phase of the reflection coefficient (for electric fields) is 180° at resonance, ensuring the effect of an equivalent perfect electric conductor around the resonance frequency. For larger angles, the reflection bandwidth increases due to the naturally larger reflection from the array in this polarization.

We do not report here the corresponding transmission coefficient, which is simply given by $|T| = \sqrt{1 - |R|^2}$ as long as the far-field is formed just by the dominant Floquet mode (as for this choice of periods) due to the absence of absorption.

Figure 3.4 reports the equivalent shunt impedance for this geometry, which is the same for all incidence angles, due to the absence of higher-order propagating Floquet modes and the small granularity of the array. This is consistent with recent findings on metasurfaces whose surface impedance is weakly dependent on the angle of incidence [27]. It is seen that the impedance is purely reactive due to the absence of absorption on the surface, and it crosses zero at the resonance frequency, ensuring total reflection. The surface impedance diverges near the transparency frequency, as expected. In Fig. 3.5 we report the magnitude of the induced dipole moment on planar arrays with $d_x/3 = d_y = 2.2a$ and $a = 10$ nm (Fig. 3.5a), $a = 20$ nm (Fig. 3.5b). On comparing Fig. 3.5(a) with Fig. 3.2, it is seen that a larger spacing in the $\hat{x}$ direction ensures a sharper resonance and stronger induced dipole moments. This is expected, and is consistent with our results for linear arrays and 3-D arrays of nanoparticles: decreasing the coupling in the direction of polarization

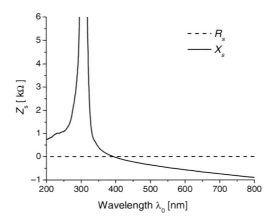

Figure 3.4. Dispersion of the effective surface impedance for the periodic array of Fig. 3.2 ($a = 10$ nm, $d_x = d_y = 22$ nm). This impedance is the same for all angles of incidence.

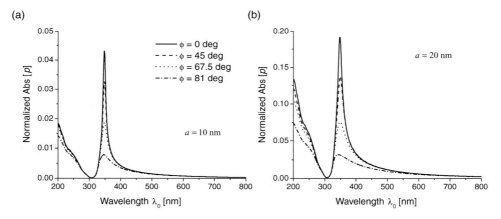

Figure 3.5. Dispersion of the magnitude of the induced dipole moments for a periodic array with $d_x/3 = d_y = 2.2a$ for two different sizes of the particles.

of the nanoparticles ensures sharper and stronger resonant properties for the whole array. Of course, the price to be paid is an inherently anisotropic response.

Figure 3.5(b) shows the effect of scaling the whole geometry by a factor of two. It is seen how the resonant peak is somehow shifted to larger frequencies due to the different size of the individual particles and the different coupling among them. The induced dipole moments are also larger, as expected due to the larger nanoparticle polarizability. The dependence on the angle of incidence is analogous in all these examples.

(a)

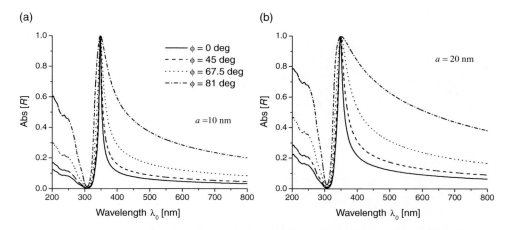

(b)

Figure 3.6. Dispersion of the magnitude of the reflection coefficient for the periodic arrays of Fig. 3.5.

(a)

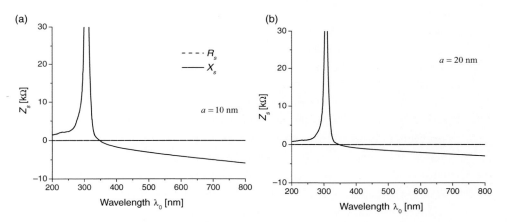

(b)

Figure 3.7. Dispersion of the effective surface impedance for the periodic arrays of Fig. 3.5. This impedance is the same for all angles of incidence.

Figure 3.6 reports the corresponding magnitude of the reflection coefficient for these arrays. It is evident that the angular response is analogous to the one reported in Fig. 3.3, but a larger distance d_x ensures a narrower bandwidth and sharper filtering properties (which is quite counterintuitive, given the strong reduction in the number of nanoparticles employed for the array). Moreover, larger nanoparticles inherently correspond to larger bandwidths of operation.

Figure 3.7 reports analogous plots for the effective surface impedance in these cases. Although the impedance is always purely reactive and crosses zero at the array's resonance frequency, the variation and slope of surface reactance versus frequency may be tuned by varying the geometry of the array, increasing or reducing

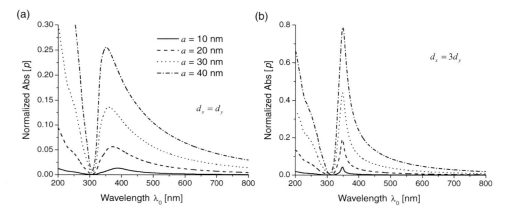

Figure 3.8. Dispersion of the magnitude of the induced dipole moments for plane
wave excitation at normal incidence in a periodic array with $d_y = 2.2a$, for several
sizes of the particles.

the resonance bandwidth. These plots may be used to tailor the dispersion properties
of the surface reactance with a proper design of the nanoparticles and their period.
Also, in these examples, the surface impedance is the same for all incidence angles.

Figure 3.8 reports the normalized induced dipole moment for normal incidence
and several sizes of the nanoparticles, for $d_y = 2.2a$ and $d_x = d_y$ (Fig. 3.8a), $d_x = 3d_y$ (Fig. 3.8b). These plots show how the induced dipole moment at resonance
grows approximately as the square of the diameter for small nanoparticles, and
the resonance frequency is fairly constant with scaling the array. As expected,
the Q-factor of the resonance increases for anisotropic arrays when the period is
increased in the direction of the polarization of the electric field.

Figure 3.9 shows the corresponding amplitude of the reflection coefficient $|R|$,
which always reaches unity at the resonance frequency, as expected, due to the
absence of absorption. It is evident that the resonance bandwidth is strongly affected
by the variation in the period of the array along the electric field polarization.

Figures 3.10 and 3.11 report similar plots, but for larger nanoparticles, in the case
$d_x = 3d_y$. It is observed how further increasing a, and correspondingly increasing the periods, causes the occurrence of Wood's anomalies or lattice resonances
associated with higher-order Floquet harmonics excited by the impinging plane
wave. These resonances correspond to the scenario in which the nanoparticles
are far from their individual resonance, but the lattice period is comparable with
the transverse wavelength and may produce an overall resonance effect. In this
scenario, condition (3.17) is not met below a certain wavelength, and additional
side lobes are produced in the far-field of the surface, on both sides of the array.
It is seen that the occurrence of these lattice resonances produces a very distinct
behavior when compared to the "quasi-static" small-radius resonances associated

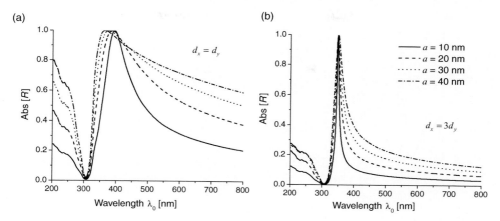

Figure 3.9. Dispersion of the reflection coefficient for the geometries of Fig. 3.8.

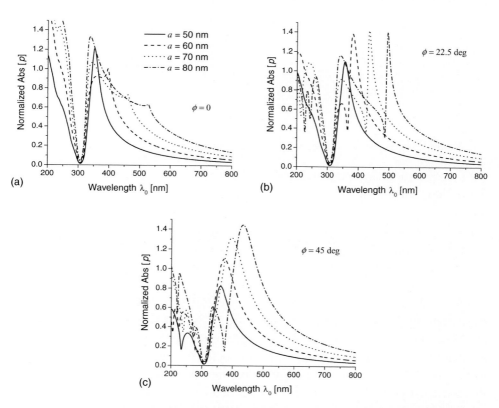

Figure 3.10. Dispersion of the magnitude of the induced dipole moments for plane wave excitation in a periodic array with $d_x/3 = d_y = 2.2a$, for several sizes of the particles.

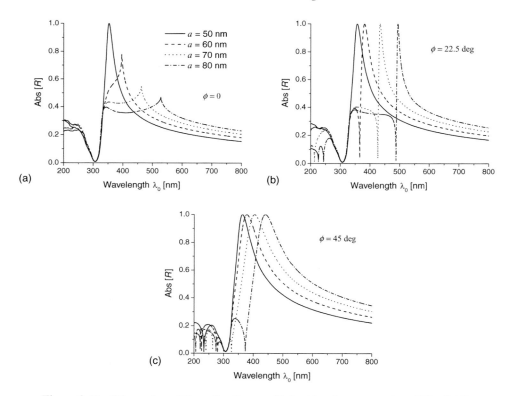

Figure 3.11. Dispersion of the reflection coefficient for the geometries of Fig. 3.10.

with the plasmonic properties of the nanoparticles for smaller sizes: as expected, these Wood's resonances are strongly dependent on the period (varied here together with the nanoparticle size) and on the incidence angle, which varies the effective transverse wavelength seen by the periodic array, consistent with our previous theoretical results. In Fig. 3.10, we observe that these lattice resonances may provide additional peaks in the induced dipole moment, distinct from the plasmonic resonances.

In Fig. 3.11, we note that the reflection coefficient dispersion behaves more irregularly in the region where condition (3.17) is not satisfied, with peaks that are lower than unity and sharp changes in the slope, even in the case in which we are neglecting Ohmic absorption. This is associated with the fact that, as predicted in Section 3.2, for larger periods additional modes can radiate away a portion of the extracted power. In this regime, the reflection and transmission coefficients, Eq. (3.12), do not provide a complete description of the array interaction, and power balance is not satisfied if we do not consider the other propagating waves beyond the dominant Floquet mode.

This is further detailed in Fig. 3.12, where we report for the examples of Fig. 3.11 the quantity $1 - |R|^2 - |T|^2$, which is identically zero for energy conservation as

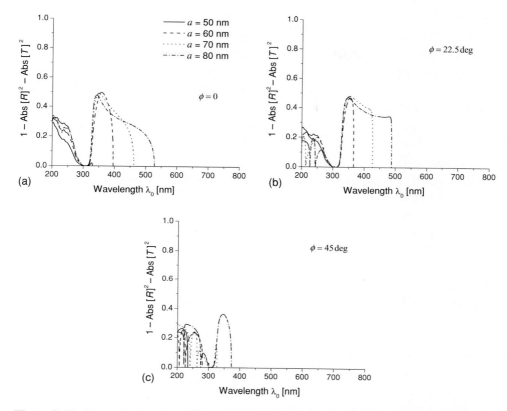

Figure 3.12. Power balance for the array of Fig. 3.11 for different angles of incidence.

long as condition (3.17) holds, i.e. as long as there is only one propagating Floquet wave. As can be seen in the figure, when the interparticle distance increases together with the particle size, higher-order propagating Floquet modes arise, implying that power is coupled into additional plane waves and that a nonzero value of the power balance quantity is obtained in Fig. 3.12, if we consider only the dominant Floquet mode in the reflection and transmission scenarios. For this scenario, due to the larger period along x, the cut-off wavelength for which higher-order modes propagate out of the array and $|R|^2 + |T|^2 < 1$, occurs for $N_x = \pm 1$ in Eq. (3.17), i.e. at the free-space wavelength

$$\lambda_0 = d_x \sqrt{1 - \sin^2 \phi}. \tag{3.53}$$

Any wavelength below this value is characterized by the presence of at least one higher-order Floquet mode carrying additional power away from the array.

Larger periods along the polarization direction, although ensuring larger Q-factors and stronger polarization vectors, imply lower cut-off frequencies for the higher-order Floquet modes and undesired side lobes in the radiation pattern of the array. These results confirm that it may be more appealing to achieve highly reflective surfaces or anomalous wave interaction properties from these planar arrays based on plasmonic resonance effects associated with smaller nanoparticles, rather than with regular Wood's anomalies and lattice resonances associated with the lattice periodicity. A metasurface based on plasmonic resonances would provide a resonant response that is independent of the angle of incidence and is not affected by the presence of side lobes and radiation in undesired directions.

It is also noticed that, together with strong reflections, lattice resonances may produce sharp transparency windows for the dominant mode (reflection coefficient R), associated with enhanced optical transmission effects [34] through such arrays. We stress the fact that these are lattice effects, associated with the array period, which is comparable with the effective incident wavelength, and that they are not related to the plasmonic features of the array. As shown in Section 3.2, for sub-wavelength features the resonance of these arrays is always associated with strong reflections, since the large polarization currents on each nanoparticle produce a maximized reflected wave and cancel out the impinging wave on the transmission side of the array.

3.3.2 TE polarization: realistic levels of absorption

After having established the main properties of such arrays for lossless nanoparticles, we consider in this section the presence of realistic material absorption. In particular, we use the extracted experimental values of the permittivity of silver given in ref. [5] for our nanoparticles. Analogous to Fig. 3.3(a), Fig. 3.13(a) reports the reflection coefficient from the array of Fig. 3.2, with isotropic periods $d_x = d_y = 22$ nm and small silver nanoparticles, $a = 10$ nm. It is seen that the reflection curves are analogous to those in the lossless case, even if the peak in reflection is slightly lower than unity, due to absorption, and the dip around $\lambda_0 = 300$ nm is larger than zero, since silver is not transparent in that range of wavelengths. The levels of reflection consistently shift down for larger angles of incidence at all wavelengths.

The corresponding transmission and absorption coefficients are also reported in Fig. 3.13. We observe that a significant level of absorption is experienced on the surface, in particular near the array resonance for normal incidence and near the transparency window for grazing angles. Due to the small periodicity, in this scenario the quantity $1 - |R|^2 - |T|^2$ directly corresponds to the absorbed power, since the cut-off frequency for higher-order modes is in the UV.

Figure 3.14, analogous to Fig. 3.4, reports the effective surface impedance for the array of Fig. 3.13. Also in this case, this quantity is independent of the angle of

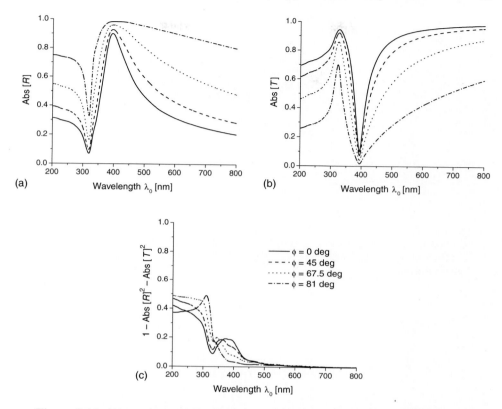

Figure 3.13. Dispersion of the reflection, transmission, and absorption for the periodic array of Fig. 3.2 ($a = 10$ nm, $d_x = d_y = 22$ nm), but now for realistic values of the absorption for silver.

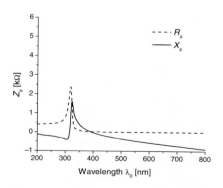

Figure 3.14. Analogous to Fig. 3.4, dispersion of the effective surface impedance for the periodic array of Fig. 3.13 ($a = 10$ nm, $d_x = d_y = 22$ nm), for realistic values of the absorption for silver.

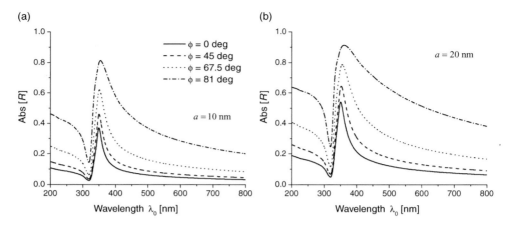

Figure 3.15. Analogous to Fig. 3.6, dispersion of the magnitude of the reflection coefficient for the periodic arrays of Fig. 3.5, but now for realistic values of the absorption for silver.

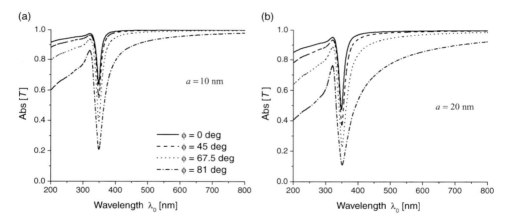

Figure 3.16. Similar to Fig. 3.15, but reporting the transmission coefficients for the same array, for realistic values of the absorption for silver.

incidence, due to the small features of the array. As expected, the absorbing nature of the nanoparticles mainly affects the surface resistance, which is not negligible for wavelengths below the array resonance. The surface impedance is still relatively small around the resonance wavelength, ensuring a low-impedance behavior with high reflectivity around this regime. However, the large surface reactance around the transparency window is strongly affected by the material absorption, increasing the reflective properties of the array in this frequency range.

Figures 3.15–3.17 report the reflection, transmission, and absorption coefficients, respectively, for the periodic arrays of Fig. 3.5, with anisotropic periods

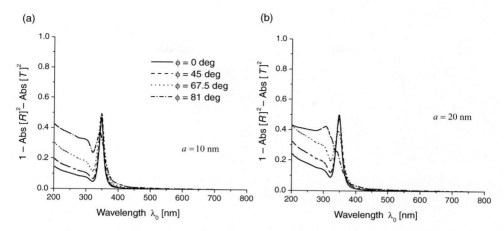

Figure 3.17. Similar to Figs. 3.15 and 3.16, but reporting the absorption loss associated with realistic values of Ohmic absorption in silver.

$d_x/3 = d_y = 2.2a$ and $a = 10$ nm (panels (a)), $a = 20$ nm (panels (b)). It is observed that in this case the high-Q and narrow bandwidth features obtained in Fig. 3.6 are strongly affected by the presence of absorption in the particles. In this case, the array is considerably less dense, and the particle absorption seriously affects the plasmonic resonances of the nanoparticles, considerably broadening the resonant features. Still, the overall coupling in the array ensures a stable reflection peak weakly dependent on the angle of incidence and a maximized reflection for larger angles of incidence. As expected, larger nanoparticles (panels (b)) ensure more robustness to loss and larger reflectivities. Also in this case, on comparing these results with Fig. 3.13, one may note how reducing the period in the direction of polarization of the nanoparticles ensures sharper resonance features.

Figures 3.16 and 3.17 report similar trends for the transmission and absorption features of the array. It is particularly interesting to verify that by reducing the array density, and removing a significant number of nanoparticles from the array, it is possible to increase the absorption at the surface. This counterintuitive effect is again associated with the improved resonant features arising from the coupling reduction in the direction of polarization of the nanoparticles, consistent with our findings in ref. [11] for linear arrays of nanoparticles.

3.3.3 *TM polarization: realistic levels of absorption*

As in the preceding subsection for TE polarization, we present here some numerical examples of wave interaction with planar arrays of nanoparticles for TM plane wave excitation. In this section, we concentrate for simplicity on arrays of nanodiscs of

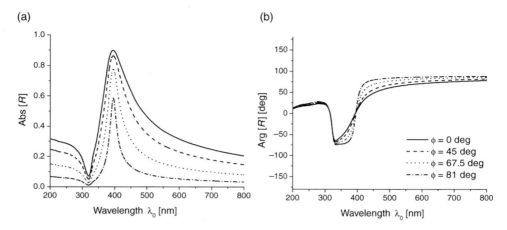

Figure 3.18. Dispersion of the amplitude and phase of the reflection coefficient for a periodic array of nanodiscs and geometry as in Fig. 3.2 ($a = 10$ nm, $d_x = d_y = 22$ nm).

radius a for which

$$\alpha = \alpha_{xx} \left(\hat{\mathbf{x}}\hat{\mathbf{x}} + \hat{\mathbf{y}}\hat{\mathbf{y}} \right). \tag{3.54}$$

For easy comparison with the TE case, we assume that the polarizability value in the xy plane is the same as used in Eq. (3.52). However, the normal component of polarizability, α_{zz}, is assumed to be negligible here. In this case, the reflection and transmission coefficients are given by Eq. (3.39) and the shunt surface impedance is given by Eq. (3.40). We consider the experimental values of absorption in silver from the literature [5], as in the preceding subsection.

Figure 3.18 reports the amplitude and phase of the reflection coefficient for an array with $a = 10$ nm, $d_x = d_y = 22$ nm, consistent with the geometry of Fig. 3.2. It is seen that for normal incidence, as expected, the results coincide with Fig. 3.13, but, for an increasing angle of incidence, the reflection coefficient decreases, rather than increasing as in the TE case. The reason is due to the fact that in TM polarization, as noted in the preceding section, the tangential component of the electric field is not constant on the surface of the array, but decreases for larger angles of incidence. Correspondingly, the polarization of the nanoparticles, and the effective averaged currents, significantly decrease as the grazing angle is approached. The phase of the reflection coefficient now crosses zero at resonance, since the reflection coefficient is calculated here for magnetic fields, consistent with the definition given in Eq. (3.38). Interestingly, in this polarization the resonance becomes sharper for larger angles of incidence, in contrast with the TE case.

Figure 3.19 reports the corresponding values of transmission and absorption for the same array. It is particularly interesting to notice the sharp absorption

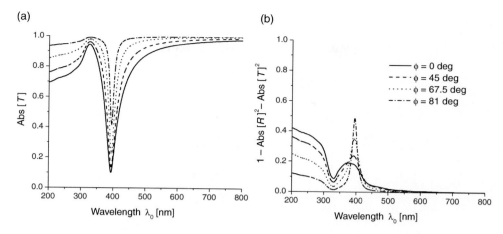

Figure 3.19. Dispersion of the transmission and absorption coefficients for the array of Fig. 3.17.

features obtained for large incidence angles, consistent with the enhanced resonant features obtained for larger angles in this polarization. The corresponding surface impedance for this polarization is consistent with Fig. 3.14. Also in this case, the variation of the surface impedance with the incidence angle is negligible, since the array period is much smaller than the wavelength of operation, and therefore in both polarizations the impedance, with good approximation, is constant with respect to the incidence angle. Of course, in the TM case we expect in general to obtain a series component to the surface impedance; this is not present here, due to the assumption of a negligible normal component of the polarizability tensor.

Figures 3.20–3.22 present the reflection, transmission, and absorption coefficients, respectively, for an array of nanodiscs as in Fig. 3.18, but with periods consistent with Fig. 3.15, i.e. $d_x/3 = d_y = 2.2a$ and $a = 10$ nm (panels (a)), $a = 20$ nm (panels (b)). It can be seen that in this case maximum wave interaction is also achieved, as expected, for normal incidence, whereas larger angles of incidence produce a weaker interaction with the impinging wave. Interestingly, increasing the distance between neighboring nanoparticles in the direction of polarization now weakens the resonant response for larger angles of incidence. This is particularly obvious in the reflection (Fig. 3.20) and absorption coefficients (Fig. 3.22), which become very small for larger angles of incidence. This is associated with the weak response of a sparser array of lossy nanoparticles, which interact only with the transverse component of the electric field.

These findings confirm that the homogenized description provided by an averaged surface impedance, as in Eq. (3.40), well describes the interaction of waves of both polarizations with such an array. Due to space limitations, the numerical

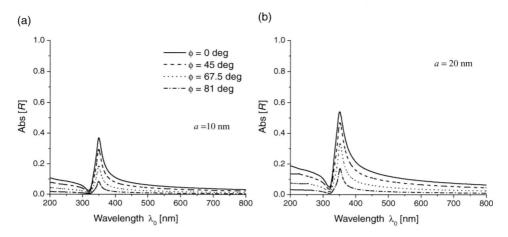

Figure 3.20. Analogous to Fig. 3.15, dispersion of the magnitude of the reflection coefficient for a periodic array of nanodiscs as in Fig. 3.18, illuminated by a TM plane wave. In this case, $d_x/3 = d_y = 2.2a$ and (a) $a = 10$ nm, (b) $a = 20$ nm.

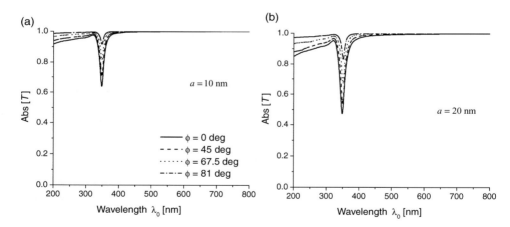

Figure 3.21. Similar to Fig. 3.20, but reporting the transmission coefficients for the same array.

results for arrays that have a nonzero normal component of the polarizability are not given here, but in light of the theoretical results reported in the preceding section, similar features may be seen. For larger periods, higher-order Floquet modes become more relevant, and the periodic features of these surfaces produce side lobes and additional higher-order Floquet plane waves, analogous to the features discussed in the preceding subsection for TE polarization.

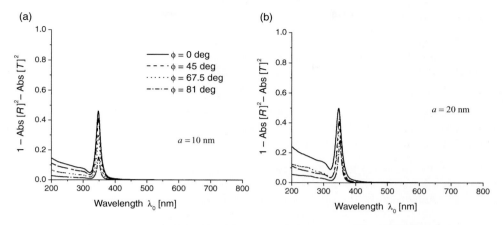

Figure 3.22. Similar to Figs. 3.20 and 3.21, but reporting the absorption loss.

3.4 Conclusions

In this chapter, we have analyzed in detail the properties of the interaction of an optical plane wave with two-dimensional planar arrays of nanoparticles. We have distinguished features of the resonant reflection and transmission properties arising from the plasmonic features of periodic arrays of nanoparticles, associated with the small value of $Re[\alpha^{-1}]$ (individual resonant properties of the nanoparticles) and lattice resonances arising due to the large value of $Re[\beta]$ (coupling among the entire lattice due to its periodic features). The plasmonic features are particularly appealing, since their resonance features are weakly dependent on the angle of incidence and on the order of the array, in contrast to the regular lattice resonances. We have shown that it may be possible to realize highly reflective and/or frequency-selective surfaces at optical frequencies for filtering, absorption, and radiation purposes. We have discussed in detail both the theoretical solution and associated power issues for plane wave excitation and several numerical examples, taking into account the effect of losses and polarization of the incident plane wave. These findings may be useful for a variety of applications at optical frequencies.

Acknowledgments

This work is supported in part by the US Air Force Office of Scientific Research (AFOSR) grant number FA9550-08-1-0220 and by the US Air Force Research Laboratory (AFRL) with contract number FA8718-09-C-0061.

References

[1] M. A. McCord and M. J. Rooks, *SPIE Handbook of Microlithography, Micromachining and Microfabrication* (Bellingham, WA: SPIE Press, 2000).

[2] R. E. Dunin-Borkowski, A. Kasama, A. Wei, S. L. Tripp, M. J. Hÿtch, E. Snoeck, R. J. Harrison, and A. Putnis, "Off-axis electron holography of magnetic nanowires and chains, rings, and planar arrays of magnetic nanoparticles," *Microsc. Res. Tech.* **64**, 390–402 (2004).

[3] M. Mayy, G. Zhu, Y. Barnakov, and M A. Noginov, "Development of composite silver-polymer metamaterials," *J. Appl. Phys.* **105**, 084318(1-6) (2009).

[4] J. F. Galisteo, F. Garcá-Santamará, D. Golmayo, B. H. Juárez, C. López, and E. Palacios, "Self-assembly approach to optical metamaterials," *J. Opt. A: Pure Appl. Opt.* **7**, S244–S254 (2005).

[5] P. B. Johnson and R. W. Christy, "Optical constants of the noble metals," *Phys. Rev. B* **6**, 4370–4379 (1972).

[6] C. F. Bohren and D. R. Huffman, *Absorption and Scattering of Light by Small Particles* (New York: Wiley, 1983).

[7] M. Quinten, A. Leitner, J. R. Krenn, and F. R. Aussenegg, "Electromagnetic energy transport via linear chains of silver nanoparticles," *Opt. Lett.* **23**, 1331–1333 (1998).

[8] S. A. Tretyakov and A. J. Vitanen, "Line of periodically arranged passive dipole scatterers," *Elec. Eng.* **82**, 353–361 (2000).

[9] M. L. Brongersma, J. W. Hartman, and H. A. Atwater, "Electromagnetic energy transfer and switching in nanoparticle chain arrays below the diffraction limit," *Phys. Rev. B* **62**, 16356–16369 (2000).

[10] R. A. Shore and A. D. Yaghjian, "Travelling electromagnetic waves on linear periodic arrays of lossless spheres," *Electron. Lett.* **41**, 578–580 (2005).

[11] A. Alú and N. Engheta, "Theory of linear chains of metamaterial/plasmonic particles as subdiffraction optical nanotransmission lines," *Phys. Rev. B* **74**, 205436(1-18) (2006).

[12] A. Alú and N. Engheta, "Three-dimensional nanotransmission lines at optical frequencies: a recipe for broadband negative-refraction optical metamaterials," *Phys. Rev. B* **75**, 024304(1-20) (2007).

[13] F. J. García de Abajo, G. Gomez-Santos, L. A. Blanco, A. G. Borisov, and S. V. Shabanov, "Tunneling mechanism of light transmission through metallic films," *Phys. Rev. Lett.* **95**, 067403(1-4) (2005).

[14] A. Alú and N. Engheta, "Optical nanotransmission lines: synthesis of planar left-handed metamaterials in the infrared and visible regimes," *J. Opt. Soc. Am. B* **23**, 571–583 (2006).

[15] C. Enkrich, M. Wegener, S. Linden *et al.* "Magnetic metamaterials at telecommunication and visible frequencies," *Phys. Rev. Lett.* **95**, 203901(1-4) (2005).

[16] J. Zhou, T. Koschny, M. Kafesaki, E. N. Economou, J. B. Pendry, and C. M. Soukoulis, "Saturation of the magnetic response of split-ring resonators at optical frequencies," *Phys. Rev. Lett.* **95**, 223902(1-4) (2005).

[17] A. Alú, A. Salandrino, and N. Engheta, "Negative effective permeability and left-handed materials at optical frequencies," *Opt. Express.* **14**, 1557–1567 (2006).

[18] G. Shvets and Y. A. Urzhumov, "Negative index meta-materials based on two-dimensional metallic structures," *J. Opt. A: Pure Appl. Opt.* **8**, S122–S130 (2006).

[19] P. J. Schuck, D. P. Fromm, A. Sundaramurthy, G. S. Kino, and W. E. Moerner, "Improving the mismatch between light and nanoscale objects with gold bowtie nanoantennas," *Phys. Rev. Lett.* **94**, 017402(1-4) (2005).

[20] P. Muhlschlegel, H. J. Eisler, O. J. F. Martin, B. Hecht, and D. W. Pohl, "Resonant optical antennas," *Science* **308**, 1607–1609 (2005).

[21] A. Alú and N. Engheta, "Tuning the scattering response of optical nanoantennas with nanocircuit loads," *Nature Photon.* **2**, 307–310 (2008).

[22] A. Alú and N. Engheta, "Input impedance, nanocircuit loading, and radiation tuning of optical nanoantennas," *Phys. Rev. Lett.* **101**, 043901(1-4) (2008).

[23] F. J. Garcia de Abajo, "Light scattering by particle and hole arrays," *Rev. Mod. Phys.* **79**, 1267–1290 (2007).

[24] A. J. Viitanen, I. Hanninen, and S. A. Tretyakov, "Analytical model for regular dense arrays of planar dipole scatterers," *Prog. Electromag. Res.*, **38**, 97–110 (2002).

[25] M. Englund and A. J. Viitanen, "A planar dipole array formed by small resonant particles," *Microwave. Opt. Technol. Lett.* **49**, 2419–2422 (2007).

[26] C. L. Holloway, A. Dienstfrey, E. F. Kuester, J. F. O'Hara, A. K. Azad, and A. J. Taylor, "A discussion on the interpretation and characterization of metafilms/metasurfaces: the two-dimensional equivalent of metamaterials," *Metamaterials* **3**, 100–112 (2009).

[27] J. A. Gordon, C. L. Holloway, and A. Dienstfrey, "A physical explanation of angle-independent reflection and transmission properties of metafilms/metasurfaces," *IEEE Antenn. Wireless Propag. Lett.* **8** 1127–1130 (2009).

[28] P. Belov and C. Simovski, "Homogenization of electromagnetic crystals formed by uniaxial resonant scatterers," *Phys. Rev. E* **72**, 026615(1-15) (2005).

[29] A. Alú, P. A. Belov, and N. Engheta, "Parallel-chain optical transmission line for a low-loss ultraconfined light beam," *Phys. Rev. B* **80**, 113101 (1-4) (2009).

[30] L. Lewin, *Polylogarithms and Associated Functions* (New York: Elsevier North-Holland, 1981).

[31] I. A. Stegun, "Miscellaneous functions," in *Handbook of Mathematical Functions*, eds. M. Abramowitz, I. A. Stegun (New York: Dover Publications, Inc., 1970).

[32] A. Alú and N. Engheta, "Polarizabilities and effective parameters for collections of spherical nanoparticles formed by pairs of concentric double-negative, single-negative, and/or double-positive metamaterial layers," *J. Appl. Phys.* **97**, 094310(1-12) (2005).

[33] J. Sipe and J. Van Kranendonk, "Macroscopic electromagnetic theory of resonant dielectrics," *Phys. Rev. A* **9**, 1806–1822 (1974).

[34] T. W. Ebbesen, H. J. Lezec, H. F. Ghaemi, T. Thio, and P. A. Wolff, "Extraordinary optical transmission through subwavelength hole arrays," *Nature* **391**, 667–669 (1998).

4

Chirality and anisotropy of planar metamaterials

ERIC PLUM AND NIKOLAY I. ZHELUDEV

4.1 Introduction

In recent years it has emerged that planar metamaterials offer a vast range of custom-designed electromagnetic functionalities. The best known are wire grid polarizers, which are established standard components for microwaves, terahertz waves, and the far-infrared. They are expected to be of increasing importance also for the near-infrared [1] and visible light [2]. Equally well developed are frequency selective surfaces [3–6], which are used as filters in radar systems, antenna technology [7], broadband communications, and terahertz technology [8, 9]. However, the range of optical effects observable in planar metamaterials and the variety of potential applications have only become clear since metamaterials research took off in 2000 [10]. Wave plate [11, 12] as well as polarization rotator and circular polarizer [13–15] functionalities have been demonstrated in metamaterials of essentially zero thickness. Traditionally, such components are large as they rely on integrating weak effects over thick functional materials. Polarization rotation has also been seen at planar chiral diffraction gratings [16, 17] and thin layered stereometamaterials [18, 19]. Electromagnetically induced transparency (EIT) [20–24] and high quality factor resonances [20] have been observed at planar structured interfaces. And finally, new fundamental electromagnetic effects leading to directionally asymmetric transmission of circularly [25–29] and linearly [30] polarized waves have been discovered in planar metamaterials.

Planar metamaterials derive their properties from artificial structuring rather than atomic or molecular resonances, and therefore appropriately scaled versions of such structures will show similar properties for radio waves, microwaves, terahertz waves, and, to some extent, in the infrared and optical spectral regions where losses are becoming more important. Planar metamaterials are compatible with

Structured Surfaces as Optical Metamaterials, ed. A. A. Maradudin. Published by Cambridge University Press. © Cambridge University Press 2011.

well-established fabrication technologies such as lithography and nanoimprint, allowing for high-throughput manufacturing and making them suitable for highly integrated applications and miniaturization.

The effects underlying the functionalities of planar metamaterials can be divided into two categories: dispersion phenomena in the forms of narrow resonances, stop bands, and EIT-like behaviors on the one hand, and polarization phenomena, such as circular and linear birefringence and dichroism and polarization-sensitive asymmetric transmission, on the other hand. The present chapter is devoted to the analysis of polarization phenomena and propagation asymmetries.

Given the huge range of potential applications of planar metamaterials, it is time to ask the following question: what are the fundamental limits of polarization func-tionalities and propagation asymmetries one can expect from planar metamaterials? In this work we aim to examine what limitations are imposed on the performance of planar metamaterials by energy conservation and fundamental symmetries.

Before we start, we need to clarify what we mean by a planar metamaterial. Idealized, a planar metamaterial is a flat two-dimensional surface of zero thick-ness that is periodically structured on the sub-wavelength scale. This ideal is best approximated by a single periodically patterned metal layer with a thickness that is comparable to the skin depth. An ideal planar metamaterial is still well approxi-mated by a single periodically patterned metal or dielectric layer that is very thin compared to the wavelength. In practice, such structures are often supported by a transparent substrate. Due to their sub-wavelength periodicity, planar metamateri-als do not diffract electromagnetic waves at normal incidence. Here we consider only nondiffracting angles of incidence.

Oblique incidence onto a planar metamaterial is considered in Section 4.2, and then the consequences of different structural symmetries and symmetries of the experimental arrangement are examined. We show that, in general, planar meta-materials can manifest circular birefringence and circular dichroism, as well as linear birefringence and dichroism and directionally asymmetric transmission; see Fig. 4.1. Following general definitions in Section 4.3, these phenomena are further discussed in Section 4.4. In Section 4.5, an analysis of polarization eigenstates for planar metamaterials is presented. In Section 4.6 the role of energy conservation is studied. The performance limits of planar metamaterials as linear polarizers, circu-lar polarizers, wave plates, or polarization rotators will be examined in Section 4.7. The final Section 4.8 is devoted to normal incidence onto planar metamaterials.

4.2 General planar metamaterials

The transmission and reflection properties of a linear planar metamaterial can be written in terms of the transmission[1] and reflection matrices, t and r, which relate

---

[1] Also known as Jones matrix.

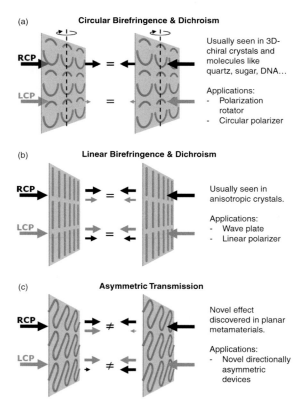

Figure 4.1. Exemplar polarization effects in planar metamaterials. (a) Optical activity in the form of polarization rotation and differential transmission for circularly polarized waves of different handedness (observable at oblique incidence onto planar metamaterials). (b) Linear birefringence and dichroism in an anisotropic metamaterial at normal incidence. (c) Asymmetric transmission arising from circular polarization conversion of the incident wave.

the transmitted and reflected electric fields, $\mathbf{E^t}$ and $\mathbf{E^r}$, to the incident field $\mathbf{E^0}$. Where the propagation direction of the incident wave, forwards or backwards, is important, it is indicated by an arrow over the matrix, vector, or scalar. For example, for a forward-propagating incident wave,

$$\overrightarrow{\mathbf{E}}^t = \overrightarrow{t}\,\overrightarrow{\mathbf{E}}^0, \tag{4.1}$$

$$\overrightarrow{\mathbf{E}}^r = \overrightarrow{r}\,\overrightarrow{\mathbf{E}}^0. \tag{4.2}$$

It is also convenient to express the transmission and reflection matrices in terms of the scattering matrix s. As a planar metamaterial is just a nondiffracting array of scatterers, the transmitted field is simply the superposition of the scattered field

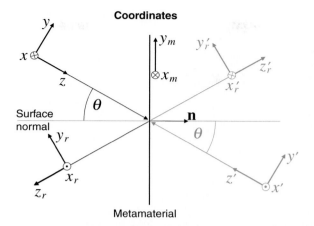

Figure 4.2. Coordinate systems. Each wave's polarization state (see Fig. 4.3(a)) is defined in its own right-handed Cartesian coordinate system defined by its propagation direction and plane of incidence; see the text. Here the coordinates are shown for a wave incident on the metamaterial's front xyz (back $x'y'z'$) and the corresponding reflected wave $x_r y_r z_r$ ($x'_r y'_r z'_r$). The coordinates $x_m y_m$ are used to define the metamaterial's in-plane orientation (see Fig. 4.3(b)) and correspond to xy projected onto the metamaterial.

and the incident wave, i.e.

$$\overrightarrow{t} = \overrightarrow{s} + 1, \tag{4.3}$$

where 1 is the unit matrix.

As illustrated by Fig. 4.2, we will define the polarization state of any electromagnetic wave in its own right-handed Cartesian coordinate system xyz, where x is perpendicular to the plane of incidence,[2] y is parallel to the plane of incidence, and z is the wave's propagation direction. Note that y is chosen consistently so that its projection onto the metamaterial is parallel for all waves within the same plane of incidence.

For example, incident and transmitted waves, as well as scattered fields (measured in the transmission direction), all have the same propagation direction and therefore the same coordinates xyz. The coordinates for an incident wave xyz and the reflected wave $x_r y_r z_r$ have anti-parallel x axes $x_r = -x$. The coordinates xyz and $x'y'z'$ for waves with opposite propagation directions are related by $x' = -x$, $y' = y$, and $z' = -z$.

---

[2] The plane of incidence contains the propagation direction and the metamaterial's surface normal **n**.

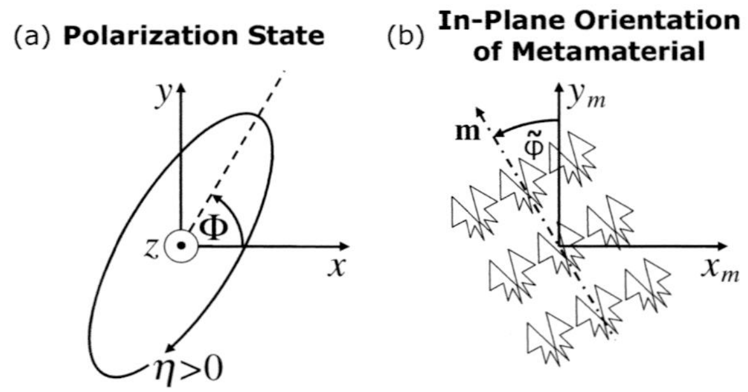

Figure 4.3. Polarization state and metamaterial orientation. (a) Each wave's polarization state is defined in its own coordinate system. Looking along the negative
z axis, i.e. into the beam, positive ellipticity η corresponds to a right-handed path
of the electric field vector at a fixed position in space. The azimuth Φ is measured from the positive x axis and increases towards the positive y axis. (b) The
metamaterial's orientation $\tilde{\varphi}$ corresponds to a preferred direction, e.g. a line of
(glide) mirror symmetry **m** or direction of anisotropy, which is measured from the
positive y_m axis, increasing towards the negative x_m axis; $x_m y_m$ correspond to xy
projected onto the metamaterial (see Fig. 4.2).

Planar structures can only couple to tangential electric fields and normal magnetic fields.[3] Therefore two electromagnetic waves with identical tangential electric fields and identical normal magnetic fields cannot be distinguished by a planar
metamaterial. In terms of the electric field – which fully defines an electromagnetic
plane wave – these waves correspond to mirror images with respect to the plane of
the metamaterial.

In particular, this implies that the electric field radiated by a planar metamaterial
(or planar current configuration) must be symmetric with respect to the metamaterial plane. However, the polarization states (see Fig. 4.3(a)) of waves scattered in the
transmission and reflection directions are defined in different coordinate systems
(see Fig. 4.2), in which they have the opposite handedness. This is why the reflection
matrix (describing scattering in the reflection direction) differs from the scattering
matrix (describing scattering in the transmission direction) by a coordinate transformation. Choosing right-handed (RCP, +) and left-handed (LCP, −) circularly
polarized waves as our basis, the reflection matrix for forward-propagating incident
waves is given by

$$\overrightarrow{r} = \sigma \overrightarrow{s}, \tag{4.4}$$

where σ_{ij} switches between the coordinate systems for the incident and reflected
waves, and is 0 for $i = j$ and 1 otherwise.

---

[3] Coupling to normal electric fields and tangential magnetic fields is not possible, as the electric charges cannot
leave the plane of the structure.

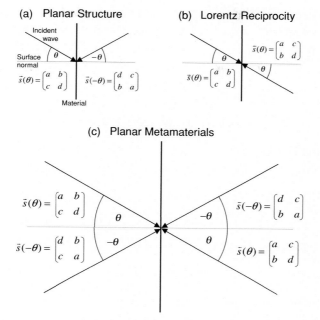

Figure 4.4. Scattering matrices for opposite angles and/or opposite directions of incidence in the same plane. (a) Form of the scattering matrices for waves incident at opposite angles $\pm\theta$ on opposite sides of a planar metamaterial according to conditions (4.5) and (4.6). (b) Form of the scattering matrices for opposite directions of incidence as required by Lorentz reciprocity; see Eqs. (4.7) and (4.8). (c) For nonmagnetized planar metamaterials, both (a) and (b) must apply, thus the scattering coefficients b, c are reversed for opposite propagation directions, while the coefficients a, d are reversed for opposite angles of incidence: $\pm\theta$.

As discussed above, planar metamaterials can only couple to tangential electric fields and normal magnetic fields. Therefore two electromagnetic waves that do not differ in these field components must excite the metamaterial in the same way. In particular, this is the case for circularly polarized waves of opposite handedness that are incident in the same plane at angles θ and $-\theta$ from the normal on the front and back of the metamaterial; see Fig. 4.4(a). These waves must cause the same scattered field. Taking into account that the scattering matrix describes the scattered field in the respective transmission direction (opposite handedness for the considered cases), we can identify pairs of identical scattering coefficients for opposite angles of incidence on opposite sides of the metamaterial:

$$\overrightarrow{s_{++}}(\theta) = \overleftarrow{s_{--}}(-\theta) \quad \text{and} \quad \overrightarrow{s_{--}}(\theta) = \overleftarrow{s_{++}}(-\theta), \tag{4.5}$$

$$\overrightarrow{s_{+-}}(\theta) = \overleftarrow{s_{-+}}(-\theta) \quad \text{and} \quad \overrightarrow{s_{-+}}(\theta) = \overleftarrow{s_{+-}}(-\theta). \tag{4.6}$$

If no static magnetic field is present, the Lorentz reciprocity lemma [31] must hold. Lorentz reciprocity requires, for opposite propagation directions, that $\overrightarrow{t_{ij}} = \overleftarrow{t_{ji}}$, which, due to Eq. (4.3), is equivalent to $\overrightarrow{s_{ij}} = \overleftarrow{s_{ji}}$. This gives us (see also Fig. 4.4(b))

$$\overrightarrow{s_{++}}(\theta) = \overleftarrow{s_{++}}(\theta) \quad \text{and} \quad \overrightarrow{s_{--}}(\theta) = \overleftarrow{s_{--}}(\theta), \tag{4.7}$$

$$\overrightarrow{s_{+-}}(\theta) = \overleftarrow{s_{-+}}(\theta) \quad \text{and} \quad \overrightarrow{s_{-+}}(\theta) = \overleftarrow{s_{+-}}(\theta). \tag{4.8}$$

By combining Eqs. (4.5)–(4.8), we arrive at four complex coefficients describing the scattering properties of a planar metamaterial for one plane of incidence:

$$a(\theta) := \overrightarrow{s_{++}}(\theta) = \overleftarrow{s_{++}}(\theta) = \overrightarrow{s_{--}}(-\theta) = \overleftarrow{s_{--}}(-\theta), \tag{4.9}$$

$$b(\theta) := \overrightarrow{s_{+-}}(\theta) = \overleftarrow{s_{-+}}(\theta) = \overrightarrow{s_{+-}}(-\theta) = \overleftarrow{s_{-+}}(-\theta), \tag{4.10}$$

$$c(\theta) := \overrightarrow{s_{-+}}(\theta) = \overleftarrow{s_{+-}}(\theta) = \overrightarrow{s_{-+}}(-\theta) = \overleftarrow{s_{+-}}(-\theta), \tag{4.11}$$

$$d(\theta) := \overrightarrow{s_{--}}(\theta) = \overleftarrow{s_{--}}(\theta) = \overrightarrow{s_{++}}(-\theta) = \overleftarrow{s_{++}}(-\theta). \tag{4.12}$$

The corresponding scattering matrices are written out explicitly in Fig. 4.4(c), where

$$a(\theta) = d(-\theta), \quad d(\theta) = a(-\theta), \tag{4.13}$$

$$b(\theta) = b(-\theta), \quad c(\theta) = c(-\theta). \tag{4.14}$$

A planar metamaterial's response depends not only on the angle of incidence θ, but also on the orientation $\tilde{\varphi} \in [0, 2\pi)$ of the metamaterial structure in its plane. With reference to Fig. 4.4(c), reversal of the angle of incidence, $\theta \to -\theta$, is equivalent to an in-plane rotation of the metamaterial by π, i.e. $\tilde{\varphi} \to \tilde{\varphi} + \pi$. Therefore $\overrightarrow{s}(\theta, \tilde{\varphi}) = \overrightarrow{s}(-\theta, \tilde{\varphi} + \pi)$ and

$$a(\theta, \tilde{\varphi}) = d(\theta, \tilde{\varphi} + \pi), \quad d(\theta, \tilde{\varphi}) = a(\theta, \tilde{\varphi} + \pi), \tag{4.15}$$

$$b(\theta, \tilde{\varphi}) = b(\theta, \tilde{\varphi} + \pi), \quad c(\theta, \tilde{\varphi}) = c(\theta, \tilde{\varphi} + \pi). \tag{4.16}$$

For reference, the scattering, transmission, and reflection matrices for opposite directions of incidence (same θ, φ) onto a planar metamaterial are given explicitly:

$$\overrightarrow{s} = \begin{pmatrix} a & b \\ c & d \end{pmatrix} \quad \text{and} \quad \overleftarrow{s} = \begin{pmatrix} a & c \\ b & d \end{pmatrix}, \tag{4.17}$$

$$\overrightarrow{t} = \begin{pmatrix} a+1 & b \\ c & d+1 \end{pmatrix} \quad \text{and} \quad \overleftarrow{t} = \begin{pmatrix} a+1 & c \\ b & d+1 \end{pmatrix}, \tag{4.18}$$

$$\overrightarrow{r} = \begin{pmatrix} c & d \\ a & b \end{pmatrix} \quad \text{and} \quad \overleftarrow{r} = \begin{pmatrix} b & d \\ a & c \end{pmatrix}. \tag{4.19}$$

4.2.1 Lossless complementary planar metamaterials

According to Babinet's principle, complementary waves ($\mathbf{E}^{\text{cpl},0} = c\mathbf{B}^0$, $\mathbf{B}^{\text{cpl},0} = -\mathbf{E}^0/c$) incident on complementary perfectly conducting planar thin screens cause transmitted fields $\mathbf{E}^t$, $\mathbf{B}^t$ and $\mathbf{E}^{\text{cpl},t}$, $\mathbf{B}^{\text{cpl},t}$, which are related by $\mathbf{E}^t - c\mathbf{B}^{\text{cpl},t} = \mathbf{E}^0$ and $\mathbf{B}^t + \mathbf{E}^{\text{cpl},t}/c = \mathbf{B}^0$ [32, 33].

This requires the transmission matrices of complementary planar metamaterials to be related as follows:

$$\overrightarrow{t} = \begin{pmatrix} a+1 & b \\ c & d+1 \end{pmatrix} \quad \text{and} \quad \overrightarrow{t^{cpl}} = \begin{pmatrix} -a & b \\ c & -d \end{pmatrix}. \tag{4.20}$$

The corresponding scattering matrices are given by

$$\overrightarrow{s} = \begin{pmatrix} a & b \\ c & d \end{pmatrix} \quad \text{and} \quad \overrightarrow{s^{cpl}} = \begin{pmatrix} -(a+1) & b \\ c & -(d+1) \end{pmatrix}. \tag{4.21}$$

Thus, apart from a phase shift by π, complementary lossless planar metamaterials have interchanged direct transmission and scattering properties for circularly polarized waves, e.g. $t_{++}^{cpl} = -s_{++}$. On the other hand, the circular polarization conversion coefficients (off-diagonal elements) are the same for such complementary structures, e.g. $\overrightarrow{t_{-+}^{cpl}} = \overrightarrow{s_{-+}^{cpl}} = \overrightarrow{t_{-+}} = \overrightarrow{s_{-+}}$.

4.2.2 Two-dimensional (2D) achiral planar metamaterials

A volume structure that cannot be superimposed with its mirror image is called chiral, or 3D-chiral. Infinitely thin planar objects can always be superimposed with their mirror image and therefore are 3D-achiral. However, the notion of chirality also exists in two dimensions: any planar object that cannot be superimposed with its mirror image without being lifted off the plane is called 2D-chiral. The consequences of 2D and 3D chirality are very different in electrodynamics, and a substantial body of literature exists on establishing continuous measures of 3D and 2D chirality [34, 35].

Like all planar periodic structures, planar metamaterials can be classified according to their wallpaper symmetry group [16, 36, 37]. For the examples of planar metamaterials presented here, we specify the wallpaper symmetry using crystallographic notation [38].

Planar metamaterials with a line of (glide) mirror symmetry **m** are achiral and will be considered further in this section. The orientation of a 2D-achiral metamaterial pattern with respect to the plane of incidence is of special significance. This orientation is most conveniently defined by an angle $\tilde{\varphi}$ between the pattern's line of (glide) mirror symmetry **m** and the y_m axis along which the plane of incidence crosses the metamaterial plane.

For planar metamaterials with only one line of (glide) mirror symmetry,[4] an in-plane rotation by π results in a different metamaterial orientation with the same line of mirror symmetry. In order to resolve this, a direction needs to be assigned to the line of mirror symmetry **m** in this case.[5] For metamaterials with multiple lines of mirror symmetry,[6] a particular mirror line needs to be chosen. Here we measure $\tilde{\varphi}$ from the y_m axis to the line of (glide) mirror symmetry **m**, increasing towards the *negative* x_m axis (compare with Fig. 4.3(b)).

Before considering general values of $\tilde{\varphi}$, we will consider special cases corresponding to the metamaterial's (glide) mirror line being either parallel ($\tilde{\varphi} = 0, \pi$) or perpendicular ($\tilde{\varphi} = \pi/2, 3\pi/2$) to the plane of incidence.

As illustrated by Fig. 4.5(a), if the metamaterial has a line of (glide) reflection symmetry **m** parallel to the plane of incidence, identical experiments result from waves incident at angles θ and $-\theta$ on opposite sides of the metamaterial. These experiments can be superimposed by rotating one of them by π around **m**. Identical experiments must have identical scattering matrices, $\overrightarrow{s}(\theta) = \overleftarrow{s}(-\theta)$, and therefore $a = d$ *and* $b = c$ *must hold for planar metamaterials with a (glide) mirror line parallel to the plane of incidence* ($\tilde{\varphi} = 0$ *or* π); compare with Fig. 4.4(c).

Figure 4.5(b) illustrates the case when the metamaterial has a line of (glide) reflection symmetry **m** perpendicular to the plane of incidence. In this case, opposite directions of incidence result in identical experiments, which can be superimposed by rotating one experiment by π around **m**. The corresponding scattering matrices must be identical, $\overrightarrow{s}(\theta) = \overleftarrow{s}(\theta)$, and therefore $b = c$ *must hold for planar metamaterials with a (glide) mirror line perpendicular to the plane of incidence* ($\tilde{\varphi} = \pi/2$ *or* $3\pi/2$); compare with Fig. 4.4(c).

Thus, the following conditions apply to achiral planar metamaterials:

$$a(\theta, n\pi) = d(\theta, n\pi), \tag{4.22}$$

$$b(\theta, n\tfrac{\pi}{2}) = c(\theta, n\tfrac{\pi}{2}) \quad \text{for} \quad n \in \mathbb{Z}. \tag{4.23}$$

Figure 4.5(c) illustrates the case of a fixed direction of incidence onto an achiral planar metamaterial for opposite metamaterial orientations $\pm\tilde{\varphi}$. These experiments are mirror images of each other and therefore they must show the same behavior

---

[4] Wallpaper symmetry groups with a single line of (glide) reflection: *pm, pg, cm*.
[5] For example, narrow end to wide end of the pattern.
[6] Wallpaper symmetry groups with multiple lines of (glide) reflection: *pmm, pmg, pgg, cmm, p4m, p4g, p3m1, p31m, p6m*.

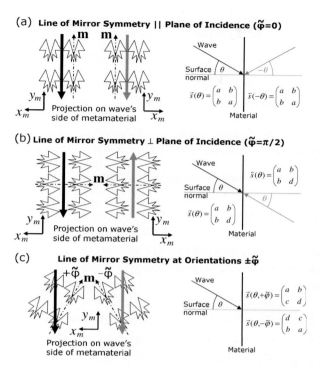

Figure 4.5. Achiral planar metamaterials: special orientations $\tilde{\varphi}$ of the metamaterial's line of (glide) mirror symmetry **m** relative to the plane of incidence (y_m axis, see Fig. 4.3(b)), illustrated for a pattern in the *pm* wallpaper group. (a) For $\tilde{\varphi} = 0$ or π, opposite angles of incidence on opposite sides of the metamaterial correspond to identical experiments with identical scattering matrices and therefore $a = d$ and $b = c$. (b) For $\tilde{\varphi} = \pi/2$ or $3\pi/2$ opposite directions of incidence correspond to identical experiments with identical scattering matrices and therefore $b = c$. (c) For the same direction of incidence, opposite metamaterial orientations $\pm\tilde{\varphi}$ result in mirrored experiments with the same properties for opposite circular polarizations.

for opposite circular polarizations:

$$a(\theta, +\tilde{\varphi}) = d(\theta, -\tilde{\varphi}), \tag{4.24}$$

$$b(\theta, +\tilde{\varphi}) = c(\theta, -\tilde{\varphi}). \tag{4.25}$$

4.2.3 Normal incidence onto achiral planar metamaterials

The special cases when the metamaterial's line of (glide) mirror symmetry is either parallel or perpendicular to the plane of incidence have particular relevance to the special case of normal incidence onto achiral planar metamaterials. At

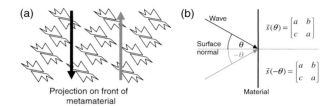

Figure 4.6. Two-fold rotational symmetry, illustrated for a pattern in the $p2$ wallpaper group. (a) Opposite angles of incidence $\pm\theta$ onto a planar metamaterial with 2-fold rotational symmetry result in identical experiments. (b) The corresponding scattering matrices must also be identical, therefore $a = d$.

normal incidence any plane containing the metamaterial's surface normal can be considered the plane of incidence, including those parallel and perpendicular to the metamaterial's line of (glide) mirror symmetry **m**. Therefore the *physical* restrictions for both of these cases must apply to normal incidence.

However, whereas at oblique incidence these special cases correspond to **m** being perpendicular or parallel to the x axis of the incident wave's coordinates, at normal incidence the plane of incidence, and therefore the orientations of xy, are not defined by the experimental geometry. No new physics should appear if we rotate the coordinate system around the z axis. Later we will find that for normal incidence a rotation of the coordinate system by some angle $-\Delta_\varphi$ (or the metamaterial by $+\Delta_\varphi$) around the z axis simply corresponds to a change in the phases of b and c, which is given explicitly by Eq. (4.80). The remaining restrictions that must hold for *normal incidence onto achiral planar metamaterials* are

$$a = d \quad \text{and} \quad |b| = |c|. \tag{4.26}$$

Note that $b = c$ must hold in the special coordinate systems in which the x axis is either perpendicular or parallel to the metamaterial's line of (glide) mirror symmetry.

4.2.4 Two-fold rotational symmetry or normal incidence

Here we will consider two common cases: 2-fold rotational symmetry[7] (for any angle of incidence) and normal incidence (for any planar metamaterial).

As illustrated by Fig. 4.6, opposite angles of incidence onto the same side of a planar metamaterial with 2-fold rotational symmetry result in the same experiment. One experiment can be superimposed on the other through a rotation by π around

---

[7] Wallpaper symmetry groups with 2-fold rotational symmetry: $p2$, pmm, pmg, pgg, cmm, $p4$, $p4m$, $p4g$, $p6$, $p6m$.

the metamaterial's normal. The corresponding scattering matrices must be identical, $\overrightarrow{s}(+\theta) = \overrightarrow{s}(-\theta)$, and therefore $a = d$.

On the other hand, for any planar metamaterial, $\overrightarrow{s}(+\theta) = \overrightarrow{s}(-\theta)$ must hold for the limiting case $\theta = 0$, and therefore $a = d$ is required at normal incidence.

Thus for (i) planar metamaterials at normal incidence and (ii) 2-fold rotationally symmetric planar metamaterials at any given angle of incidence $\pm\theta$, the most general form the scattering matrices can take is

$$\overrightarrow{s} = \begin{pmatrix} a & b \\ c & a \end{pmatrix} \quad \text{and} \quad \overleftarrow{s} = \begin{pmatrix} a & c \\ b & a \end{pmatrix}. \tag{4.27}$$

For the corresponding transmission and reflection matrices see Eqs. (4.18) and (4.19), with $a = d$.

4.3 Definitions

4.3.1 Alternative variables for the elements of scattering and transmission matrices

It turns out to be convenient to make the following definitions for the diagonal elements of the scattering matrix:

$$a = Ae^{i(\xi - 2\delta)} - \frac{1}{2} \quad \text{and} \quad d = De^{i(\xi + 2\delta)} - \frac{1}{2}. \tag{4.28}$$

These definitions allow us to specify a and d in terms of the real parameters A, D, ξ, δ, which we define as follows:

$$A = |a + \tfrac{1}{2}| \quad \text{and} \quad D = |d + \tfrac{1}{2}|, \tag{4.29}$$

$$\xi = \frac{\arg(a+1/2)+\arg(d+1/2)}{2}, \qquad \xi \in (-\pi, \pi], \tag{4.30}$$

$$\delta = \frac{\arg(d+1/2)-\arg(a+1/2)}{4}, \qquad \delta \in (-\tfrac{\pi}{2}, \tfrac{\pi}{2}). \tag{4.31}$$

Importantly, some later arguments regarding the meaning of δ require us to choose a phase convention. We choose ξ, δ, so that $\xi \pm 2\delta \in (-\pi, \pi]$. This convention is automatically satisfied if we choose $\arg(a + 1/2)$, $\arg(d + 1/2) \in (-\pi, \pi]$ and follow the above definitions.

Similarly it is convenient to define alternative parameters κ, φ for the phases of the off-diagonal elements of the scattering matrix b and c,

$$b = |b|e^{i(\kappa - 2\varphi)} \quad \text{and} \quad c = |c|e^{i(\kappa + 2\varphi)}, \tag{4.32}$$

where κ and φ are defined as

$$\kappa = \frac{\arg(b) + \arg(c)}{2}, \qquad \kappa \in [0, 2\pi), \tag{4.33}$$

$$\varphi = \frac{\arg(c) - \arg(b)}{4}, \qquad \varphi \in [0, \tfrac{\pi}{2}). \tag{4.34}$$

It will be shown that in the case of normal incidence, or of 2-fold rotationally symmetric planar metamaterials, φ is the azimuth of one eigenstate for forward propagation, and that in these cases the eigenpolarizations are independent of κ. We will also find that for lossless metamaterials without linear birefringence/dichroism, δ specifies the polarization azimuth rotation of the transmitted field.

4.3.2 Polarization states

In a circular basis, the azimuth Φ of a polarization state $\mathbf{E} = (E_+, E_-)$ is defined as

$$\Phi = -\frac{1}{2}[\arg(E_+) - \arg(E_-)], \tag{4.35}$$

and its ellipticity angle η is given by

$$\eta = \frac{1}{2} \arcsin\left(\frac{|E_+|^2 - |E_-|^2}{|E_+|^2 + |E_-|^2}\right). \tag{4.36}$$

Clearly, any polarization can be represented as a vector $\mathbf{E} = (E_+, E_-)$, where any collinear vectors $\mathbf{E}' = \tau\mathbf{E}$, $\tau \in \mathbb{C}$ represent the same polarization state. The squared magnitude $|\mathbf{E}|^2$ represents the intensity of the electromagnetic wave. As we only consider linear materials, none of the phenomena discussed here depends on the intensity of the incident wave, and therefore it is sufficient to consider incident waves of normalized amplitude $\hat{\mathbf{u}} = \mathbf{E}/|\mathbf{E}|$. All normalized polarization states can be written as

$$\hat{\mathbf{u}} = \begin{pmatrix} \sin(\beta)e^{i(\gamma-\alpha)} \\ \cos(\beta)e^{i(\gamma+\alpha)} \end{pmatrix}, \quad \alpha \in (-\tfrac{\pi}{2}, \tfrac{\pi}{2}], \ \ \beta \in [0, \tfrac{\pi}{2}], \ \ \gamma \in [0, 2\pi). \tag{4.37}$$

From Eqs. (4.35) and (4.36) it follows that the azimuth and ellipticity of $\hat{\mathbf{u}}$ are

$$\Phi(\hat{\mathbf{u}}) = \alpha,$$

$$\eta(\hat{\mathbf{u}}) = \beta - \tfrac{\pi}{4}.$$

Thus $\hat{u}$ represents a polarization state of azimuth α and ellipticity angle $\beta - \pi/4$. Different values of γ correspond to the same polarization state at different times.

4.4 Polarization effects

Planar metamaterials may exhibit several polarization effects. As shown in Section 4.4.1, optical activity is present when $a \neq d$. It will be demonstrated in Section 4.4.2 that $|b| \neq |c|$ corresponds to directionally asymmetric transmission, reflection and absorption of circularly polarized waves (circular conversion dichroism). As derived in Section 4.4.3, corresponding directional asymmetries for linearly polarized waves (linear conversion dichroism) may accompany optical activity in lossy planar metamaterials. Finally, it will be shown in Section 4.4.4 that linear birefringence and dichroism are present in their pure form if $a = d$ and $|b| = |c| \neq 0$. Here we will also find that polarization-azimuth-independent transmission and reflection properties correspond to $b = c = 0$.

Note that for a general lossy planar metamaterial without 2-fold rotational symmetry at oblique incidence all of the above effects should be expected to occur, if the structure does not have a line of (glide) mirror symmetry either parallel or perpendicular to the plane of incidence. In these cases the scattering matrix takes its most general form, which allows the magnitudes and phases of the scattering coefficients a, b, c, d to be all different.

4.4.1 Optical activity at oblique incidence $(a \neq d)$

For oblique incidence onto a planar metamaterial that does not have 2-fold rotational symmetry,[8] we found that generally $\overrightarrow{t_{++}} \neq \overrightarrow{t_{--}}$, or $a \neq d$ should be expected, if the metamaterial does not have a line of (glide) mirror symmetry in the plane of incidence. We will find that $a \neq d$ corresponds to optical activity, i.e. circular birefringence and circular dichroism. These effects are normally associated with intrinsic 3D chirality.

Optical activity due to extrinsic 3D chirality (see Fig. 4.8) was first predicted by Bunn [39] and later detected in liquid crystals [40, 41]. However, the topic attracted hardly any attention until the observation of large circular birefringence and circular dichroism due to extrinsic 3D chirality in planar metamaterials [13–15].

In order to see how $a \neq d$ leads to circular birefringence, consider the scattered field for an incident wave $\hat{u}$ with azimuth α as defined in Eq. (4.37). The scattered

---

[8] Wallpaper symmetry groups without 2-fold rotational symmetry: $p1$, pm, pg, cm, $p3$, $p3m1$, $p31m$.

field is given by (for forward-propagation)

$$\overrightarrow{\mathbf{E}}^s = \overrightarrow{s}\,\hat{\mathbf{u}}(\alpha)$$

$$= \begin{pmatrix} a & b \\ c & d \end{pmatrix} \begin{pmatrix} \sin(\beta)e^{i(\gamma-\alpha)} \\ \cos(\beta)e^{i(\gamma+\alpha)} \end{pmatrix}$$

$$= \begin{pmatrix} a\sin(\beta)e^{i(\gamma-\alpha)} + b\cos(\beta)e^{i(\gamma+\alpha)} \\ c\sin(\beta)e^{i(\gamma-\alpha)} + d\cos(\beta)e^{i(\gamma+\alpha)} \end{pmatrix}. \tag{4.38}$$

Now we will calculate the azimuth rotation, i.e. the difference between the azimuth of scattered and incident waves. As the azimuth of a polarization state **E** is $\Phi = -(1/2)[\arg(E_+) - \arg(E_-)]$, see Eq. (4.35), the azimuth rotation can be written as follows:

$$\Delta\Phi^s = \Phi(\overrightarrow{\mathbf{E}}^s) - \Phi(\hat{\mathbf{u}})$$

$$= -\frac{1}{2}\left[\arg(a\sin(\beta)e^{i(\gamma-\alpha)} + b\cos(\beta)e^{i(\gamma+\alpha)})\right.$$

$$\left. - \arg(c\sin(\beta)e^{i(\gamma-\alpha)} + d\cos(\beta)e^{i(\gamma+\alpha)})\right]$$

$$+ \frac{1}{2}\left[\arg(\sin(\beta)e^{i(\gamma-\alpha)}) - \arg(\cos(\beta)e^{i(\gamma+\alpha)})\right]$$

$$= -\frac{1}{2}\left[\arg(a + b\cot(\beta)e^{+i2\alpha})\right.$$

$$\left. - \arg(c\tan(\beta)e^{-i2\alpha} + d)\right]$$

$$= -\frac{1}{2}\left[\arg(a) - \arg(d)\right.$$

$$+ \arg(1 + \tfrac{b}{a}\cot(\beta)e^{+i2\alpha})$$

$$\left. - \arg(1 + \tfrac{c}{d}\tan(\beta)e^{-i2\alpha})\right]. \tag{4.39}$$

We can clearly see that for $b = c = 0$ the azimuth rotation does not depend on the azimuth of the incident wave, as we would expect for an ideal, circularly birefringent medium. The eigenstates corresponding to this case of pure optical activity will be derived in Section 4.5.1. In the general case, however, the planar metamaterial may also show linear birefringence/dichroism or circular/linear conversion dichroism, and the rotation can depend on the azimuth α of the incident wave. Therefore it is more meaningful to consider the average rotation experienced by all waves with the same ellipticity. Clearly, when averaging over all $\alpha \in (-\pi/2, \pi/2]$, the last two terms must vanish, giving us the *average azimuth rotation for the scattered field*

$$\langle\Delta\Phi^s\rangle = -\frac{1}{2}\left[\arg(a) - \arg(d)\right]. \tag{4.40}$$

Similarly, it can be shown that the *average azimuth rotation for the transmitted field* is given by

$$\langle \Delta \Phi^t \rangle = -\tfrac{1}{2} \left[\arg(a + 1) - \arg(d + 1) \right]. \tag{4.41}$$

Note that the azimuth rotations for scattered and reflected fields have opposite signs, $\Delta \Phi^r = -\Delta \Phi^s$, as scattering rotation is measured looking into the incident beam while reflection rotation is measured looking into the reflected beam.

It can be easily seen that generally $a \neq d$ leads not only to circular birefringence, but also to circular dichroism. Transmission circular dichroism corresponds to different direct transmission levels for opposite circular polarizations:

$$\Delta T := |t_{++}|^2 - |t_{--}|^2 = |a + 1|^2 - |d + 1|^2, \tag{4.42}$$

and for planar metamaterials it is accompanied by analogous phenomena in scattering and reflection, i.e.

$$\Delta S := |s_{++}|^2 - |s_{--}|^2 = |a|^2 - |d|^2, \tag{4.43}$$

$$\Delta R := |r_{-+}|^2 - |r_{+-}|^2 = |a|^2 - |d|^2. \tag{4.44}$$

Note that polarization states are always measured looking into the beam, thus the reflected beam must be measured in a different coordinate system, which gives rise to different indices in Eq. (4.44).

As the diagonal elements of the scattering and transmission matrices do not depend on reversal of the propagation direction, Eqs. (4.40)–(4.44) apply to forward and backward propagation. For the opposite angle of incidence $-\theta$ and in-plane rotations of the metamaterial by π, however, the roles of a and d are reversed (see Eqs. (4.13) and (4.15)), and these equations change sign. In other words, *circular birefringence and circular dichroism are the same for opposite propagation directions, but the effects change sign for opposite angles of incidence and in-plane rotations of the metamaterial by π*. It follows that polarization rotation and circular dichroism in planar metamaterials without 2-fold rotational symmetry are tunable via the angle of incidence. The effects are "switched off" at normal incidence and can be "switched on" with one sign for angles of incidence $\theta > 0$ and the other sign for $\theta < 0$.

It was derived in Section 4.2.1 that lossless complementary planar metamaterials have interchanged direct scattering and transmission coefficients for circularly polarized waves (apart from an overall phase shift). Therefore *transmission and*

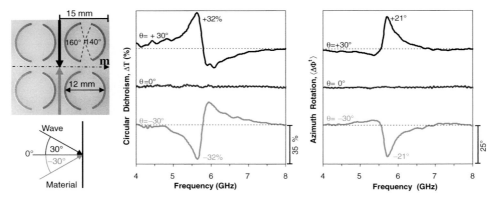

Figure 4.7. Experimental demonstration of optical activity due to extrinsic 3D chirality. Spectra of transmission circular dichroism, ΔT, and polarization rotation, $\langle \Delta \Phi^t \rangle$, measured for a planar metamaterial based on achiral asymmetrically split wire rings (wallpaper group *pm*). The respective zero-level is indicated by a dashed line. The 3D-chiral phenomena have reversed signs for opposite angles of incidence $\theta = \pm 30°$ and vanish at normal incidence. The structure's line of mirror symmetry **m** is oriented perpendicular to the plane of incidence ($\tilde{\varphi} = \pi/2$), which ensures the absence of the asymmetric effect discussed in Section 4.4.2. How extrinsic 3D chirality controls the properties of this metamaterial is discussed in detail in ref. [13].

scattering optical activity are interchanged for lossless complementary planar metamaterials.

For achiral planar metamaterials,[9] it follows from Eq. (4.22) that optical activity is absent, if the metamaterial has a line of (glide) mirror symmetry in the plane of incidence. Thus, for the orientations $\tilde{\varphi} = 0, \pi$, circular birefringence and circular dichroism are absent independent of θ. Equation (4.24) implies that for achiral planar metamaterials the signs of circular birefringence and circular dichroism will be reversed for in-plane metamaterial orientations $\pm\tilde{\varphi}$. (Of course, as we found for the general case, the signs will also be reversed for orientations $\tilde{\varphi}, \tilde{\varphi} + \pi$ or alternatively angles of incidence $\pm\theta$.) Importantly, if the achiral planar metamaterial has a line of (glide) mirror symmetry perpendicular to the plane of incidence, i.e. $\tilde{\varphi} = \pi/2, 3\pi/2$, the asymmetric phenomena for circularly polarized waves ($|b| \neq |c|$) that will be discussed in Section 4.4.2 are absent, while circular birefringence and circular dichroism are still permitted. Experimental results illustrating this case are shown in Fig. 4.7.

We will find in Section 4.4.3 that optical activity in lossy planar metamaterials is usually accompanied by directionally asymmetric transmission, reflection, and absorption of linearly polarized waves. Importantly, such directional asymmetries,

---

[9] Achiral wallpaper symmetry groups without 2-fold rotational symmetry: *pm, pg, cm, p3m1, p31m*.

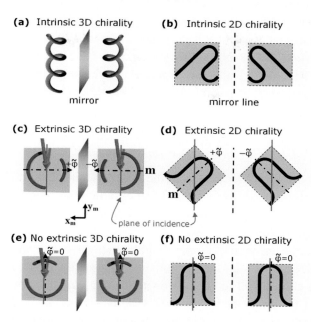

Figure 4.8. Intrinsic and extrinsic chirality. Panels (a) and (b) show examples of intrinsically 3D-chiral and 2D-chiral elements, respectively. (c) Extrinsic 3D chirality: the arrangement of structure and incident wave cannot be superimposed with its mirror image. (d) Extrinsic 2D chirality: the structure combined with the direction introduced by the plane of incidence cannot be superimposed with its mirror image without being lifted off the plane. Panels (e) and (f) show configurations where extrinsic chirality is absent.

both for linearly and circularly polarized waves, are not allowed for lossless planar metamaterials.

Optical activity is generally associated with 3D chirality. One might correctly argue that a planar structure, which can always be superimposed with its mirror image via an out-of-the-plane rotation by π, cannot be 3D-chiral. However, it turns out that for oblique incidence onto a structure without 2-fold rotational symmetry, the metamaterial and incident beam form a 3D-chiral experimental arrangement, if the metamaterial does not have a line of (glide) mirror symmetry in the plane of incidence; see Figs. 4.8(c) and (e). Chirality arising from the mutual orientation of metamaterial and incident beam has been called *extrinsic chirality* [13] in order to distinguish it from *intrinsic chirality*, which refers to the symmetry of a material; see Fig. 4.8(a). Inherently tunable optical activity in planar metamaterials allows the realization of tunable planar metamaterial polarization rotators and circular polarizers. The potential performance of such devices will be discussed in Sections 4.7.4 and 4.7.5.

For any planar metamaterial at normal incidence, and for planar metamaterials with 2-fold rotational symmetry[10] in general, equality of the diagonal elements of the transmission matrix is required, i.e. $\overrightarrow{t_{++}} = \overleftarrow{t_{--}}$, and thus optical activity is prohibited. This should also be expected, as the metamaterial and incident beam cannot form an extrinsically 3D-chiral experimental arrangement in these cases.

4.4.2 Circular conversion dichroism ($|b| \neq |c|$)

Intriguing behavior may also arise from the in general nonequal off-diagonal terms of the transmission matrix. We will find that circular conversion dichroism, $|b| \neq |c|$, leads to directionally asymmetric transmission, reflection, and absorption of circularly polarized waves.

Asymmetric transmission of circularly polarized waves was discovered in 2006 for normal incidence of microwaves onto an intrinsically 2D-chiral metamaterial [25], which is shown in Fig. 4.9. Soon thereafter the effect was also seen in optics [26] and for terahertz waves [42]. Corresponding asymmetries were first numerically predicted [43] and then measured [27] for reflection and absorption. Asymmetric transmission due to extrinsic 2D chirality was discovered in 2009 for oblique incidence onto an achiral metamaterial [28]; see Fig. 4.10. The effect has since even been observed for oblique incidence onto a highly symmetric, achiral, and isotropic lossy planar metamaterial [29], providing experimental evidence that circular conversion dichroism should be expected for extrinsically 2D-chiral directions of incidence onto any lossy periodically structured interface. Furthermore, the observation of circular conversion dichroism in planar metamaterials has led to the discovery of a similar phenomenon in plasmonics [44] and has generated theoretical interest [45].

Consider right-handed circularly polarized waves $\mathbf{E}^0 = (1, 0)^{tr}$ with opposite propagation directions incident on the front and back of a planar metamaterial:

$$\overrightarrow{\mathbf{E}}^t = \overrightarrow{t}(\theta, \tilde{\varphi})\overrightarrow{\mathbf{E}}^0 = \begin{pmatrix} a+1 \\ c \end{pmatrix},$$

$$\overleftarrow{\mathbf{E}}^t = \overleftarrow{t}(\theta, \tilde{\varphi})\overleftarrow{\mathbf{E}}^0 = \begin{pmatrix} a+1 \\ b \end{pmatrix}.$$

$$(4.45)$$

Assuming the case of $|b(\theta, \tilde{\varphi})| > |c(\theta, \tilde{\varphi})|$, this implies that the considered planar structure would be less transparent for right-handed circularly polarized waves that are incident on its front than for those on its back. One can easily see that

---

[10] Wallpaper symmetry groups with 2-fold rotational symmetry: *p2, pmm, pmg, pgg, cmm, p4, p4m, p4g, p6, p6m*.

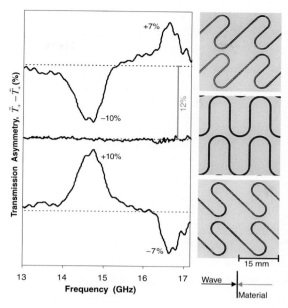

Figure 4.9. Experimental demonstration of asymmetric transmission at normal incidence due to intrinsic 2D chirality. The spectral dependence of the directional transmission asymmetry, $\overrightarrow{T}_+ - \overleftarrow{T}_+$, for right-handed circularly polarized waves is shown for right-handed, achiral and left-handed forms of the metamaterial (wallpaper groups $p2$, pmg, $p2$). The respective zero-level is indicated by a dashed line. The insets show the front side of the structures. See ref. [25] for a detailed discussion of the experimental work.

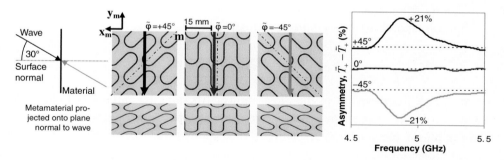

Figure 4.10. Experimental demonstration of asymmetric transmission due to extrinsic 2D chirality. The spectral dependence of the directional transmission asymmetry, $\overrightarrow{T}_+ - \overleftarrow{T}_+$, for right-handed circularly polarized waves is shown for orientations $\tilde{\varphi} = 0, \pm\pi/4$ of the achiral metamaterial (wallpaper group pmg). The angle of incidence is $\theta = 30°$ in all cases. Note that the metamaterial's projection onto the plane normal to the direction of incidence is 2D-chiral when extrinsic 2D chirality is present. See ref. [28] for a detailed discussion of the experimental work.

the transmission asymmetry has to be reversed for left-handed waves. The asymmetry arises from different circular polarization conversion efficiencies $\overleftarrow{t}_{-+} = b$ and $\overrightarrow{t}_{-+} = c$ for the same circular polarization in opposite propagation directions (circular conversion dichroism).

If we calculate the reflectivity for right-handed circularly polarized waves, we find

$$\overrightarrow{\mathbf{E}}^{\mathbf{r}} = \overrightarrow{r}(\theta, \tilde{\varphi}) \overrightarrow{\mathbf{E}}^{\mathbf{0}} = \begin{pmatrix} c \\ a \end{pmatrix},$$

$$\overleftarrow{\mathbf{E}}^{\mathbf{r}} = \overleftarrow{r}(\theta, \tilde{\varphi}) \overleftarrow{\mathbf{E}}^{\mathbf{0}} = \begin{pmatrix} b \\ a \end{pmatrix}. \tag{4.46}$$

Assuming again that $|b(\theta, \tilde{\varphi})| > |c(\theta, \tilde{\varphi})|$, this implies that the considered planar structure would be simultaneously less transparent and less reflective for right-handed waves that are incident on its front than when they are incident on its back. The asymmetry is reversed for left-handed waves. As a fraction of incident power, the transmission asymmetry and reflection asymmetry are given by

$$\overrightarrow{T}_{\pm} - \overleftarrow{T}_{\pm} = \overrightarrow{R}_{\pm} - \overleftarrow{R}_{\pm} = \pm(|c|^2 - |b|^2), \tag{4.47}$$

where $\overrightarrow{T}_{\pm}, \overrightarrow{R}_{\pm}, \overleftarrow{T}_{\pm}, \overleftarrow{R}_{\pm}$ correspond to the transmitted (T) and reflected (R) power fractions for circularly polarized (right, +, or left, −) incident waves with opposite propagation directions (arrows). For example, the transmitted power fraction for a forward-propagating right-handed circularly polarized incident wave is $\overrightarrow{T}_{+} = |\overrightarrow{t_{++}}|^2 + |\overrightarrow{t_{-+}}|^2$.

This corresponds to an asymmetry of losses, L, of

$$\overrightarrow{L}_{\pm} - \overleftarrow{L}_{\pm} = \mp 2(|c|^2 - |b|^2), \tag{4.48}$$

where $\overrightarrow{L}_{\pm}, \overleftarrow{L}_{\pm}$ correspond to the "lost" power fractions for circularly polarized (right, +, or left, −) incident waves with opposite propagation directions (arrows). Thus, a planar metamaterial for which $|b| \neq |c|$ will have larger losses for one circular polarization for forward propagation and for the other circular polarization for backward propagation. The eigenstates associated with these asymmetric phenomena are discussed in Section 4.5.2.

In particular, the asymmetric phenomena, which occur when $|b| \neq |c|$, imply that at least one circular polarization experiences losses. For metamaterials, which do not diffract, the only loss mechanism is absorption. Thus, *for lossless planar metamaterials the asymmetric phenomena for circularly polarized waves cannot occur and $|b| = |c|$ must hold*. However, note that in principle diffraction losses could play the same role as absorption losses, and thus circular conversion dichroism ($|b| \neq |c|$) may be possible in nonabsorbing planar diffraction gratings, i.e. structures that would be lossless planar metamaterials at longer wavelengths.

For any planar metamaterial the angular dependence of the scattering coefficients b and c must obey Eqs. (4.14) and (4.16), which imply that the difference $|c|^2 - |b|^2$ takes the same value for opposite angles of incidence $\pm\theta$ and for in-plane rotations of the metamaterial by π, i.e.

$$(|c|^2 - |b|^2)(\theta, \tilde{\varphi}) = (|c|^2 - |b|^2)(-\theta, \tilde{\varphi})$$
$$= (|c|^2 - |b|^2)(\theta, \tilde{\varphi} + \pi). \tag{4.49}$$

This indicates that the asymmetric properties are the same for opposite angles of incidence $\pm\theta$. For in-plane rotation of the metamaterial, circular conversion dichroism is periodic with period π, i.e. the metamaterial must show the same asymmetric response for orientations $\tilde{\varphi}$ and $\tilde{\varphi} + \pi$.

For achiral planar metamaterials[11] the scattering coefficients b and c must also obey Eqs. (4.25) and (4.23), from which we can conclude that the difference $|c|^2 - |b|^2$ is an odd function of the metamaterial orientation $\tilde{\varphi}$ that is zero for multiples of $\pi/2$:

$$(|c|^2 - |b|^2)(\theta, +\tilde{\varphi}) = -(|c|^2 - |b|^2)(\theta, -\tilde{\varphi}),$$
$$(|c|^2 - |b|^2)(\theta, n\tfrac{\pi}{2}) = 0 \quad \text{for} \quad n \in \mathbb{Z}. \tag{4.50}$$

This means that for achiral planar metamaterials the directional asymmetries are reversed for in-plane orientations $\pm\tilde{\varphi}$ of the metamaterial, and that circular conversion dichroism is absent if the metamaterial has a line of (glide) mirror symmetry either parallel ($\tilde{\varphi} = 0, \pi$) or perpendicular ($\tilde{\varphi} = \pi/2, 3\pi/2$) to the plane of incidence; see Fig. 4.10. In particular, it follows that achiral planar metamaterials do not allow circular conversion dichroism for the special case of normal incidence.

Importantly, cases that allow circular conversion dichroism, in which optical activity ($a \neq d$) and linear conversion dichroism, see Sections 4.4.1 and 4.4.3, cannot occur, are (i) lossy planar metamaterials with 2-fold rotational symmetry[12] and (ii) normal incidence onto lossy 2D-chiral planar metamaterials[13] (that are anisotropic, as explained in Section 4.4.4).

The asymmetric transmission, reflection, and absorption phenomena arise from reversed right-to-left and left-to-right circular polarization conversion efficiencies

---

[11] Achiral wallpaper symmetry groups: *pm, pg, cm, pmm, pmg, pgg, cmm, p4m, p4g, p3m1, p31m, p6m.*

[12] Wallpaper symmetry groups with 2-fold rotational symmetry: *p2, pmm, pmg, pgg, cmm, p4, p4m, p4g, p6, p6m.*

[13] Anisotropic 2D-chiral wallpaper symmetry groups: *p1, p2.*

for opposite directions of wave propagation. Such properties resemble the perceived handedness of 2D-chiral patterns (e.g. flat spirals), which is reversed for opposite directions of observation. In fact, the abovementioned symmetry requirements are equivalent to stating that circular conversion dichroism requires 2D chirality. This can be either intrinsic 2D chirality of the planar metamaterial itself (see Fig. 4.8(b)) or it can be extrinsic 2D chirality of the experimental arrangement (see Figs. 4.8(d) and (f)).

4.4.3 Linear conversion dichroism

Based on the results of the preceding section, one may expect that also directionally asymmetric transmission, reflection and absorption of linearly polarized waves should be observable in planar metamaterials.

In fact asymmetric transmission of linearly polarized waves was reported in 2010 for a lossy extrinsically 3D-chiral planar metamaterial [30] as well as an intrinsically 3D-chiral metamaterial which was not planar [46].

Consider two linearly polarized waves $\overrightarrow{\mathbf{E}}^0$, $\overleftarrow{\mathbf{E}}^0$ with opposite propagation directions and parallel polarization states incident on front and back of a planar metamaterial. Here we must take into account that the polarization of these waves is defined in their own mutually rotated coordinate systems (see Figs. 4.2 and 4.3), therefore a forward-propagating wave of azimuth α corresponds to a backward-propagating wave of azimuth $-\alpha$. Using Eq. (4.37) with $\beta = \frac{\pi}{4}$ for linear polarization we get the normalized incident waves

$$\overrightarrow{\mathbf{E}}^0(\alpha) = \frac{e^{i\gamma}}{\sqrt{2}} \begin{pmatrix} e^{-i\alpha} \\ e^{i\alpha} \end{pmatrix} \quad \text{and} \quad \overleftarrow{\mathbf{E}}^0(\alpha) = \frac{e^{i\gamma}}{\sqrt{2}} \begin{pmatrix} e^{i\alpha} \\ e^{-i\alpha} \end{pmatrix}, \tag{4.51}$$

where γ corresponds to the same polarization state at different times. The corresponding scattered field is given by

$$\overrightarrow{\mathbf{E}}^s(\alpha) = \overrightarrow{s}\,\overrightarrow{\mathbf{E}}^0(\alpha) = \frac{e^{i\gamma}}{\sqrt{2}} \begin{pmatrix} ae^{-i\alpha} + be^{i\alpha} \\ ce^{-i\alpha} + de^{i\alpha} \end{pmatrix},$$

$$\overleftarrow{\mathbf{E}}^s(\alpha) = \overleftarrow{s}\,\overleftarrow{\mathbf{E}}^0(\alpha) = \frac{e^{i\gamma}}{\sqrt{2}} \begin{pmatrix} ae^{i\alpha} + ce^{-i\alpha} \\ be^{i\alpha} + de^{-i\alpha} \end{pmatrix}. \tag{4.52}$$

As the reflected field corresponds to the scattered field described in different coordinates, the reflected power fraction is given by $|\mathbf{E}^s|^2$ in each case and therefore the directional reflection asymmetry as a fraction of incident power is

$$(\overrightarrow{R} - \overleftarrow{R})(\alpha) = |\overrightarrow{\mathbf{E}}^s(\alpha)|^2 - |\overleftarrow{\mathbf{E}}^s(\alpha)|^2 \tag{4.53}$$

$$= [Re(a-d)Re(b-c) + Im(a-d)Im(b-c)]\cos(2\alpha)$$

$$+ [Im(a-d)Re(b+c) - Re(a-d)Im(b+c)]\sin(2\alpha).$$

In order to derive the corresponding transmission asymmetry, we need to consider the transmitted fields rather than the scattered fields, which requires us to replace a, d with $a + 1, d + 1$ throughout this derivation. However, as the above expression only depends on $a - d$ it remains unchanged by this substitution. Therefore the directional transmission and reflection asymmetries are identical and must be compensated by a corresponding asymmetry of losses:

$$(\overrightarrow{T} - \overleftarrow{T})(\alpha) = (\overrightarrow{R} - \overleftarrow{R})(\alpha) = -\tfrac{1}{2}(\overrightarrow{L} - \overleftarrow{L})(\alpha) \tag{4.54}$$

It is clear from Eq. (4.53) that these asymmetries change sign for the orthogonal incident polarization. Thus if the metamaterial is *more* transparent (and reflective) for forward-propagating than backward-propagating **x**-polarized waves, it will be *less* transparent (and reflective) for forward-propagating than backward-propagating **y**-polarized waves.

It follows from Eq. (4.53) that the directional transmission, reflection and absorption asymmetries will only vanish for all linear polarizations if

$$Re(a-d)Re(b-c) + Im(a-d)Im(b-c) = 0,$$

$$Im(a-d)Re(b+c) - Re(a-d)Im(b+c) = 0.$$

If we interpret the scattering coefficients a, b, c, d as vectors in the complex plane, then these conditions are equivalent to $a - d \perp b - c$ and $a - d \parallel b + c$, respectively. This has the trivial solutions (4.55) and (4.56), while the remaining nontrivial solutions can be written in a simplified form (4.57):

$$a = d \tag{4.55}$$

$$b = c = 0 \tag{4.56}$$

$$a - d \perp b - c \quad \text{and} \quad |b| = |c| \tag{4.57}$$

The directional asymmetries for transmission, reflection and absorption of linearly polarized waves will be absent if and only if one of the above conditions (4.55)–(4.57) is satisfied.

In particular the *presence of these phenomena requires losses and optical activity* $(a \neq d)$. Just like optical activity, it may only occur at oblique incidence onto planar metamaterials without 2-fold rotational symmetry[14] if the structure does not have a line of (glide) mirror symmetry in the plane of incidence, see Section 4.4.1. The effect is reversed for opposite angles of incidence $\pm\theta$ and linked to *extrinsic 3D chirality*.

If no static magnetic field is present the Lorentz reciprocity lemma [31] must hold. Lorentz reciprocity requires the component of the transmitted wave which is polarized parallel to the incident wave to be the same for opposite propagation directions. In case of planar metamaterials the same holds for reflection. Therefore asymmetric transmission/reflection of linearly polarized waves can only result from a directional asymmetry of the linear polarization conversion efficiency from the incident wave to the orthogonal polarization, i.e. *linear conversion dichroism*.

4.4.4 Linear birefringence and linear dichroism

In their pure form, linear birefringence and dichroism are associated with orthogonal linearly polarized eigenpolarizations (see Section 4.5.2), which experience different phase delays (linear birefringence) or different transmission/reflection/absorption levels (linear dichroism).

Before focusing on the above special case, let's begin by asking more generally under which conditions the transmission/reflection properties of a planar metamaterial can depend on the polarization azimuth of the incident wave.

At normal incidence, the electromagnetic properties can only depend on the azimuth of the incident wave for anisotropic planar metamaterials, where for achiral anisotropic structures[15] the intrinsic preferred directions (i.e. eigenpolarizations) are parallel and perpendicular to the pattern's line of (glide) mirror symmetry.

At oblique incidence, additional extrinsic preferred directions parallel and perpendicular to the plane of incidence are associated with the experimental arrangement. Therefore isotropic planar metamaterials[16] can show linear birefringence/dichroism at oblique incidence. For achiral anisotropic planar metamaterials the eigenpolarizations become frequency-dependent at oblique incidence if the metamaterial pattern's line of (glide) mirror symmetry is neither parallel nor perpendicular to the plane of incidence: the intrinsic preferred directions of the metamaterial pattern compete with the extrinsic preferred directions of the experimental arrangement in this case.

We note that at oblique incidence isotropic structures can exhibit linear birefringence and/or linear dichroism. Also, at specific frequencies linear birefringence/dichroism may vanish for anisotropic planar metamaterials, as illustrated by Fig. 4.11.

---

[14] Wallpaper symmetry groups without 2-fold rotational symmetry: *p*1, *pm*, *pg*, *cm*, *p*3, *p*3*m*1, *p*31*m*.

[15] Achiral anisotropic wallpaper symmetry groups: *pm*, *pg*, *cm*, *pmm*, *pmg*, *pgg*, *cmm*.

[16] Isotropic wallpaper symmetry groups: *p*3, *p*3*m*1, *p*31*m*, *p*4, *p*4*m*, *p*4*g*, *p*6, *p*6*m*.

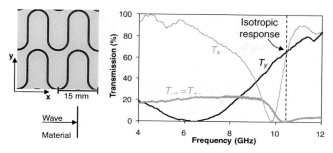

Figure 4.11. Experimental example of a metamaterial with linear dichroism. Normal incidence transmission spectra for waves polarized perpendicular (x) and parallel (y) to the metamaterial's line of mirror symmetry. Consistent with its highly anisotropic structure, the planar metamaterial (wallpaper group *pmg*) generally shows substantial linear dichrism, $T_x - T_y$. This is reflected by nonzero circular polarization conversion T_{-+}. However, at 10.5 GHz linear dichroism vanishes: transmission levels for x and y polarizations are identical and circular polarization conversion is absent. Refer to ref. [11] for more details on this experimental work.

In order to see how the scattering coefficients reveal a dependence of the scattering properties on the azimuth of the incident wave, let's consider the azimuth rotation and ellipticity angle of the scattered field for some incident wave $\hat{u}$ of azimuth α. Any polarization state with azimuth α can be written as (4.37)

$$\hat{u} = \begin{pmatrix} \sin(\beta)e^{i(\gamma-\alpha)} \\ \cos(\beta)e^{i(\gamma+\alpha)} \end{pmatrix},$$

$$\alpha \in (-\tfrac{\pi}{2}, \tfrac{\pi}{2}], \qquad \beta \in [0, \tfrac{\pi}{2}], \qquad \gamma \in [0, 2\pi).$$

The azimuth rotation of the scattered field is given by Eq. (4.39),

$$\Delta\Phi^s = -\tfrac{1}{2}\left[\arg(a) - \arg(d) + \arg(1 + \tfrac{b}{a}\cot(\beta)e^{+i2\alpha}) - \arg(1 + \tfrac{c}{d}\tan(\beta)e^{-i2\alpha})\right].$$

Thus $\Delta\Phi^s$ depends on the azimuth α of the incident wave if and only if at least one of the scattering coefficients b, c is nonzero.

To see whether all diagonal scattering matrices correspond to polarization-azimuth-independent scattering properties, we must check whether the ellipticity angle and amplitude of the scattered field are also independent of the incident wave's azimuth α when $b = c = 0$. The scattered field for $b = c = 0$ is (for forward and backward propagation since $b = c$)

$$\mathbf{E}^s = s\,\hat{u}$$

$$= \begin{pmatrix} a & 0 \\ 0 & d \end{pmatrix} \begin{pmatrix} \sin(\beta)e^{i(\gamma-\alpha)} \\ \cos(\beta)e^{i(\gamma+\alpha)} \end{pmatrix}$$

$$= \begin{pmatrix} a\sin(\beta)e^{i(\gamma-\alpha)} \\ d\cos(\beta)e^{i(\gamma+\alpha)} \end{pmatrix}. \tag{4.58}$$

It can be easily seen that the amplitude $|\mathbf{E}^s|$ does not depend on α. Using Eq. (4.36) we obtain the ellipticity angle of the scattered field as follows:

$$
\begin{aligned}
\eta^s &= \frac{1}{2} \arcsin \left(\frac{|E_+^s|^2 - |E_-^s|^2}{|E_+^s|^2 + |E_-^s|^2} \right) \\
&= \frac{1}{2} \arcsin \left(\frac{|a \sin(\beta)|^2 - |d \cos(\beta)|^2}{|a \sin(\beta)|^2 + |d \cos(\beta)|^2} \right).
\end{aligned}
$$

Clearly also the ellipticity angle of the scattered field does not depend on the azimuth α of the incident field, if $b = c = 0$. By replacing a, d with $a + 1, d + 1$, we could write the same proof for the transmitted field.

Thus the transmission and reflection properties are polarization-azimuth-dependent if and only if $|b| \neq 0$ and/or $|c| \neq 0$. Note that – apart from linear birefringence and linear dichroism – this also includes the lossy chiral effects of circular conversion dichroism (see Section 4.4.2) and linear conversion dichroism (see Section 4.4.3). On the other hand, simultaneous absence of these azimuth-dependent phenomena corresponds to $b = c = 0$.

We will speak of pure linear birefringence/dichroism if no other polarization effects are present. This requires the simultaneous absence of circular/linear conversion dichroism and optical activity ($a = d$ and $|b| = |c| \neq 0$). Provided that preferred directions are defined by the structure (anisotropy) or the experimental arrangement (oblique incidence), pure linear birefringence/dichroism occurs for (i) lossless planar metamaterials without extrinsic 3D chirality and (ii) lossy planar metamaterials without 2D or 3D chirality. Linear dichroism in planar metamaterials can be exploited for the realization of linear polarizers, and linear birefringence allows planar metamaterial wave plates to be developed. The potential performance of such devices will be discussed in Sections 4.7.2 and 4.7.3.

4.5 Eigenstates

To gain a better understanding of polarization effects in planar metamaterials, we study the associated eigenstates. The transmission eigenstates simply correspond to those polarization states that are transmitted without any change in azimuth and ellipticity; see Fig. 4.12 (top). For reflection the situation is more complex. As polarization states must be measured looking into the beam, the reflection matrix relates incident and reflected waves that are measured in mutually rotated coordinate systems; compare the top of Fig. 4.12 with the bottom. Thus the eigen-states of the reflection matrix are polarization states for which the azimuth and ellipticity of the incident and reflected waves are the same in their respective coor-dinate systems. This, however, is not physically meaningful, as it depends on the

Eigenstates

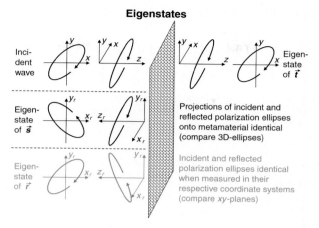

Figure 4.12. Eigenstates of the transmission (top), scattering (middle), and reflection (bottom) matrices. For scattering eigenstates the projections of the polarization ellipses of the incident and reflected waves onto the metamaterial are the same, whereas for eigenstates of the reflection matrix, the incident and reflected waves have the same parameters in their mutually rotated coordinate systems.

choice of coordinates (orientation of x and x_r in the plane normal to the propagation direction). For reflection the eigenstates of the scattering matrix, which describes reflection in the coordinates of the incident wave, are *physically meaningful eigenstates*. For eigenstates of the scattering matrix, the projection of the polarization ellipses of the incident and reflected waves onto the metamaterial are the same.

For the scattering eigenstates, the eigenvalues $\lambda_{1,2}^s$ are given by the eigenvalue equation

$$\det\left[\overrightarrow{s} - \lambda^s\right] = \det\begin{pmatrix} a - \lambda^s & b \\ c & d - \lambda^s \end{pmatrix}$$

$$= (a - \lambda^s)(d - \lambda^s) - bc = 0, \qquad (4.59)$$

which, for both forward and backward propagation, has the following solutions:

$$\lambda_{1,2}^s = \frac{a+d}{2} \pm \sqrt{\left(\frac{a-d}{2}\right)^2 + bc}$$

$$= \frac{a+d}{2} \pm \sqrt{\left(\frac{a-d}{2}\right)^2 + |bc|e^{i2\kappa}}. \qquad (4.60)$$

Similarly, the eigenvalues of the transmission matrix are given by

$$\lambda_{1,2}^t = 1 + \frac{a+d}{2} \pm \sqrt{\left(\frac{a-d}{2}\right)^2 + |bc|e^{i2\kappa}} \tag{4.61}$$

for both forward and backward propagation. Thus the simple relation

$$\lambda_{1,2}^t = 1 + \lambda_{1,2}^s \tag{4.62}$$

applies for transmission and scattering eigenvalues in planar metamaterials. Note that the eigenvalues do not depend on φ.

The scattering eigenstates (or eigenvectors) for the general case can be calculated from the eigenvector equation for forward propagation:

$$\overrightarrow{s}\,\overrightarrow{v} = \lambda \overrightarrow{v}$$

$$\begin{pmatrix} a & b \\ c & d \end{pmatrix} \begin{pmatrix} \overrightarrow{v}_+ \\ \overrightarrow{v}_- \end{pmatrix} = \lambda \begin{pmatrix} \overrightarrow{v}_+ \\ \overrightarrow{v}_- \end{pmatrix}$$

$$\begin{pmatrix} a\overrightarrow{v}_+ + b\overrightarrow{v}_- \\ c\overrightarrow{v}_+ + d\overrightarrow{v}_- \end{pmatrix} = \lambda \begin{pmatrix} \overrightarrow{v}_+ \\ \overrightarrow{v}_- \end{pmatrix}, \tag{4.63}$$

where $\overrightarrow{v}$ is an eigenstate. By eliminating λ from the two conditions contained in the last line, we arrive at the eigenstate condition

$$a + b\frac{\overrightarrow{v}_-}{\overrightarrow{v}_+} = c\frac{\overrightarrow{v}_+}{\overrightarrow{v}_-} + d. \tag{4.64}$$

Note that by replacing a and d with $a + 1$ and $d + 1$, we get the corresponding condition for the transmission eigenstates. Clearly Eq. (4.64) remains unchanged by this substitution, and thus the eigenstate conditions for both forward transmission and forward scattering are identical. It follows that *planar metamaterials have identical transmission and scattering eigenstates*. This becomes obvious when considering that for transmission eigenstates the superposition of the scattered field and the incident wave must have the same polarization state as the incident wave on its own. This can only be achieved if transmission eigenstates are simultaneously also scattering eigenstates. Note, however, that we have to distinguish between eigenstates for forward-propagating and backward-propagating incident waves. The eigenstate condition for backward propagation corresponds to Eq. (4.64) with reversed values of b and c, i.e.

$$a + c\frac{\overleftarrow{v}_-}{\overleftarrow{v}_+} = b\frac{\overleftarrow{v}_+}{\overleftarrow{v}_-} + d. \tag{4.65}$$

In the general case, when the polarization effects of optical activity, circular/ linear conversion dichroism, linear birefringence, and linear dichroism may occur simultaneously, any pair of polarization states is the pair of eigenstates of some scattering matrix. Thus it is more instructive to derive the eigenstates associated with specific polarization effects. Here we do this for the cases of pure optical activity ($b = c = 0$) and the absence of optical activity ($a = d$). Note that the latter case corresponds to the general eigenstates for (i) normal incidence onto planar metamaterials and (ii) any angle of incidence onto 2-fold rotationally symmetric planar metamaterials, as in these cases $a = d$ always applies.

4.5.1 Eigenstates for pure optical activity ($b = c = 0$)

Optical activity has been introduced in Section 4.4.1. Pure optical activity corresponds to the special situation when no other polarization effects are present. In this case $b = c = 0$ is required, which results in diagonal scattering and transmission matrices and simple solutions for eigenvalues and eigenstates. The eigenvalues given by Eqs. (4.60) and (4.61) simplify to the diagonal elements of the corresponding matrices:

$$\lambda_1^s = a, \qquad \lambda_2^s = d, \tag{4.66}$$

$$\lambda_1^t = a + 1, \qquad \lambda_2^t = d + 1, \tag{4.67}$$

while the eigenstates $v_{1,2}$ simply correspond to the polarization states that form our basis, i.e. right-handed and left-handed circular polarizations,

$$v_1 = \begin{pmatrix} 1 \\ 0 \end{pmatrix}, \qquad v_2 = \begin{pmatrix} 0 \\ 1 \end{pmatrix}. \tag{4.68}$$

Note that eigenvalues and eigenstates with the same indices correspond to each other, and that in this case eigenvalues and eigenstates are the same for opposite propagation directions. For the opposite angle of incidence $-\theta$ the roles of a and d are reversed, which means in this case that the same eigenstates correspond to the opposite eigenvalues.

So in the case of pure optical activity, i.e. $b = c = 0$, the eigenstates are counterrotating circular polarizations.

4.5.2 Eigenstates in the absence of optical activity ($a = d$)

Here we derive the eigenstates for planar metamaterials not showing optical activity. Such structures can show the directionally asymmetric transmission, reflection, and absorption effects for circularly polarized waves (see Section 4.4.2) as well as linear birefringence and dichroism (see Section 4.4.4) in their purest forms. Note that $a = d$ is automatically satisfied in two important cases for which optically active behavior cannot occur. These are (i) normal incidence onto any planar metamaterial and (ii) any angle of incidence onto 2-fold rotationally symmetric planar metamaterials. Therefore the eigenvalues and eigenstates derived here are the complete set for cases (i) and (ii).

For $a = d$ the scattering and transmission eigenvalues given by Eqs. (4.60) and (4.61), respectively, simplify to

$$\lambda_{1,2}^s = \quad a \pm \sqrt{bc} \quad = a \pm \sqrt{|bc|}e^{i\kappa}, \tag{4.69}$$

$$\lambda_{1,2}^t = a + 1 \pm \sqrt{bc} = a + 1 \pm \sqrt{|bc|}e^{i\kappa}, \tag{4.70}$$

which apply to both forward and backward propagation. Solutions λ_1 and λ_2 correspond to "+" and "−", respectively. Note that the eigenvalues do not depend on φ.

The transmission and scattering eigenstates for forward propagation are given by Eq. (4.64), which simplifies for $a = d$ to

$$\frac{(\vec{v}_+)^2}{(\vec{v}_-)^2} = \frac{b}{c}. \tag{4.71}$$

Note that the equivalent expression for the backward-propagation eigenstates has reversed values of b and c. From the magnitude of the terms in Eq. (4.71) we can determine the ellipticity angle η of the eigenstates, while the azimuth follows from their phases. Thus the eigenstates are fully defined by Eq. (4.71).

Generally the ellipticity angle of a polarization state is given by Eq. (4.36):

$$\eta = \frac{1}{2} \arcsin \left(\frac{|E_+|^2 - |E_-|^2}{|E_+|^2 + |E_-|^2} \right).$$

By considering the magnitudes of all terms in Eq. (4.71) we can easily see that for the forward-propagation eigenstates, $\mathbf{E} = \vec{v}_{1,2}$, the ellipticity angle must be given by

$$\vec{\eta} = \frac{1}{2} \arcsin \left(\frac{|b| - |c|}{|b| + |c|} \right). \tag{4.72}$$

This indicates that both eigenstates must have the same ellipticity angle. In particular, if they are not linear polarizations, they must be co-rotating.

For backward propagation the roles of b and c are reversed and therefore

$$\overleftarrow{\eta} = -\vec{\eta}, \tag{4.73}$$

which indicates that the eigenstates for backward propagation are also co-rotating if they are not linear; however, they have the opposite handedness compared to the eigenstates for forward propagation.

The azimuth Φ of a polarization state is generally given by Eq. (4.35), i.e.

$$\Phi = -\tfrac{1}{2}[\arg(E_+) - \arg(E_-)].$$

By considering the phases of all terms in Eq. (4.71) it can easily be seen that for the forward-propagation eigenstates $\mathbf{E}^0 = \overrightarrow{\mathbf{v}}_{1,2}$ the following phase condition must be satisfied:

$$\arg(\overrightarrow{v}_+) - \arg(\overrightarrow{v}_-) = \begin{cases} \frac{\arg(b) - \arg(c)}{2} \\ \frac{\arg(b) - \arg(c)}{2} - \pi. \end{cases} \tag{4.74}$$

With Eq. (4.35) the azimuths $\Phi_{1,2}$ of the forward-propagation eigenstates are found to be

$$\overrightarrow{\Phi_1} = \frac{\arg(c) - \arg(b)}{4} = \varphi, \tag{4.75}$$

$$\overrightarrow{\Phi_2} = \varphi + \tfrac{\pi}{2}. \tag{4.76}$$

For backward propagation, the azimuths of the eigenstates are

$$\begin{aligned} \overleftarrow{\Phi_1} &= -\varphi, \\ \overleftarrow{\Phi_2} &= -\varphi + \tfrac{\pi}{2}, \end{aligned} \tag{4.77}$$

as in this case the roles of b and c are switched.

Thus for normal incidence or 2-fold rotational symmetry, planar metamaterials have eigenstates with orthogonal azimuths $\Phi_{1,2}$. The parameter φ, defined in Eq. (4.34), simply corresponds to the azimuth of one of the eigenstates for forward propagation in these cases. Note that, in their respective coordinate systems, the eigenstates for backward propagation are rotated relative to those for forward propagation by -2φ.

By combining the magnitudes of the terms in Eq. (4.71) with the phase conditions (4.74) and normalizing, we can write down the eigenstates for forward propagation for normal incidence or 2-fold rotational symmetry:

$$\begin{aligned} \overrightarrow{v}_1 &= \frac{e^{i\gamma}}{\sqrt{|b| + |c|}} \begin{pmatrix} \sqrt{|b|}e^{-i\varphi} \\ \sqrt{|c|}e^{i\varphi} \end{pmatrix}, \\ \overrightarrow{v}_2 &= \frac{e^{i\gamma}}{\sqrt{|b| + |c|}} \begin{pmatrix} \sqrt{|b|}e^{-i(\varphi+\pi/2)} \\ \sqrt{|c|}e^{i(\varphi+\pi/2)} \end{pmatrix}, \end{aligned} \tag{4.78}$$

where the eigenstate azimuth φ is defined by Eq. (4.34). Different values of $\gamma \in [0, 2\pi)$, which has been included for completeness, correspond to the same polarization state at different times.

The eigenstates for backward propagation, where b and c are switched, are

$$\overleftarrow{\mathbf{v}}_1 = \frac{e^{i\gamma}}{\sqrt{|b| + |c|}} \begin{pmatrix} \sqrt{|c|}e^{-i(-\varphi)} \\ \sqrt{|b|}e^{i(-\varphi)} \end{pmatrix},$$

$$\overleftarrow{\mathbf{v}}_2 = \frac{e^{i\gamma}}{\sqrt{|b| + |c|}} \begin{pmatrix} \sqrt{|c|}e^{-i(-\varphi+\pi/2)} \\ \sqrt{|b|}e^{i(-\varphi+\pi/2)} \end{pmatrix}.$$

(4.79)

The eigenpolarizations do not depend on the scattering coefficients $a = d$ and the phase κ. Note that the eigenvalues λ_i and the eigenvectors v_i with the same indices do correspond to each other.

Importantly, we found that planar metamaterials not showing optical activity ($a = d$) have eigenstates with orthogonal azimuths. In the general case, when $|b| \neq |c|$, the directionally asymmetric transmission, reflection, and absorption effects for circularly polarized waves (see Section 4.4.2) are present, and the eigenstates are co-rotating with the same ellipticity angle. In the special case of $|b| = |c| \neq 0$, which corresponds to linear birefringence and dichroism without the presence of other polarization effects (see Section 4.4.4), the eigenstates are simply orthogonal linear polarizations.

Note that, for *normal incidence*, the azimuths of the eigenstates, if well-defined, must correspond to preferred directions of the metamaterial. For 2D-chiral metamaterials these preferred directions may be frequency-dependent, while for metamaterials with a single line of (glide) mirror symmetry they must be oriented parallel and perpendicular to the line of mirror symmetry. The orientation of this line of mirror symmetry is given by $\tilde{\varphi}$, which is measured in the coordinates for forward propagation at normal incidence. Thus the azimuth of one normal incidence forward-propagation eigenstate, φ or $\varphi + \pi/2$ (measured from the $\mathbf{x}$-axis) must correspond to $\tilde{\varphi}$ (measured from the $\mathbf{y_m}$-axis). So for normal incidence onto metamaterials with a single line of (glide) mirror symmetry, $\varphi = \tilde{\varphi} + n\pi/2$ must hold for some $n \in \mathbb{Z}$.

Obviously, at normal incidence, if φ is well-defined, rotation of the metamaterial by some angle $+\Delta_\varphi$ must result in the same rotation of the eigenstates, without affecting the eigenvalues or the ellipticity of the eigenstates. Thus at normal incidence the rotation $\tilde{\varphi} \rightarrow \tilde{\varphi} + \Delta_\varphi$ must simply translate into a change of the parameter $\varphi \rightarrow \varphi + \Delta_\varphi$. From Eqs. (4.32) it follows that an in-plane rotation of a planar metamaterial at normal incidence by Δ_φ corresponds to

$$b \rightarrow be^{-i2\Delta_\varphi} \quad \text{and} \quad c \rightarrow ce^{+i2\Delta_\varphi}. \tag{4.80}$$

4.6 Energy conservation

Energy conservation implies that the combined power of the transmitted and reflected waves plus losses, $L \geq 0$, must equal the power of the incident wave. Here we consider the case of forward propagation; the opposite propagation direction corresponds to reversed values of b and c:

$$|\mathbf{E^0}|^2 = |\mathbf{E^t}|^2 + |\mathbf{E^r}|^2 + L|\mathbf{E^0}|^2$$

$$= |\overrightarrow{t} \, \mathbf{E^0}|^2 + |\overrightarrow{r} \, \mathbf{E^0}|^2 + L|\mathbf{E^0}|^2. \tag{4.81}$$

Division by $|\mathbf{E^0}|^2$ allows us to introduce the unit vector $\hat{u} = \mathbf{E^0}/|\mathbf{E^0}|$:

$$1 - L = |\overrightarrow{t} \, \hat{u}|^2 + |\overrightarrow{r} \, \hat{u}|^2. \tag{4.82}$$

This condition must be satisfied for any incident polarization, i.e. for any complex unit vector $\hat{u}$. All possible unit vectors or polarization states can be written as Eq. (4.37),

$$\hat{u} = \begin{pmatrix} \sin(\beta)e^{i(\gamma-\alpha)} \\ \cos(\beta)e^{i(\gamma+\alpha)} \end{pmatrix},$$

$$\alpha \in (-\tfrac{\pi}{2}, \tfrac{\pi}{2}], \qquad \beta \in [0, \tfrac{\pi}{2}], \qquad \gamma \in [0, 2\pi).$$

Thus energy conservation for forward propagation written out explicitly is given by

$$1 - L = \left| \begin{pmatrix} a+1 & b \\ c & d+1 \end{pmatrix} \begin{pmatrix} \sin(\beta)e^{i(\gamma-\alpha)} \\ \cos(\beta)e^{i(\gamma+\alpha)} \end{pmatrix} \right|^2$$

$$+ \left| \begin{pmatrix} c & d \\ a & b \end{pmatrix} \begin{pmatrix} \sin(\beta)e^{i(\gamma-\alpha)} \\ \cos(\beta)e^{i(\gamma+\alpha)} \end{pmatrix} \right|^2$$

$$\forall \, \alpha, \beta, \gamma. \tag{4.83}$$

This simplifies to

$$0 \geq -L = p_1 \sin^2 \beta + p_2 \cos^2 \beta$$

$$+ \sin 2\beta [p_4 \cos(2\alpha) - p_5 \sin(2\alpha)]$$

$$\forall \, \alpha \in (-\tfrac{\pi}{2}, \tfrac{\pi}{2}], \beta \in [0, \tfrac{\pi}{2}], \tag{4.84}$$

with the parameters

$$\tfrac{1}{2}p_1 := |a|^2 + |c|^2 + Re(a), \tag{4.85}$$

$$\tfrac{1}{2}p_2 := |d|^2 + |b|^2 + Re(d), \tag{4.86}$$

$$\tfrac{1}{2}p_4 := Re(a + \tfrac{1}{2})Re(b) + Im(a + \tfrac{1}{2})Im(b)$$
$$+ Re(d + \tfrac{1}{2})Re(c) + Im(d + \tfrac{1}{2})Im(c), \tag{4.87}$$

$$\tfrac{1}{2}p_5 := Re(a + \tfrac{1}{2})Im(b) - Im(a + \tfrac{1}{2})Re(b)$$
$$- Re(d + \tfrac{1}{2})Im(c) + Im(d + \tfrac{1}{2})Re(c). \tag{4.88}$$

Equation (4.84) is equivalent to

$$0 \geq p_1 \sin^2 \beta + p_2 \cos^2 \beta + p_3 \sin 2\beta$$
$$\forall \beta \in [0, \tfrac{\pi}{2}], \tag{4.89}$$

where the parameter p_3 corresponds to

$$p_3 := \sqrt{(p_4)^2 + (p_5)^2}$$
$$= 2\left\{|a + \tfrac{1}{2}|^2|b|^2 + |d + \tfrac{1}{2}|^2|c|^2\right.$$
$$+ 2|a + \tfrac{1}{2}||d + \tfrac{1}{2}||b||c|$$
$$\left. \cdot \cos[\arg((a + \tfrac{1}{2})(d + \tfrac{1}{2})) - \arg(bc)]\right\}^{1/2}$$
$$= 2\left\{A^2|b|^2 + D^2|c|^2 + 2AD|b||c|\cos[2\xi - 2\kappa]\right\}^{1/2}. \tag{4.90}$$

The parameters A, D, ξ, κ used in the last line are defined by Eqs. (4.29), (4.30), and (4.33).

The right-hand side of Eq. (4.89) reaches its largest value for β_0 given by

$$\beta_0 = \begin{cases} \tfrac{1}{2}\arctan\left(2\tfrac{p_3}{p_2-p_1}\right) & \text{for } p_1 < p_2 \\ \tfrac{\pi}{4} & \text{for } p_1 = p_2 \\ \tfrac{1}{2}\arctan\left(2\tfrac{p_3}{p_2-p_1}\right) + \tfrac{\pi}{2} & \text{for } p_1 > p_2. \end{cases} \tag{4.91}$$

Therefore energy conservation is satisfied, if Eq. (4.89) is satisfied for β_0 both in the cases of forward and backward propagation, where backward propagation corresponds to reversed roles of b and c in all formulas. (If this is satisfied, energy conservation for the opposite angle of incidence $-\theta$ is also satisfied.)

Although this does define the scattering coefficients that satisfy energy conservation, it is not very intuitive. One can easily see that the energy conservation equation (4.89) requires $p_{1,2} \leq 0$. As we will find for the lossless case, $p_3 = 0$ corresponds to a constraint on the possible values of the phase κ. We will derive that for lossless planar metamaterials κ has two solutions for each allowed pair of values of a, d (except for trivial cases). The constraint on κ is less tight for lossy planar metamaterials. Note that $p_{1,2} \leq 0$ can be written as

$$\left|a + \tfrac{1}{2}\right|^2 + |c|^2 \leq \left(\tfrac{1}{2}\right)^2, \tag{4.92}$$

$$\left|d + \tfrac{1}{2}\right|^2 + |b|^2 \leq \left(\tfrac{1}{2}\right)^2, \tag{4.93}$$

which must also hold for reversed values of b and c, due to energy conservation for the opposite propagation direction.

From this we can obtain for a, d and the magnitudes of $|b|$ and $|c|$ that

$$\left|a + \tfrac{1}{2}\right|, \left|d + \tfrac{1}{2}\right| \leq \tfrac{1}{2}, \tag{4.94}$$

$$|b|, |c| \leq \sqrt{\tfrac{1}{4} - \max\left(\left|a + \tfrac{1}{2}\right|^2, \left|d + \tfrac{1}{2}\right|^2\right)} \leq \tfrac{1}{2}. \tag{4.95}$$

Note that by using Eqs. (4.29) these conditions can rewritten in terms of the alternative parameters A, D resulting in the simplified forms

$$A, D \leq \tfrac{1}{2},$$

$$|b|, |c| \leq \sqrt{\tfrac{1}{4} - \max\left(A^2, D^2\right)} \leq \tfrac{1}{2}.$$

From Eqs. (4.94) and (4.95) it follows that, in the complex plane, the solutions for a, b, c, d must always fall in the regions indicated in Fig. 4.13(a).

For the lossless case the conditions $|a + 1/2| = |d + 1/2|$ and the left equality of Eq. (4.95) are required. Compared to any chosen lossless solution, $|a + 1/2|$, $|d + 1/2|$, $|b|$, $|c|$ may take smaller values for a lossy planar metamaterial; see Fig. 4.13(b).

4.6.1 Lossless planar metamaterials

For lossless planar metamaterials, the equality in Eq. (4.89) must hold for all β. Thus the expressions p_1, p_2, p_3 must all be zero. The conditions $p_{1,2} = 0$ are equivalent to the equality in Eqs. (4.92) and (4.93), which must also hold for reversed values of b, c due to lossless behavior for the opposite propagation direction. From this it follows that the following relations must hold for lossless planar metamaterials:

$$\left|a + \tfrac{1}{2}\right| = \left|d + \tfrac{1}{2}\right| \leq \tfrac{1}{2}, \tag{4.96}$$

$$|b| = |c| \quad \leq \tfrac{1}{2}, \tag{4.97}$$

$$\left|a + \tfrac{1}{2}\right|^2 + |b|^2 = \left(\tfrac{1}{2}\right)^2. \tag{4.98}$$

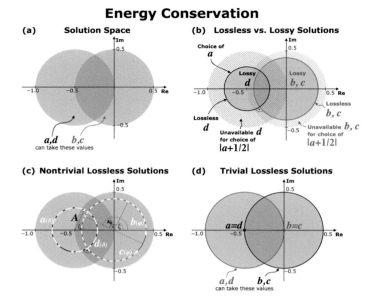

Figure 4.13. Constraints on the scattering coefficients a, b, c, d from energy conservation. (a) The solution space for a, b, c, d. (b) Lossy vs. lossless solutions for b, c, d for a particular choice of $|a + 1/2| = const$ (here $1/\sqrt{3}$). The roles of a, d can be switched. (c) The nontrivial solutions (4.104) for the lossless case illustrated for an arbitrary choice of A, ξ. Solutions for a, d are corresponding points on the dashed arrows indicating a and d as functions of δ. Solutions b, c are corresponding points on the dashed arrows indicating b and c as functions of the parameter φ. (d) Trivial solutions for the lossless case: $a = d = -1/2$ is shown in black (Eq. (4.105)) and $b = c = 0$ is plotted in gray (Eq. (4.106)).

Rewritten in terms of the alternative parameters A, D defined by Eqs. (4.29) these conditions simplify to

$$A = D \ \leq \tfrac{1}{2},$$

$$|b| = |c| \ \leq \tfrac{1}{2},$$

$$A^2 + |b|^2 = \left(\tfrac{1}{2}\right)^2.$$

Note that condition (4.97) implies that lossless planar metamaterials cannot show the asymmetric transmission, reflection, and absorption effects for circularly polarized waves, which were identified in Section 4.4.2. Equation (4.98) describes two circles in the complex plane that intersect orthogonally (see Fig. 4.13(b)). In the lossless case a and d lie on a circle of radius $A = D$ centered at $-1/2$, while b and c lie on a circle of radius $|b| = |c|$ centered at the origin. The sum of the squared radii of both circles is $1/4$. In particular, this implies that the maximum radius of either circle is $1/2$.

For lossless planar metamaterials, p_3, defined by Eq. (4.90), must also be zero. Using Eqs. (4.96) and (4.97), $p_3 = 0$ simplifies to

$$|a + \tfrac{1}{2}||b| \left\{ 1 + \cos[\arg((a + \tfrac{1}{2})(d + \tfrac{1}{2})) - 2\kappa] \right\}$$
$$= A|b| \left\{ 1 + \cos[2\xi - 2\kappa] \right\}$$
$$= 0, \tag{4.99}$$

where A, ξ, and κ are defined by Eqs. (4.29), (4.30), and (4.33). Thus the nontrivial solutions of $p_3 = 0$ correspond to $\cos[2\xi - 2\kappa] = -1$, which requires κ to take one of the values

$$\kappa_0 := \xi \pm \tfrac{\pi}{2}. \tag{4.100}$$

Condition (4.99) also has two trivial solutions corresponding to the cases (see Fig. 4.13(d))

$$a = d = -\frac{1}{2} \tag{4.101}$$

or

$$b = c = 0. \tag{4.102}$$

Note that we defined κ_0 as the allowed values for κ in the lossless nontrivial case.

So any lossless planar metamaterial must satisfy Eqs. (4.96), (4.97), and (4.98), and additionally one of Eqs. (4.100), (4.101), and (4.102). The latter three are the nontrivial and trivial solutions of $p_3 = 0$.

For the lossless case, Eq. (4.98) allows us to write the values $|b| = |c|$ as functions of A:

$$|b| = |c| = \sqrt{\tfrac{1}{4} - |a + \tfrac{1}{2}|^2} = \sqrt{\tfrac{1}{4} - A^2} \leq \tfrac{1}{2}. \tag{4.103}$$

Using Eqs. (4.28) and (4.32) to express a, b, c, d in terms of the alternative parameters, we can easily write down the lossless nontrivial scattering matrices (see Fig. 4.13(c)):

$$\vec{s} = \begin{pmatrix} a & b \\ c & d \end{pmatrix} = \begin{pmatrix} Ae^{i(\xi-2\delta)} - \tfrac{1}{2} & |b|e^{i(\kappa_0-2\varphi)} \\ |b|e^{i(\kappa_0+2\varphi)} & Ae^{i(\xi+2\delta)} - \tfrac{1}{2} \end{pmatrix}, \tag{4.104}$$

$$A \in (0, \tfrac{1}{2}), \quad \xi \in (-\pi, \pi], \quad \delta \in (-\tfrac{\pi}{2}, \tfrac{\pi}{2}), \quad \varphi \in [0, \tfrac{\pi}{2}),$$

where κ_0 and $|b|$ are given by Eqs. (4.100) and (4.103), respectively. Note that there are four independent parameters for the lossless nontrivial solutions, compared to eight free parameters for general planar metamaterials.

For the trivial cases Eqs. (4.101) and (4.102), we get the following trivial solutions (see Fig. 4.13(d)):

$$\overrightarrow{s} = \frac{1}{2} \begin{pmatrix} -1 & e^{i(\kappa-2\varphi)} \\ e^{i(\kappa+2\varphi)} & -1 \end{pmatrix}, \tag{4.105}$$

$$\kappa \in [0, 2\pi), \qquad \varphi \in [0, \tfrac{\pi}{2}),$$

$$s = \frac{1}{2} \begin{pmatrix} e^{i(\xi-2\delta)} - 1 & 0 \\ 0 & e^{i(\xi+2\delta)} - 1 \end{pmatrix}, \tag{4.106}$$

$$\xi \in (-\pi, \pi], \qquad \delta \in (-\tfrac{\pi}{2}, \tfrac{\pi}{2}).$$

All scattering matrices allowed for lossless planar metamaterials are given by Eqs. (4.104), (4.105), and (4.106).

How can we interpret these solutions? Clearly, $|b| = |c|$ must be satisfied for all lossless planar metamaterials and thus circular conversion dichroism (see Section 4.4.2), cannot be observed for lossless planar metamaterials. Also linear conversion dichroism (see Section 4.4.3) is absent for the above solutions, which satisfy the conditions (4.57), (4.55) and (4.56), respectively. Therefore directionally asymmetric transmission, reflection, and absorption of circularly or linearly polarized waves cannot occur for lossless planar metamaterials.

For Eqs. (4.104) and (4.106) the diagonal elements of the scattering matrix, a and d, are allowed to take different values, which corresponds to the effects of circular birefringence and circular dichroism; see Section 4.4.1. Here $\xi = 0$ corresponds to the special case of circular birefringence without circular dichroism.

Solutions (4.104) and (4.105) correspond to responses with linear dichroism and/or linear birefringence, as they have $|b| = |c| \neq 0$; see Section 4.4.4. In the absence of optical activity, $a = d$, the eigenstates are linearly polarized with azimuths φ and $\varphi + \pi/2$ for forward propagation; see Section 4.5.2. Special cases of these linearly birefringent/dichroic lossless solutions without optical activity correspond to lossless linear polarizers and lossless wave plates. These important cases will be identified and discussed in Sections 4.7.2 and 4.7.3.

Lossless solution (4.106), which has $b = c = 0$, corresponds to a response that does not depend on the azimuth of the incident wave; see Section 4.4.4. This family of solutions corresponds to lossless planar metamaterials without linear birefringence/dichroism that may exhibit optical activity. Section 4.6.2 studies such lossless pure optical activity in more detail, while Sections 4.7.4

and 4.7.5 will discuss the special cases of lossless polarization rotators and circular polarizers.

4.6.2 Lossless planar metamaterials without linear birefringence/dichroism
$$(L = 0, b = c = 0)$$

We now study optical activity for lossless planar metamaterials without linear birefringence/dichroism. The complete set of scattering matrices for this case is given by Eq. (4.106), and the corresponding eigenstates are derived in Section 4.5.1. Note that in general optical activity can only occur at oblique incidence onto planar metamaterials without 2-fold rotational symmetry. Quantitatively the properties discussed here can only be realized for planar metamaterials that are also lossless. Importantly, absence of linear birefringence/dichroism at some particular frequency does not require the metamaterial itself to be isotropic. First we will derive polarization rotation and circular dichroism of lossless planar metamaterials without linear birefringence/dichroism in general, and then in Sections 4.7.4 and 4.7.5 the special cases of an ideal lossless rotator and circular polarizer will be discussed.

Since in our case the metamaterial properties do not depend on the azimuth of the incident wave, it follows from Eqs. (4.40) and (4.41) that the polarization azimuths of the scattered and transmitted waves will be rotated by

$$\Delta\Phi^s = \tfrac{1}{2}\left[\arg(d) - \arg(a)\right],$$
$$\Delta\Phi^t = \tfrac{1}{2}\left[\arg(d + 1) - \arg(a + 1)\right]. \tag{4.107}$$

Note that for lossless planar metamaterials without linear birefringence/dichroism both a and d are of the form $(1/2)e^{i(\xi \mp 2\delta)} - 1/2$, where ξ, δ are chosen to satisfy the phase convention $\xi \pm 2\delta \in (-\pi, \pi]$. As illustrated by Fig. 4.14, $\arg(d + 1) - \arg(a + 1) = 2\delta$ follows for this case. The phase difference $\arg(d) - \arg(a)$ is 2δ if a, d lie in the same half-plane (upper or lower) and $\pi - 2\delta$ otherwise. From these results we obtain the polarization azimuth rotations for lossless planar metamaterials without linear birefringence/dichroism for the scattered and transmitted fields as follows:

$$\Delta\Phi^s = \begin{cases} \delta \\ \tfrac{\pi}{2} - \delta \end{cases} \quad \text{for} \quad \begin{matrix} Im(a)Im(d) \geq 0 \\ Im(a)Im(d) < 0, \end{matrix} \tag{4.108}$$
$$\Delta\Phi^t = \delta.$$

Lossless Pure Optical Activity

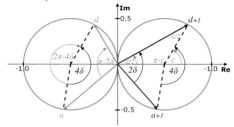

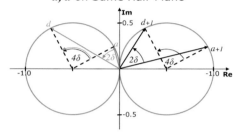

Figure 4.14. For lossless metamaterials without linear birefringence/dichroism (shown in black), $a + 1$ and $d + 1$ lie on a circle through the origin centered at $+1/2$. The angle from $a + 1$ to $d + 1$ around the circle's center that faces away from the origin is $\arg(d + 1/2) - \arg(a + 1/2) = 4\delta$; see Eq. (4.31). The angle $\arg(d + 1) - \arg(a + 1)$ that the same points form with the origin is just half as large. (We use basic geometry to sketch the proof (in gray) on the right-hand side of the top panel; the same idea is used in all cases.) For the angle $\arg(d) - \arg(a)$ from a to d (shown in gray) we must distinguish two cases. The top panel shows that if a and d fall in different half-planes (lower/upper), $\arg(d) - \arg(a)$ is just half as large as $2\pi - 4\delta$. The bottom panel shows that $\arg(d) - \arg(a) = \arg(d + 1) - \arg(a + 1) = 2\delta$ if a and d fall in the same half-plane (upper/lower).

Circular dichroism for transmission and scattering are generally given by Eqs. (4.42) and (4.43), which in our lossless case without linear birefringence/dichroism take the form

$$\Delta S = |a|^2 - |d|^2$$
$$= |\tfrac{1}{2}e^{i(\xi-2\delta)} - \tfrac{1}{2}|^2 - |\tfrac{1}{2}e^{i(\xi+2\delta)} - \tfrac{1}{2}|^2,$$
$$\Delta T = |a + 1|^2 - |d + 1|^2$$
$$= |\tfrac{1}{2}e^{i(\xi-2\delta)} + \tfrac{1}{2}|^2 - |\tfrac{1}{2}e^{i(\xi+2\delta)} + \tfrac{1}{2}|^2.$$

After simplifying we find that for lossless planar metamaterials without linear birefringence/dichroism, circular dichroism for scattering and transmission is given by

$$\Delta S = - \sin(\xi) \sin(2\delta),$$ (4.109)

$$\Delta T = + \sin(\xi) \sin(2\delta).$$ (4.110)

4.7 Applications and limitations

We are used to a variety of optical components, some of which are polarization-insensitive, such as ideal mirrors, beam splitters, and attenuators, while others, such as wave plates, linear and circular polarizers, and polarization rotators are used to manipulate polarization states. In this section, we explore the potential of planar metamaterial realizations of these components.

Planar metamaterials acting as mirrors, beam splitters, or reflective attenuators should not be surprising, as the usual realizations of these components can be understood as trivial cases of planar metamaterials. The same applies to wire grid linear polarizers commonly used in the microwave, terahertz, and mid-infrared parts of the spectrum.

We show that planar metamaterials can also be used as wave plates, which can be very efficient for small phase delays. Large phase delays, however, come at the cost of large insertion losses, and $\lambda/2$-plates cannot be realized.

We find that, at oblique incidence, planar metamaterials can act as polarization rotators or circular polarizers. Polarization rotators can perform well for small rotation angles, but large rotation angles come at the cost of large insertion losses, and rotation angles of $\pm\pi/2$ cannot be realized. Circular polarizers are not limited in efficiency.

Importantly, planar metamaterials allow the realization of wave plates, linear and circular polarizers and polarization rotators that operate in transmission and/or reflection. Thus planar metamaterials provide not only an opportunity to miniaturize existing polarization optics for transmission, but they also allow the development of novel components such as reflection wave plates, reflection circular polarizers, and reflection polarization rotators.

It will be shown that for all functionalities listed here, apart from attenuators, the best performance can be achieved with lossless planar metamaterials.

4.7.1 Attenuators, beam splitters, mirrors, and empty space

Attenuators, beam splitters, mirrors, and empty space exhibit neither polarization-azimuth-dependent properties ($b = c = 0$) nor optical activity ($a = d$). Thus they correspond to scattering matrices

$$s = \begin{pmatrix} a & 0 \\ 0 & a \end{pmatrix},$$ (4.111)

$$|a + \tfrac{1}{2}| \le \tfrac{1}{2}.$$

The corresponding scattered (reflected) and transmitted power fractions S and T are given by

$$\begin{aligned} S &= |a|^2, \\ T &= |a + 1|^2. \end{aligned}$$ (4.112)

The equality $|a + 1/2| = 1/2$ corresponds to all lossless planar metamaterials without linear birefringence/dichroism and optical activity. Using Eqs. (4.28), all lossless a can be written as $a = (1/2)(e^{i\xi} - 1)$, which allows us to rewrite the scattered and transmitted power fractions for the *lossless case* as

$$\begin{aligned} S &= \sin^2\left(\tfrac{\xi}{2}\right), \\ T &= \cos^2\left(\tfrac{\xi}{2}\right). \end{aligned}$$ (4.113)

Lossless planar metamaterials showing neither linear birefringence/dichroism nor optical activity are beam splitters. The transmitted and scattered power fractions are given by Eqs. (4.113). A simple $50 - 50$ beam splitter corresponds to $\xi = \pm\pi/2$, while the two limiting cases are a perfect mirror for $\xi = \pi$ ($a = -1$) and empty space with $\xi = 0$ ($a = 0$).

All lossy planar metamaterials without polarization-azimuth-dependent characteristics and without optical activity, i.e. $|a + 1/2| < 1/2$, correspond to attenuating beam splitters.

4.7.2 Linear polarizer

A transmission linear polarizer transmits one specific linearly polarized component of the incident wave without changing its polarization state, while it is opaque for the orthogonal linearly polarized component. A reflection linear polarizer exhibits corresponding behavior for the scattered field.

Thus, linear polarizers must have orthogonal linearly polarized eigenstates, where, for reflection linear polarizers one scattering eigenvalue is zero, and for transmission linear polarizers one transmission eigenvalue is zero. Orthogonal, linearly polarized eigenstates correspond to linearly dichroic (and/or birefringent) behavior without optical activity or circular/linear conversion dichroism, i.e. $a = d$ and $|b| = |c|$. The additional requirement that one eigenvalue must be zero allows us to write down linear polarizer conditions based on the eigenvalue equations (4.69) and (4.70). A reflection linear polarizer must satisfy

$$a = \pm|b|e^{i\kappa} \tag{4.114}$$

for one sign, and a transmission linear polarizer must satisfy

$$a = \pm|b|e^{i\kappa} - 1 \tag{4.115}$$

for one sign.

Linear polarizers that operate in transmission and reflection are of particular interest. It can be easily seen that Eqs. (4.114) and (4.115) can only be simultaneously satisfied for $|b| = 1/2$, $\kappa \in \{0, \pi\}$, and $a = -1/2$.

Thus *all* simultaneous transmission and reflection linear polarizers are given by

$$\overrightarrow{s} = \tfrac{1}{2}\begin{pmatrix} -1 & \pm e^{-i2\varphi} \\ \pm e^{i2\varphi} & -1 \end{pmatrix},$$
$$\varphi \in [0, \tfrac{\pi}{2}). \tag{4.116}$$

These scattering matrices correspond to the lossless planar metamaterial solution (4.105) with $\kappa \in \{0, \pi\}$. It follows that energy conservation is satisfied and that simultaneous reflection and transmission planar metamaterial linear polarizers, like good wire grid polarizers [1, 2], *must be lossless*. They completely transmit one linear polarization and completely reflect the orthogonal linear polarization. Note that the parameter φ specifies the azimuth of either the transmitted or the reflected polarization state. At normal incidence φ simply corresponds to the orientation of the linear polarizer.

Importantly, all lossy (and many lossless) planar metamaterial linear polarizers only work for either reflection or transmission, as they do not reflect/transmit the wanted polarization completely. To see this, consider a reflection linear polarizer that reflects the wanted linear polarization completely. It must have scattering eigenvalues $|\lambda^s_{1,2}| = 1, 0$. Due to $\lambda^t_{1,2} = \lambda^s_{1,2} + 1$ (see Eq. (4.62)) and energy conservation, $|\lambda^s_i|^2 + |\lambda^t_i|^2 \leq 1$, the scattering eigenvalues must be $\lambda^s_{1,2} = -1, 0$, and the corresponding transmission eigenvalues are $\lambda^t_{1,2} = 0, 1$. Thus a planar metamaterial reflection linear polarizer that reflects the wanted polarization completely must also work for transmission (and vice versa). As discussed in the preceding paragraph, this is only possible if no absorption losses are present.

It follows that *all* planar metamaterial linear polarizers with 100% efficiency are lossless, work simultaneously for both reflection and transmission, and are given by Eq. (4.116). Here efficiency is the reflected or transmitted power fraction for the desired polarization state.

4.7.3 Wave plates

A wave plate introduces some phase delay ρ between two orthogonal, linearly polarized components of an electromagnetic wave. For any wave plate, the polarization state of these orthogonal, linearly polarized components will not be affected by the wave plate, and therefore these components must be eigenstates. As wave plates have identical transmission levels for both eigenstates, the corresponding eigenvalues must have the same magnitude, i.e. they only differ by the desired phase difference ρ.

Wave plates do not show optical activity or circular/linear conversion dichroism and thus $a = d$ and $|b| = |c|$. The behavior of wave plates is simply linearly birefringent.

It follows from Section 4.5.2 that orthogonal linearly polarized eigenstates result automatically from $a = d$ and $|b| = |c|$. Thus all scattering matrices that additionally have eigenvalues of the same magnitude correspond to wave plates.

Here we generalize the concept of wave plates to the scattered field. Thus, we consider scattering wave plates and transmission wave plates. Scattering wave plates act as wave plates for the reflected wave. In order to simplify this derivation, we introduce a parameter $\tilde{a}$, which is defined differently for scattering and transmission wave plates, as follows:

$$\begin{aligned} \text{scattering:} \quad & \tilde{a} := a, \\ \text{transmission:} \quad & \tilde{a} := a + 1 \end{aligned} \tag{4.117}$$

In the absence of optical activity, the eigenvalues are given by Eqs. (4.69) and (4.70) for scattering and transmission, respectively. Here, with $|b| = |c|$ and the definition of $\tilde{a}$, these equations simplify to

$$\lambda_{1,2} = \tilde{a} \pm |b|e^{i\kappa},$$

for both scattering and transmission eigenstates. Clearly, $|\lambda_1| = |\lambda_2|$ is only satisfied when the vectors represented by $\tilde{a}$ and $e^{i\kappa}$ in the complex plane are perpendicular to each other, i.e.

$$\kappa = \arg(\tilde{a}) \pm \tfrac{\pi}{2}. \tag{4.118}$$

The angle formed by the sum and difference of two perpendicular vectors is twice the arctangent of the ratio of their magnitudes. Therefore the phase difference $\rho \in [0, \pi]$ between the eigenvalues corresponds to

$$\rho = |\arg \lambda_2 - \arg \lambda_1| = 2 \arctan \left| \frac{b}{\tilde{a}} \right|. \tag{4.119}$$

Note that all scattering matrices with $a = d$ and $|b| = |c|$, which additionally satisfy condition (4.118) and energy conservation, correspond to wave plates. For scattering wave plates $\tilde{a} = a$, while for transmission wave plates $\tilde{a} = a + 1$.

The phase difference induced by a scattering wave plate is given by

$$\rho^s = 2\arctan\left|\frac{b}{a}\right| \leq 2\arctan\sqrt{\frac{Re(a)+1}{-Re(a)}}, \tag{4.120}$$

where we used inequality (4.95) to write the largest possible $|b|$ in terms of a. Similarly the phase difference induced by a transmission wave plate is

$$\rho^t = 2\arctan\left|\frac{b}{a+1}\right| \leq 2\arctan\sqrt{\frac{-Re(a)}{Re(a)+1}}. \tag{4.121}$$

Planar metamaterials have identical scattering and transmission eigenstates. As wave plates have orthogonal linearly polarized eigenstates, the corresponding scattering and transmission matrices can always be written in the orthogonal eigenstate basis. Written in the eigenbasis, both scattering and transmission matrices become diagonal with the corresponding eigenvalues as entries. For a scattering wave plate both scattering eigenvalues have the same magnitude, and therefore the scattered power fraction S for any incident polarization is simply given by the squared magnitude of its scattering eigenvalues, i.e.

$$S = |\lambda^s|^2 = |a|^2 + |b|^2$$
$$\leq |a|^2 + \tfrac{1}{4} - |a + \tfrac{1}{2}|^2 = -Re(a). \tag{4.122}$$

Similarly the transmission level of a transmission wave plate is given by the squared magnitude of its transmission eigenvalues, i.e.

$$T = |\lambda^t|^2 = |a + 1|^2 + |b|^2$$
$$\leq |a + 1|^2 + \tfrac{1}{4} - |a + \tfrac{1}{2}|^2 = Re(a) + 1. \tag{4.123}$$

Importantly, a planar metamaterial $\lambda/2$-plate cannot be realized. A phase delay of π in transmission requires $a = -1$ (see Eq. (4.121)), but for $a = -1$ the wave plate must be opaque. Similarly, for a scattering $\lambda/2$-plate $a = 0$ would be required, but for $a = 0$ the wave plate cannot reflect.

Lossy planar metamaterial wave plates [11, 12] typically work for either transmission or scattering. However, there is one important exception: if $a \in \mathbb{R}$, and thus also $\tilde{a} \in \mathbb{R}$, the wave plate condition (4.118) will be satisfied for both scattering and transmission by $\kappa = \pm\pi/2$. It follows that all wave plates that simultaneously work for transmission and scattering are given by

$$\vec{s} = \begin{pmatrix} a & b \\ c & d \end{pmatrix} = \begin{pmatrix} a & |b|e^{i(\pm\frac{\pi}{2}-2\varphi)} \\ |b|e^{i(\pm\frac{\pi}{2}+2\varphi)} & a \end{pmatrix},$$

(4.124)

$$-1 < a < 0, \qquad |b| \leq \sqrt{\tfrac{1}{4} - (a + \tfrac{1}{2})^2}, \qquad \varphi \in [0, \tfrac{\pi}{2}),$$

where one preferred direction is given by φ. All solutions (4.124) satisfy energy conservation (4.89) as $p_3 = 0$ for $A = D$, $|b| = |c|$, and $\kappa - \xi = \pm\pi/2$, and as the conditions (4.94) and (4.95), which arise from $p_{1,2}$, are obviously satisfied. The limiting cases $a \in \{-1, 0\}$, which lead to $|b| = 0$, correspond to a phase delay of $\rho = 0$ and the absence of linear birefringence. As a and $|b|$ are both free parameters, the possible choices for phase delays for transmitted and reflected waves are largely independent in lossy planar metamaterial wave plates that work simultaneously in transmission and reflection.

Lossless wave plates

All scattering matrices allowed by energy conservation for lossless planar meta-materials are given by Eqs. (4.104), (4.105), and (4.106). As shown earlier in this section, wave plates must satisfy $a = d$ (i.e. $\delta = 0$).

In the nontrivial case, Eq. (4.104), lossless planar metamaterials must meet the condition $\kappa = \arg(a + 1/2) \pm \pi/2$, Eq. (4.100), while wave plates must satisfy $\kappa = \arg(\tilde{a}) \pm \pi/2$, Eq. (4.118). Clearly these conditions can only be simultaneously satisfied if $a \in \mathbb{R}$ and thus $\arg(a + 1/2)$, $\arg(\tilde{a}) \in \{0, \pi\}$. The only linearly birefringent trivial lossless solution, Eq. (4.105), also satisfies $a \in \mathbb{R}$.

Thus *all* lossless planar metamaterial wave plates are given by

$$\vec{s} = \begin{pmatrix} a & |b|e^{i(\pm\frac{\pi}{2}-2\varphi)} \\ |b|e^{i(\pm\frac{\pi}{2}+2\varphi)} & a \end{pmatrix},$$

(4.125)

$$-1 < a < 0, \qquad |b| = \sqrt{\tfrac{1}{4} - (a + \tfrac{1}{2})^2}, \qquad \varphi \in [0, \tfrac{\pi}{2}),$$

where φ specifies one preferred direction. All lossless wave plates are simultaneously wave plates for scattered and transmitted waves. Importantly, in the lossless case there is only one free parameter a, apart from the orientation of the anisotropic direction φ. Therefore the phase delays ρ^s and ρ^t for scattering and transmission are not independent in lossless planar metamaterial wave plates, and Eqs. (4.120) and (4.121) become equalities:

$$\rho^s = 2\arctan\left|\frac{b}{a}\right| = 2\arctan\sqrt{\frac{a+1}{-a}}, \tag{4.126}$$

$$\rho^t = 2\arctan\left|\frac{b}{a+1}\right| = 2\arctan\sqrt{\frac{-a}{a+1}}. \tag{4.127}$$

It follows that

$$\rho^s + \rho^t = \pi. \tag{4.128}$$

Equations (4.122) and (4.123) for the scattered power fraction S and the transmitted power fraction T for any incident polarization also become equalities, and can be written in terms of the phase delay using Eqs. (4.126) and (4.127) as follows:

$$S = -a = cos^2\left(\frac{\rho^s}{2}\right), \tag{4.129}$$

$$T = a + 1 = cos^2\left(\frac{\rho^t}{2}\right). \tag{4.130}$$

It follows that a large phase delay comes at the cost of low efficiency. For example, a transmission phase delay of $3\pi/4$ can only be achieved with 15% transmission, while the corresponding $\pi/4$ scattering phase delay is achieved with 85% reflection. The special case of $\rho^s = \rho^t = \pi/2$, i.e. the lossless $\lambda/4$-plate, has transmission and reflection levels of 50% each.

From Eqs. (4.120)–(4.123) and (4.126)–(4.130) it follows that lossless planar metamaterial wave plates introduce larger phase delays combined with higher transmission and scattering levels than lossy planar metamaterial wave plates. Simply put, lossless wave plates perform better.

4.7.4 Polarization rotators

Here we examine the special case of polarization rotators. Note that rotators require oblique incidence onto planar metamaterials without 2-fold rotational symmetry. Polarization rotators rotate the azimuth of any polarization state by some fixed angle $\Delta\Phi$ without affecting the ellipticity of the polarization state.

As shown in Section 4.4.4, the same azimuth rotation for any polarization state will be seen if and only if $b = c = 0$. For this case energy conservation, Eq. (4.89), reduces to the simple condition $|a + 1/2|, |d + 1/2| \le 1/2$, where the equalities correspond to the lossless case; see the gray circles in Fig. 4.15.

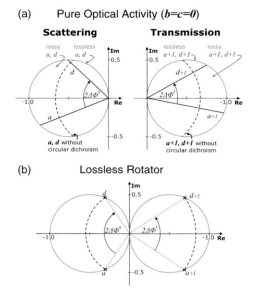

Figure 4.15. Pure optical activity and lossless rotators. As shown in Section 4.4.4, absence of linear birefringence/dichroism and circular/linear conversion dichroism requires $b = c = 0$. For this case energy conservation, Eq. (4.89), allows all a, d that satisfy $|a + 1/2|, |d + 1/2| \leq 1/2$. The equalities, corresponding to the lossless case, are indicated by gray circles. The relevant coefficients for scattering are a, d, while the relevant transmission coefficients are $a + 1, d + 1$. (a) For lossless planar metamaterials the scattering and transmission coefficients must fall on the gray circles, while for lossy planar metamaterials all values within the gray circles are allowed. Circular dichroism is only absent if the relevant coefficients have the same magnitude, i.e. fall on the same (dashed) circle around the origin. The angle that both coefficients form with the origin is twice as large as the polarization azimuth rotation $\Delta\Phi$ experienced by incident waves. (b) The lossless rotator is the special case where circular dichroism and losses are absent.

From Eqs. (4.40) and (4.41) it follows that for the absence of linear birefringence/dichroism and circular/linear conversion dichroism the azimuth rotations $\Delta\Phi^s$ and $\Delta\Phi^t$ are given by

$$\Delta\Phi^s = -\tfrac{1}{2}\left[\arg(a) - \arg(d)\right], \tag{4.131}$$

$$\Delta\Phi^t = -\tfrac{1}{2}\left[\arg(a + 1) - \arg(d + 1)\right]. \tag{4.132}$$

Thus, for scattering the rotation is half as large as the angle formed by a and d with the origin; and for transmission the rotation is half as large as the angle formed by $a + 1$ and $d + 1$ with the origin; see the black lines in Fig. 4.15.

Rotators do not change the ellipticity of any polarization state, thus a scattering rotator must not have scattering circular dichroism (4.43), and a transmission rotator must not have transmission circular dichroism (4.42). For scattering rotators this requires $|a| = |d|$, while for transmission rotators $|a + 1| = |d + 1|$ is needed; see the dashed arcs in Fig. 4.15.

Due to the absence of circular dichroism, the scattering matrix of a scattering rotator is diagonal with entries a, d (= eigenvalues) of equal magnitude. Therefore the scattered power fraction S of a scattering rotator is given by

$$S = |a|^2,$$

independent of the incident polarization state. Similarly the transmitted power fraction of a transmission rotator is given by

$$T = |a + 1|^2.$$

It can easily be seen from Fig. 4.15(a) that the largest reflectivity S for any choice of $\Delta\Phi^s$ (angle between the black lines in the left panel) without scattering circular dichroism (a and d on same dashed circle) corresponds to the lossless case (gray circle). Similarly, the largest transmission T for any choice of $\Delta\Phi^t$ without transmission circular dichroism corresponds to the lossless case; see the right panel. Thus, for any azimuth rotation, the most transparent transmission rotator and the most reflective scattering rotator are lossless. We will find on the following pages that for lossless scattering/transmission rotators the reflected/transmitted power fraction is given by $\cos^2(\Delta\Phi)$. Therefore the scattered power fraction of a scattering rotator is given by

$$S = |a|^2 \leq \cos^2(\Delta\Phi^s), \tag{4.133}$$

and similarly the transmitted power fraction of a transmission rotator is given by

$$T = |a + 1|^2 \leq \cos^2(\Delta\Phi^t). \tag{4.134}$$

Typically lossy planar metamaterial rotators [13] will only work for scattered or transmitted fields. However, there are rotator solutions ($a \neq d$) that satisfy $|a| = |d|$ and $|a + 1| = |d + 1|$ simultaneously. All of these scattering and transmission rotators are given by $d = \bar{a}$, where $\bar{a}$ is the complex conjugate of a. The corresponding scattering matrices are

$$s = \begin{pmatrix} a & 0 \\ 0 & \bar{a} \end{pmatrix}, \tag{4.135}$$

$$|a + \tfrac{1}{2}| \leq \tfrac{1}{2}.$$

The lossless rotator

Here we examine the special case of lossless rotators. Note that lossless rotators require oblique incidence onto lossless planar metamaterials without 2-fold rotational symmetry.

Lossless rotators must correspond to lossless scattering matrices without any linear birefringence/dichroism ($b = c = 0$) or circular dichroism for scattering ($|a| = |d|$) or transmission ($|a + 1| = |d + 1|$). All lossless scattering matrices without linear birefringence/dichroism are given by Eq. (4.106). From Eq. (4.109) and (4.110) it follows that circular dichroism is generally absent simultaneously for scattered and transmitted fields. For our phase convention, $\xi \pm 2\delta \in (-\pi, \pi]$, all circularly birefringent solutions without circular dichroism correspond to $\xi = 0$. Thus the scattering matrices of all lossless rotators are given by Eq. (4.106) with $\xi = 0$:

$$s = \tfrac{1}{2} \begin{pmatrix} e^{-i2\delta} - 1 & 0 \\ 0 & e^{+i2\delta} - 1 \end{pmatrix}, \tag{4.136}$$

$$\delta \in (-\tfrac{\pi}{2}, \tfrac{\pi}{2}).$$

Note that this just corresponds to Eq. (4.135) with $|a + 1/2| = 1/2$.

We have found that for any given azimuth rotation the most transparent transmission rotator and the most reflective scattering rotator are lossless. In order to assess the potential of planar metamaterials as rotators, we will calculate transmission, reflection, and the associated polarization rotation for the lossless case. As the diagonal elements have the same magnitude, in both the scattering matrix and the transmission matrix, the scattered power fraction S and transmitted power fraction T are given by

$$\begin{aligned} S &= \quad |d|^2 \quad = \tfrac{1}{4}|e^{i2\delta} - 1|^2, \\ T &= |d + 1|^2 = \tfrac{1}{4}|e^{i2\delta} + 1|^2, \end{aligned}$$

which simplify to

$$\begin{aligned} S &= \sin^2 \delta, \\ T &= \cos^2 \delta. \end{aligned} \tag{4.137}$$

In the lossless case without linear birefringence/dichroism or circular dichroism the polarization rotation for scattered and transmitted fields is given by Eq. (4.108), where $Im(a)Im(d) \leq 0$. It follows that lossless rotators rotate the azimuth of

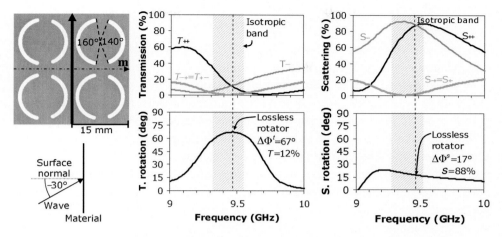

Figure 4.16. Experimental demonstration of an almost ideal lossless rotator response for a 1 mm thick aluminum film perforated with asymmetrically split ring apertures (wallpaper group *pm*). At 9.46 GHz, linear birefringence/dichroism ($T_{-+} = T_{+-} = 0$ and $S_{-+} = S_{+-} = 0$) and circular dichroism ($T_{++} = T_{--}$ and $S_{++} = S_{--}$) are absent, both in transmission and scattering. Transmission rotation of 67° is achieved at 12% transmission (limit $\cos^2 67° = 15\%$), and scattered field rotation of 17° is achieved at 88% scattering (limit $\cos^2 17° = 91\%$). Transmission and scattering rotation add up to 84° (ideal case 90°). All data shown correspond to oblique incidence at $\theta = -30°$ with the metamaterial's line of mirror symmetry **m** perpendicular to the plane of incidence. The structure's transmission properties are discussed in detail in ref. [14].

scattered and transmitted fields by

$$\Delta\Phi^s = \tfrac{\pi}{2} - \delta,$$
$$\Delta\Phi^t = \delta. \tag{4.138}$$

Thus the scattering matrices of all lossless rotators are given by Eq. (4.136). Their transmission and reflection levels are described by Eqs. (4.137) and their rotary power is given by Eqs. (4.138). As illustrated by Fig. 4.16, real planar metamaterials can come very close to ideal lossless rotators [14]. Importantly, a large polarization rotation of the transmitted wave comes at the expense of reduced transmission, while a large rotation of the scattered field comes at the expense of reduced reflection. Thus planar metamaterial rotators can be efficient only for small rotation angles. For example, a transmission rotator that rotates up to 6° can be $\geq 99\%$ transparent, while one rotating by 45° can only transmit up to 50%. In general, lossless planar metamaterial rotators have the highest efficiency that can be achieved with planar metamaterial rotators. Due to being measured in different coordinate systems, the rotation for scattering and that for reflection have opposite signs.

4.7.5 Circular polarizers

The behavior discussed here can only occur at oblique incidence onto planar metamaterials without 2-fold rotational symmetry.

A transmission circular polarizer transmits one circular polarization without changing its polarization state, while it is opaque for the other circular polarization. A reflection circular polarizer shows analogous behavior for the scattered field.

Thus circular polarizers must have counter-rotating circularly polarized eigenstates, where for reflection polarizers one scattering eigenvalue is zero, and for transmission polarizers one transmission eigenvalue is zero. Counter-rotating circularly polarized eigenstates correspond to pure optical activity, i.e. $b = c = 0$ (see Section 4.5.1). This leaves us with diagonal scattering matrices which have scattering eigenvalues as their entries. For reflection circular polarizers, exactly one scattering eigenvalue is zero, thus all reflection circular polarizers are given by

$$s(\theta) = \begin{pmatrix} 0 & 0 \\ 0 & \lambda \end{pmatrix} \quad \text{and} \quad s(-\theta) = \begin{pmatrix} \lambda & 0 \\ 0 & 0 \end{pmatrix}, \tag{4.139}$$

$$|\lambda + \tfrac{1}{2}| \leq \tfrac{1}{2}, \lambda \neq 0,$$

which correspond to reflection of left-handed and right-handed incident circular components, respectively. The reflected power fraction of the desired circular component is $|\lambda|^2$.

Similarly, for transmission circular polarizers exactly one transmission eigenvalue must be zero. From $\lambda_{1,2}^t = \lambda_{1,2}^s + 1$ (see Eq. (4.62)) it follows that exactly one diagonal entry of the scattering matrix must be -1, and thus all transmission circular polarizers are given by

$$s(\theta) = \begin{pmatrix} \lambda & 0 \\ 0 & -1 \end{pmatrix} \quad \text{and} \quad s(-\theta) = \begin{pmatrix} -1 & 0 \\ 0 & \lambda \end{pmatrix}, \tag{4.140}$$

$$|\lambda + \tfrac{1}{2}| \leq \tfrac{1}{2}, \lambda \neq -1,$$

which correspond to transmission of right-handed and left-handed incident circular components, respectively. The transmitted power fraction of the desired circular component is $|\lambda + 1|^2$.

As circular polarizers rely on optical activity, which is reversed for opposite angles of incidence $\pm\theta$, a planar metamaterial circular polarizer for one circular polarization $s(\theta)$ can always be turned into a circular polarizer for the other circular polarization $s(-\theta)$ by reversing the angle of incidence.

Obviously, circular polarizers that work simultaneously for transmission and reflection are of particular interest. It can easily be seen that the only matrices that satisfy Eqs. (4.139) and (4.140) are

$$\overrightarrow{s}(\theta) = \begin{pmatrix} 0 & 0 \\ 0 & -1 \end{pmatrix} \quad \text{and} \quad s(-\theta) = \begin{pmatrix} -1 & 0 \\ 0 & 0 \end{pmatrix}. \tag{4.141}$$

The corresponding eigenvalues have magnitudes 0, 1, for both scattering and transmission, and thus one circular polarization is completely transmitted while the other is completely reflected. In fact, Eqs. (4.141) correspond to the lossless solution (4.106) with $\xi = \pi/2$ and $\delta = \pm\pi/4$. It follows that simultaneous transmission and reflection planar metamaterial circular polarizers must be lossless.

All lossy planar metamaterial circular polarizers only work for either reflection or transmission. Note that there are also lossless circular polarizers that only work either in reflection or transmission.

It can be easily seen that scattering circular polarizers, Eqs. (4.139), and transmission circular polarizers, Eqs. (4.140), which reflect or transmit 100% of the desired circular component, must correspond to the special case given by Eqs. (4.141), which is lossless and works simultaneously for reflection and transmission. Therefore all planar metamaterial circular polarizers with 100% efficiency are lossless, work simultaneously for reflection and transmission and are given by Eqs. (4.141).

In this case, $s(\theta)$ corresponds to 100% transmission of right-handed circular polarization and 100% reflection of left-handed circular polarization, while the properties for $s(-\theta)$ are reversed. Interestingly, as the reflected wave changes handedness, $s(\theta)$ splits any beam into two right-handed circularly polarized beams, while $s(-\theta)$ splits any beam into two left-handed circularly polarized waves.

Note that Eqs. (4.141) can also be found by maximizing circular dichroism in lossless planar metamaterials without linear birefringence/dichroism; see Eqs. (4.109) and (4.110).

4.8 Normal incidence

Here we explore the properties of planar metamaterials under normal incidence conditions. We found in Section 4.2.4 that for normal incidence $a = d$ must hold, and that therefore optical activity and linear conversion dichroism cannot be observed in this case. We also found that the asymmetric transmission, reflection, and absorption effects for circularly polarized waves ($|b| \neq |c|$) can occur at normal incidence onto planar metamaterials only if they are simultaneously 2D-chiral, lossy, and anisotropic. The arguments for this were presented in Sections 4.2.3, 4.4.2, and 4.4.4, respectively.

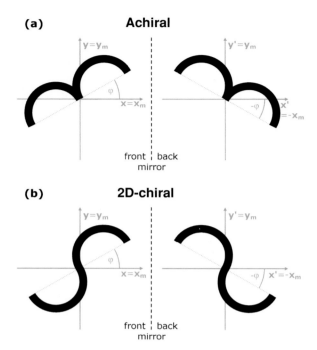

Figure 4.17. Achiral and planar chiral patterns and their mirror images, which correspond to the pattern seen from the other side. (a) Achiral patterns and their mirror images are not different from each other as they can be superimposed by an in-plane rotation of the mirror image by 2φ. (b) 2D-chiral patterns are different from their mirror image, as the pattern and its mirror image cannot be superimposed by an in-plane rotation. In the 2D-chiral case the pattern and its mirror image have opposite senses of twist.

The following three sections examine normal incidence for the cases of achiral, isotropic, and lossless planar metamaterials. For each of these cases the scattering matrices allowed by symmetry and energy conservation will be derived. Furthermore, it will be shown that for normal incidence onto lossless or isotropic planar metamaterials 2D chirality does not lead to any polarization effect. For normal incidence the scattering matrices of all lossless ($a = d$, $|b| = |c|$) or isotropic ($a = d$, $b = c = 0$) planar 2D-chiral metamaterials could also correspond to achiral planar metamaterials ($a = d$, $|b| = |c|$).

4.8.1 Achiral planar metamaterials at normal incidence

For planar metamaterials, waves normally incident on the front and back observe mirror images of the structure; see Fig. 4.17. If the structure and its mirror image, i.e.

the patterns seen by forward- and backward-propagating waves, cannot be superimposed without being lifted off the plane, the metamaterial is 2D-chiral. In the case of normal incidence on achiral structures,[17] forward- and backward-propagating waves see the same pattern. However, the reflected pattern will be rotated by an angle -2φ; see Fig. 4.17 and Section 4.5.2 (at normal incidence eigenstate azimuth = metamaterial orientation seen by the wave). Thus in the achiral case we must be able to overlap the mirror-image structure with the original by a $+2\varphi$ in-plane rotation. Taking into account that mirror images rotate in opposite directions and using Eqs. (4.80) this rotation corresponds to multiplying the scattering coefficients b and c by the phase factors $e^{+i(4\varphi)}$ and $e^{-i(4\varphi)}$, respectively. After this rotation the rotated scattering matrix $\overleftarrow{s}^{rot}$ must be equal to the original scattering matrix $\overrightarrow{s}$:

$$\overrightarrow{s} = \begin{pmatrix} a & b \\ c & a \end{pmatrix}, \tag{4.142}$$

$$\overleftarrow{s}^{rot} = \begin{pmatrix} a & c\,e^{-i(4\varphi)} \\ b\,e^{i(4\varphi)} & a \end{pmatrix}. \tag{4.143}$$

For the achiral case at normal incidence, $\overrightarrow{s} = \overleftarrow{s}^{rot}$ must hold, and therefore

$$c = b\,e^{i(4\varphi)}. \tag{4.144}$$

By using $b = |b|e^{i(\kappa-2\varphi)}$, from Eqs. (4.32), this gives us $c = |b|e^{i(\kappa+2\varphi)}$. Written in this form the scattering matrices for normal incidence on achiral planar metamaterials are

$$\overrightarrow{s} = \begin{pmatrix} a & |b|e^{i(\kappa-2\varphi)} \\ |b|e^{i(\kappa+2\varphi)} & a \end{pmatrix},$$

$$\overleftarrow{s} = \begin{pmatrix} a & |b|e^{i(\kappa+2\varphi)} \\ |b|e^{i(\kappa-2\varphi)} & a \end{pmatrix}, \tag{4.145}$$

$$\varphi \in [0, \tfrac{\pi}{2}), \qquad |a + \tfrac{1}{2}| \le \tfrac{1}{2},$$

where the allowed values for κ are derived below (see Eq. (4.150)) and $|b|$ must satisfy Eq. (4.95), which simplifies in this case to

$$|b| \le \sqrt{\tfrac{1}{4} - |a + \tfrac{1}{2}|^2} \le \tfrac{1}{2}. \tag{4.146}$$

This set of scattering matrices implies that achiral planar metamaterials can serve as normal incidence linear polarizers or wave plates operating in transmission and/or reflection. Obviously normal incidence attenuators, beam splitters, and mirrors can also be realized using achiral planar metamaterials; see Section 4.7.

---

[17] Achiral wallpaper symmetry groups: *pm, pg, cm, pmm, pmg, pgg, cmm, p4m, p4g, p3m*1, *p*31*m, p*6*m.*

From the results obtained in Section 4.5.2, we can conclude that achiral planar metamaterials at normal incidence have linear, orthogonal eigenstates, where the azimuth of one eigenstate for forward (backward) propagation is $+\varphi$ $(-\varphi)$. Note that for anisotropic achiral metamaterials, the orientation $\tilde{\varphi}$ of the structure's line of (glide) mirror symmetry must correspond to the orientation of a polarization eigenstate. As $\tilde{\varphi}$ is measured in the coordinates of a normally incident forward-propagating wave, the relationship $\tilde{\varphi} = \varphi + n\pi/2$ must be satisfied for some $n \in \mathbb{Z}$.

Since $|b| = |c|$, circular conversion dichroism cannot occur for normal incidence onto achiral planar metamaterials. Thus at normal incidence circular conversion dichroism requires 2D-chiral metamaterials.

In order to find the allowed values for the parameter κ, we must examine the general energy conservation condition (4.89). For normal incidence onto achiral planar metamaterials, we found $a = d$ and $|b| = |c|$, and therefore $p_1 = p_2$ (see Eqs. (4.85) and (4.86)) must hold, which simplifies the energy conservation condition to

$$p_3 \leq -p_1. \tag{4.147}$$

Note that Eq. (4.85) is equivalent to $(1/2)p_1 = |a + 1/2|^2 + |c|^2 - 1/4$, and that p_3 is defined by Eq. (4.90). Using this and $a = d$, $|b| = |c|$, and the alternative parameters A, ξ, κ defined in Eqs. (4.29), (4.30), and (4.33), Eq. (4.147) can be written as

$$\sqrt{2}A|b|\sqrt{1 + \cos[2(\xi - \kappa)]} \leq \tfrac{1}{4} - A^2 + |b|^2. \tag{4.148}$$

Using $(1 + \cos 2\alpha) = 2\cos^2 \alpha$ this is equivalent to

$$|\cos[\xi - \kappa]| \leq \frac{1/4 - A^2 + |b|^2}{2A|b|}. \tag{4.149}$$

This expression describes how much the parameter κ can deviate from $\kappa_0 = \xi \pm \pi/2$, the values allowed in the lossless case; see Eq. (4.100). We note that because $a = d$, in our case the simplified expression $\xi = \arg(a + 1/2)$ holds,

$$|\kappa - \kappa_0| \leq \left| \arcsin\left(\frac{1/4 - A^2 - |b|^2}{2|b|A} \right) \right|,$$

or, in terms of the scattering coefficient a,

$$|\kappa - \kappa_0| \leq \left| \arcsin\left(\frac{1/4 - |a + 1/2|^2 - |b|^2}{2|b|\,|a + 1/2|} \right) \right| \tag{4.150}$$

applies for the nontrivial solutions of Eq. (4.147). For the trivial solutions, which have either $a = -1/2$ or $b = 0$, all values of $\kappa \in [0, 2\pi)$ are allowed.

4.8.2 Isotropic planar metamaterials at normal incidence

At normal incidence, isotropic planar metamaterials do not have a preferred direction and therefore their response cannot depend on the azimuth of the incident wave. In Section 4.4.4, we have shown that such an absence of linear birefringence/dichroism and circular/linear conversion dichroism is equivalent to $b = c = 0$. Together with the energy conservation condition (4.94), this gives us the set of scattering matrices allowed for normal incidence onto isotropic planar metamaterials,

$$s = \begin{pmatrix} a & 0 \\ 0 & a \end{pmatrix}, \tag{4.151}$$

$$|a + \tfrac{1}{2}| \leq \tfrac{1}{2}.$$

This set of scattering matrices implies that isotropic planar metamaterials can serve as normal incidence attenuators, beam splitters, and mirrors; see Section 4.7.1. Note that these scattering matrices are a subset of those allowed for normal incidence onto achiral planar metamaterials, Eqs. (4.145). Thus for normal incidence onto isotropic planar metamaterials 2D-chiral symmetry cannot lead to any polarization effect.

Importantly, any planar metamaterial with 3-fold or higher rotational symmetry[18] is isotropic. This can be seen from the fact that for any chosen direction in a pattern with 3-fold or higher rotational symmetry there is at least one different direction that is absolutely equivalent. Therefore such structures do not have any preferred direction, and their transmission and reflection properties must be independent of the azimuth of normally incident waves.

4.8.3 Lossless planar metamaterials: normal incidence or two-fold rotational symmetry

Here we determine the scattering matrices and associated properties for the cases of (i) any lossless planar metamaterials at normal incidence and (ii) 2-fold rotationally symmetric lossless planar metamaterials[19] at any angle of incidence.

All lossless planar metamaterials must satisfy energy conservation without losses. The complete set of scattering matrices meeting this requirement is given by Eqs. (4.104), (4.105), and (4.106). In Section 4.2.4 we found that $a = d$ holds for any planar metamaterial at normal incidence and that the same constraint applies to 2-fold rotationally symmetric planar metamaterials at any angle of incidence.

---

[18] Isotropic wallpaper symmetry groups: *p3*, *p3m*1, *p31m*, *p4*, *p4m*, *p4g*, *p6*, *p6m*.
[19] Wallpaper symmetry groups with 2-fold rotational symmetry: *p2*, *pmm*, *pmg*, *pgg*, *cmm*, *p4*, *p4m*, *p4g*, *p6*, *p6m*.

Thus the scattering matrices for cases (i) and (ii) are the subset of Eqs. (4.104), (4.105), and (4.106) for which $a = d$ is satisfied. This subset is

$$\overrightarrow{s} = \begin{pmatrix} a & |b|e^{i(\kappa_0 - 2\varphi)} \\ |b|e^{i(\kappa_0 + 2\varphi)} & a \end{pmatrix},$$ (4.152)

$$|a + \tfrac{1}{2}| \in (0, \tfrac{1}{2}), \qquad \varphi \in [0, \tfrac{\pi}{2}),$$

with with κ_0 and $|b|$ given by Eqs. (4.100) and (4.103), respectively, and

$$\overrightarrow{s} = \tfrac{1}{2} \begin{pmatrix} -1 & e^{i(\kappa - 2\varphi)} \\ e^{i(\kappa + 2\varphi)} & -1 \end{pmatrix},$$ (4.153)

$$\kappa \in [0, 2\pi), \qquad \varphi \in [0, \tfrac{\pi}{2}),$$

$$s = \begin{pmatrix} a & 0 \\ 0 & a \end{pmatrix},$$ (4.154)

$$|a + \tfrac{1}{2}| = \tfrac{1}{2}.$$

For (i) normal incidence onto any lossless planar metamaterial and (ii) any angle of incidence onto a 2-fold rotationally symmetric lossless planar metamaterial, all allowed scattering matrices are given by Eqs. (4.152), (4.153), and (4.154).

It follows from Eqs. (4.72), (4.75), and (4.76) that the eigenstates are orthogonal linear polarizations, where the azimuth of one forward-propagation eigenstate is given by φ. Solutions (4.152) and (4.153) have $|b| = |c| \neq 0$ and thus φ is well-defined by Eq. (4.34). These solutions correspond to lossless planar metamaterials exhibiting linear birefringence and/or linear dichroism. Importantly, these scattering matrices include the special cases of lossless linear polarizers and wave plates for transmission and/or reflection, which are discussed in Sections 4.7.2 and 4.7.3, respectively.

Solution (4.154) corresponds to lossless planar metamaterials without linear birefringence/dichroism, i.e. mirrors ($a = -1$), empty space ($a = 0$), and lossless beam splitters; see Section 4.7.1.

Importantly, 2D-chiral circular conversion dichroism ($|b| \neq |c|$) or 3D-chiral optical activity ($a \neq d$) or linear conversion dichroism cannot occur in the cases (i) and (ii) considered. We found in Sections 4.4.2 and 4.4.3 that directionally asymmetric transmission, reflection, and absorption phenomena, which rely on absorption losses, cannot be observed for lossless planar metamaterials in general, not even in the presence of chirality. Such a generalization cannot be made for 3D-chiral optical activity, which can occur for oblique incidence onto a lossless planar metamaterial without 2-fold rotational symmetry.

Note that the scattering matrices in Eqs. (4.152), (4.153), and (4.154) are a subset of those allowed for normal incidence onto achiral planar metamaterials, Eqs. (4.145). Thus for normal incidence onto lossless planar metamaterials 2D-chiral symmetry cannot lead to any polarization effect. The same holds for 2-fold rotationally symmetric lossless planar metamaterials at any angle of incidence.

4.9 Summary

We have found that planar metamaterials can show distinctly different polarization effects of 3D-chiral and 2D-chiral nature. Importantly, these effects do not require the metamaterial itself to be chiral.

The first 3D-chiral effect corresponds to optical activity, i.e. circular birefringence and circular dichroism. In terms of the scattering or transmission matrices for circularly polarized waves, optical activity corresponds to nonequal diagonal elements. Conventionally, optical activity has been associated with 3D-chiral structures. Just as 3D-chiral structures have the same sense of twist when observed from opposite sides, optical activity is the same for opposite propagation directions. Even though planar metamaterials cannot have 3D-chiral symmetry, we found that circular birefringence and circular dichroism *can* occur in planar metamaterials if the following conditions are met:

- oblique incidence,
- no 2-fold rotational symmetry,[20]
- no (glide) mirror line parallel to the plane of incidence.

Under these conditions, the experimental arrangement consisting of the metamaterial combined with the direction of incidence has 3D-chiral symmetry (extrinsic 3D chirality). Due to its dependence on the angle of incidence, optical activity in planar metamaterials is inherently tunable. In this respect it is particularly useful that circular birefringence and circular dichroism are each absent at normal incidence and have opposite signs for opposite angles of incidence.

The second 3D-chiral phenomenon is linear conversion dichroism, which leads to asymmetric transmission, reflection, and absorption of linearly polarized waves with parallel polarization states for opposite propagation directions. Linear conversion dichroism can accompany optical activity in lossy planar metamaterials, i.e. its observation requires absorption losses and the conditions for optical activity, which are listed above.

The 2D-chiral phenomenon is circular conversion dichroism, which leads to asymmetric transmission, reflection, and absorption of circularly polarized waves

---

[20] Wallpaper symmetry groups without 2-fold rotational symmetry: $p1$, pm, pg, cm, $p3$, $p3m1$, $p31m$.

of the same handedness for opposite directions of propagation. The asymmetric behavior arises from reversed right-to-left and left-to-right circular polarization conversion efficiencies for opposite propagation directions. Thus, in terms of scattering or transmission matrices for circular polarization, the phenomenon corresponds to different magnitudes of the off-diagonal terms. At oblique incidence circular conversion dichroism requires structures that have

- losses,
- no (glide) mirror line parallel or perpendicular to the plane of incidence.

As these conditions can be satisfied by any lossy planar periodic interface, circular conversion dichroism at oblique incidence appears to be a generic property of lossy planar metamaterials. At normal incidence[21] circular conversion dichroism requires planar metamaterials that have

- losses,
- 2D chirality,
- anisotropy.

Note that at normal incidence the reversed circular polarization conversion efficiencies for opposite propagation directions correspond to the reversed sense of twist of 2D-chiral structures for opposite directions of observation. At oblique incidence, the above criteria require that the metamaterial combined with the direction introduced by the plane of incidence is 2D-chiral (extrinsic 2D chirality). Thus, while optical activity and linear conversion dichroism are 3D-chiral phenomena, circular conversion dichroism is of 2D-chiral nature.

We found that both linear and circular conversion dichroism rely on asymmetric absorption losses, and that therefore the asymmetric effects cannot occur in lossless planar metamaterials.

Optical activity and circular conversion dichroism have remarkably different polarization eigenstates. In the case of pure optical activity, the eigenstates are the same counter-rotating circular polarizations for opposite propagation directions. On the other hand, circular conversion dichroism is associated with co-rotating elliptically polarized eigenstates of orthogonal azimuth, which are left-handed for one propagation direction and right-handed for the opposite direction.

Apart from chiral polarization effects, planar metamaterials can also show linear birefringence and linear dichroism. In terms of scattering or transmission matrices for circularly polarized waves, pure linear birefringence and dichroism, i.e. without the presence of chiral polarization effects, correspond to identical diagonal elements and non-zero off-diagonal terms of the same magnitude. Provided

---

[21] 2D-chiral anisotropic wallpaper symmetry groups: $p1$, $p2$.

that preferred directions are provided by the structure (anisotropy) or the experimental arrangement (oblique incidence), pure linear birefringence and dichroism can be seen in lossless planar metamaterials when extrinsic 3D chirality is absent and in lossy planar metamaterials when both 2D and 3D chirality are absent. The eigenstates associated with pure linear birefringence and dichroism, are orthogonal linear polarization states.

Importantly, planar metamaterials can act as linear or circular polarizers, wave plates, or polarization rotators. Each of these functionalities can be realized for transmission and/or reflection. In particular this allows the realization of reflection circular polarizers, reflection wave plates, and reflection rotators. The potential efficiency of planar metamaterial wave plates and rotators decreases with increasing phase delay and rotation angle, respectively. Low phase delay wave plates and weak rotators can be very efficient, while planar metamaterial λ/2-plates and ±π/2-rotators cannot be realized. Planar metamaterial rotators and circular polarizers require oblique incidence, and have the interesting property that their rotation and polarizing properties, respectively, are reversed for opposite angles of incidence.

References

[1] H. Tamada, T. Doumuki, T. Yamaguchi, and S. Matsumoto, "Al wire-grid polarizer using the s-polarization resonance effect at the 0.8-μm-wavelength band," *Opt. Lett.* **22**, 419–421 (1997).

[2] S. W. Ahn, K. D. Lee, J. S. Kim, S. H. Kim, J. D. Park, S. H. Lee, and P. W. Yoon, "Fabrication of a 50 nm half-pitch wire grid polarizer using nanoimprint lithography," *Nanotechnol.* **16**, 1874–1877 (2005).

[3] R. Ott, R. Kouyoumjian, and L. Peters, Jr., "Scattering by a two-dimensional periodic array of narrow plates," *Radio Sci.* **2**, 1347–1359 (1967).

[4] C. Chen, "Scattering by a two-dimensional periodic array of conducting plates," *IEEE Trans. Antenn. Propag.* **AP-18**, 660–665 (1970).

[5] B. Munk, R. Kouyoumjian, and L. Peters Jr., "Reflection properties of periodic surfaces of loaded dipoles," *IEEE Trans. Antenn. Propag.* **AP-19**, 612–617 (1971).

[6] B. A. Munk, *Frequency Selective Surfaces: Theory and Design*, 1st edn (New York: Wiley-Interscience, 2000).

[7] J. Huang, T. K. Wu, and S. W. Lee, "Tri-band frequency-selective surface with circular ring elements," *IEEE Trans. Antenn. Propag.* **AP-42**, 166–175 (1994).

[8] R. Ulrich, "Far infrared properties of metallic mesh and its complementary structure," *Infrared Phys.* **7**, 37–55 (1967).

[9] V. P. Tomaselli, D. C. Edewaard, P. Gillan, and K. D. Möller, "Far infrared bandpass filters from cross shaped grids," *Appl. Opt.* **20**, 1361–1366 (1981).

[10] D. R. Smith, W. J. Padilla, D. C. Vier, S. C. Nemat-Nasser, and S. Schultz, "Composite medium with simultaneously negative permeability and permittivity," *Phys. Rev. Lett.* **84**, 4184–4187 (2000).

[11] V. A. Fedotov, P. L. Mladyonov, S. L. Prosvirnin, and N. I. Zheludev, "Planar electromagnetic metamaterial with a fish scale structure," *Phys. Rev. E* **72**, 056613(1-4) (2005).

[12] X. G. Peralta, E. I. Smirnova, A. K. Azad, H.-T. Chen, A. J. Taylor, I. Brener, and J. F. O'Hara, "Metamaterials for THz polarimetric devices," *Opt. Express* **17**, 773–783 (2009).

[13] E. Plum, V. A. Fedotov, and N. I. Zheludev, "Optical activity in extrinsically chiral metamaterial," *Appl. Phys. Lett.* **93**, 191911(1-3) (2008).

[14] E. Plum, X.-X. Liu, V. A. Fedotov, Y. Chen, D. P. Tsai, and N. I. Zheludev, "Metamaterials: optical activity without chirality," *Phys. Rev. Lett.* **102**, 113902(1-4) (2009).

[15] R. Singh, E. Plum, W. Zhang, and N. I. Zheludev, "Highly tunable optical activity in planar achiral terahertz metamaterials," *Opt. Exp.* **18**, 13425 (2010).

[16] A. Papakostas, A. Potts, D. M. Bagnall, S. L. Prosvirnin, H. J. Coles, and N. I. Zheludev, "Optical manifestations of planar chirality," *Phys. Rev. Lett.* **90**, 107404(1-4) (2003).

[17] S. L. Prosvirnin and N. I. Zheludev, "Polarization effects in the diffraction of light by a planar chiral structure," *Phys. Rev. E* **71**, 037603(1-4) (2005).

[18] Y. Svirko, N. Zheludev, and M. Osipov, "Layered chiral metallic microstructures with inductive coupling," *Appl. Phys. Lett.* **78**, 498–500 (2001).

[19] E. Plum, V. A. Fedotov, A. S. Schwanecke, N. I. Zheludev, and Y. Chen, "Giant optical gyrotropy due to electromagnetic coupling," *Appl. Phys. Lett.* **90**, 223113(1-3) (2007).

[20] V. A. Fedotov, M. Rose, S. L. Prosvirnin, N. Papasimakis, and N. I. Zheludev, "Sharp trapped-mode resonances in planar metamaterials with a broken structural symmetry," *Phys. Rev. Lett.* **99**, 147401(1-4) (2007).

[21] S. Zhang, D. A. Genov, Y. Wang, M. Liu, and X. Zhang, "Plasmon-induced transparency in metamaterials," *Phys. Rev. Lett.* **101**, 047401(1-4) (2008).

[22] B. Luk'yanchuk, N. I. Zheludev, S. A. Maier, N. J. Halas, P. Nordlander, H. Giessen, and C. T. Chong, "The Fano resonance in plasmonic nanostructures and metamaterials," *Nat. Mater.* **9**, 707 (2010).

[23] P. Tassin, L. Zhang, T. Koschny, E. N. Economou, and C. M. Soukoulis, "Low-loss metamaterials based on classical electromagnetically induced transparency," *Phys. Rev. Lett.* **102**, 053901(1-4) (2009).

[24] N. Papasimakis and N. I. Zheludev, "Metamaterial-induced transparency," *Opt. Photon. News* **20**, 22 (2009).

[25] V. A. Fedotov, P. L. Mladyonov, S. L. Prosvirnin, A.V. Rogacheva, Y. Chen, and N. I. Zheludev, "Asymmetric propagation of electromagnetic waves through a planar chiral structure," *Phys. Rev. Lett.* **97**, 167401(1-4) (2006).

[26] A. S. Schwanecke, V. A. Fedotov, V. V. Khardikov, S. L. Prosvirnin, Y. Chen, and N. I. Zheludev, "Nanostructured metal film with asymmetric optical transmission," *Nano Lett.* **8**, 2940–2943 (2008).

[27] E. Plum, V. A. Fedotov, and N. I. Zheludev, "Planar metamaterial with transmission and reflection that depend on the direction of incidence," *Appl. Phys. Lett.* **94**, 131901(1-3) (2009).

[28] E. Plum, V. A. Fedotov, and N. I. Zheludev, "Extrinsic electromagnetic chirality in metamaterials," *J. Opt. A: Pure Appl. Opt.* **11**, 074009(1-7) (2009).

[29] E. Plum, V. A. Fedotov, and N. I. Zheludev, "Asymmetric transmission: a generic property of two-dimensional periodic patterns," *J. Opt.* **13**, 024006 (2011).

[30] E. Plum, V. A. Fedotov, and N. I. Zheludev, "Metamaterial optical diodes for linearly and circularly polarized light," *arXiv.org.* 1006.0870 (2010).

[31] J. A. Kong, *Electromagnetic Wave Theory* (Cambridge, MA: EMW Publishing, 2005).

[32] F. Falcone, T. Lopetegi, M. A. G. Laso *et al.* "Babinet principle applied to the design of metasurfaces and metamaterials," *Phys. Rev. Lett.* **93**, 197401(1-4) (2004).

[33] J. D. Jackson, *Classical Electrodynamics* (New York: Wiley, 1999).

[34] A. Potts, D. M. Bagnall, and N. I. Zheludev, "A new model of geometric chirality for two-dimensional continuous media and planar meta-materials," *J. Opt. A: Pure Appl. Opt.* **6**, 193–203 (2004).

[35] M. A. Osipov, B. T. Pickup, M. Fehervari, and D. A. Dunmur, "Chirality measure and chiral order parameter for a two-dimensional system," *Mol. Phys.* **94**, 283–287 (1998).

[36] W. J. Padilla, "Group theoretical description of artificial electromagnetic metamaterials," *Opt. Express* **15**, 1639–1646 (2007).

[37] C. M. Bingham, H. Tao, X. Liu, R. D. Averitt, X. Zhang, and W. J. Padilla, "Planar wallpaper group metamaterials for novel terahertz applications," *Opt. Express* **16**, 18565–18575 (2008).

[38] D. Schattschneider, "The plane symmetry groups: their recognition and notation," *Am. Math. Mon.* **85**, 439–450 (1978).

[39] C. W. Bunn, *Chemical Crystallography* (New York: Oxford University Press, 1945).

[40] R. Williams, "Optical rotatory effect in the nematic liquid phase of p-azoxyanisole," *Phys. Rev. Lett.* **21**, 342–344 (1968).

[41] R. Williams, "Optical-rotary power and linear electro-optic effect in nematic liquid crystals of p-azoxyanisole," *J. Chem. Phys.* **50**, 1324–1332 (1969).

[42] R. Singh, E. Plum, C. Menzel *et al.* "Terahertz metamaterial with asymmetric transmission," *Phys. Rev. B* **80**, 153104(1-4) (2009).

[43] V. A. Fedotov, A. S. Schwanecke, N. I. Zheludev, V. V. Khardikov, and S. L. Prosvirnin, "Asymmetric transmission of light and enantiomerically sensitive plasmon resonance in planar chiral nanostructures," *Nano Lett.* **7**, 1996–1999 (2007).

[44] A. Drezet, C. Genet, J.-Y. Laluet, and T. W. Ebbesen, "Optical chirality without optical activity: how surface plasmons give a twist to light," *Opt. Express* **16**, 12559–12570 (2008).

[45] S. V. Zhukovsky, A. V. Novitsky, and V. M. Galynsky, "Elliptical dichroism: operating principle of planar chiral metamaterials," *Opt. Lett.* **34**, 1988–1991 (2009).

[46] C. Menzel, C. Helgert, C. Rockstuhl, E.-B. Kley, A. Tünnermann, T. Pertsch, and F. Lederer, "Asymmetric transmission of linearly polarized light at optical metamaterials," *Phys. Rev. Lett.* **104**, 253902 (2010).

5

Novel optical devices using negative refraction of light by periodically corrugated surfaces

WENTAO TRENT LU AND SRINIVAS SRIDHAR

5.1 Introduction

Negative refraction (NR) has been theoretically predicted [1, 2] and experimentally realized [3–7] in three types of materials. One is a material with a simultaneously negative permittivity and permeability [8–12], leading to a negative refractive index for the medium. The second consists of a photonic crystal (PhC) [13–21], which is a periodic arrangement of scatterers in which the group and phase velocities can be in different directions leading to NR. The third is the indefinite medium [22–28], whose permittivity and/or permeability tensor is an indefinite matrix. In all cases, the bulk properties of the medium, which is inherently inhomogeneous at a subwavelength scale, can be described as having an effective negative refractive index. The active research in these artificial materials has opened doors to a plethora of unusual electromagnetic properties and new applications such as a perfect lens [29], subwavelength imaging [30], cloaking [31], slow light, and optical data storage [32, 33], that cannot be obtained with naturally occurring materials. The holy grail of manufacturing these artificial photonic metamaterial structures is to manipulate light at the nanoscale level for optical information processing and high-resolution imaging.

In order to achieve NR, engineering the bulk electromagnetic properties is normally needed such that the group velocity and phase velocity be at an obtuse angle or even anti-parallel to each other. However, refraction is a surface phenomenon. A bulk-engineered material will have certain inherent surface properties. Negative refraction can be realized in positive index materials by special orientation [34] or by engineering the interface properties [35–37]. Although the bulk materials have a positive refractive index, and thus will not lead to subwavelength imaging resolution, the far-field properties are the same as those of negative index materials.

Structured Surfaces as Optical Metamaterials, ed. A. A. Maradudin. Published by Cambridge University Press.
© Cambridge University Press 2011.

In this chapter we describe a new mechanism of achieving NR by utilizing surface corrugation. Negative refraction through diffraction is realized in a step-by-step fashion. We will show that by adding a surface grating to a smooth surface, the forward diffracted beam can be bent positively or negatively, with respect to the transverse wave vector direction. By tuning the periodicity of diffraction, we will suppress all orders of diffraction except the -1 order. Even though true negative refraction is not yet achieved at this point, this negatively diffracted beam will be utilized to make a grating lens that is realized in both microwaves and in optics. We will show that one can further eliminate the -1 order reflection by applying surface corrugation to photonic band gap materials, which will lead to true negative refraction. Surface corrugation is applied to one-dimensional (1D) and two-dimensional (2D) photonic band gap structures to realize all-angle negative refraction (ANNR) [18] by folding the band structures. This leads to new frequency windows and lattice orientations for AANR. Flat lens imaging will be demonstrated. Unlike previously realized flat lens in PhCs, these flat lenses can have very large object–image distances.

5.2 Negative refraction with visible light and microwaves by selective diffraction

No homogeneous materials have yet been found to exhibit negative refraction. Causality prevents the existence of materials with both negative static permeability and permittivity. Negative refraction is demonstrated only in inhomogeneous systems [38] or low-dimensional structures with wave confinement [39].

We start with an initially smooth interface that is totally reflecting. Consider a plane wave incident on the corrugated surface of a material with bulk refractive index $n > 1$ at an angle of incidence θ as shown in Fig. 5.1(a). A single beam of light can be *refracted* negatively or positively by (i) making the angle of incidence greater than the critical angle,

$$\theta > \theta_c = \sin^{-1}(1/n), \tag{5.1}$$

which suppresses the zeroth and all positive orders, and then (ii) tuning the corrugation wave vector to select the -1 order.

We demonstrate this concept experimentally using grisms (grating prisms) at visible and microwave frequencies. The presence of the surface grating changes the wave vector in the bulk medium and gives a new handle to control the light emerging from the interface. The corrugation provides a momentum kick to the incident light, enabling it to cross the interface and emerge refractively at angles that can be controlled.

The component of the the wave vector **k** of the incident light parallel to the surface is

$$k_{||} = nk_0 \sin \theta. \qquad (5.2)$$

Here $k_0 = 2\pi/\lambda$ is the wave number and λ is the wavelength in free space. Due to the surface corrugation of periodicity a_s, the wave vector along the grating surface is not conserved. The parallel components of the transmitted and reflected wave vectors along the grating surface are

$$k_{||m} = nk_0 \sin \theta + 2m\pi/a_s, \qquad (5.3)$$

according to the Floquet theorem [40]. Here m is the order of the so-called Bragg waves. The radiating Bragg waves into the air have

$$-1 < n \sin \theta + m\lambda/a_s < 1. \qquad (5.4)$$

Otherwise, they will be evanescent, constituting surface waves. For angles of incidence larger than the critical angle, all the radiating Bragg waves have negative orders, $m < 0$. Within the wavelength range

$$a_s(1 + n \sin \theta)/2 < \lambda < a_s(1 + n \sin \theta), \qquad (5.5)$$

only the $m = -1$ Bragg wave will radiate from the grating surface into the air, which we call the refracted beam with wave vector $\mathbf{k}_f$ and $k_{f||} = nk_0 \sin \theta - 2\pi/a_s$. For light within this range, an effective refractive index can be defined as

$$n_{\text{eff}} = n - \lambda/(a_s \sin \theta), \qquad (5.6)$$

and Snell's law applies. If $a_s(1 + n \sin \theta)/2 < \lambda < na_s \sin \theta$, $n_{\text{eff}} > 0$ and the refraction will be positive, while for $a_s n \sin \theta < \lambda < a_s(1 + n \sin \theta)$, $n_{\text{eff}} > 0$ and the refraction will be negative. This is illustrated in Fig. 5.1(a).

An experimental demonstration of NR at visible light using this mechanism is shown in Fig. 5.1(b). A holographic transmission grating with ruling density 2400 lines/mm and estimated groove depth $h \sim 130$ nm was replicated on one of the sides of an equilateral right-angle BK7 prism of size 2 cm. A collimated laser beam is incident on the hypotenuse and passes through the grating surface of the grism. The angle of incidence at the grating is the prism angle $\theta = \pi/4$, which is greater than the critical angle for the BK7 glass. Theoretical analysis indicates that only the $m = -1$ order beam will radiate into the air at negative angles if the incident light is within the 435–860 nm wavelength range. Photographs of the experiments clearly indicate that the incident He–Ne (632.8 nm) laser beam "refracts" negatively through an angle $\phi = 27°$, as shown in Fig. 5.1(b). Indeed, this is indistinguishable from refraction by a prism made of a negative refractive

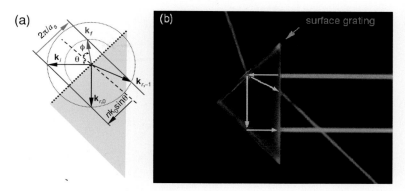

Figure 5.1. (a) Wave vector diagram for NR from a corrugated surface with grating period a_s. The semicircle is the equi-frequency surface (EFS) in the air, while the circle is the EFS in the dielectric. Here $\mathbf{k}_i$ is the incident wave vector in the glass, $\mathbf{k}_f$ is the refracted one of -1 order in the air; $\mathbf{k}_{r,0}$ and $\mathbf{k}_{r,-1}$ are the reflected wave vectors in the dielectric of the zeroth and -1 orders, respectively. (b) Optical experiment demonstrating NR using a grism of size 2 cm with a grating density of 2400 lines/mm on the upper short surface. The He–Ne laser beam is normally incident to the hypotenuse. To visualize the beam path in air, the grism was placed inside a glass enclosure that was sparsely filled with smoke. The solid lines with arrows indicate the propagation of the beams inside the grism.

index material with $n_{\text{eff}} = -0.63$ for the red light. A sketch of the main beam trajectories inside the glass is also shown in Fig. 5.1(b). All the beams can and have been explained by diffraction theory; NR was also observed on this grism for the green laser (532 nm), in which case $n_{\text{eff}} = -0.29$.

These experiments were repeated on a 1800 lines/mm grism with an estimated groove depth $h \sim 150$ nm. In the case of red light with normal incidence to the hypotenuse, only the $m = -1$ beam will emerge negatively from the grating surface, whereas for the green light the $m = -1$ order is diffracted positively with the additional appearance of the $m = -2$ order. Control experiments performed on regular prisms (without the grating) show positive refraction or complete reflection without any transmitted beam, depending on the angle of incidence, as is to be expected.

For a fixed wavelength and groove geometry, the fraction of light diffracted into the $m = -1$ order depends strongly on the polarization state of the incident light. The intensity transmission efficiency $\eta = I_{-1}/I_{in}$ for the $m = -1$ order was measured at different polarization orientations of the incident light. A half-wave plate inserted between the grism and a polaroid is used to rotate the orientation of the linearly polarized light. For P polarization, the electric-field vector was parallel to the grooves, whereas for S polarization it was perpendicular to the grooves. A

Wentao Trent Lu and Srinivas Sridhar

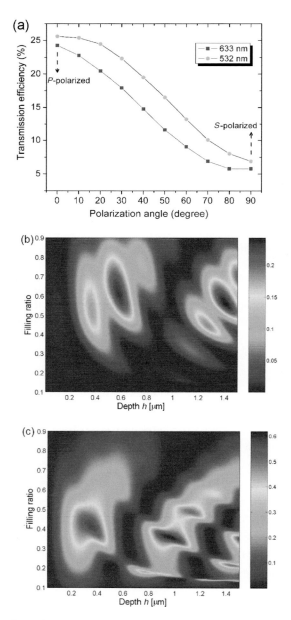

Figure 5.2. (a) Experimentally measured transmission efficiency of the beam negatively refracted through a BK7 grism with 2400 lines/mm grating. The electric vector is parallel to the groove for a $0°$ orientation (P polarization) and perpendicular for a $90°$ orientation of the polarizer. Calculated transmission efficiency in S polarization (b) and in P polarization (c) with an angle of incidence $\pi/4$ for $\lambda = 532$ nm through a lamellar grating on BK7 glass. The period of the lamellar grating has a density of 2400 lines/mm ($a_s = 416.7$ nm). See color plates section.

maximum efficiency of 25% is attained for the 2400 lines/mm grism with the P-polarized green light. The detailed transmission curves for different polarizations are shown in Fig. 5.2(a).

For practical applications, the efficiency of the power transmission is of primary concern. Since the direction of the refracted beam is determined only by the surface periodicity and not by the detailed geometry of the grating, there is considerable freedom allowed to design the grating surface to maximize the transmission. To exploit this freedom for transmission enhancement, we consider a specific grating, the lamellar grating. The transmission and reflection of waves were calculated using the Bloch wave expansion method. The transmission efficiency is plotted as a function of the groove depth h and the filling ratio for both the S and P polarizations at $\lambda = 532$ nm, as shown in Figs. 5.2(b) and (c), respectively. For a groove depth $h < 50\,\mu$m, an efficiency of over 60% can be reached in P polarization for $h \sim 1\,\mu$m and a filling ratio 0.34. For $h \sim 0.4\,\mu$m and a filling ratio of 0.4, the efficiency is 50%.

The concept of NR by using selective diffraction is applicable over a wide range of frequencies. This is further exemplified by the grism experiments performed using microwaves, as shown in Fig. 5.3. The grism consists of a right-angled polystyrene ($\varepsilon = 2.56$) prism with a surface grating of alumina rods next to the hypotenuse. The alumina rods have a diameter of 0.635 cm with grating periodicity $a_s = 2$ cm. The experiments were carried out in a parallel-plate waveguide. The distance between the two plates is 1.26 cm. The excitation in the parallel-plate waveguide is a transverse magnetic (TM) mode of up to 12 GHz such that the electric field is vertical and the magnetic field is within the plane. The collimated microwave beam is incident normally on the shortest side of the prism and hits the hypotenuse with an angle of incidence $\theta = \pi/3$. A dipole antenna attached to an X–Y robot maps the electric field. As shown in Fig. 5.3, at 9 GHz the beam emerges as if it were refracted negatively at an angle $\phi = -16°$, leading to $n_{\text{eff}} = -0.32$; NR was observed between 6.3 and 10.8 GHz. Experimental data are in excellent agreement with theory and numerical simulations. For all the microwave experiments and the corresponding numerical simulations carried out in this section, only the TM modes are considered.

5.3 Focusing microwaves by a plano-concave grating lens

A unique feature of a negative index material is that it leads to focusing by a plano-concave lens [41, 42]. Focusing by plano-concave lenses was realized in 2D and 1D PhCs [43–45]. We show that the NR mechanism demonstrated in Section 5.2 can be used to design a plano-concave grating lens. For a plano-concave lens with a circular curved surface of radius R, and if the grating is placed such that the

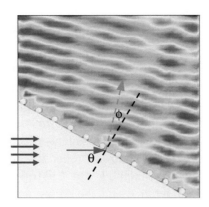

Figure 5.3. Microwave experiment demonstrating NR using a polystyrene grism with a surface grating period $a_s = 2$ cm and angle of incidence $\theta = \pi/3$ at 9 GHz. Plotted is the electric field (real part of the measured transmission coefficient S_{21}). The solid arrows on the left indicate the direction of the incident microwave beam. The dashed line is the surface normal, and the dashed arrow indicates the direction of propagation of the refracted beam. See color plates section.

groove distance along the optical axis is a fixed number a, the surface periodicity will be $a_s = a/\sin\theta$. Here the angle θ is the angle of incidence toward the curved surface. The effective refractive index is given by

$$n_{\text{eff}} = n - \lambda/a, \tag{5.7}$$

which is independent of θ. A focus is expected with a focal length $f(\theta) = R[1 + \sin\phi/\sin(\theta - \phi)]$. The focal length depends on the angle θ, leading to aberration, which is present even in conventional lenses. The image quality is mainly impacted by (a) the variation of the focal length for a large angle of incidence θ and (b) the zeroth order diffraction, which is present when $\theta < \sin^{-1}(1/n)$. The strategy to improve the image quality is discussed next.

A good quality focus can be observed for the plano-concave grating lens with circular surface if $\lambda/a \sim n$, $n_{\text{eff}} \sim 0$, in which case the focal length $f(\theta)$ is flat. For $|n_{\text{eff}}| < 1$, one can use a noncircular curve instead of a circular one to minimize spherical aberration. This curve assumes an elliptical form

$$y^2/b^2 + x^2 = R^2, \tag{5.8}$$

where $b = (1 - n_{\text{eff}})^{1/2}$ and

$$f = R/(1 - n_{\text{eff}}) \tag{5.9}$$

is the desired focal length. On this elliptical curve one places the grating such that the distance along the optical axis is a constant a as in the circular case.

In order to eliminate the diverging beam around the optical axis of the plano-concave lens due to the zeroth order diffraction, one can simply block this part of the lens. Even if the interference from the zeroth order diffraction could not be eliminated, it can be reduced. For a plano-concave lens with higher refractive index, this effect is smaller. For certain gratings on the plano-concave lens, the zeroth order diffraction can also be suppressed. For example, for the staggered cut, as in a 1D PhC, the part of the lens around the optical axis is flat, as shown in Fig. 5.4. For this grating one can choose the thickness of the lens around the optical axis such that the transmission through this part is a minimum. This is confirmed in our numerical simulations by using the finite-difference time-domain (FDTD) method [46].

An elliptical plano-concave grating lens made of alumina with a semimajor axis $R = 15$ cm and a semiminor axis 12.7 cm is shown in Fig. 5.4(c). The grating is made by staggered cuts on the concave surface, such that the horizontal distance of consecutive cuts is 1 cm. The effective index is $n_{eff} = -0.53$ at 8.5 GHz. A high quality focus of a microwave beam is observed at 8.4 GHz, as shown in Fig. 5.4(a). The inverse experiment was also performed in which a point source placed at the focal point will radiate a plane wave beam at 8.4 GHz. The plano-concave grating lens was placed inside the parallel-plate waveguide [43]. Numerical simulation (Fig. 5.4(b)) verifies both of the abovementioned focusing experiments at 8.5 GHz.

5.4 Realization of a plano-concave grating lens in optics

In this section we demonstrate how this binary-staircase optical element can be tailor-made to have an effective negative refractive index at optical frequencies, thus bringing a new approach to negative index optical elements.

The binary-staircase lens [47] we consider here consists of a sequence of zones configured as flat parallel steps each having an annular shape. The binary-staircase lens is a plano-concave lens. Proof-of-concept experiments have been carried out at microwave frequencies, and were described in Section 5.3. However the plano-concave lens used in the microwave range consisted of an assembly of commercial alumina bars, placed in a parallel-plate waveguide, which are not suitable for integration in optoelectronic circuits.

Geometrical parameters of the binary-staircase lens were determined by considering the transverse size of the lens, the focal length, the wavelength of the incoming radiation, the index of the material used to fabricate the lens itself, and mainly the surface periodicity. For a binary-staircase lens with a plano-concave shape, and in the case that the grating period is much smaller than the incident wavelength, an effective refractive index n_{eff} can be used to describe the refraction

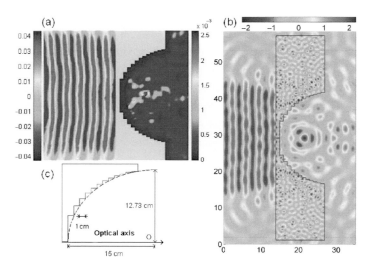

Figure 5.4. Demonstration of plano-concave grating lens focusing. (a) Composite figure of the microwave focusing experiment at 8.4 GHz using a plano-concave grating lens made of alumina with a grating on the curved surface. The electric field of the incident beam measured without the presence of the grating lens is plotted on the left. The intensity of the electric field is plotted on the right. In the middle is a photo of the lens. The grating lens behaves like a smooth plano-concave lens made of negative index material with $n_{\text{eff}} = -0.57$ at 8.4 GHz. (b) FDTD simulations at a plano-concave lens without aberration made with $n = 3$, $R = 15$ cm, and $a = 1$ cm at 8.5 GHz. Plotted is the electric field. The size of the system is measured in centimeters. (c) Details of the plano-concave lens (half of which is shown). The dashed curve is an ellipse with a semimajor axis of 15 cm and a semiminor axis of 12.73 cm. The horizontal length of the grooves is 1 cm. See color plates section.

at the modified concave surface. The effective index is related to the bulk refractive index of the medium n_{med}, the step size a, and the free space wavelength λ through Eq. (5.7) (with $a < \lambda$). The number of steps N_{steps} or zones is then R/a, where $2R$ is the transverse size of the binary-staircase lens. The focal length f is calculated by using Eq. (5.9). To obtain a good focus, $a \sim \lambda/n_{\text{med}}$ (Abbe's diffraction limit). In the present case, $\lambda = 1550$ nm and a was chosen as 450 nm with $n_{\text{med}} = 3.231$ for TE modes and $n_{\text{med}} = 3.216$ for TM modes; a has been given an arbitrary value close to λ/n_{med}; $N_{\text{steps}} = 11$, so that R is 20 μm. Thus n_{eff} is -0.2133 for TE modes and -0.2889 for TM modes.

The actual lens has been nanofabricated by a combination of electron beam lithography and reactive ion etching in an InP/InGaAsP heterostructure. The fabrication platform consisted of a 400 nm InGaAsP core layer on an InP substrate with a 200 nm InP top cladding layer. The waves are trapped and propagate within

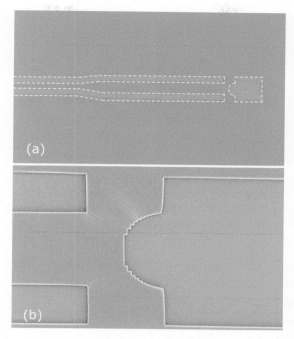

Figure 5.5. (a) Bird's eye view of the tapered waveguide and the binary-staircase lens. (b) Close-up view of the binary-staircase lens.

the core layer plane with an effective permittivity of 3.231 (TE modes) and 3.216 (TM modes). The final structure for optical measurements consisted of three sub units (shown in Fig. 5.5(a)). (i) A 0.5 mm long waveguide, laterally tapered, having 5 μm wide trenches on each side; the taper starts at a distance of 100 μm from the edge of the waveguide, with a core width varying from 5 μm to 10 μm. (ii) Binary-staircase plano-concave lens with ten zones on the optical axis, having a step height of 450 nm and a transverse size of 10 μm, located at a distance of 5 μm from the tapered end of the waveguide, as shown in Fig. 5.5(b). (iii) Finally, an open cavity (semicircle attached to a 20 μm × 20 μm square) at the end of the binary-staircase lens.

An analogous structure, having the same geometrical dimensions but bearing no steps (or zones), was also fabricated. The purpose of the analogous design was to prove that the periodicity of the steps is a decisive structural element in realizing a negative index prototype. The structures were written using electron beam lithography on polymethylmethacrylate (PMMA) resist. Pattern transfers to a silicon nitride working mask and subsequently to the InP/InGaAsP layers were achieved with a reactive ion etching (RIE) method.

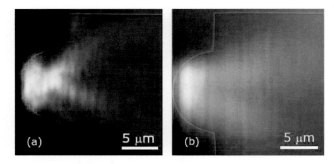

Figure 5.6. Optical images from an optical scanning microscope, obtained at
$\lambda = 1550$ nm around the focal point of (a) the binary-staircase lens and (b) of the
analogous structure with no zones/steps (semicircle with smooth walls). Note that
focusing is observed only with the binary-staircase lens.

In the characterization experiment, a continuous wave (CW) tunable semiconductor laser (1550 nm–1580 nm) was used as the input light source. The laser
light was coupled into the cleaved end of the input waveguides using a monomode
lensed fiber (working distance ≈ 14 µm and FWHM ≈ 2.5 µm in air) mounted on
a five-axis positioning stage. An infrared (IR) camera (Hamamatsu Model C2741)
connected to a microscope port aids the initial alignment to optimize the IR light
coupling from the optical fiber to the waveguide. In the FDTD simulation, a 10 µm
wide plane parallel, Gaussian beam was chosen as the incident field for the grating
lens. In the actual sample, the 5 µm wide input facet of the waveguide was inversely
tapered to a 10 µm width (see Fig. 5.5(a)), so that the propagating Gaussian beam
is expanded sufficiently inside the guiding channel before reaching the device end.
The planar wavefront after emerging from the binary-staircase lens is expected to
focus in the air cavity.

Subsequently, the focusing properties of the device were experimentally verified
using a scanning probe optical technique. A tapered fiber probe (250 nm aperture
diameter) metalized with a thin chromium and gold layer was raster scanned
just above the sample surface. The output end of the fiber probe was connected
to a nitrogen cooled germanium detector (North Coast Scientific Corp. Model #
EO-817L). Additionally, a typical lock-in amplifier was utilized to optimize the
detection scheme. Scanning the fiber tip at a constant height about 500 nm above
the sample surface allowed us to probe the optical intensity distribution over a
grid of 256×256 points spanning a 15×15 µm^2 area. The reconstructed image
is shown in Fig. 5.6(a).

The intensity distribution near the cavity center clearly shows the light focusing
from the binary-staircase lens. Identical focusing patterns were observed when
the experiment was repeated over a range of wavelengths varying from 1510 nm

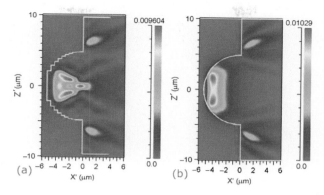

Figure 5.7. (a) Three-dimensional FDTD simulation of the plano-concave binary-staircase lens. (b) Three-dimensional FDTD simulation of the lens having the same geometrical dimensions as the binary-staircase one, but bearing no steps (or zones). See color plates section.

to 1580 nm. Another controlled experiment was performed in which the binary-staircase lens was replaced by an analogous structure (having the same geometrical features) with no steps. In the latter case, as shown in Fig. 5.6(b), no beam focusing was observed. Nevertheless, we can distinguish a bright spot near the device's edge, which is attributed to a sudden beam divergence as it propagates into open space from its initial confinement in the InGaAsP core waveguide layer (diffraction).

The numerical simulations were performed by using in-house 3D FDTD codes with perfectly matched layer boundary conditions that minimize reflections at the edges. The chosen input field excitation for the FDTD simulation was a TE polarized Gaussian beam which closely resembles the beam shape of the fiber source in the actual experiment. The energy density of the propagating H-field was mapped at different plane heights. Figures 5.7(a) and (b) show the simulated H-field density of the binary-staircase lens and the analog structure at about 800 nm above the center of the core layer, respectively. No focusing is achieved if the zones are removed, reinforcing the fact the steps are the decisive structural elements.

5.5 AANR and a negative lateral shift through a multilayered structure with surface gratings

So far we have demonstrated NR through the combination of total internal reflection and selective negative diffraction by using a surface grating on a homogeneous isotropic bulk material. Focal grating lenses have also been fabricated and realized. Strictly speaking, the negative refraction we have demonstrated in previous sections

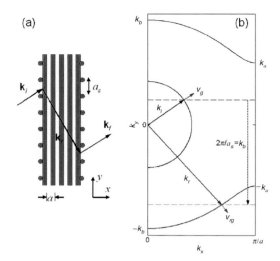

Figure 5.8. All-angle negative refraction using a surface grating. (a) A slab of 1D multilayer PhC of lattice spacing a, with a surface grating of period a_s on both surfaces. The surface grating gives rise to NR for the 1D PhC operating within the band gap. (b) Illustration of NR using a surface grating. The EFS of S-polarized modes in a 1D PhC made of alumina bars with lattice spacing $a = 0.9$ cm and a bar thickness $d = 0.5$ cm, at 6.85 GHz, is shown in the curves at top and bottom. The semicircle is the corresponding EFS in the air.

is not refraction at all due to the fact that, at the interface, two diffraction beams were reflected.

Apparently, an additional requirement is needed on the bulk properties to demonstrate true negative refraction in the presence of surface modification. In this section, we apply surface corrugation to one-dimensional photonic crystals. A negative lateral shift of an incident microwave beam by a flat multilayered structure with a surface grating [18] will be demonstrated. We further show that all-angle negative refraction (AANR) can also be achieved.

To illustrate this principle to achieve AANR, we consider a multilayered structure which behaves as a 1D PhC. For a 1D PhC as shown in Fig. 5.8(a), there will be a band gap for normally incident plane waves within a certain frequency range. For these frequencies, transmission may be allowed for oblique angles of incidence. For example, for the equi-frequency surface (EFS) of the 1D PhC shown in Fig. 5.8(b), waves with an angle of incidence θ, such as $k_a < k_0 \sin\theta < k_b$, will propagate. If, for some frequencies, $k_0 < k_a$, then, for all the incident plane waves, there is total external reflection. So this 1D PhC behaves as an omnidirectional mirror [48] for these S-polarized waves. If a grating with period a_s is introduced on the flat surface of the 1D PC, for example with $2\pi/a_s = k_a$, then a plane wave with an angle of incidence θ will receive a positive momentum kick along the surface. Thus the incident wave will couple to the Bloch wave with $k_y = k_0 \sin \theta + 2\pi/a_s$ and propagate inside the 1D PhC. However, if $2\pi/a_s = k_b$, the incident wave will receive a negative momentum kick along the surface and couple to the Bloch wave with $k_y = k_0 \sin \theta - 2\pi/a_s$. With proper design it is possible that only the Bloch wave with $k_y = k_0 \sin \theta - 2\pi/a_s$ will propagate inside the 1D PhC. This refraction is well-defined and negative. Furthermore, if $k_0 = k_b - k_a$, all the propagating waves will be transmitted into the 1D PhC. Since k_y of the Bloch wave is negative for every positive angle of incidence θ, this leads to a single-beam AANR. In this case, both the wave vector and the group velocity refraction is negative. This scenario for NR is illustrated in Fig. 5.8(b). For 1D PhCs, different k_y correspond to different modes. With the introduction of a surface grating, Bloch states with k_y and $k_y + 2m\pi/a_s$ are identical. Thus the 1D PhC effectively becomes a 2D PhC due to band structure folding, resulting in a finite-sized first Brillouin zone of rectangular shape. This is the simplest all-dielectric structure to achieve AANR.

A microwave experiment carried out in a parallel-plate waveguide confirms this mechanism for NR. A negative lateral shift was observed experimentally in the range 6.65–7.74 GHz for a multilayered grating structure. The 1D PhC is made of six layers of alumina bars with thickness $d = 0.5$ cm, lattice spacing $a = 0.9$ cm, and surface grating $a_s = 1.8$ cm (see Fig. 5.9). The angle of incidence of the 10 cm wide microwave beam was 13.5°. A 5.6 cm negative lateral shift is observed at 6.96 GHz, as shown in Fig. 5.9. Numerical simulations confirm AANR and a negative lateral shift for a large range of angles of incidence for frequencies around 6.85 GHz, as shown in Fig. 5.10.

Though negative refraction has been shown in 1D PhC [44], the range of negative refraction is quite limited. No AANR has been demonstrated in 1D all-dielectric PhCs. However, with surface engineering, AANR can be realized in 1D PhCs.

Even though AANR can be realized in 1D PhCs, these devices cannot be used to demonstrate flat lens imaging, which requires elliptic dispersion with negative group refraction [49]. This requirement will lead all negatively refracted light rays to focus at a single point. However, for the 1D PhC with a surface grating, different rays will focus at different points and the focal point of the paraxial rays is at infinity. Thus severe aberrations will render focusing impossible. Nevertheless these structures may be used to converge beams.

5.6 Surface corrugation approach to AANR in 2D photonic crystals

In this section, we will obtain new windows of AANR in two-dimensional (2D) PhCs using this new mechanism of surface modification.

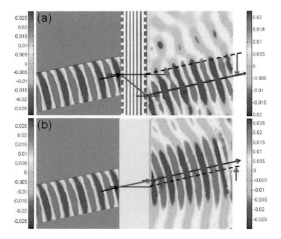

Figure 5.9. (a) Experimental demonstration of a negative lateral shift by a 1D PhC with a surface grating, at 6.96 GHz. A 5.6 cm negative lateral shift was observed. The 1D PhC is made of six layers of alumina bars with width $d = 0.5$ cm and spacing $a = 0.9$ cm. The surface grating was formed by rods of the same material, alumina, with diameter 0.63 cm and spacing $a_s = 1.8$ cm. The width of the incident beam is 10 cm and the angle of incidence is 13.5°. The incident and outgoing beams are plotted as the real parts of the measured transmission coefficient S_{21}. (b) A positive lateral shift for a microwave beam at 6.96 GHz by a slab of polystyrene with thickness 7.5 cm. See color plates section.

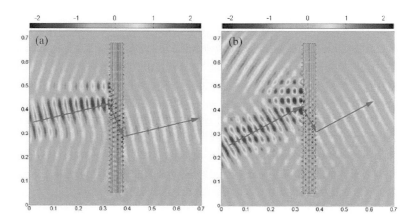

Figure 5.10. FDTD simulation of a negative lateral shift of microwave beams through a 1D PhC with surface gratings as specified in Fig. 5.8 at 6.96 GHz. (a) Microwave beam with an angle of incidence 13.5°. (b) Microwave beam with an angle of incidence 30°. The arrows indicate the energy flows of the incident and refracted beams. Lengths are measured in meters. See color plates section.

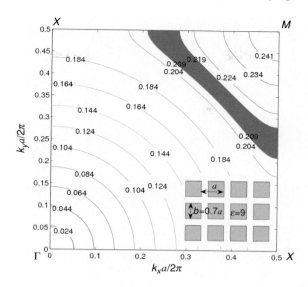

Figure 5.11. EFS of the TM modes of a square lattice PhC. The lattice is made of square rods of alumina ($\varepsilon = 9$) with a filling ratio 0.49 (see inset). The shaded area is the window of AANR for the PhC with surface grating $a_s = 2a$ (see Fig. 5.13). A slab of this PhC is oriented such that the surface normal is along the ΓX direction.

In a pioneering paper [18], Luo *et al.* showed that AANR can be achieved within certain frequency windows in the first band of a PhC. Specifically, within the first band, AANR is possible along the ΓM direction for a square lattice PhC. We will show that with an appropriate surface grating, NR and AANR are also possible along the ΓX direction in the first band of a square lattice PhC.

In the main text of this section, we consider only TM modes of a square lattice PhC of square rods. A square lattice of circular rods, or even of rods whose cross section is a rhombus, can be treated similarly. The generalization to TE modes and lattice structures other than the square lattice is straightforward. As a specific example, we consider a square lattice of rhombus rods with the ratio $b/a = 0.7$, and thus a filling ratio of 0.49. The EFS of this PhC is calculated by using the plane-wave expansion method [16] with 5041 plane waves. The EFS of the first band is shown in Fig. 5.11. The frequencies at the X point and the M point are $\omega_X = 0.1943(2\pi c/a)$ and $\omega_M = 0.2446(2\pi c/a)$, respectively.

Consider a slab of this PhC with surface normal along the ΓX direction. If one increases the frequency, $\omega > \omega_X$, there will be a partial band gap for waves incident on the air–PhC interface since the Bloch waves have $k_y \geq k_a$ and the incident plane wave with $k_y < k_a$ will be completely reflected. Here k_a is the k_y value of the crossing point of the EFS with the XM boundary of the first Brillouin zone,

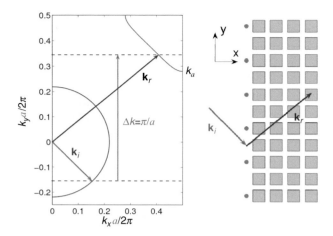

Figure 5.12. Mechanism for NR and AANR for a square lattice PhC with surface grating $a_s = 2a$. The curve is the EFS in the PhC and the semicircle is that in the air for $\omega = 0.219(2\pi c/a)$. With a vertical momentum kick $\Delta k = \pi/a$, an incident plane wave with $\mathbf{k}_i$ will be refracted negatively into a Bloch wave with $\mathbf{k}_r$.

as shown in Fig. 5.11 and Fig. 5.12. For certain frequencies $\omega_l < \omega < \omega_M$ with $\omega_l = 0.2089(2\pi c/a)$ when $k_a = \omega/c$, a flat slab of such a PhC is an omnidirectional reflector [48]. For example, for $\omega = 0.219(2\pi c/a)$, as shown in Fig. 5.12, there will be total external reflection of any incident plane wave. However, for these frequencies, a surface grating with period

$$a_s = 2a \qquad\qquad (5.10)$$

will give a momentum boost along the surface to the incident plane wave with angle of incidence θ such that it will be coupled to the Bloch waves inside the PhC with transverse momentum $k_y = \pi/a + (\omega/c)\sin\theta$, if θ is negative, and $k_y = -\pi/a + (\omega/c)\sin\theta$ if θ is positive. The refracted wave will propagate on the opposite side of the surface normal with respect to the incident beam. Thus NR is achieved. This is illustrated in Fig. 5.12. The effect of this surface grating is equivalent to bringing down the EFS around the M point to the X point for $\omega_l < \omega < \omega_M$. As we pointed out in ref. [35], it is the surface periodicity which determines the size of the EFS and the folding of the band structure. Furthermore, if $\pi/a - k_a \geq \omega/c$, AANR can be achieved. The upper limit for AANR is $\omega_u = 0.2192(2\pi c/a)$. Thus we obtained a 4.7% AANR around ω_u.

The above approach to NR and AANR is confirmed in our FDTD simulations. Here we consider the lateral shift of an incident beam by a slab made of a square lattice PhC. The details of the slab are shown in Fig. 5.13. Negative lateral shifts

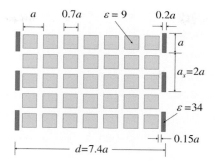

Figure 5.13. Details of slab made of a square lattice PhC with surface grating $a_s = 2a$. For simplicity, the thickness d of the slab is defined as the distance from the first surface to the last surface of the structure.

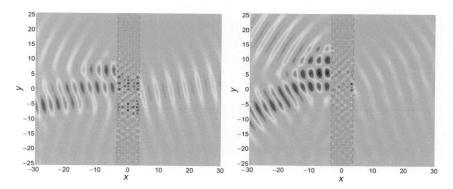

Figure 5.14. Negative lateral shift by a slab of PhC given in Fig. 5.13 for an incident Gaussian beam with angles of incidence 15° (left) and 30° (right) at $\omega = 0.219(2\pi c/a)$. The distance is measured in units of the lattice spacing a. See color plates section.

are observed for different angles of incidence, as shown in Fig. 5.14 for beams at $\omega = 0.219(2\pi c/a)$. It can be verified that, for this slab, AANR can be achieved for $0.2089 \leq \omega a/2\pi c \leq 0.2192$. Note that the details of the surface grating are not essential, except for its period, $a_s = 2a$. The grating can be holographic or an array of circular rods, as long as it is not too thick. For the specific surface grating shown in Fig. 5.13, the energy transmissions are 99.2% and 4.7% for plane waves with angles of incidence 15° and 30°, respectively.

Our new approach gives a much larger window of AANR than previously realized. For example, for a square lattice of air holes in $\varepsilon = 12$ with $r/a = 0.35$ studied in ref. [18], our approach gives a lower limit $\omega_l = 0.183(2\pi c/a)$ and an upper limit $\omega_u = 0.206(2\pi c/a)$, hence a fraction of AANR frequency range of 11% around $0.206(2\pi c/a)$. This range is much larger than the 6.1% AANR range

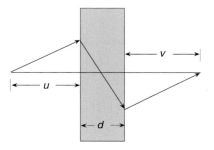

Figure 5.15. Flat lens with the lens equation $u + v = \sigma d$.

along the ΓM direction for the TE modes. This window of AANR is easier to locate. The two limits are obtained from the crossing of the band with the light lines around the X point and the M point, respectively. For the determination of the lower limit ω_l, there is no need to compute the frequency at which the radius of curvature of the contours along ΓM diverges [18].

This approach to AANR can also be extended to three-dimensional PhCs.

5.7 Flat lens imaging with large σ

One prominent application of negative refraction is the Veselago–Pendry perfect lens [29]. A flat slab of thickness d can focus an object with distance u on one side to a distance v on the other side with $u + v = d$ if the refractive index $n = -1$. For a generalized flat lens without an optical axis [49], the lens equation takes the form

$$u + v = \sigma d, \tag{5.11}$$

where σ is a material property, depending on the dispersion characteristics of the flat lens. This is illustrated in Fig. 5.15. This lens equation requires the following form of the EFS at the operating frequency:

$$k_{rx} = \kappa - \sigma \sqrt{\omega^2/c^2 - k_y^2}. \tag{5.12}$$

The lens surface is in the y-direction. The surface normal is along the x-axis. Here k_{rx} is the longitudinal component of the wave vector in the lens medium and κ is the center of the EFS ellipse [49].

Even though AANR can be realized in the first band along the ΓM direction for a square lattice PhC [18], the EFS is very flat around the lens normal. Thus one has $\sigma \ll 1$ [49]. Although $\sigma \sim 1$ has been reported in PCs with other structures [50–52], the focusing is still limited to the vicinity of the lens surface [53]. For practical applications, we need a large σ so that the object and image can be far away from the lens. We also show that a flat lens made of a photonic crystal with a surface

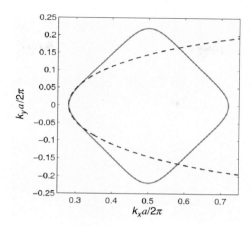

Figure 5.16. Fitting of the EFS of the TM modes at $\omega a/2\pi c = 0.219$ by Eq. (5.12). Here $\sigma_0 = 4$ and $\kappa a/2\pi c = 1.16$. Note that the center of the EFS is shifted from the M point to the X point due to the surface grating with $a_s = 2a$.

grating can have $u + v = \sigma d$ [49] with $\sigma \gg 1$, while for the Veselago–Pendry flat lens [29] $\sigma = 1$. Thus a flat lens can focus large and far away objects.

As we have stated in Section 5.6, the first Brillouin zone of a square lattice PhC with surface grating $a_s = 2a$ takes the shape of a rectangle instead of a square and its vertical size is reduced to $-\pi/a \le k_y \le \pi/a$. The center of the EFS for $\omega_l \le \omega \le \omega_M$ of the original PhC is moved from the M point to the X point. The fitting of the modified EFS for this frequency by Eq. (5.12) will give the lens property σ. An inspection of the band structure reveals that the EFS is not elliptical. This results in a σ that depends on the angle of incidence [49]. Nevertheless, the EFS can be fitted well with a constant σ_0 for small k_y, as shown in Fig. 5.16. For the square lattice PhC we have designed (Fig. 5.13), one has $\sigma_0 \sim 4$ for $\omega_l \le \omega \le \omega_u$.

Focusing by such a flat lens is shown in Fig. 5.17. For a point source with $u = 13.6a$, a clear focused image is obtained at $v = 12.4a$ for the operating frequency $\omega = 0.219(2\pi c/a)$, which is consistent with the lens equation $u + v = \sigma_{\text{eff}} d$ with $\sigma_{\text{eff}} = 3.5$ and $d = 7.4a$. There are two reasons for $\sigma_{\text{eff}} < \sigma_0$. First, the EFS is elliptical only for small $k_y = (\omega/c)\sin\theta$, and $\sigma \equiv -dk_{rx}/dk_x$ decreases with increasing angle of incidence θ. The effective σ_{eff} is an average and thus smaller than $\sigma_0 = -dk_{rx}/dk_x|_{k_y=0}$. Second, the thickness of a PhC slab is not a well-defined quantity. Here we simply define the lens thickness as the distance from the first surface to the last surface, as shown in Fig. 5.13. This may overestimate the effective thickness of the lens.

To check the performance of this flat lens further, we vary the object distance u. In Fig. 5.18 we show that the ratio $(u + v)/d$ is almost constant and very close to σ_0 for different object distances u.

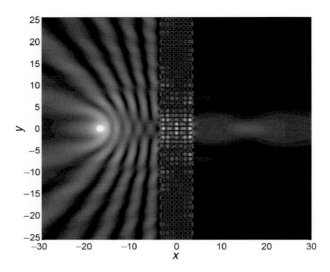

Figure 5.17. FDTD simulation of flat lens focusing of a point source. For better contrast effect, the field intensity at the point source is suppressed. The details of the lens are given in Fig. 5.13. The distance is measured in units of the lattice spacing a.

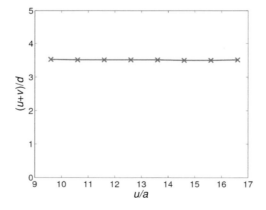

Figure 5.18. Ratio of the object–image distance to the slab thickness $(u + v)/d$ vs. the object distance u for the flat lens shown in Fig. 5.13 at the operating frequency $\omega = 0.219(2\pi c/a)$.

The primary concern for practical applications is the power transmission through the lens. As expected, the transmission through the flat lens is low due to the impedance mismatch. However, since the details of the grating on the PhC will not alter the scenario for NR, the power transmission can be enhanced through careful

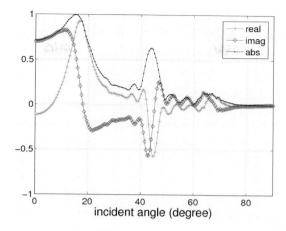

Figure 5.19. Transmission coefficient for a plane wave incident on the flat lens shown in Fig. 5.13.

engineering of the grating. In our simulation, we find that the grating on the 2D PhC with large dielectric constant has strong power transmission. The parameters of the surface grating shown in Fig. 5.13 are not optimized. Further improvement of transmission may be possible. For the grating parameters given in Fig. 5.13, the transmission coefficient is calculated and plotted in Fig. 5.19.

5.8 Discussions and conclusions

In this chapter we have shown that negative refraction of light can be achieved using surface corrugation, leading to novel microwave and optical devices.

Although the phenomena presented in this chapter are due to diffraction, ray optics does apply, as we have shown in the design of the plano-concave grating lens. The mechanism of plano-concave lens focusing is different from that of the zone plate, where concentric rings are carved to give each ray the corrected phase and ray optics does not apply. Previous use of diffractive optics has been limited to reflection gratings. The grism used in astronomy [54] satisfies the condition $n \sin \theta < 1$, which allows the zeroth order diffraction. Our approach is different from the suppression of zeroth and enhancement of -1 order transmission through surface grating depth modification [55]. By removing the zeroth order Bragg diffraction completely, our work opens the door for new phenomena and applications such as plano-concave lens focusing and flat lens imaging.

We have also achieved NR using a new approach: photonic band gap with surface grating. This approach enables us to demonstrate NR at the interface between two homogeneous positive media, achieve AANR in 1D PhC, and discover new windows for AANR in 2D PhCs. This approach also enables us to design a flat lens made of these PCs with a large object–image distance. Through the process of achieving these, we have provided a new perspective on the phenomenon of negative refraction. The NR can be seen as attributed to the folding of the band structure and EFS of the bulk material by a proper surface grating. Under certain conditions, surface periodicity alone can be sufficient to achieve NR even with homogeneous positive index materials, as we have demonstrated. For a 2D or higher dimensional PhC, the bulk periodic structure naturally introduces a surface periodicity at the interface between a PhC and another medium. Improper surface modification of the PhC may suppress or even diminish NR. Here we have shown that NR can be achieved by combining a surface grating with a multilayer 1D PhC structure that is relatively easier to fabricate. Previous approaches towards creating NR materials with multilayered structures required the use of alternating layers of negative permittivity and negative permeability materials [56, 57], and have not yet been realized experimentally. The realization of AANR in 1D all-dielectric PhCs with surface corrugation opens new realms of NR applications. Many structures, such as the photonic band gap materials currently used to guide waves or form cavities as photon insulators [13–15, 58–60] can be modified to have NR and AANR through surface engineering [36].

We have shown that negative refraction can be realized without changing the bulk properties. Our approach helps to deepen our understanding of negative refraction; it also gives us an extra handle on engineering the electromagnetic properties of materials. Surface engineering will enable us to manipulate waves further. Refraction is an interfacial phenomenon. Throughout this chapter we have shown that by engineering the interface, a totally reflecting surface can be made to refract negatively or positively, even though the materials utilized do not possess bulk negative refractive indices. The importance of surface modification has been previously recognized [61, 62], but has not been used as a mechanism to achieve negative refraction. Surface modification can also lead to the formation of surface states, which can be used to enhance subwavelength imaging.

The concepts discussed in this chapter are particularly suitable for integrated optical circuits, where the device dimension is about the size of the free space wavelength. Our work provides new ideas to harness diffraction to produce focusing devices and other optical elements.

Acknowledgments

We would like to thank our collaborators P. Vodo, Y. J. Huang, R. K. Banyal, and B. D. F. Casse for their contributions in this project. Work supported by the Air

Force Research Laboratory, Hanscom, MA through contract FA8718-06-C-0045 and the National Science Foundation through PHY-0457002.

References

[1] V. Veselago, "The electrodynamics of substances with simultaneously negative values of ε and μ," *Sov. Phys. Usp.* **10**, 509–514 (1968).

[2] V. Veselago and E. E. Narimanov, "The left hand of brightness: past, present and future of negative index materials," *Nature Mater.* **5**, 759–762 (2006).

[3] R. A. Shelby, D. R. Smith, and S. Schultz, "Experimental verification of a negative index of refraction," *Science* **292**, 77–79 (2001).

[4] C. G. Parazzoli, R. B. Greegor, K. Li, B. E. C. Koltenbah, and M. Tanielian, "Experimental verification and simulation of negative index of refraction using Snell's law," *Phys. Rev. Lett.* **90**, 107401(1-4) (2003).

[5] E. Cubukcu, K. Aydin, E. Ozbay, S. Foteinopoulou, and C. M. Soukoulis, "Negative refraction by photonic crystals," *Nature* **423**, 604–605 (2003).

[6] P. V. Parimi, W. T. Lu, P. Vodo, J. Sokoloff, J. S. Derov, and S. Sridhar, "Negative refraction and left-handed electromagnetism in microwave photonic crystals," *Phys. Rev. Lett.* **92**, 127401(1-4) (2004).

[7] P. V. Parimi, W. T. Lu, P. Vodo, and S. Sridhar, "Imaging by flat lens using negative refraction," *Nature* **426**, 404 (2003).

[8] J. B. Pendry, A. J. Holden, W. J. Stewart, and I. Youngs, "Extremely low frequency plasmons in metallic mesostructures," *Phys. Rev. Lett.* **76**, 4773–4776 (1996).

[9] J. B. Pendry, A. J. Holden, D. J. Robbins, and W. J. Stewart, "Magnetism from conductors and enhanced nonlinear phenomena," *IEEE Trans. Microwave Theory Tech.* **47**, 2075–2084 (1999).

[10] D. R. Smith, J. B. Pendry, and M. C. K. Wiltshire, "Metamaterials and negative refractive index," *Science* **305**, 788–792 (2004).

[11] V. M. Shalaev, "Optical negative-index metamaterials," *Nature Photonics* **1**, 41–48 (2007).

[12] B. D. F. Casse, H. O. Moser, J. W. Lee, M. Bahou, S. Inglis, and L. K. Jian, "Towards three-dimensional and multilayer rod-split-ring metamaterial structures by means of deep x-ray lithography," *Appl. Phys. Lett.* **90**, 254106(1-3) (2007).

[13] E. Yablonovitch, "Inhibited spontaneous emission in solid-state physics and electronics," *Phys. Rev. Lett.* **58**, 2059–2062 (1987).

[14] E. Yablonovitch, "Photonic band-gap structures," *J. Opt. Soc. Am. B* **10**, 283–296 (1993).

[15] S. John, "Strong localization of photons in certain disordered dielectric superlattices," *Phys. Rev. Lett.* **58**, 2486–2489 (1987).

[16] J. D. Joannopoulos, R. Meade, and J. N. Winn, *Photonic Crystals: Modeling the Flow of Light* (Princeton, NJ: Princeton University Press, 1995).

[17] M. Notomi, "Theory of light propagation in strongly modulated photonic crystals: refraction like behavior in the vicinity of the photonic band gap," *Phys. Rev. B* **62**, 10696–10705 (2000).

[18] C. Luo, S. G. Johnson, J. D. Joannopoulos, and J. B. Pendry, "All-angle negative refraction without negative effective index," *Phys. Rev. B* **65**, 201104(1-4) (2002).

[19] B. Gralak, S. Enoch, and G. Tayeb, "Anomalous refractive properties of photonic crystals," *J. Opt. Soc. Am. A* **17**, 1012–1020 (2000).

[20] A. Berrier, M. Mulot, M. Swillo, M. Qiu, L. Thylén, A. Talneau, and S. Anand, "Negative refraction at infrared wavelengths in a two-dimensional photonic crystal," *Phys. Rev. Lett.* **93**, 073902(1-4) (2004).

[21] B. D. F. Casse, W. T. Lu, R. K. Banyal, Y. J. Huang, S. Selvarasah, M. R. Dokmeci, C. H. Perry, and S. Sridhar, "Imaging with subwavelength resolution by a generalized superlens at infrared wavelengths," *Opt. Lett.* **34**, 1994–1996 (2009).

[22] D. R. Smith and D. Schurig, "Electromagnetic wave propagation in media with indefinite permittivity and permeability tensors," *Phys. Rev. Lett.* **90**, 077405(1-4) (2003).

[23] D. R. Smith, P. Kolinko, and D. Schurig, "Negative refraction in indefinite media," *J. Opt. Soc. Am. B* **21**, 1032–1043 (2004).

[24] A. J. Hoffman, L. Alekseyev, S. S. Howard *et al.*, "Negative refraction in semiconductor metamaterials," *Nature Mater.* **6**, 946–950 (2007).

[25] W. T. Lu and S. Sridhar, "Superlens imaging theory for anisotropic nanostructured metamaterials with broadband all-angle negative refraction," *Phys. Rev. B* **77**, 233101(1-4) (2008).

[26] L. Menon, W. T. Lu, A. L. Friedman, S. Bennett, D. Heiman, and S. Sridhar, "Negative index metamaterials based on metal-dielectric nanocomposites for imaging applications," *Appl. Phys. Lett.* **93**, 123117(1-3) (2008).

[27] J. Valentine, S. Zhang, T. Zentgraf, E. Ulin-Avila, D. A. Genov, G. Bartal, and X. Zhang, "Three-dimensional optical metamaterial with a negative refractive index," *Nature* **455**, 376–380 (2008).

[28] B. D. F. Casse, W. T. Lu, Y. J. Huang, E. Gultepe, L. Menon, and S. Sridhar, "Super-resolution imaging using a three-dimensional metamaterials nanolens," *Appl. Phys. Lett.* **96**, 023114(1-3) (2010).

[29] J. B. Pendry, "Negative refraction makes a perfect lens," *Phys. Rev. Lett.* **85**, 3966–3969 (2000).

[30] C. Luo, S. G. Johnson, J. D. Joannopoulos, and J. B. Pendry, "Subwavelength imaging in photonic crystals," *Phys. Rev. B* **68**, 045115(1-15) (2003).

[31] D. Schurig, J. J. Mock, B. J. Justice, S. A. Cummer, J. B. Pendry, A. F. Starr, and D. R. Smith, "Metamaterial electromagnetic cloak at microwave frequencies," *Science* **314**, 977–980 (2006).

[32] K. L. Tsakmakidis, A. D. Boardman, and O. Hess, "'Trapped rainbow' storage of light in metamaterials," *Nature* **450**, 397–401 (2007).

[33] W. T. Lu, S. Savo, B. D. F. Casse, and S. Sridhar, "Slow microwave waveguide made of negative permeability metamaterials," *Microwave Opt. Tech. Lett.* **51**, 2705 (2009).

[34] Y. Zhang, B. Fluegel, and A. Mascarenhas, "Total negative refraction in real crystals for ballistic electrons and light," *Phys. Rev. Lett.* **91**, 157404(1-4) (2003).

[35] W. T. Lu, Y. J. Huang, P. Vodo, R. K. Banyal, C. H. Perry, and S. Sridhar, "A new mechanism for negative refraction and focusing using selective diffraction from surface corrugation," *Opt. Express* **15**, 9166–9175 (2007).

[36] Y. J. Huang, W. T. Lu, and S. Sridhar, "Alternative approach to all-angle negative refraction in two-dimensional photonic crystals," *Phys. Rev. A* **76**, 013824(1-5) (2007).

[37] B. D. F. Casse, R. K. Banyal, W. T. Lu, Y. J. Huang, S. Selvarasah, M. Dokmeci, and S. Sridhar, "Nanoengineering of a negative-index binary-staircase lens for the optics regime," *Appl. Phys. Lett.* **92**, 243122(1-3) (2008).

[38] Z. Feng, X. Zhang, Y. Wang, Z.-Y. Li, B. Cheng, and D.-Z. Zhang, "Negative refraction and imaging using 12-fold-symmetry quasicrystals," *Phys. Rev. Lett.* **94**, 247402(1-4) (2005).

[39] H. J. Lezec, J. A. Dionne, and H. A. Atwater, "Negative refraction at visible frequencies," *Science* **316**, 430–432 (2007).

[40] M. Neviere and E. Popov, *Light Propagation in Periodic Media: Differential Theory and Design* (New York: Marcel Dekker, Inc., 2003), p. 3.

[41] S. Enoch, G. Tayeb, and B. Gralak, "The richness of the dispersion relation of electromagnetic bandgap materials," *IEEE Trans. Antennas Propag.* **51**, 2659–2666 (2003).

[42] C. G. Parazzoli, R. B. Greegor, J. A. Nielsen, M. A. Thompson, K. Li, A. M. Vetter, M. H. Tanielian, and D. C. Vier, "Performance of a negative index of refraction lens," *Appl. Phys. Lett.* **84**, 3232(1-3) (2004).

[43] P. Vodo, P. V. Parimi, W. T. Lu, and S. Sridhar, "Focusing by plano-concave lens using negative refraction," *Appl. Phys. Lett.* **86**, 201108(1-3) (2005).

[44] P. Vodo, W. T. Lu, Y. Huang, and S. Sridhar, "Negative refraction and plano-concave lens focusing in one-dimensional photonic crystals," *Appl. Phys. Lett.* **89**, 084104(1-3) (2006).

[45] B. D. F. Casse, W. T. Lu, Y. J. Huang, and S. Sridhar, "Nano-optical microlens with ultrashort focal length using negative refraction," *Appl. Phys. Lett.* **93**, 053111(1-3) (2008).

[46] A. Taflove and S. C. Hagness, *Computational Electrodynamics: The Finite-Difference Time-Domain Method*, 3rd edn (Norwood, MA: Artech House Publishers, 2005).

[47] J. Alda, J. M. Rico-García, J. M. López-Alonso, B. Lail, and G. Boreman, "Design of Fresnel lenses and binary-staircase kinoforms of low value of the aperture number," *Opt. Commun.* **260**, 454–461 (2005).

[48] Y. Fink, J. N. Winn, S. Fan, C. Chen, J. Michel, J. D. Joannopoulos, and E. L. Thomas, "A dielectric omnidirectional reflector," *Science* **282**, 1679–1682 (1998).

[49] W. T. Lu and S. Sridhar, "Flat lens without optical axis: theory of imaging," *Opt. Express* **13**, 10673–10680 (2005).

[50] X. Zhang, "Absolute negative refraction and imaging of unpolarized electromagnetic waves by two-dimensional photonic crystals," *Phys. Rev. B* **70**, 205102(1-6) (2004).

[51] X. Zhang, "Subwavelength far-field resolution in a square two-dimensional photonic crystal" *Phys. Rev. E* **71**, 037601(1-4) (2005).

[52] R. Gajić, R. Meisels, F. Kuchar, and K. Hingerl, "All-angle left-handed negative refraction in Kagomé and honeycomb lattice photonic crystals," *Phys. Rev. B* **73**, 165310(1-5) (2006).

[53] Z.-Y. Li and L.-L. Lin, "Evaluation of lensing in photonic crystal slabs exhibiting negative refraction," *Phys. Rev. B* **68**, 245110(1-7) (2003).

[54] E. G. Loewen and E. Popov, *Diffraction Gratings and Applications* (New York: Marcel Dekker, Inc., 1997).

[55] E. Noponen, *Electromagnetic Theory of Diffractive Optics*. Ph.D. thesis, Helsinki University of Technology, Espoo, Finland, 1994.

[56] D. R. Fredkin and A. Ron, "Effectively left-handed (negative index) composite material," *Appl. Phys. Lett.* **81**, 1753(1-3) (2002).

[57] A. Alù and N. Engheta, "Pairing an epsilon-negative slab with a mu-negative slab: resonance, tunneling and transparency," *IEEE Trans. Antenn. Propag.* **51**, 2558–2571 (2003).

[58] S. Y. Lin, J. G. Fleming, D. L. Hetherington *et al.*, "A three-dimensional photonic crystal operating at infrared wavelengths," *Nature* **394**, 251–253 (1998).

[59] M. Ibanescu, Y. Fink, S. Fan, E. L. Thomas, and J. D. Joannopoulos, "An all-dielectric coaxial waveguide," *Science* **289**, 415–419 (2000).

[60] M. Qi, E. Lidorikis, P. T. Rakich, S. G. Johnson, J. D. Joannopoulos, E. P. Ippen, and H. I. Smith, "A three-dimensional optical photonic crystal with designed point defects," *Nature* **429**, 538–542 (2004).

[61] T. Decoopman, G. Tayeb, S. Enoch, D. Maystre, and B. Gralak, "Photonic crystal lens: from negative refraction and negative index to negative permittivity and permeability," *Phys. Rev. Lett.* **97**, 073905(1-4) (2006).

[62] D. R. Smith, P. M. Rye, J. J. Mock, D. C. Vier, and A. F. Starr, "Enhanced diffraction from a grating on the surface of a negative-index metamaterial," *Phys. Rev. Lett.* **93**, 137405(1-4) (2004).

6

Transformation of optical fields by structured surfaces

A. A. MARADUDIN, E. R. MÉNDEZ, AND T. A. LESKOVA

6.1 Introduction

A reader of this book will quickly see that structured surfaces, whether deterministic or random, can reflect, transmit, refract, and amplify volume or surface electromagnetic waves in ways that naturally occurring surfaces cannot. They can also change the nature of an electromagnetic field incident on them. For example, they can change a beam with one intensity distribution into a beam with a different intensity distribution, or they can convert a plane wave into a beam. The use of structured surfaces, specifically randomly rough surfaces, to effect such transformations of optical fields is the subject of this chapter, where two examples of this use are presented, namely beam shaping and the formation of pseudo-nondiffracting beams.

The creation of optical elements that transform an electromagnetic beam with a specified transverse intensity distribution into a beam with a different specified transverse intensity distribution, especially those that transform a laser beam with a Gaussian intensity profile into a beam with a constant intensity profile – a flat top beam, has been studied theoretically and experimentally for many years [1–38]. The interest in beam shaping is due to a wide range of applications for beams with a variety of non-Gaussian intensity distributions. These applications include laser surgery [39], laser radar [40], laser microstructuring of materials [41], metal hardening [42], optical communication [43], and optical scanning [44], among others. Some of them and other applications of beam shaping are discussed in the recent book by Dickey *et al.* [45].

A variety of experimental methods has been developed for shaping a laser beam into a beam with a different transverse intensity profile. These include the use of a binary diffractive optical element [6], a distributed phase plate [11],

Structured Surfaces as Optical Metamaterials, ed. A. A. Maradudin. Published by Cambridge University Press.
© Cambridge University Press 2011.

a refractive optical element [3, 13], a polarization-holographic optical element [5, 8], aspherical refractive lenses [22], microlens arrays [7, 12], and holograms [21], for example. Many of these methods are described in the review article by Shealy [16].

There is also a great deal of interest at the present time in a different type of transformation of an optical field, namely one that produces a nondiffracting beam. It is well known that a bounded optical field propagating in free space ordinarily undergoes diffractive spreading. However, there are exceptions to this general statement if the field is unbounded. In 1987, in a study of the scalar wave equation in cylindrical coordinates,

$$\left(\nabla^2 + k^2\right) \Psi(\rho, \phi, z) = 0, \tag{6.1}$$

Durnin [46] obtained an exact, cylindrically symmetric, solution:

$$\Psi(\rho, z) = \exp(i\beta z) J_0(\alpha\rho), \tag{6.2}$$

where $\alpha^2 + \beta^2 = k^2$, $J_0(x)$ is the Bessel function of the first kind and zero order, and (ρ, ϕ, z) are the cylindrical coordinates. This beam has an infinite extent in the transverse plane, and is capable of propagating to infinity in the z direction without spreading. It has been named a *diffraction-free beam*. Durnin's discovery stimulated a great number of studies of diffraction-free beams. It has been shown that there exists an infinite number of diffraction-free beams with different transverse field profiles [47]. Much of this work is cited in a recent book by Ostrovsky [48].

However, the ideal diffraction-free beams considered by Durnin and by other authors possess wave functions $\Psi(\rho, \phi, z)$ that are not square integrable and that contain an infinite amount of energy. They are therefore impossible to realize in practice. Consequently, studies of diffraction-free beams have shifted to studies of *pseudo-nondiffracting beams*. These beams have a finite beam aperture, and exhibit the main propagation features of true diffraction-free beams, namely a constant intensity along the direction of propagation, and a beam-like profile in any transverse plane, over a finite propagation range [49–81]. This propagation length can extend to several tens of centimeters, or more, which is long enough for many applications, which include optical interconnection [76], laser Doppler velocimetry [72], precision alignment [51, 65, 72], laser machining [60], and laser surgery [60].

Several approaches to the production of three-dimensional [49, 51, 57–60, 62, 64, 65, 68–70, 72, 73, 76–78, 81] and even two-dimensional [75] pseudo-nondiffracting beams have been developed. The latter are characterized by a constant intensity along the direction of propagation z, and a beam-like shape in one of the transverse directions, say x.

The great majority of the methods proposed for the transformation of optical beams into different optical beams, and the methods actually realized for performing such transformations, have been based on diffractive optical elements. These are deterministic periodic structures that produce, essentially, an array of diffractive orders over the desired region in the far field [82]. To produce the desired intensity distribution of light scattered from or transmitted through such a diffuser, the diffracted beams have to overlap to some extent. This means that the cells, whose periodic repetition in two dimensions forms the diffuser, must be larger than about 500 wavelengths of the incident light. The design of these diffusers is usually based on scalar diffraction theory and the paraxial approximation [83]. The smallest features impressed in the diffuser plate must then be larger than the wavelength of the incident light, and the angles of scattering must be small, so that the cosine of the polar scattering angle can be replaced by unity. These requirements impose some limitations on the design of a diffuser. It is generally not possible to obtain the structure within a cell analytically, which then has to be found through some kind of optimization algorithm.

In addition, many of the approaches to the design of optical diffractive elements are based on the use of binary phase distributions. These are two-dimensional periodic structures fabricated from horizontal cells whose heights above a base plane can have only two values. The resulting diffusers work well only for wavelengths of the incident light in the vicinity of the design wavelength, i.e. they are chromatic. Therefore they are not useful in applications that use broadband illumination. Consequently, although diffractive optical elements have been successful in modifying laser beams to produce desired transverse intensity profiles, or to produce desired axial intensity distributions, for example, there are situations for which they would not be appropriate.

In this chapter we describe an alternative approach to the design of optical elements that effect desired transformations of optical fields that is based on refraction rather than diffraction. This kind of element is well suited for applications in which broadband illumination is required. The approach described here is based on the use of random surfaces and, as we will see, provides simple and well-defined procedures for designing these optical elements, without having to resort to iterative techniques that might not converge for some designs. In contrast with the diffractive approach, the methods described here are well suited for cases involving partially coherent broadband beams.

We show in particular that beam shaping and the generation of pseudo-nondiffracting beams can be effected by the transmission of an incident electromagnetic field through a suitably designed random surface. The random surfaces we consider are the surfaces (microreliefs) of a thin random phase screen. A thin random phase screen can be thought of as a layer of negligible thickness in the plane $x_3 = 0$ that in transmission produces a phase shift $\Phi(\mathbf{x}_\parallel)$ that is a random function of the coordinates $\mathbf{x}_\| = (x_1, x_2, 0)$ in the plane $x_3 = 0$ [84], but does not change the amplitude of the field. We assume that $\Phi(\mathbf{x}_\|)$ is a single-valued function of $\mathbf{x}_\|$ that is differentiable with respect to x_1 and x_2, and that constitutes a random process, but not necessarily a stationary one. This phase function can be related to variations in the thickness of a dielectric film, i.e. to the surface profile function $\zeta(\mathbf{x}_\|)$ of the rough illuminated surface of the film. In the case of small polar angles of incidence and transmission, this relation is given by [85–87]

$$\Phi(\mathbf{x}_\|) = \frac{\omega}{c} \Delta n \zeta(\mathbf{x}_\|), \tag{6.3}$$

where Δn is the difference between the refractive index of the material from which the film is fabricated and the refractive index of the surrounding medium, ω is the frequency of the incident light, and c is the speed of light in a vacuum. In more sophisticated models [87, 88], the relation between the phase function $\Phi(\mathbf{x}_\|)$ and the surface profile function $\zeta(\mathbf{x}_\|)$ has the form

$$\Phi(\mathbf{x}_\|) = \frac{\omega}{c} \mu(\mathbf{x}_\|) \zeta(\mathbf{x}_\|), \tag{6.4}$$

where $\mu(\mathbf{x}_\|)$ depends on the slope of the surface profile function and the angles of incidence and observation. However, the use of such a relation would complicate the analysis presented in this chapter enough to make it unworkable, and we do not consider it further.

The two examples of field transformations by transmission through deliberately structured surfaces considered in this chapter suggest that additional types of such transformations can be effected in this way.

6.2 Beam shaping

In this section we show how the transformation of a beam with a specified transverse intensity distribution into a beam with a different specified transverse intensity distribution can be effected by its transmission through a suitably designed random surface.

6.2.1 The transmitted field

The system we consider is depicted in Fig. 6.1. It consists of a thin random phase screen in the plane $x_3 = 0$ that fills the aperture A in an opaque screen in the plane $x_3 = 0$. The region $x_3 < 0$ is a vacuum, as is the region $x_3 > 0$. We assume that the phase screen is illuminated at normal incidence from the region $x_3 < 0$ by a scalar beam of frequency ω, written as a superposition of incoming plane waves,

$$\psi(\mathbf{x}|\omega)_{inc} = \int \frac{d^2 k_\|}{(2\pi)^2} F(\mathbf{k}_\|) \exp[i\mathbf{k}_\| \cdot \mathbf{x}_\| + i\alpha_0(\mathbf{k}_\|, \omega)x_3], \tag{6.5}$$

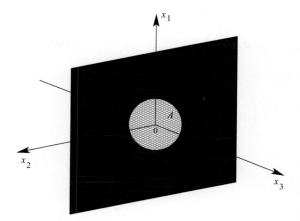

Figure 6.1. The system studied in Section 6.2.

where $\mathbf{x}_\parallel = (x_1, x_2, 0)$ is an arbitrary vector in the plane $x_3 = 0$, while $\mathbf{k}_\parallel = (k_1, k_2, 0)$ is a two-dimensional wave vector. The weight function $F(\mathbf{k}_\parallel)$ at this point is arbitrary, subject to the condition that the integral converges. The function $\alpha_0(k_\parallel, \omega)$ is defined by

$$\alpha_0(k_\parallel, \omega) = [(\omega/c)^2 - k_\parallel^2]^{1/2}, \qquad Re\,\alpha_0(k_\parallel, \omega) > 0, \qquad Im\,\alpha_0(k_\parallel, \omega) > 0. \quad (6.6)$$

The field transmitted through the phase screen can be written formally as follows:

$$\psi(\mathbf{x}|\omega)_{tr} = \int \frac{d^2q_\parallel}{(2\pi)^2} \, T(\mathbf{q}_\parallel) \exp[i\mathbf{q}_\parallel \cdot \mathbf{x}_\parallel + i\alpha_0(q_\parallel, \omega)x_3]. \quad (6.7)$$

From the result that

$$\psi(\mathbf{x}_\parallel, 0+ |\omega)_{tr} = \int \frac{d^2q_\parallel}{(2\pi)^2} \, T(\mathbf{q}_\parallel) \exp(i\mathbf{q}_\parallel \cdot \mathbf{x}_\parallel), \quad (6.8)$$

where the notation "0+" emphasizes that this is the field just after its passage through the phase screen, we find that the transmission amplitude $T(\mathbf{q}_\parallel)$ is given by

$$T(\mathbf{q}_\parallel) = \int d^2x_\parallel' \exp(-i\mathbf{q}_\parallel \cdot \mathbf{x}_\parallel')\psi(\mathbf{x}_\parallel', 0+ |\omega)_{tr}. \quad (6.9)$$

When this result is substituted into Eq. (6.7) the transmitted field in the region $x_3 > 0+$ takes the following form:

$$\psi(\mathbf{x}|\omega)_{tr} = \int d^2x_\parallel' \, \psi(\mathbf{x}_\parallel', 0+ |\omega)_{tr}$$

$$\times \int \frac{d^2q_\parallel}{(2\pi)^2} \exp[i\mathbf{q}_\parallel \cdot (\mathbf{x}_\parallel - \mathbf{x}_\parallel') + i\alpha_0(q_\parallel, \omega)x_3]. \quad (6.10)$$

At this point we make the parabolic approximation

$$\alpha_0(q_\parallel, \omega) = [(\omega/c)^2 - q_\parallel^2]^{1/2} \simeq (\omega/c) - (c/2\omega)q_\parallel^2, \qquad (6.11)$$

with which Eq. (6.10) becomes

$$\psi(\mathbf{x}|\omega)_{tr} = \left(\frac{\omega}{2\pi i c x_3}\right) \exp\left(i\frac{\omega}{c}x_3\right)$$

$$\times \int d^2 x_\parallel' \exp\left[i\frac{\omega}{2cx_3}(\mathbf{x}_\parallel - \mathbf{x}_\parallel')^2\right] \psi(\mathbf{x}_\parallel', 0+ |\omega)_{tr}. \qquad (6.12)$$

The mean intensity of the transmitted field in the region $x_3 > 0+$ is obtained from Eq. (6.12) in the form

$$\langle |\psi(\mathbf{x}|\omega)_{tr}|^2 \rangle = \left(\frac{\omega}{2\pi c x_3}\right)^2$$

$$\times \int d^2 x_\parallel' \int d^2 x_\parallel'' \exp\left\{i\frac{\omega}{2cx_3}[(\mathbf{x}_\parallel - \mathbf{x}_\parallel')^2 - (\mathbf{x}_\parallel - \mathbf{x}_\parallel'')^2]\right\}$$

$$\times \langle \psi(\mathbf{x}_\parallel', 0+ |\omega)_{tr} \psi^*(\mathbf{x}_\parallel'', 0+ |\omega)_{tr} \rangle, \qquad (6.13)$$

where the angle brackets denote an average over the ensemble of realizations of the random phase screen.

The correlation function $\langle \psi(\mathbf{x}_\parallel', 0+ |\omega)_{tr} \psi^*(\mathbf{x}_\parallel'', 0+ |\omega)_{tr} \rangle$ is called the *cross-spectral density* of the transmitted field in the plane $x_3 = 0+$, and is denoted by $W^{(0)}(\mathbf{x}_\parallel, 0+ |\mathbf{x}_\parallel', 0+)$ [89]. We assume that it has the form corresponding to a Schell-model source in this plane [90], namely

$$W^{(0)}(\mathbf{x}_\parallel', 0+ |\mathbf{x}_\parallel'', 0+) = \langle \psi(\mathbf{x}_\parallel', 0+ |\omega)_{tr} \psi^*(\mathbf{x}_\parallel'', 0+ |\omega)_{tr} \rangle$$

$$= [S^{(0)}(\mathbf{x}_\parallel')]^{1/2} g^{(0)}(\mathbf{x}_\parallel' - \mathbf{x}_\parallel'') [S^{(0)}(\mathbf{x}_\parallel'')]^{1/2}. \qquad (6.14)$$

In this expression $g^{(0)}(\mathbf{x}_\parallel' - \mathbf{x}_\parallel'')$ is the *spectral degree of coherence* of the source. It is a Hermitian function of $\mathbf{x}_\parallel'$ and $\mathbf{x}_\parallel''$,

$$g^{(0)}(\mathbf{x}_\parallel'' - \mathbf{x}_\parallel') = g^{(0)}(\mathbf{x}_\parallel' - \mathbf{x}_\parallel'')^*, \qquad (6.15)$$

where the star denotes complex conjugation and has the following properties [89]:

$$0 \leq |g^{(0)}(\mathbf{x}_\parallel' - \mathbf{x}_\parallel'')| \leq 1 \qquad (6.16)$$

and

$$g^{(0)}(\mathbf{0}) = 1. \qquad (6.17)$$

It is assumed to be nonzero only for values of $|\mathbf{x}'_{\parallel} - \mathbf{x}''_{\parallel}|$ smaller than a characteristic (coherence) length σ_g.

The function $S^{(0)}(\mathbf{x}_{\parallel})$ is the *spectral density* (intensity) of the transmitted field in the plane $x_3 = 0+$, and is given by

$$S^{(0)}(\mathbf{x}_{\parallel}) = W^{(0)}(\mathbf{x}_{\parallel}, 0+ |\mathbf{x}_{\parallel}, 0+)$$

$$= \langle |\psi(\mathbf{x}_{\parallel}, 0+ |\omega)_{tr}|^2 \rangle. \tag{6.18}$$

It is a real non-negative function of $\mathbf{x}_{\parallel}$,

$$S^{(0)}(\mathbf{x}_{\parallel}) \geq 0, \tag{6.19}$$

and is assumed to be nonzero only for values of $|\mathbf{x}_{\parallel}|$ smaller than a characteristic length σ_s.

The square root $[S^{(0)}(\mathbf{x}_{\parallel})]^{1/2}$ is also a real non-negative function of $\mathbf{x}_{\parallel}$, and has the Fourier representation

$$[S^{(0)}(\mathbf{x}_{\parallel})]^{1/2} = \int \frac{d^2 p_{\parallel}}{(2\pi)^2} \, \hat{S}(\mathbf{p}_{\parallel}) \exp(i\mathbf{p}_{\parallel} \cdot \mathbf{x}_{\parallel}), \tag{6.20}$$

where the Fourier coefficient $\hat{S}(\mathbf{p}_{\parallel})$ has the property

$$\hat{S}(-\mathbf{p}_{\parallel}) = \hat{S}^*(\mathbf{p}_{\parallel}). \tag{6.21}$$

When Eq. (6.14) is substituted into Eq. (6.13) and use is made of Eq. (6.20), we obtain the mean intensity of the transmitted field in the following form:

$$\langle |\psi(\mathbf{x}|\omega)_{tr}|^2 \rangle = \left(\frac{\omega}{2\pi c x_3}\right)^2 \int d^2 u_{\parallel} \, g^{(0)}(\mathbf{u}_{\parallel}) \exp\left[-i\frac{\omega}{c x_3}\mathbf{x}_{\parallel} \cdot \mathbf{u}_{\parallel}\right]$$

$$\times \int \frac{d^2 p_{\parallel}}{(2\pi)^2} \exp(i\mathbf{p}_{\parallel} \cdot \mathbf{u}_{\parallel}) \hat{S}\left(\mathbf{p}_{\parallel} - \frac{\omega}{2c x_3}\mathbf{u}_{\parallel}\right) \hat{S}^*\left(\mathbf{p}_{\parallel} + \frac{\omega}{2c x_3}\mathbf{u}_{\parallel}\right). \tag{6.22}$$

In obtaining the result given by Eq. (6.22), we have not made use of any properties of the phase function $\Phi(\mathbf{x}_{\parallel}) = \Delta n(\omega/c)\zeta(\mathbf{x}_{\parallel})$. However, as our aim is to determine the function $\zeta(\mathbf{x}_{\parallel})$ that produces the mean intensity given by Eq. (6.22), we now have to obtain an expression for the mean intensity of the transmitted beam that depends explicitly on $\zeta(\mathbf{x}_{\parallel})$. On equating the latter expression with Eq. (6.22), we will obtain an equation from which $\zeta(\mathbf{x}_{\parallel})$ can be determined.

Our starting point is Eq. (6.10). The transmitted field just beyond the phase screen, $\psi(\mathbf{x}_{\parallel}, 0+ |\omega)_{tr}$, is related to the incident field just before the phase screen,

$\psi(\mathbf{x}_\parallel, 0- |\omega)_{inc}$, as follows:

$$\psi(\mathbf{x}_\parallel, 0+ |\omega)_{tr} = \psi(\mathbf{x}_\parallel, 0- |\omega)_{inc} \exp\left[i\frac{\omega}{c}\Delta n \zeta(\mathbf{x}_\parallel)\right]$$

$$= \exp\left[i\frac{\omega}{c}\Delta n \zeta(\mathbf{x}_\parallel)\right] \int \frac{d^2 k_\parallel}{(2\pi)^2} F(\mathbf{k}_\parallel) \exp(i\mathbf{k}_\parallel \cdot \mathbf{x}_\parallel) \qquad \mathbf{x}_\parallel \in A \quad (6.23a)$$

$$= 0 \qquad\qquad\qquad\qquad\qquad\qquad\qquad\qquad\qquad\qquad \mathbf{x}_\parallel \notin A, \quad (6.23b)$$

where we have used Eq. (6.5). When Eq. (6.23) is combined with Eq. (6.10), the mean intensity of the transmitted field in the region $x_3 > 0+$ can be written as follows:

$$\langle |\psi(\mathbf{x}|\omega)_{tr}|^2 \rangle = \left(\frac{\omega}{2\pi c x_3}\right)^2 \int \frac{d^2 k'_\parallel}{(2\pi)^2} F(\mathbf{k}'_\parallel) \int \frac{d^2 k''_\parallel}{(2\pi)^2} F^*(\mathbf{k}''_\parallel)$$

$$\times \int_A d^2 x'_\parallel \exp\left\{i\left[\mathbf{k}'_\parallel - \frac{\omega}{c x_3}\mathbf{x}_\parallel\right] \cdot \mathbf{x}'_\parallel + i\frac{\omega}{2 c x_3}x'^2_\parallel\right\}$$

$$\times \int_A d^2 x''_\parallel \exp\left\{-i\left[\mathbf{k}''_\parallel - \frac{\omega}{c x_3}\mathbf{x}_\parallel\right] \cdot \mathbf{x}''_\parallel - i\frac{\omega}{2 c x_3}x''^2_\parallel\right\}$$

$$\times \left\langle \exp\left\{i\frac{\omega}{c}\Delta n[\zeta(\mathbf{x}'_\parallel) - \zeta(\mathbf{x}''_\parallel)]\right\}\right\rangle. \qquad (6.24)$$

As it stands, this equation is too difficult to invert to obtain $\zeta(\mathbf{x}_\parallel)$ in terms of $\langle |\psi(\mathbf{x}|\omega)_{tr}|^2 \rangle$. To obtain an expression that can be inverted, we first extend the integration over $\mathbf{x}'_\parallel$ and $\mathbf{x}''_\parallel$ to the entire $x_3 = 0$ plane. This is allowable if the linear dimensions of the aperture A are larger than the width of the intensity distribution of the incident beam. We then pass to the geometrical optics limit of Eq. (6.24) by making the change of variable $\mathbf{x}''_\parallel = \mathbf{x}'_\parallel - \mathbf{u}_\parallel$, expanding the difference $\zeta(\mathbf{x}'_\parallel) - \zeta(\mathbf{x}'_\parallel - \mathbf{u}_\parallel)$ in powers of $\mathbf{u}_\parallel$, and keeping only the linear term. In this way we obtain the following result:

$$\langle |\psi(\mathbf{x}|\omega)_{tr}|^2 \rangle = \left(\frac{\omega}{2\pi c x_3}\right)^2 \int \frac{d^2 k'_\parallel}{(2\pi)^2} F(\mathbf{k}'_\parallel) \int \frac{d^2 k''_\parallel}{(2\pi)^2} F^*(\mathbf{k}''_\parallel)$$

$$\times \int d^2 u_\parallel \exp\left\{i\left[\mathbf{k}''_\parallel - \frac{\omega}{c x_3}\mathbf{x}_\parallel\right] \cdot \mathbf{u}_\parallel - i\frac{\omega}{2 c x_3}u^2_\parallel\right\}$$

$$\times \int d^2 x'_\parallel \exp\left\{i\left[\mathbf{k}'_\parallel - \mathbf{k}''_\parallel + \frac{\omega}{c x_3}\mathbf{u}_\parallel\right] \cdot \mathbf{x}'_\parallel\right\}$$

$$\times \left\langle \exp\left[i\frac{\omega}{c}\Delta n \mathbf{u}_\parallel \cdot \nabla \zeta(\mathbf{x}'_\parallel)\right]\right\rangle. \qquad (6.25)$$

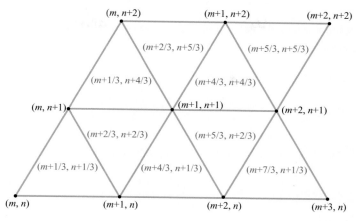

Figure 6.2. A segment of the x_1x_2 plane showing the equilateral triangles above which the triangular facets that generate the two-dimensional rough surface defined by Eq. (6.26) are placed.

This is the equation that we can invert to obtain the surface profile function $\zeta(\mathbf{x}_\parallel)$ that produces a specified form for $\langle |\psi(\mathbf{x}|\omega)_{tr}|^2 \rangle$.

6.2.2 The inverse problem

To evaluate the integral over $\mathbf{x}'_\parallel$ in Eq. (6.25) we begin by covering the x_1x_2 plane by equilateral triangles of edge b (Fig. 6.2). The vertices of these triangles are given by the vectors $\{\mathbf{x}_\parallel(m, n)\}$ that are defined by $\mathbf{x}_\parallel(m, n) = m\mathbf{a}_1 + n\mathbf{a}_2$, where $m, n = 0, \pm1, \pm2\ldots$, and the basis vectors $\mathbf{a}_1$ and $\mathbf{a}_2$ are $\mathbf{a}_1 = (b, 0)$, $\mathbf{a}_2 = (b/2, \sqrt{3}b/2)$ [91]. Each triangle is labeled by the coordinates of its center of gravity. These are given by the mean values of the coordinates of its three vertices. Thus, the triangle defined by the vertices (m, n), $(m + 1, n)$, and $(m, n + 1)$ is the $(m + 1/3, n + 1/3)$ triangle, while the triangle whose vertices are $(m + 1, n)$, $(m + 1, n + 1)$, and $(m, n + 1)$ is the $(m + 2/3, n + 2/3)$ triangle. As m and n take the values $0, \pm1, \pm2, \ldots$, the $(m + 1/3, n + 1/3)$ triangles and the $(m + 2/3, n + 2/3)$ triangles generated cover the x_1x_2 plane.

For $\mathbf{x}_\parallel$ contained in the triangle $(m + 1/3, n + 1/3)$ the surface profile function is represented by

$$\zeta(\mathbf{x}_\parallel) = b^{(0)}_{m+1/3,n+1/3} + a^{(1)}_{m+1/3,n+1/3}x_1 + a^{(2)}_{m+1/3,n+1/3}x_2. \tag{6.26a}$$

For $\mathbf{x}_\parallel$ contained within the triangle $(m + 2/3, n + 2/3)$ the surface profile function is represented by

$$\zeta(\mathbf{x}_\parallel) = b^{(0)}_{m+2/3,n+2/3} + a^{(1)}_{m+2/3,n+2/3}x_1 + a^{(2)}_{m+2/3,n+2/3}x_2. \tag{6.26b}$$

The coefficients $a^{(1,2)}_{m+1/3,n+1/3}$ and $a^{(1,2)}_{m+2/3,n+2/3}$ are assumed to be independent identically distributed random deviates. Therefore, the joint probability density functions (pdf) of the two coefficients associated with a given triangle,

$$f(\gamma_1, \gamma_2) = \langle \delta(\gamma_1 - a^{(1)}_{m+1/3,n+1/3})\delta(\gamma_2 - a^{(2)}_{m+1/3,n+1/3})\rangle$$

$$= \langle \delta(\gamma_1 - a^{(1)}_{m+2/3,n+2/3})\delta(\gamma_2 - a^{(2)}_{m+2/3,n+2/3})\rangle, \qquad (6.27)$$

are independent of the coordinates labeling the triangles.

With the use of the representation of $\zeta(\mathbf{x}_\parallel)$ given by Eqs. (6.26) we find that

$$\int d^2 x'_\parallel \exp\left\{i\left[\mathbf{k}'_\parallel - \mathbf{k}''_\parallel + \frac{\omega}{cx_3}\mathbf{u}_\parallel\right] \cdot \mathbf{x}'_\parallel\right\} \left\langle \exp\left[i\frac{\omega}{c}\Delta n \mathbf{u}_\parallel \cdot \nabla \zeta(\mathbf{x}'_\parallel)\right]\right\rangle$$

$$= \sum_{m=-\infty}^{\infty}\sum_{n=-\infty}^{\infty}\left\{\int_{(m+1/3,n+1/3)} d^2 x'_\parallel \exp(i\mathbf{a}_\parallel \cdot \mathbf{x}'_\parallel)\right.$$

$$\times \left\langle \exp\left[i\frac{\omega}{c}\Delta n \left(u_1 a^{(1)}_{m+1/3,n+1/3} + u_2 a^{(2)}_{m+1/3,n+1/3}\right)\right]\right\rangle$$

$$+ \int_{(m+2/3,n+2/3)} d^2 x'_\parallel \exp(i\mathbf{a}_\parallel \cdot \mathbf{x}'_\parallel)$$

$$\left.\times \left\langle \exp\left[i\frac{\omega}{c}\Delta n \left(u_1 a^{(1)}_{m+2/3,n+2/3} + u_2 a^{(2)}_{m+2/3,n+2/3}\right)\right]\right\rangle\right\}, \qquad (6.28)$$

where, for example, the notation $\int_{(m+1/3,n+1/3)} d^2 x_\parallel$ indicates that the integration is carried out over the area of the triangle $(m + 1/3, n + 1/3)$, and where, to simplify the notation, we have defined

$$\mathbf{a}_\parallel = \mathbf{k}'_\parallel - \mathbf{k}''_\parallel + (\omega/cx_3)\mathbf{u}_\parallel. \qquad (6.29)$$

Using Eqs. (6.27), the averages in Eq. (6.28) can be evaluated, with the result that the integral becomes

$$\int d^2\gamma_\parallel \, f(\gamma_\parallel)\exp\left(i\frac{\omega}{c}\Delta n\gamma_\parallel \cdot \mathbf{u}_\parallel\right)$$

$$\times \sum_{m=-\infty}^{\infty}\sum_{n=-\infty}^{\infty}\left\{\int_{(m+1/3,n+1/3)} d^2 x'_\parallel \exp(i\mathbf{a}_\parallel \cdot \mathbf{x}'_\parallel) + \int_{(m+2/3,n+2/3)} d^2 x'_\parallel \exp(i\mathbf{a}_\parallel \cdot \mathbf{x}'_\parallel)\right\}$$

$$= \int d^2\gamma_\parallel \, f(\gamma_\parallel)\exp\left(i\frac{\omega}{c}\Delta n\gamma_\parallel \cdot \mathbf{u}_\parallel\right)\int d^2 x'_\parallel \exp(i\mathbf{a}_\parallel \cdot \mathbf{x}'_\parallel)$$

$$= (2\pi)^2 \int d^2\gamma_\parallel \, f(\gamma_\parallel)\exp\left(i\frac{\omega}{c}\Delta n\mathbf{u}_\parallel \cdot \gamma_\parallel\right)\delta(\mathbf{k}'_\parallel - \mathbf{k}''_\parallel + (\omega/cx_3)\mathbf{u}_\parallel). \qquad (6.30)$$

When the result given by Eq. (6.30) is substituted into Eq. (6.25), with some changes of variables we find that

$$\langle |\psi(\mathbf{x}|\omega)_{tr}|^2 \rangle = \left(\frac{\omega}{2\pi c x_3} \right)^2 \int d^2 \gamma_\| \, f(\gamma_\|)$$

$$\times \int d^2 u_\| \exp \left(i \frac{\omega}{c} \Delta n \gamma_\| \cdot \mathbf{u}_\| \right) \exp \left[-i \frac{\omega}{c x_3} (\mathbf{x}_\| \cdot \mathbf{u}_\|) \right]$$

$$\times \int \frac{d^2 p_\|}{(2\pi)^2} \exp(i \mathbf{p}_\| \cdot \mathbf{u}_\|) F(\mathbf{p}_\| - (\omega/2 c x_3) \mathbf{u}_\|) F^*(\mathbf{p}_\| + (\omega/2 c x_3) \mathbf{u}_\|). \quad (6.31)$$

On equating the right-hand sides of Eqs. (6.22) and (6.31), we obtain an equation that relates the spectral degree of coherence $g^{(0)}(\mathbf{u}_\|)$ of the phase screen to the joint pdf $f(\gamma_\|)$ of the coefficients $\{a^{(1,2)}_{m+1/3,n+1/3}\}$ and $\{a^{(1,2)}_{m+2/3,n+2/3}\}$ defining the surface profile function $\zeta(\mathbf{x}_\|)$:

$$\int d^2 u_\| \, g^{(0)}(\mathbf{u}_\|) \exp \left[-i \frac{\omega}{c x_3} \mathbf{x}_\| \cdot \mathbf{u}_\| \right]$$

$$\times \int \frac{d^2 p_\|}{(2\pi)^2} \exp(i \mathbf{p}_\| \cdot \mathbf{u}_\|) \hat{S}(\mathbf{p}_\| - (\omega/2 c x_3) \mathbf{u}_\|) \hat{S}^*(\mathbf{p}_\| + (\omega/2 c x_3) \mathbf{u}_\|)$$

$$= \int d^2 \gamma_\| \, f(\gamma_\|) \int d^2 u_\| \exp \left(i \frac{\omega}{c} \Delta n \gamma_\| \cdot \mathbf{u}_\| \right) \exp \left[-i \frac{\omega}{c x_3} \mathbf{x}_\| \cdot \mathbf{u}_\| \right]$$

$$\times \int \frac{d^2 p_\|}{(2\pi)^2} \exp(i \mathbf{p}_\| \cdot \mathbf{u}_\|) F(\mathbf{p}_\| - (\omega/2 c x_3) \mathbf{u}_\|) F^*(\mathbf{p}_\| + (\omega/2 c x_3) \mathbf{u}_\|). \quad (6.32)$$

We see immediately that if

$$\hat{S}(\mathbf{k}_\|) = A F(\mathbf{k}_\|), \quad (6.33)$$

where A is a constant, Eq. (6.32) is satisfied if

$$|A|^2 \int d^2 u_\| \, g^{(0)}(\mathbf{u}_\|) \exp \left[-i \frac{\omega}{c x_3} \mathbf{x}_\| \cdot \mathbf{u}_\| \right]$$

$$= \int d^2 \gamma_\| \, f(\gamma_\|) \int d^2 u_\| \exp \left(i \frac{\omega}{c} \Delta n \gamma_\| \cdot \mathbf{u}_\| \right) \exp \left[-i \frac{\omega}{c x_3} \mathbf{x}_\| \cdot \mathbf{u}_\| \right]. \quad (6.34)$$

On equating the $\mathbf{x}_\|$ Fourier coefficients on both sides of this equation, we find that

$$|A|^2 g^{(0)}(\mathbf{u}_\|) = \int d^2 \gamma_\| \, f(\gamma_\|) \exp \left(i \frac{\omega}{c} \Delta n \gamma_\| \cdot \mathbf{u}_\| \right), \quad (6.35)$$

which can be inverted to yield

$$f(\pmb{\gamma}_\parallel) = |A|^2 \int \frac{d^2 u_\parallel}{(2\pi)^2} \, g^{(0)} \left(\frac{c}{\omega \Delta n} \mathbf{u}_\parallel \right) \exp(-i\pmb{\gamma}_\parallel \cdot \mathbf{u}_\parallel). \tag{6.36}$$

From the normalization of $f(\pmb{\gamma}_\parallel)$,

$$\int d^2\gamma_\parallel \, f(\pmb{\gamma}_\parallel) = |A|^2 \int \frac{d^2 u_\parallel}{(2\pi)^2} \, g^{(0)} \left(\frac{c}{\omega \Delta n} \mathbf{u}_\parallel \right) (2\pi)^2 \delta(\mathbf{u}_\parallel)$$

$$= |A|^2 g^{(0)}(\mathbf{0}) = |A|^2$$

$$= 1, \tag{6.37}$$

where we have used Eq. (6.17), we find that

$$|A|^2 = 1. \tag{6.38}$$

With no loss of generality, because $\hat{S}(\mathbf{p}_\parallel)$ always appears multiplied by a complex conjugate $\hat{S}^*(\mathbf{p}'_\parallel)$ in a calculation of a physical property, we can assume that

$$A = 1. \tag{6.39}$$

Thus we have obtained the result that in order for the beam transmitted through the random phase screen to be described by the cross-spectral density, Eq. (6.14), the amplitudes $\{a^{(1,2)}_{m+1/3,n+1/3}\}$ and $\{a^{(1,2)}_{m+2/3,n+2/3}\}$ defining the phase screen through Eqs. (6.26) have to be drawn from the joint pdf given in terms of the spectral degree of coherence by

$$f(\pmb{\gamma}_\parallel) = \int \frac{d^2 u_\parallel}{(2\pi)^2} \, g^{(0)} \left(\frac{c}{\omega \Delta n} \mathbf{u}_\parallel \right) \exp(-i\pmb{\gamma}_\parallel \cdot \mathbf{u}_\parallel), \tag{6.40}$$

and the phase screen must be illuminated from the region $x_3 < 0$ by a beam defined in terms of the spectral density by

$$\psi(\mathbf{x}|\omega)_{inc} = \int \frac{d^2 k_\parallel}{(2\pi)^2} \, \hat{S}(\mathbf{k}_\parallel) \exp[i\mathbf{k}_\parallel \cdot \mathbf{x}_\parallel + i\alpha_0(k_\parallel, \omega)x_3]. \tag{6.41}$$

For the generation of the surface profile function $\zeta(\mathbf{x}_\parallel)$ we also need the marginal pdf given by

$$f(\gamma_1) = \int f(\gamma_1, \gamma_2)d\gamma_2, \tag{6.42}$$

and the conditional pdf of $a^{(2)}_{m+1/3,n+1/3}$ $(a^{(2)}_{m+2/3,n+2/3})$ given $a^{(1)}_{m+1/3,n+1/3}$ $(a^{(1)}_{m+2/3,n+2/3})$,

$$f(\gamma_2|\gamma_1) = f(\gamma_1, \gamma_2)/f(\gamma_1). \tag{6.43}$$

In addition, if we denote the value of $\zeta(\mathbf{x}_\|)$ at the vertex (m, n) by $h_{m,n}$, we find that $b^{(0)}_{m+1/3,n+1/3} = (m + n + 1)h_{m,n} - mh_{m+1,n} - nh_{m,n+1}$, while $b^{(0)}_{m+2/3,n+2/3} = (n + 1)h_{m+1,n} - (m + n + 1)h_{m+1,n+1} + (m + 1)h_{m,n+1}$.

The preceding results, together with the rejection method [92] for generating random deviates from a given pdf, enable us to generate an ensemble of N_p realizations of the two-dimensional random surface profile function $\zeta(\mathbf{x}_\|)$ that are continuous functions of $\mathbf{x}_\|$, and whose statistical properties are defined by the joint pdf $f(\boldsymbol{\gamma}_\|)$ derived from the cross-spectral density, Eq. (6.14), we wish the random phase screen to produce. The precise manner in which this generation is carried out is described in detail elsewhere [91]. Here we focus on its use in beam shaping.

6.2.3 Beam shaping

We now return to Eq. (6.22) and replace $g^{(0)}(\mathbf{u}_\|)$ through the use of Eq. (6.35) (with $|A|^2 = 1$), and find that the mean intensity of the transmitted beam becomes

$$
\langle |\psi(\mathbf{x}|\omega)_{tr}|^2 \rangle = \left(\frac{\omega}{2\pi c x_3}\right)^2 \int d^2 u_\| \int d^2 \gamma_\| \, f(\boldsymbol{\gamma}_\|)
$$

$$
\times \exp\left\{ i \left[\frac{\omega}{c} \Delta n \boldsymbol{\gamma}_\| - \frac{\omega}{c x_3} \mathbf{x}_\| \right] \cdot \mathbf{u}_\| \right\} \int \frac{d^2 p_\|}{(2\pi)^2} \exp(i\mathbf{p}_\| \cdot \mathbf{u}_\|)
$$

$$
\times \hat{S}\left(\mathbf{p}_\| - \frac{\omega}{2c x_3} \mathbf{u}_\|\right) \hat{S}^*\left(\mathbf{p}_\| + \frac{\omega}{2c x_3} \mathbf{u}_\|\right). \tag{6.44}
$$

In the far field this expression takes the form

$$
\langle |\psi(\mathbf{x}|\omega)_{tr}|^2 \rangle = \left(\frac{\omega}{2\pi c x_3}\right)^2 \int d^2 \gamma_\| \, f(\boldsymbol{\gamma}_\|) \left| \hat{S}\left(\frac{\omega}{c} \Delta n \boldsymbol{\gamma}_\| - \frac{\omega}{c x_3} \mathbf{x}_\|\right) \right|^2
$$

$$
= \frac{1}{(2\pi \Delta n x_3)^2} \int d^2 p_\| \, f\left(\frac{c}{\omega \Delta n} \mathbf{p}_\| + \frac{1}{\Delta n x_3} \mathbf{x}_\|\right) |\hat{S}(\mathbf{p}_\|)|^2. \tag{6.45}
$$

If $S^{(0)}(\mathbf{x}_\|)$ is a slowly varying function of $\mathbf{x}_\|$, $\hat{S}(\mathbf{p}_\|)$ is sharply peaked at $\mathbf{p}_\| = 0$. In this case we can make the approximation

$$
|\hat{S}(\mathbf{p}_\|)|^2 = B\delta(\mathbf{p}_\|), \tag{6.46}
$$

where

$$
B = \int d^2 p_\| \, |\hat{S}(\mathbf{p}_\|)|^2 = (2\pi)^2 \int d^2 x_\| \, S^{(0)}(\mathbf{x}_\|). \tag{6.47}
$$

It follows, then, that in the far field

$$\langle |\psi(\mathbf{x}|\omega)_{tr}|^2 \rangle \cong \frac{B}{(2\pi \, \Delta n x_3)^2} \, f\left(\frac{1}{\Delta n x_3} \mathbf{x}_{\|}\right). \tag{6.48}$$

The intensity profile of the mean intensity of the transmitted beam at a distance x_3 from the phase screen is thus a scaled version of the joint pdf $f(\boldsymbol{\gamma}_{\|})$. This result shows how to design a random phase screen, i.e. a random surface in view of Eq. (6.3), which transforms an incident beam, Eq. (6.41), with an intensity profile $S^{(0)}(\mathbf{x}_{\|})$ into a transmitted beam with a specified intensity profile $\langle |\psi(\mathbf{x}|\omega)_{tr}|^2 \rangle$ at some fixed value of x_3 in the far field: one simply introduces a scaled version of $\langle |\psi(\mathbf{x}|\omega)_{tr}|^2 \rangle$,

$$f(\gamma_1, \gamma_2) = \frac{(2\pi \, \Delta n x_3)^2}{B} \left\langle |\psi \, (\Delta n x_3 \gamma_1, \, \Delta n x_3 \gamma_2, \, x_3|\omega)_{tr}|^2 \right\rangle, \tag{6.49}$$

which is normalized to unity when integrated over $\boldsymbol{\gamma}_{\|}$, and uses it to generate the surface profile function $\zeta(\mathbf{x}_{\|})$ in the manner described in ref. [91]. We note that the intensity profile of the incident beam is unimportant here provided that the function $\hat{S}(\mathbf{k}_{\|})$ that defines it through Eq. (6.41) is sufficiently sharply peaked at $\mathbf{k}_{\|} = 0$ for Eq. (6.46) to be a good approximation. This intensity profile then enters Eq. (6.48) only through the amplitude B.

6.2.4 Example

To illustrate the result in Eq. (6.49), we consider the transformation of a laser beam whose intensity profile in the plane $x_3 = 0$ has the Gaussian form

$$S^{(0)}(\mathbf{x}_{\|}) = \exp(-x_{\|}^2/2\sigma_s^2) \tag{6.50}$$

into a flat-top beam of circular cross section in the far field. From Eq. (6.49) we therefore find that $f(\boldsymbol{\gamma}_{\|})$ must have the form

$$f(\boldsymbol{\gamma}_{\|}) = \frac{1}{\pi b^2} \theta(b - \gamma_{\|}), \tag{6.51}$$

where $\theta(z)$ is the Heaviside unit step function. It follows from Eq. (6.35) that the spectral degree of coherence $g^{(0)}(\mathbf{u}_{\|})$ is given by

$$g^{(0)}(\mathbf{u}_{\|}) = 2\frac{J_1(\Delta n(\omega/c)bu_{\|})}{\Delta n(\omega/c)bu_{\|}}, \tag{6.52}$$

where $J_1(z)$ is the Bessel function of the first kind and first order. If we wish $g^{(0)}(\mathbf{u}_{\|})$ to have significant values only for $u_{\|}$ smaller than a characteristic length σ_g, we can set the parameter b equal to $3.83c/\Delta n\omega\sigma_g$ (the first zero of $J_1(z)/z$ is $z = 3.83$),

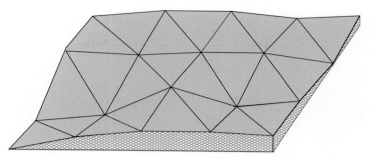

Figure 6.3. Segment of one numerically generated realization of the surface profile function $\zeta(\mathbf{x}_\parallel)$ calculated on the basis of the joint pdf $f(\boldsymbol{\gamma}_\parallel)$ given by Eq. (6.53). The values of the parameters used in obtaining this result were $\omega = 2\pi c/\lambda$, where $\lambda = 1.55$ μm, $\Delta n = 0.6$, $b = 20$ μm, and $\sigma_g = 31$ μm.

so that

$$f(\boldsymbol{\gamma}_\parallel) = \frac{\sigma_g^2}{(3.83)^2 \pi}(\Delta n(\omega/c))^2 \theta\left(\frac{3.83}{\Delta n}\frac{c}{\omega\sigma_g} - \gamma_\parallel\right) \tag{6.53}$$

and

$$g^{(0)}(\mathbf{u}_\parallel) = \frac{2\sigma_g}{3.83 u_\parallel} J_1\left(\frac{3.83 u_\parallel}{\sigma_g}\right). \tag{6.54}$$

In Fig. 6.3 we plot a segment of a single realization of the surface profile function $\zeta(\mathbf{x}_\parallel)$ for which the joint pdf $f(\boldsymbol{\gamma}_\parallel)$ is given by Eq. (6.53). The values of the parameters used in obtaining this result are $\omega = 2\pi c/\lambda$, where $\lambda = 1.55$ μm, $b = 20$ μm, $\sigma_g = 31$ μm, and $\Delta n = 0.6$.

In Fig. 6.4(a) we plot the intensity profile of the incident field in the plane $x_3 = 0-$. It is a Gaussian, $\exp(-x_\parallel^2/w^2)$, with a $1/e$ half width $w = 2\sigma_s$. In Fig. 6.4(b) we plot the intensity profile of the field in the far zone transmitted through a random phase screen that has been designed to produce a cross-spectral density in the plane $x_3 = 0+$ defined by Eq. (6.14) and Eqs. (6.50) and (6.54). It is seen that the incident Gaussian beam is transformed into a flat-top beam with a circular cross-section. The values of the parameters assumed in obtaining these results are $\sigma_s = 155$ μm, $\sigma_g = 31$ μm, $\omega = 2\pi c/\lambda$ with $\lambda = 1.55$ μm, $b = 20$ μm, $N_p = 20\,000$, and $x_3 = 50$ cm. The index of refraction of the phase screen was assumed to be $n = 1.6$.

6.2.5 Fabrication of surfaces formed from triangular facets

The design procedure presented in the preceding discussion would be of limited interest if the designed surfaces could not be fabricated. It is therefore important to discuss at least one possible fabrication scheme.

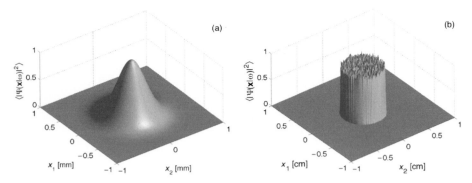

Figure 6.4. (a) Intensity profile of the incident beam in the plane $x_3 = 0-$. It is a Gaussian, $\exp(-x_\parallel^2/w^2)$, with $1/e$ half width $w = 2\sigma_s$. (b) Intensity profile of the field in the far zone transmitted through a random phase screen that has been designed to produce a cross-spectral density in the plane $x_3 = 0+$, defined by Eqs. (6.14), (6.50), and (6.54). The values of the parameters assumed in obtaining these results were $\omega = 2\pi c/\lambda$ with $\lambda = 1.55\,\mu\text{m}$, $\Delta n = 0.6$, $\sigma_s = 155\,\mu\text{m}$, $\sigma_g = 31\,\mu\text{m}$, $b = 20\,\mu\text{m}$, $N_p = 20\,000$, and $x_3 = 50\,\text{cm}$.

A few years ago, the fabrication of surfaces according to the designs presented here would have been a major undertaking. Fabrication by optical methods, for instance, would involve an elaborate setup with computer controlled scanning and beam intensity modulation. One would also have to deal with issues like the linearity of the recording medium (e.g. photoresist).

Fortunately, due to the availability of three-dimensional (3D) printers, things are much simpler now. The fabrication of 3D structures from their computer models is a problem that has relevance in fields as diverse as paleontology, archaeology, architecture, and medicine. Three-dimensional printing techniques are of central importance to the emerging field of rapid prototyping and manufacturing.

The most common additive fabrication techniques synthesize the 3D object by breaking it into slices and printing layer by layer onto some suitable material. Subtractive techniques, on the other hand, start from a solid block from which material is removed until the desired shape is reached. Typical examples of additive fabrication techniques are selective sintering, fused deposition modeling, and selective photocuring.

In selective laser sintering, one starts with a thin layer of powder of some suitable material. A focussed laser beam is used to scan and selectively sinter the medium to form a solid layer with the cross-section of the desired object. Another layer of powder is added and the procedure is repeated. One possible variation consists of using an inkjet printing system instead of a laser. In this case, the layers of powder are selectively bonded by depositing an adhesive with the inkjet printhead.

In fused deposition modeling, a nozzle is used to deposit a molten polymer onto a support structure, layer by layer. An alternative is to deposit a curable photopolymer and use an ultraviolet (UV) flood lamp mounted in the print head to cure each layer as it is deposited.

Selective photocuring includes linear and nonlinear stereolithography. In stereolithography one builds a layer at a time using a liquid UV-curable photopolymer and a scanning UV laser. For each layer, the laser beam traces the desired cross-section on the surface of the liquid resin. Exposure to the UV laser light solidifies the pattern traced on the resin, which adheres to the solid layer below. After a pattern has been traced, the platform descends by a single layer thickness and the exposed material is re-coated with fresh material. On this new liquid surface, the subsequent layer pattern is traced; the layer solidifies and adheres to the previous layer. A complete 3D part is formed by this process.

In general, the resolution is determined by the thickness of the deposited layers and by the lateral resolution of the printing technology. Resolutions of 100 μm or better in the three directions are not uncommon with stereolithography. Ultra-small features may be made using an intense infrared laser beam (instead of a UV one) which photopolymerizes the material by two-photon absorption. Feature sizes of about 100 nm can be produced in this way. For the fabrication of samples based on the designs presented here, normal stereolithography is viable and adequate.

It is also important to consider the format for data transfer to the writing device. There are two widely used formats: *NC code* (Numerical Control code) for subtractive fabricators and *StL* (StereoLithography) [93, 94] for additive ones. Fortunately, the specification of our surfaces is quite natural in this scripting language.

To specify an object in StL format, its surface is tessellated into a series of small triangular facets. Each triangular facet is specified by its outward pointing unit normal and the coordinates of the three vertices in a 3D Cartesian coordinate system. The printer uses these data and a slicing algorithm to determine the cross-section of the 3D object that needs to be printed on each layer. The vertices must be ordered counterclockwise (right-hand rule) when the object is viewed from the outside (see Fig. 6.5). The orientation of the facets is specified redundantly in these two ways, which must be consistent. The syntax for specifying a triangle in an ASCII StL file is shown in Table 6.1.

To illustrate the generation and specification of a surface in this format, let us begin by considering the equilateral triangles of edge b that cover the $x_1 x_2$ plane

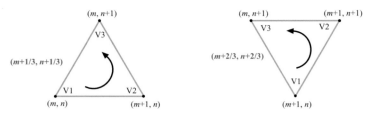

Figure 6.5. Triangles $(m + 1/3, n + 1/3)$ and $(m + 2/3, n + 2/3)$. Note that the vertices are listed in a counterclockwise manner.

Table 6.1. *Syntax for describing a triangular facet in an ASCII StL file*

The numbers n1, n2, and n3 represent the three components of the outward pointing unit vector, and the V numbers represent the coordinates of the three vertices of the facet

---

facet normal n1 n2 n3
  outer loop
    V11 V12 V13
    V21 V22 V23
    V31 V32 V33
  end loop
endfacet

(see Fig. 6.2). These triangles represent the projection of the triangular facets of the surface on the $x_1 x_2$ plane.

For the triangle $(m + 1/3, n + 1/3)$, the information required by the StL scripting language is given by the position of the vertices on the plane $x_1 x_2$, and the constants $b^{(0)}_{m+1/3,n+1/3}$, $a^{(1)}_{m+1/3,n+1/3}$, and $a^{(2)}_{m+1/3,n+1/3}$. To simplify the notation we define $\mu = (m + 1/3)$ and $\nu = (n + 1/3)$. The coordinates of the vertices in three-dimensional space are given by

$$V^{(11)}_{\mu,\nu} = \left(m + \frac{n}{2}\right)b, \qquad V^{(12)}_{\mu,\nu} = \left(\sqrt{3}\,\frac{n}{2}\right)b, \qquad V^{(13)}_{\mu,\nu} = h_{m,n}, \qquad (6.55a)$$

$$V^{(21)}_{\mu,\nu} = \left(m + 1 + \frac{n}{2}\right)b, \quad V^{(22)}_{\mu,\nu} = \left(\sqrt{3}\,\frac{n}{2}\right)b, \qquad V^{(23)}_{\mu,\nu} = h_{m+1,n}, \quad (6.55b)$$

$$V^{(31)}_{\mu,\nu} = \left(m + \frac{n+1}{2}\right)b, \quad V^{(32)}_{\mu,\nu} = \left(\sqrt{3}\,\frac{n+1}{2}\right)b, \quad V^{(33)}_{\mu,\nu} = h_{m,n+1}, \quad (6.55c)$$

where $h_{m,n}$, $h_{m,(n+1)}$, and $h_{(m+1),n}$ represent the heights of the three vertices of the triangle that, from Eqs. (6.26a), are found to be

$$V_{\mu,\nu}^{(13)} = b_{\mu,\nu}^{(0)} + a_{\mu,\nu}^{(1)} V_{\mu,\nu}^{(11)} + a_{\mu,\nu}^{(2)} V_{\mu,\nu}^{(12)}, \tag{6.56a}$$

$$V_{\mu,\nu}^{(23)} = b_{\mu,\nu}^{(0)} + a_{\mu,\nu}^{(1)} V_{\mu,\nu}^{(21)} + a_{\mu,\nu}^{(2)} V_{\mu,\nu}^{(22)}, \tag{6.56b}$$

$$V_{\mu,\nu}^{(33)} = b_{\mu,\nu}^{(0)} + a_{\mu,\nu}^{(1)} V_{\mu,\nu}^{(31)} + a_{\mu,\nu}^{(2)} V_{\mu,\nu}^{(32)}. \tag{6.56c}$$

To complete the set of equations, we note that the parameter $b_{\mu,\nu}^{(0)}$ satisfies the relation

$$b_{\mu,\nu}^{(0)} = (m + n + 1)V_{\mu,\nu}^{(13)} - mV_{\mu,\nu}^{(23)} - nV_{\mu,\nu}^{(33)}. \tag{6.57}$$

The components of the unit normal to the facet are given by

$$n_{\mu,\nu}^{(1)} = -\frac{a_{\mu,\nu}^{(1)}}{\phi_{\mu,\nu}}, \qquad n_{\mu,\nu}^{(2)} = -\frac{a_{\mu,\nu}^{(2)}}{\phi_{\mu,\nu}}, \qquad n_{\mu,\nu}^{(3)} = \frac{1}{\phi_{\mu,\nu}}, \tag{6.58}$$

where

$$\phi_{\mu,\nu} = \left[1 + \left(a_{\mu,\nu}^{(1)}\right)^2 + \left(a_{\mu,\nu}^{(2)}\right)^2\right]^{1/2}. \tag{6.59}$$

On the other hand, for the triangle $(m + 2/3, n + 2/3)$, we write $\mu = (m + 2/3)$ and $\nu = (n + 2/3)$. The coordinates of the vertices are given by

$$V_{\mu,\nu}^{(11)} = \left(m + 1 + \frac{n}{2}\right)b, \qquad\qquad V_{\mu,\nu}^{(12)} = \left(\sqrt{3}\frac{n}{2}\right)b,$$

$$V_{\mu,\nu}^{(13)} = h_{m+1,n}, \tag{6.60a}$$

$$V_{\mu,\nu}^{(21)} = \left(m + 1 + \frac{n+1}{2}\right)b, \qquad V_{\mu,\nu}^{(22)} = \left(\sqrt{3}\frac{n+1}{2}\right)b,$$

$$V_{\mu,\nu}^{(23)} = h_{m+1,n+1}, \tag{6.60b}$$

$$V_{\mu,\nu}^{(31)} = \left(m + \frac{n+1}{2}\right)b, \qquad\qquad V_{\mu,\nu}^{(32)} = \left(\sqrt{3}\frac{n+1}{2}\right)b,$$

$$V_{\mu,\nu}^{(33)} = h_{m,n+1}, \tag{6.60c}$$

where the heights and the components of the unit normal can be written as in Eqs. (6.56) and (6.58), but with $\mu = (m + 2/3)$ and $\nu = (n + 2/3)$. In this case, the parameter $b_{\mu,\nu}^{(0)}$ satisfies the relation

$$b_{\mu,\nu}^{(0)} = (n + 1)V_{\mu,\nu}^{(13)} - (m + n + 1)V_{\mu,\nu}^{(23)} + (m + 1)V_{\mu,\nu}^{(33)}. \tag{6.61}$$

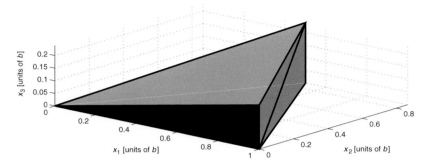

Figure 6.6. Example of a solid structure supporting a triangular facet.

As an example, we consider the triangle $(1/3, 1/3)$. That is, we take $m = n = 0$ with $V^{(13)}_{1/3,1/3} = 0$, and consider the fabrication of a solid structure that supports a facet with a sloping angle of $10°$ in both the x_1 and x_2 directions. The solid model of this structure is shown in Fig. 6.6. The ASCII StL code that generates this structure is presented in the Appendix to this chapter.

6.2.6 Replacement of ensemble averaging by frequency averaging

In concluding this section we note that in obtaining the mean intensity of the transmitted field we have assumed a monochromatic incident beam, and have performed averages over the ensemble of realizations of the surface profile function $\zeta(\mathbf{x}_\parallel)$, as in Eqs. (6.24) and (6.25). Without this averaging the intensity of the field transmitted through a single realization of the random phase screen would consist of an array of bright and dark spots – a speckle pattern – as a function of the transverse coordinates on the plane of observation, rather than the smooth function of these coordinates produced by the ensemble averaging, and displayed in Fig. 6.4(b). However, under normal experimental conditions, carrying out an ensemble average is not a practical consideration. An experimentalist has to work with a single realization of the phase screen. To obtain the kind of smooth curve produced by ensemble averaging the experimentalist has to average over the resulting speckles in some way. One way of doing this is to move the random surface, for example to rotate it or dither it. However, in some applications moving the surface may not be an option.

A speckle pattern depends on the wavelength of a monochromatic incident beam: change the wavelength, and the positions of the bright and dark spots on the plane of observation change. This means that it is possible to average over speckles by using a polychromatic, broadband, beam instead of a monochromatic beam to illuminate the phase screen. In what follows we show that this is the case.

We begin by generalizing the expression for the incident beam given by Eq. (6.41) to the case where it is time dependent:

$$\psi(\mathbf{x}; t)_{inc} = \int_{-\infty}^{\infty} \frac{d\omega}{2\pi} \, G(\omega) \int \frac{d^2k_{\parallel}}{(2\pi)^2} \, \hat{S}(\mathbf{k}_{\parallel})$$

$$\times \exp[i\mathbf{k}_{\parallel} \cdot \mathbf{x}_{\parallel} + i\alpha_0(k_{\parallel}, \omega)x_3 - i\omega t], \tag{6.62}$$

where $\alpha_0(k_{\parallel}, \omega)$ is defined in Eq. (6.6). The function $G(\omega)$ is a random function with the properties

$$\langle G(\omega)G^*(\omega')\rangle_F = 2\pi \delta(\omega - \omega')S_0(\omega), \tag{6.63a}$$

$$\langle G(\omega)G(\omega')\rangle_F = 0, \tag{6.63b}$$

where the angle brackets $\langle \ldots \rangle_F$ denote an average over the ensemble of realizations of the incident field. The fluctuation of $|G(\omega)|^2$ as a function of ω, which depends on the integration time of the detector, is below the resolution of a normal spectrograph, so, for all practical purposes, $G(\omega)$ can be considered to be delta correlated [95].

For specificity we will assume that the spectral density of the incident light $S_0(\omega)$ has a Gaussian form centered at a frequency ω_0 with $1/e$ half width $\Delta\omega$,

$$S_0(\omega) = \frac{1}{2\sqrt{\pi}\Delta\omega} \exp[-(\omega - \omega_0)^2/(\Delta\omega)^2]. \tag{6.64}$$

The half width $\Delta\omega$ is assumed to be small enough that $S_0(\omega)$ can be regarded as zero when $\omega < 0$. Broadband beams of this nature are produced, for example, by superluminescent diodes [96].

Due to the linearity of the transmission problem, the transmitted field in the region $x_3 > 0+$ can be written in the form

$$\psi(\mathbf{x}; t)_{tr} = \int_{-\infty}^{\infty} \frac{d\omega}{2\pi} \, G(\omega) \int \frac{d^2q_{\parallel}}{(2\pi)^2} \, T(\mathbf{q}_{\parallel})$$

$$\times \exp[i\mathbf{q}_{\parallel} \cdot \mathbf{x}_{\parallel} + i\alpha_0(q_{\parallel}, \omega)x_3 - i\omega t], \tag{6.65}$$

where

$$T(\mathbf{q}_{\parallel}) = \int \frac{d^2k_{\parallel}}{(2\pi)^2} \, T(\mathbf{q}_{\parallel}|\mathbf{k}_{\parallel})\hat{S}(\mathbf{k}_{\parallel}) \tag{6.66}$$

and

$$T(\mathbf{q}_{\parallel}|\mathbf{k}_{\parallel}) = \int d^2x_{\parallel} \exp\left[-i(\mathbf{q}_{\parallel} - \mathbf{k}_{\parallel}) \cdot \mathbf{x}_{\parallel} + i\frac{\omega}{c}\Delta n \zeta(\mathbf{x}_{\parallel})\right]. \tag{6.67}$$

In obtaining Eq. (6.65) we have combined Eqs. (6.7), (6.9), and (6.23), but we have not used the parabolic approximation, Eq. (6.11).

The intensity of the transmitted beam, averaged over the ensemble of realizations of the incident field, then becomes

$$\langle |\psi(\mathbf{x};t)_{tr}|^2 \rangle_F = \int_{-\infty}^{\infty} \frac{d\omega}{2\pi} \, S_0(\omega)|\psi(\mathbf{x}|\omega)|^2, \tag{6.68}$$

where

$$\psi(\mathbf{x}|\omega) = \int \frac{d^2q_{\parallel}}{(2\pi)^2} \, T(\mathbf{q}_{\parallel}) \exp[i\mathbf{q}_{\parallel} \cdot \mathbf{x}_{\parallel} + i\alpha_0(q_{\parallel}, \omega)x_3]. \tag{6.69}$$

The use of the representation for $\zeta(\mathbf{x}_{\parallel})$ given by Eqs. (6.26), and the results

$$\int_{(m+\frac{1}{3},n+\frac{1}{3})} d^2x_{\parallel} = \int_{\frac{\sqrt{3}}{2}nb}^{\frac{\sqrt{3}}{2}(n+1)b} dx_2 \int_{mb+\frac{x_2}{\sqrt{3}}}^{(m+n+1)b-\frac{x_2}{\sqrt{3}}} dx_1, \tag{6.70a}$$

$$\int_{(m+\frac{2}{3},n+\frac{2}{3})} d^2x_{\parallel} = \int_{\frac{\sqrt{3}}{2}nb}^{\frac{\sqrt{3}}{2}(n+1)b} dx_2 \int_{(m+n+1)b-\frac{x_2}{\sqrt{3}}}^{(m+1)b+\frac{x_2}{\sqrt{3}}} dx_1, \tag{6.70b}$$

enable the integral in Eq. (6.67) to be calculated straightforwardly, with the result that

$$T(\mathbf{q}_{\parallel}|\mathbf{k}_{\parallel}) = t_1(\mathbf{q}_{\parallel}|\mathbf{k}_{\parallel}) + t_2(\mathbf{q}_{\parallel}|\mathbf{k}_{\parallel}), \tag{6.71}$$

where

$t_1(\mathbf{q}_{\parallel}|\mathbf{k}_{\parallel})$

$$= i\frac{\sqrt{3}}{2}b \sum_{m=-N}^{N-1} \sum_{n=-N}^{N-1} \frac{\exp(-iab_{m+1/3,n+1/3}^{(0)})}{q_1 - k_1 + aa_{m+1/3,n+1/3}^{(1)}}$$

$$\times \exp\left[-i\left(q_1 - k_1 + aa_{m+1/3,n+1/3}^{(1)}\right)\left(m + \frac{1}{2}n + \frac{1}{2}\right)b\right.$$

$$\left. - i\frac{\sqrt{3}}{2}\left(q_2 - k_2 + aa_{m+1/3,n+1/3}^{(2)}\right)\left(n + \frac{1}{2}\right)b\right]$$

$$\times \left\{\exp\left[-i\left(q_1 - k_1 + aa_{m+1/3,n+1/3}^{(1)}\right)\frac{b}{4}\right]\right.$$

$$\times \text{sinc}\left[\frac{\sqrt{3}b}{4}\left(q_2 - k_2 + aa^{(2)}_{m+1/3,n+1/3}\right) - \frac{b}{4}\left(q_1 - k_1 + aa^{(1)}_{m+1/3,n+1/3}\right)\right]$$

$$- \exp\left[i\left(q_1 - k_1 + aa^{(1)}_{m+1/3,n+1/3}\right)\frac{b}{4}\right]$$

$$\times \text{sinc}\left[\frac{\sqrt{3}b}{4}\left(q_2 - k_2 + aa^{(2)}_{m+1/3,n+1/3}\right) + \frac{b}{4}\left(q_1 - k_1 + aa^{(1)}_{m+1/3,n+1/3}\right)\right]\Biggr\}$$

$$(6.72a)$$

and

$$t_2(\mathbf{q}_\parallel | \mathbf{k}_\parallel)$$

$$= i\frac{\sqrt{3}}{2}\sum_{n=-N}^{N-1}\sum_{n=-N}^{N-1}\frac{\exp(-iab^{(0)}_{m+2/3,n+2/3})}{q_1 - k_1 + aa^{(1)}_{m+2/3,n+2/3}}$$

$$\times \exp\left[-i\left(q_1 - k_1 + aa^{(1)}_{m+2/3,n+2/3}\right)\left(m + \frac{1}{2}n + 1\right)b\right.$$

$$\left. - i\frac{\sqrt{3}}{2}\left(q_2 - k_2 + aa^{(2)}_{m+2/3,n+2/3}\right)\left(n + \frac{1}{2}\right)b\right]$$

$$\times \left\{\exp\left[-i\left(q_1 - k_1 + aa^{(1)}_{m+2/3,n+2/3}\right)\frac{b}{4}\right]\right.$$

$$\times \text{sinc}\left[\frac{\sqrt{3}b}{4}\left(q_2 - k_2 + aa^{(2)}_{m+2/3,n+2/3}\right) + \frac{b}{4}\left(q_1 - k_1 + aa^{(1)}_{m+2/3,n+2/3}\right)\right]$$

$$- \exp\left[i\left(q_1 - k_1 + aa^{(1)}_{m+2/3,n+2/3}\right)\frac{b}{4}\right]$$

$$\times \text{sinc}\left[\frac{\sqrt{3}b}{4}\left(q_2 - k_2 + aa^{(2)}_{m+2/3,n+2/3}\right) - \frac{b}{4}\left(q_1 - k_1 + aa^{(1)}_{m+2/3,n+2/3}\right)\right]\Biggr\},$$

$$(6.72b)$$

where $\text{sinc}\, x = \sin x / x$ and $a = \Delta n(\omega/c)$.

To illustrate the preceding results, we again assume for $S^{(0)}(\mathbf{x}_\parallel)$ the Gaussian form (6.50) with $\sigma_s = 155\,\mu$m, and generate a single realization of $\zeta(\mathbf{x}_\parallel)$ by the use of the joint pdf $f(\boldsymbol{\gamma}_\parallel)$ given by Eq. (6.53). For the central frequency of the incident beam we choose $\omega_0 = 2\pi c/\lambda_0$, with $\lambda_0 = 1.55\,\mu$m, and its $1/e$ half width is $\Delta\omega = 0.1\omega_0$. The index of refraction of the phase screen is assumed to be $n = 1.6$, so that $\Delta n = n - 1 = 0.6$. In generating a realization of $\zeta(\mathbf{x}_\parallel)$, we assume that $b = 20\,\mu$m. Finally, x_3 is taken to be $x_3 = 50$ cm.

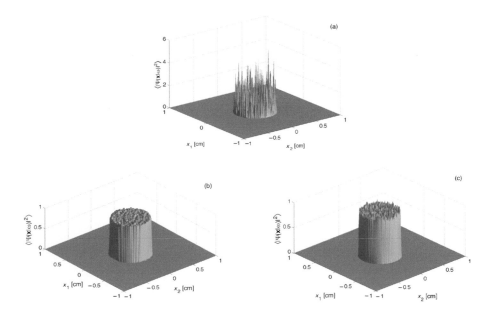

Figure 6.7. (a) The intensity distribution in the far zone when a single realization of the surface profile function, calculated on the basis of the joint pdf $f(\boldsymbol{\gamma}_\parallel)$ given by Eq. (6.53), is illuminated at normal incidence by a monochromatic Gaussian beam of frequency ω_0. (b) The intensity profile given by Eq. (6.68) when the same realization of the surface profile is illuminated at normal incidence by a broadband Gaussian beam, Eq. (6.62), centered at frequency ω_0 with $\Delta\omega = 0.1\omega_0$. (c) The intensity profile obtained by averaging the intensities of the field produced by scattering a monochromatic Gaussian beam from an ensemble of N_p realizations of the surface profile function, all drawn from the same joint pdf $f(\boldsymbol{\gamma}_\parallel)$. In all these calculations $f(\boldsymbol{\gamma}_\parallel)$ is given by Eq. (6.53), with $\omega_0 = 2\pi c/\lambda_0$ ($\lambda_0 = 1.55\,\mu\text{m}$), $\Delta\omega = 0.1\omega_0$, $\Delta n = 0.6$, $\sigma_s = 155\,\mu\text{m}$, $\sigma_g = 31\,\mu\text{m}$, $b = 20\,\mu\text{m}$, $N_p = 20\,000$, and $x_3 = 50$ cm.

The intensity profile of the incident field in the plane $x_3 = 0-$ is the same as the one plotted in Fig. 6.4(a). It is a Gaussian, $\exp(-x_\parallel^2/w^2)$, with a $1/e$ half width $w = 2\sigma_s$. The procedure now is to generate a single realization of the surface profile function $\zeta(\mathbf{x}_\parallel)$ on the basis of the joint pdf $f(\gamma_1, \gamma_2)$ of the slopes $a^{(1,2)}_{m+1/3,n+1/3}$ and $a^{(1,2)}_{m+2/3,n+2/3}$ in the representation of $\zeta(\mathbf{x}_\parallel)$ given by Eqs. (6.26a) and (6.26b) and to use it in calculating the speckle pattern produced at a distance $x_3 = 50$ cm from it when it is illuminated at normal incidence by a monochromatic Gaussian beam of frequency ω_0 ($G(\omega) = 2\pi\,\delta(\omega - \omega_0)$ in Eq. (6.62)). In Fig. 6.7(a) we present a plot of this speckle pattern. In Fig. 6.7(b) we present the intensity profile given by Eq. (6.68) when the same realization of the surface profile is illuminated at normal incidence by a broadband Gaussian beam with central wavelength $\lambda_0 = 1.55\,\mu\text{m}$ and $\Delta\lambda = 0.1\lambda_0$. Finally, in Fig. 6.7(c) we plot the intensity profile obtained by averaging the intensities of the fields produced by scattering a monochromatic Gaussian beam of frequency ω_0 from an ensemble of $N_p = 20\,000$ realizations of the surface profile function, all drawn from the same joint pdf $f(\gamma_1, \gamma_2)$. The joint pdf $f(\gamma_1, \gamma_2)$ assumed in carrying out all of the calculations whose results are presented in Figs. 6.7(a)–(c) is the one given by Eq. (6.53).

From the results presented in Fig. 6.7(a)–(c) we see that the use of a broadband beam to illuminate a single realization of the random phase screen averages over the speckles produced by a monochromatic incident beam. It thereby produces an intensity of the transmitted field that closely matches the one produced by a monochromatic incident beam when the intensity of the transmitted field is averaged over the ensemble of realizations of the random phase screen.

6.3 Pseudo-nondiffracting beams

In this section we present an approach to the production of a three-dimensional pseudo-nondiffracting beam that is based on the transmission of a scalar plane wave incident from the region $x_3 < 0$ normally on a two-dimensional circularly symmetric random phase screen situated in a circular aperture A of radius R in an opaque screen in the plane $x_3 = 0$. The region $x_3 < 0$ is a vacuum, as is the region $x_3 > 0$. We assume that the phase fluctuations introduced by the circularly symmetric random phase screen are much larger than 2π, so that the coherent component of the transmitted field is negligible.

In carrying out this program we first solve the problem of designing a two-dimensional circularly symmetric random phase screen that, when illuminated at normal incidence by a scalar plane wave, produces a transmitted field that has a specified distribution of intensity along the cylindrical axis (the positive x_3 axis). In particular, we design a random phase screen that produces a transmitted field whose intensity is a constant along a finite segment of the positive x_3 axis and vanishes along the remainder of that axis. We then show that the intensity of the field transmitted through this phase screen is a rapidly decreasing function of the radial coordinate r transverse to the x_3 axis for each value of x_3 for which the intensity along the x_3 axis is a nonzero constant. Thus, the beam transmitted through the random phase screen is a pseudo-nondiffracting beam.

In the analysis that follows we will make use of some results obtained in Section 6.2.

6.3.1 The transmitted field

We assume in the present case that the phase screen is illuminated at normal incidence from the region $x_3 < 0$ by a plane wave of frequency ω,

$$\Psi(\mathbf{x}|\omega)_{inc} = \exp\left(i\frac{\omega}{c}x_3\right). \tag{6.73}$$

The field transmitted through the phase screen is given in the parabolic approximation by Eq. (6.12):

$$\Psi(\mathbf{x}|\omega)_{tr} = \left(\frac{\omega}{2\pi i c x_3}\right) \exp\left(i\frac{\omega}{c}x_3\right)$$

$$\times \int d^2x'_{\|} \exp\left[i\frac{\omega}{2cx_3}(\mathbf{x}_{\|} - \mathbf{x}'_{\|})^2\right] \Psi(\mathbf{x}'_{\|}, 0+|\omega)_{tr}, \qquad (6.74)$$

where $\Psi(\mathbf{x}_{\|}, 0+|\omega)_{tr}$ is the field just after its passage through the phase screen.
From Eq. (6.73) and (6.23) we find that $\Psi(\mathbf{x}_{\|}, 0+|\omega)_{tr}$ is given by

$$\Psi(\mathbf{x}_{\|}, 0+|\omega)_{tr} = \Psi(\mathbf{x}_{\|}, 0-|\omega)_{inc} \exp\left[i\Delta n\frac{\omega}{c}\zeta(\mathbf{x}_{\|})\right], \qquad |\mathbf{x}_{\|}| < R \qquad (6.75a)$$

$$= 0, \qquad\qquad\qquad\qquad |\mathbf{x}_{\|}| > R. \quad (6.75b)$$

Thus the transmitted field in Eq. (6.74) becomes

$$\Psi(\mathbf{x}|\omega)_{tr} = \left(\frac{\omega}{2\pi i c x_3}\right) \exp\left(i\frac{\omega}{c}x_3\right)$$

$$\times \int_A d^2x'_{\|} \exp\left[i\frac{\omega}{2cx_3}(\mathbf{x}_{\|} - \mathbf{x}'_{\|})^2\right] \exp\left[i\Delta n\frac{\omega}{c}\zeta(\mathbf{x}'_{\|})\right]. \quad (6.76)$$

We now consider the transmitted field along the x_3 axis,

$$\Psi(0, 0, x_3|\omega)_{tr} = \left(\frac{\omega}{2\pi i c x_3}\right) \exp\left(i\frac{\omega}{c}x_3\right) \int_A d^2x'_{\|} \exp\left[i\frac{\omega}{2cx_3}(\mathbf{x}'_{\|})^2\right]$$

$$\times \exp\left[i\Delta n\frac{\omega}{c}\zeta(\mathbf{x}'_{\|})\right], \qquad (6.77)$$

make the assumption that the profile function $\zeta(\mathbf{x}_{\|})$ is a function of $\mathbf{x}_{\|}$ only through
its magnitude $|\mathbf{x}_{\|}| = r$, and write

$$\zeta(\mathbf{x}_{\|}) = H(r). \qquad (6.78)$$

This assumption, together with the assumption of an incident field in the form of
a plane wave, Eq. (6.73), ensures that the intensity of the transmitted field is also
circularly symmetric. With these assumptions Eq. (6.77) becomes

$$\Psi(0, 0, x_3|\omega)_{tr} = \left(\frac{\omega}{i c x_3}\right) \exp\left(i\frac{\omega}{c}x_3\right)$$

$$\times \int_0^R dr\, r \exp\left\{i\frac{\omega}{c}\left[\frac{r^2}{2x_3} + \Delta n H(r)\right]\right\}. \qquad (6.79)$$

We now make the change of variable $r^2 = t$, and introduce the definition

$$H(\sqrt{t}) = h(t). \tag{6.80}$$

As a result we obtain

$$\Psi(0, 0, x_3|\omega)_{tr} = \left(\frac{\omega}{2icx_3}\right) \exp\left(i\frac{\omega}{c}x_3\right)$$

$$\times \int_0^{R^2} dt \exp\left\{i\frac{\omega}{c}\left[\frac{t}{2x_3} + \Delta nh(t)\right]\right\}. \tag{6.81}$$

The mean intensity of the transmitted field along the x_3 axis is therefore given by

$$\langle I(x_3|\omega)\rangle = \langle |\Psi(0, 0, x_3|\omega)_{tr}|^2\rangle$$

$$= \left(\frac{\omega}{2cx_3}\right)^2 \int_0^{R^2} dt \int_0^{R^2} dt' \exp\left[i\frac{\omega}{2cx_3}(t - t')\right]$$

$$\times \left\langle \exp\left\{i\frac{\omega}{c}\Delta n[h(t) - h(t')]\right\}\right\rangle. \tag{6.82}$$

Our goal, now, is to find the function $h(t)$ that produces a specified form for $\langle I(x_3|\omega)\rangle$. As it stands, the expression for $\langle I(x_3|\omega)\rangle$ given by Eq. (6.82) is too difficult to invert to obtain $h(t)$. We therefore make an approximation that is analogous to passing to the geometrical optics limit of Eq. (6.82). Namely, we expand $h(t)$ about $t = t'$,

$$h(t) = h(t') + (t - t')h'(t') + \cdots, \tag{6.83}$$

where $h'(t)$ is the derivative of $h(t)$ with respect to t, and retain only the first two terms on the right-hand side of this expansion. With this approximation the expression for $\langle I(x_3|\omega)\rangle$ becomes

$$\langle I(x_3|\omega)\rangle = \left(\frac{\omega}{2cx_3}\right)^2 \int_0^{R^2} dt \int_0^{R^2} dt' \exp\left[i\frac{\omega}{2cx_3}(t - t')\right]$$

$$\times \left\langle \exp\left\{i\frac{\omega}{c}\Delta n(t - t')h'(t')\right\}\right\rangle. \tag{6.84}$$

This is the equation we can invert to obtain $h(t)$ in terms of $\langle I(x_3|\omega)\rangle$.

6.3.2 The inverse problem

To invert Eq. (6.84) we assume the following representation of $h(t)$:

$$h(t) = \frac{a_n}{b}(t - nb^2), \quad nb^2 < t < (n+1)b^2, \quad n = 0, 1, 2, \ldots, N-1, \quad (6.85)$$

where the $\{a_n\}$ are independent identically distributed random deviates and b is a characteristic length. Because the $\{a_n\}$ are independent and identically distributed random deviates, the probability density function (pdf) of a_n,

$$\langle \delta(\gamma - a_n) \rangle = f(\gamma), \quad (6.86)$$

is independent of n.

In the representation (6.85) N is a large integer. It is then convenient to define the characteristic length b through the relation $R = \sqrt{N}b$. The function $h(t)$ is not a continuous function of t: it has jump discontinuities at $t = nb^2$. Therefore the surface profile function $H(r)$,

$$H(r) = \frac{a_n}{b}(r^2 - nb^2), \quad \sqrt{n}b < r < \sqrt{n+1}b, \quad n = 0, 1, 2, \ldots, N-1, \quad (6.87)$$

also has jump discontinuities at $r = \sqrt{n}b$. This contradicts our starting assumption that the surface profile is a single-valued function of $x_\parallel$. However, as we will see, the surface profile function $h(t)$ defined by Eq. (6.85) will produce a transmitted field with the properties we seek.

With the representation for $h(t)$ given by Eq. (6.85) Eq. (6.84) becomes

$$\langle I(x_3|\omega) \rangle = \left(\frac{\omega}{2cx_3}\right)^2 \int_0^{Nb^2} dt \sum_{n=0}^{N-1} \int_{nb^2}^{(n+1)b^2} dt' \exp\left[i\frac{\omega}{2cx_3}(t - t')\right]$$

$$\times \left\langle \exp\left[i\frac{\omega}{cb}\Delta n(t - t')a_n\right]\right\rangle$$

$$= \left(\frac{\omega}{2cx_3}\right)^2 \int_0^{Nb^2} dt \sum_{n=0}^{N-1} \int_{nb^2}^{(n+1)b^2} dt' \exp\left[i\frac{\omega}{2cx_3}(t - t')\right]$$

$$\times \int_{-\infty}^{\infty} d\gamma \, f(\gamma) \exp\left[i\frac{\omega}{cb}\Delta n(t - t')\gamma\right]$$

$$= \left(\frac{\omega}{2cx_3}\right)^2 \int_{-\infty}^{\infty} d\gamma f(\gamma) \left| \int_0^{Nb^2} dt \exp\left\{ i \left[\frac{\omega}{2cx_3} + \Delta n \frac{\omega\gamma}{cb} \right] t \right\} \right|^2$$

$$= \left(\frac{\omega}{2cx_3}\right)^2 N^2 b^4 \int_{-\infty}^{\infty} d\gamma f(\gamma) \mathrm{sinc}^2 \left(N \frac{\Delta n}{2} \frac{\omega b}{c} \left(\frac{b}{2\Delta n x_3} + \gamma \right) \right).$$

$$(6.88)$$

In the limit as $N \to \infty$,

$$(\mathrm{sinc}\, Nx)^2 \underset{N\to\infty}{\to} \frac{\pi}{N} \delta(x),$$

$$(6.89)$$

where $\delta(x)$ is the Dirac delta function. With the use of this result, Eq. (6.88) simplifies to

$$\langle I(x_3|\omega) \rangle = \frac{\pi}{2\Delta n} \frac{\omega b}{c} N \frac{b^2}{x_3^2} f \left(\frac{-b}{2\Delta n x_3} \right).$$

$$(6.90)$$

Thus, the mean intensity of the transmitted field along the x_3 axis is given in terms of the pdf of a_n. With the change of variable $b/(2\Delta n x_3) = -\gamma$, we obtain the result that

$$f(\gamma) = \frac{1}{2\pi \Delta n} \frac{c}{\omega} \frac{1}{Nb} \frac{\langle I((-b/2\Delta n\gamma)|\omega) \rangle}{\gamma^2}.$$

$$(6.91)$$

Since x_3 has been assumed to be positive, we see that $f(\gamma)$ is nonzero only for negative values of γ. Therefore all the slopes $\{a_n\}$ are negative.

With the aim of obtaining a transmitted field that is a pseudo-nondiffracting beam, we seek to design a surface that produces a field with a constant intensity within the interval $z_1 < x_3 < z_2$ ($0 < z_1 < z_2$) of the x_3 axis, and a vanishing intensity along the rest of this axis, so that

$$\langle I(x_3|\omega) \rangle = I_0 \theta(x_3 - z_1) \theta(z_2 - x_3).$$

$$(6.92)$$

On combining Eqs. (6.91) and (6.92) we find that

$$f(\gamma) = I_0 \frac{1}{2\pi \Delta n} \frac{c}{\omega} \frac{1}{Nb} \frac{1}{\gamma^2} \theta \left(\gamma + \frac{b}{2\Delta n z_1} \right) \theta \left(-\frac{b}{2\Delta n z_2} - \gamma \right).$$

$$(6.93)$$

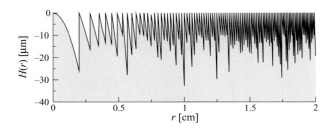

Figure 6.8. Segment of one numerically generated realization of the surface profile function $H(r)$ with 100 zones calculated on the basis of the pdf $f(\gamma)$ given by Eq. (6.97). The values of the experimental and material parameters assumed in generating this segment were $z_1 = 5$ cm, $z_2 = 50$ cm, $b = 0.2$ cm, $\Delta n = 0.6$.

The constant I_0 is determined from the normalization condition for $f(\gamma)$,

$$\int_{-\infty}^{\infty} d\gamma \, f(\gamma) = 1 = I_0 \frac{1}{2\pi \Delta n} \frac{c}{\omega} \frac{1}{Nb} \int_{\frac{-b}{2\Delta n z_1}}^{\frac{-b}{2\Delta n z_2}} d\gamma \, \frac{1}{\gamma^2}$$

$$= I_0 \frac{1}{\pi} \frac{c}{\omega} \frac{1}{Nb^2} (z_2 - z_1), \tag{6.94}$$

so that

$$I_0 = \pi \frac{\omega}{c} \frac{Nb^2}{z_2 - z_1}. \tag{6.95}$$

It therefore follows that

$$\langle I(x_3|\omega)\rangle = \pi \frac{\omega}{c} \frac{Nb^2}{z_2 - z_1} \theta(x_3 - z_1)\theta(z_2 - x_3) \tag{6.96}$$

and

$$f(\gamma) = \frac{b}{2\Delta n(z_2 - z_1)} \frac{1}{\gamma^2} \theta\left(\gamma + \frac{b}{2\Delta n z_1}\right) \theta\left(\frac{-b}{2\Delta n z_2} - \gamma\right). \tag{6.97}$$

From the result given by Eq. (6.97) a long sequence of $\{a_n\}$ is obtained, for example by the rejection method [92], and the surface profile function $H(r)$ is then constructed on the basis of Eq. (6.87). In Fig. 6.8 we present a segment of one numerically generated realization of the surface profile function $H(r)$ calculated on the basis of the pdf $f(\gamma)$ given by Eq. (6.97). The values of the experimental and material parameters employed in generating this segment were $z_1 = 5$ cm, $z_2 = 50$ cm, $b = 0.2$ cm, and $\Delta n = 0.6$.

6.3.3 *Three-dimensional distribution of the mean intensity in the radial direction from the optical axis*

The distribution of intensity along the x_3 axis has served its purpose in providing us with the pdf of slopes $\{a_n\}$. However, the axial intensity distribution is not the only function of interest if we wish to produce a pseudo-nondiffracting beam: it is also important to know the intensity distribution in the radial direction away from the x_3 axis. Consequently, we turn to the calculation of the mean intensity as a function of r for values of x_3 that include the interval $z_1 < x_3 < z_2$.

From Eqs. (6.76) and (6.78) we find that the transmitted field is given by

$$
\Psi(\mathbf{x}|\omega)_{tr} = \Psi(r, x_3|\omega)_{tr}
$$
$$
= \left(\frac{\omega}{icx_3} \right) \exp\left\{ i\frac{\omega}{c} \left[x_3 + \frac{r^2}{2x_3} \right] \right\}
$$
$$
\times \int_0^R dr'\, r'\, J_0((\omega r/cx_3)r') \exp\left\{ i\frac{\omega}{c} \left[\frac{r'^2}{2x_3} + \Delta n H(r') \right] \right\}, \quad (6.98)
$$

where $J_0(x)$ is a Bessel function of the first kind of zero order. With the form of $H(r)$ given by Eq. (6.87), this expresssion can be written in the form

$$
\Psi(r, x_3|\omega)_{tr} = \left(\frac{\omega}{icx_3} \right) \exp\left\{ i\frac{\omega}{c} \left[x_3 + \frac{r^2}{2x_3} \right] \right\}
$$
$$
\times \sum_{n=0}^{N-1} \exp\left[-i\frac{\omega b}{c} \Delta n a_n n \right] \Psi_n(r, x_3; a_n), \quad (6.99)
$$

where

$$
\Psi_n(r, x_3; a_n) = \int_{\sqrt{n}b}^{\sqrt{n+1}b} dr'\, r'\, J_0\left(\frac{\omega r}{cx_3}r' \right)
$$
$$
\times \exp\left\{ i\frac{\omega}{c} \left[\frac{1}{2x_3} + \Delta n\frac{a_n}{b} \right] r'^2 \right\}. \quad (6.100)
$$

The field $\Psi_n(r, x_3; a_n)$ gives the diffraction pattern of an annular pupil function with defocus $(\omega/c)[(1/2x_3) + \Delta n(a_n/b)]$, where a_n is a random quantity.

Diffraction integrals of the kind given by Eq. (6.100) have been extensively studied in the past [97, 98]. They are evaluated here by the Nijboer expansion [98, 99].

The mean intensity obtained from Eq. (6.99) is given by

$$\langle I(r, x_3|\omega)\rangle = \langle|\Psi(r, x_3|\omega)_{tr}|^2\rangle$$

$$= \left(\frac{\omega}{cx_3}\right)^2 \sum_{m=0}^{N-1}\sum_{n=0}^{N-1} \langle\exp[-i(\omega b/c)\Delta n(ma_n - na_n)]$$

$$\times \Psi_m(r, x_3; a_m)\Psi_n^*(r, x_3; a_n)\rangle. \tag{6.101}$$

In deriving the preceding results we have assumed that the slopes $\{a_n\}$ are independent random deviates. If we assume in addition that the coherent contribution to $\langle I(r, x_3|\omega)\rangle$ is negligible, i.e. that $\langle\Psi_n(r, x_3; a_n)\rangle = 0$, the expression for the mean intensity of the transmitted field simplifies to

$$\langle I(r, x_3|\omega)\rangle = \left(\frac{\omega}{cx_3}\right)^2 \sum_{n=0}^{N-1} \langle|\Psi_n(r, x_3; a_n)|^2\rangle. \tag{6.102}$$

The results of calculations not presented here show that the relative error made in using Eq. (6.102) instead of Eq. (6.101) is of the order of one part in 10^4 for all the parameters and values of r and x_3 assumed in this work. Consequently, in what follows we use the simple expression in Eq. (6.102) to calculate the mean intensity $\langle I(r, x_3|\omega)\rangle$.

The result given by Eq. (6.102) allows us to reach a useful conclusion about the approach we have used here to obtain the pdf of a_n. Thus, let us consider the mean intensity of the transmitted field along the x_3 axis obtained from Eq. (6.102). The amplitude along this axis produced by the nth ring of the random phase screen is

$$\Psi_n(0, x_3; a_n) = \int_{\sqrt{nb}}^{\sqrt{n+1}b} dr\, r \exp\left\{i\frac{\omega}{c}\left(\frac{1}{2x_3} + \Delta n\frac{a_n}{b}\right)r^2\right\}$$

$$= \frac{b^2}{2} \exp\left\{i\frac{\omega}{c}\left(\frac{1}{2x_3} + \Delta n\frac{a_n}{b}\right)b^2\left(n + \frac{1}{2}\right)\right\}$$

$$\times \text{sinc}\left[\frac{\omega}{c}\left(\frac{1}{2x_3} + \Delta n\frac{a_n}{b}\right)\frac{b^2}{2}\right]. \tag{6.103}$$

The average of the squared modulus of this expression,

$$\left(\frac{\omega}{cx_3}\right)^2 \langle|\Psi_n(0, x_3; a_n)|^2\rangle$$

$$= \left(\frac{\omega}{cx_3}\right)^2 \frac{b^4}{4} \left\langle\text{sinc}^2\left[\frac{\omega}{c}\left(\frac{1}{2x_3} + \Delta n\frac{a_n}{b}\right)\frac{b^2}{2}\right]\right\rangle, \tag{6.104}$$

is the contribution to the mean intensity along the x_3 axis from the nth annular ring of the random phase screen. However, because the pdf of a_n is independent of n, so is $\langle |\Psi_n(0, x_3; a_n)|^2 \rangle$. Therefore all of the annular rings produce exactly the same axial intensity distribution. The random phase screen consists of N annular rings. Therefore, the total intensity distribution along the positive x_3 axis is N times the result given by Eq. (6.104), which we write as follows:

$$\langle I(0, x_3 | \omega) \rangle = \left(\frac{\omega}{cx_3} \right)^2 \frac{R^4}{4N} \left\langle \operatorname{sinc}^2 \left[\frac{1}{N} \left(\frac{\omega R^2}{4cx_3} + \frac{\omega}{c} \Delta n \frac{R^2}{2b} a_n \right) \right] \right\rangle, \quad (6.105)$$

where we have used the relation $R^2 = Nb^2$.

It is now interesting to compare the result given by Eq. (6.105) with the expression for the mean intensity given by Eqs. (6.88) and (6.89), namely

$$\langle I(0, x_3 | \omega) \rangle = \left(\frac{\omega}{cx_3} \right)^2 \frac{R^4}{4} \left\langle \operatorname{sinc}^2 \left[\left(\frac{\omega R^2}{4cx_3} + \frac{\omega}{c} \Delta n \frac{R^2}{2b} a_n \right) \right] \right\rangle. \quad (6.106)$$

We see that the integer N that appears in Eq. (6.105) is absent from Eq. (6.106). This is due to the fact that the geometrical optics approximation employed in the derivation of Eq. (6.106) neglects diffraction effects by the small scale of the random phase screen, and retains only the diffraction effects of the complete aperture. Therefore the diffraction pattern obtained is related to the complete aperture rather than to the individual annular rings, as should be the case. The use of Eq. (6.89) removes the remaining diffraction effects and produces a final result, Eq. (6.90), that is consistent with geometrical optics. Thus, the results obtained in Section 6.3.2 are useful in the design of the phase screen, but cannot be used for diffraction calculations.

6.3.4 Pseudo-nondiffracting beam

We now seek to produce a phase screen that produces a uniform distribution of intensity in the range $z_1 < x_3 < z_2$ of the x_3 axis. The pdf of a_n in this case is given by Eq. (6.97), so that from Eq. (6.102) we obtain

$$\langle I(r, x_3 | \omega) \rangle = \left(\frac{\omega}{cx_3} \right)^2 \frac{b}{2\Delta n(z_2 - z_1)} \sum_{n=0}^{N-1} \int_{-\frac{b}{2\Delta n z_1}}^{-\frac{b}{2\Delta n z_2}} \frac{d\gamma}{\gamma^2} |\Psi_n(r, x_3; \gamma)|^2. \quad (6.107)$$

The result for the rotationally symmetric mean intensity $\langle I(r, x_3 | \omega) \rangle$ is plotted in Fig. 6.9 for the following values of the experimental and material parameters: $\omega = 2\pi c/\lambda$, where $\lambda = 632.8$ nm, $\Delta n = 0.6$, $z_1 = 10$ cm, $z_2 = 50$ cm, $R = 2$ cm,

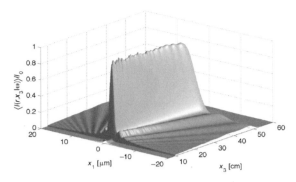

Figure 6.9. Plot of the rotationally symmetric mean intensity $\langle I(r, x_3|\omega)\rangle$ for the following values of the experimental and material parameters: $\omega = 2\pi c/\lambda$, where $\lambda = 632.8$ nm, $\Delta n = 0.6$, $z_1 = 10$ cm, $z_2 = 50$ cm, $R = 2$cm, and $N = 100$ ($b = 0.2$ cm).

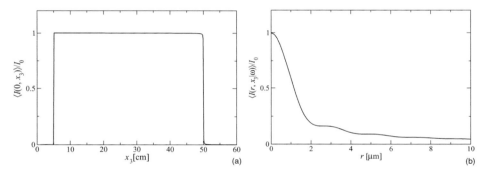

Figure 6.10. (a) The mean intensity distribution along the x_3 axis $\langle I(0, x_3|\omega)\rangle$, calculated on the basis of Eq. (6.107). (b) The mean intensity in the transverse direction $\langle I(r, x_3|\omega)\rangle$, calculated on the basis of Eq. (6.107), as a function of r for $x_3 = 27.5$ cm. The parameter values used in obtaining Fig. 6.9 were used in obtaining these results.

and $N = 100$ ($b = 0.2$ cm). It is seen to be fairly constant in the region (z_1, z_2) of the x_3 axis, and to decrease rapidly outside it and for off-axis points.

In Fig. 6.10(a) we present the mean intensity distribution along the x_3 axis, and in Fig. 6.10(b) we present the mean intensity in the transverse direction for the plane $x_3 = 27.5$ cm. The same data used in calculating the result presented in Fig. 6.9 were used in obtaining these results. From Eq. (6.105) we see that the mean intensity along the x_3 axis is the convolution of a sinc^2 function with $f(\gamma)$ which, according to Eq. (6.97), reproduces the desired intensity distribution. Therefore, the axial intensity distribution is not perfectly rectangular, but is smoothed by diffraction effects.

Several comments need to be made about the preceding results. In an earlier study [100] of the creation of a pseudo-nondiffracting beam by the scattering (reflection) of light from a random circularly symmetric Dirichlet surface a different representation of the function $h(t)$ was used, namely

$$h(t) = \frac{a_n}{b}t + b_n, \quad nb^2 \leq t \leq (n+1)b^2, \quad n = 0, 1, \ldots, N-1. \quad (6.108)$$

In this representation $\{a_n\}$ are independent identically distributed random deviates, and b is a characteristic length. The $\{b_n\}$ were determined from the condition that $h(t)$ be a continuous function of t, and were found to be

$$b_n = b_0 + (a_0 + a_1 + \cdots + a_{n-1} - na_n)b, \quad n \geq 1. \quad (6.109)$$

Because the $\{a_n\}$ were independent and identically distributed random deviates, the probability density function of a_n, $f(\gamma) = \langle \delta(\gamma - a_n) \rangle$, was also independent of n. The $\{a_n\}$ were all found to be positive for each realization of $h(t)$, so that the resulting surface profile function $h(r)$ had a scalloped bowl shape. We have chosen to use the representation of $h(t)$ given by Eq. (6.85) in the present work because it produces a surface profile function that is much closer to being planar, and hence should be easier to fabricate by the approach described in ref. [101] than a phase screen based on Eqs. (6.108) and (6.109).

All of the preceding results have been obtained on the basis of calculations that neglect the contribution to the transmitted field from the vertical segments of the surface profile function $H(r)$ at $r = \sqrt{n}b$, $n = 1, 2, \ldots, N$. However, in the present case this omission should have a small impact on the results obtained for two reasons. The first is that, since the illuminating plane wave is incident normally on the phase screen, it does not "feel" these vertical segments of the surface profile function. The second is that, since the $\{a_n\}$ are all negative and the index n of the dielectric material from which the phase screen is constructed is greater than unity, the refraction of the light through the segment of $H(r)$ in the interval $\sqrt{n}b < r < \sqrt{n+1}b$ is toward the x_3 axis and away from the vertical segments of the surface profile function. Thus, in a ray optics sense the refracted light also does not "feel" the vertical segments of the surface.

6.3.5 Fabrication of circularly symmetric radially random surfaces

To fabricate the circularly symmetric radially random surfaces studied in this section one could use the approach described in Section 6.2.5. However, this is not the best option, as it would be desirable to exploit the circular symmetry of these surfaces. One possible way in which to proceed is to use an approach based on diamond turning techniques [102]. Diamond turning is a process of mechanical

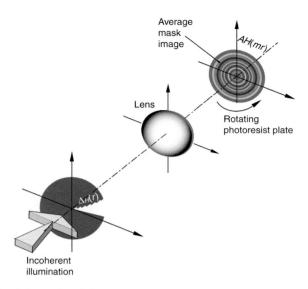

Figure 6.11. Schematic of the setup employed for the fabrication of a circularly symmetric radially random surface.

machining that uses a diamond-tipped cutting element to shape a rotating solid block of material. It is a subtractive computer numerical control (CNC) technique. The process is widely used to manufacture precision optical elements such as aspheric lenses and mirrors for the infrared. A variety of materials, such as plastics, crystals, and metals, can be used. Blazed-like structures, such as Fresnel lenses [103], can be fabricated by diamond turning processes, for example the Fresnel lenses used in overhead projectors.

An alternative approach, which we review below, has been described in ref. [101]. The rotationally symmetric surfaces discussed here can be fabricated by exposing photoresist-coated plates to blue light ($\lambda = 442$ nm) from a He–Cd laser transmitted through a rotating ground glass to reduce its coherence. A schematic depiction of the experimental setup employed in the fabrication is shown in Fig. 6.11. An incoherent image of a disk-shaped mask is formed on the rotating photoresist-coated plate by a well-corrected imaging system with magnification m. The photoresist plate is exposed for a time T_e, during which it executes a large number of revolutions. As explained below, this setup produces a total exposure of the plate that is a scaled version of the profile function used in the generation of the mask.

To produce a suitable mask one first needs to generate a realization of a profile function $H(r)$. An example has been shown in Fig. 6.8. We next introduce a function $\Delta_H(r) = K H(r)$ that, by a suitable choice of the units of K, can be interpreted as an angle. For a given radius r the angles θ that fall in the transparent

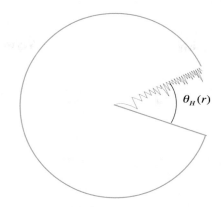

Figure 6.12. Mask generated from the random profile of Fig. 6.8.

section of the mask are defined by $\Delta_0 < \theta < \Delta_H(r)$, where Δ_0 is a constant that is smaller than the minimum value of $\Delta_H(r)$ (see Fig. 6.12).

An incoherent image of the mask is formed on the surface of the rotating photoresist plate. The resulting exposure is circularly symmetric with a radial dependence of the form $E(r) = I_e T_e [\Delta_H(mr) - \Delta_0]/2\pi$, where I_e is a constant related to the intensity of the illumination. This expression can be rewritten in the form $E(r) = E_0 + AH(mr)$, where $E_0 = -I_e T_e \Delta_0/2\pi$ and $A = I_e T_e K/2\pi$. The values of the constants E_0 and A can be adjusted by varying the intensity of the light reaching the photoresist plate, the aperture of the mask, and the exposure time. If we assume a linear relation between exposure and the resulting height of the surface, the developed surface will have the desired property of being proportional to $AH(mr)$.

The theory developed in this section has been based on a random phase screen defined by a surface profile function $H(r)$. The question therefore arises of how the mean intensity produced by a random phase screen with a surface profile function $AH(mr)$ differs from the mean intensity obtained with the original function $H(r)$. From Eq. (6.85) we see that the transformation is equivalent to choosing new random deviates $a'_n = Am^2 a_n$. As a result, the new region of constant intensity along the x_3 axis becomes $z'_1 < x_3 < z'_2$, where $z'_j = Am^2 z_j$ ($j = 1, 2$). In other words, by scaling the profile function $H(r)$ in the vertical or horizontal direction one changes the length of the region of constant intensity along the x_3 axis. Such scaling is almost inevitable in the fabrication process, and is defined by the exposure of the plate and by the magnification of the optical system.

The approach described in this section has been used successfully in the fabrication of optical diffusers that can be used to extend the depth of focus of an imaging

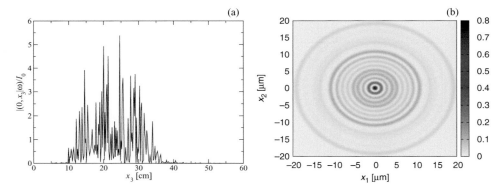

Figure 6.13. (a) The intensity distribution along the x_3 axis when a single realization of the phase screen is illuminated by monochromatic light of wavelength
$\lambda = 632.8$ nm. (b) Transverse intensity image at $x_3 = 20$ cm. The parameter values assumed in obtaining these results are $\omega = 2\pi c/\lambda$, where $\lambda = 632.8$ nm,
$\Delta n = 0.6$, $z_1 = 10$ cm, $z_2 = 35$ cm, $R = 5$cm, and $N = 2500$ ($b = 0.1$ cm).

system [101]. It has yet to be used in fabricating diffusers for use in producing
pseudo-nondiffracting beams.

6.3.6 Replacement of ensemble averaging by frequency averaging

The average over the ensemble of realizations of the surface profile function that
yields the mean intensity of the transmitted beam is readily carried out in theoreti-
cal/numerical calculations. A theorist generates a large number N_p of realizations
of the surface profile function, and calculates the intensity of the transmitted field
for each realization. An arithmetic average of these N_p intensities gives the ensemble average sought. In contrast, an experimentalist ordinarily has only a single
realization of the random phase screen to work with. For a particular realization
of the surface profile function the different annular rings focus the light on different points on the x_3 axis, with equal probability within the region (z_1, z_2). When
monochromatic illumination is used, the interference between all of these randomly
phased contributions produces speckle, which manifests itself as random variations
of the intensity along the x_3 axis. Simultaneously, the rotational symmetry of the
system leads to a transverse intensity pattern that contains rings that change rapidly
as one moves along the x_3 axis. As an example, we show in Fig. 6.13(a) the intensity
distribution of the transmitted field along the x_3 axis when a single realization of
the phase screen is illuminated by monochromatic light. Random variations of this
intensity distribution are observed. Similarly, in Fig. 6.13(b) we show a calculated
transverse intensity image at $x_3 = 20$ cm. A ring structure is clearly seen.

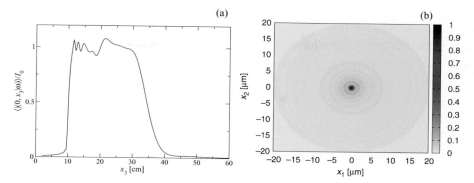

Figure 6.14. (a) The intensity distribution along the x_3 axis when a single realization of the phase screen is illuminated by a broadband source whose spectral density is given by Eq. (6.110), where $\omega_0 = 2\pi c/\lambda_0$ with $\lambda_0 = 632.8$ nm, $\Delta\omega_0 = 0.1\omega_0$. (b) The transverse intensity image at $x_3 = 20$ cm, obtained from the same realization of the phase screen when it is illuminated by the same broadband source.

Both types of intensity fluctuations can be smoothed through the use of broadband illumination instead of monochromatic illumination. In Fig. 6.14(a) we show the intensity distribution of the transmitted field along the x_3 axis when a single realization of the phase screen is illuminated by a broadband source whose spectral density has the Gaussian form

$$S_0(\omega) = \frac{1}{\sqrt{\pi}\Delta\omega} \exp\left[-\frac{(\omega - \omega_0)^2}{(\Delta\omega)^2}\right]. \tag{6.110}$$

The central frequency $\omega_0 = 2\pi c/\lambda_0$ corresponds to a wavelength $\lambda = 632.8$ nm and $\Delta\omega = 0.1\omega_0$. The random intensity fluctuations present in the results depicted in Fig. 6.13(a) have almost completely disappeared. In Fig. 6.14(b) we show calculated transverse intensity images obtained at $x_3 = 20$ cm from the same realization of the phase screen when it is illuminated by the same broadband source. It can be seen that the ring structure present in the results presented in Fig. 6.13(b) has practically disappeared. The smoothing of the axial intensity and the transverse intensity images improves as the number of rings N forming the random phase screen is increased, and the results approach closely those obtained by ensemble averaging. Thus, it can be said that the performance of the phase screen designed here improves with the replacement of monochromatic light by polychromatic light.

In this section we have used scalar diffraction theory to design a random phase screen that, when illuminated at normal incidence by a scalar plane wave, produces a transmitted beam with an intensity that is constant along a segment of the optical axis, and decays rapidly with increasing radial distance from this axis in any transverse plane within this region. Such a random phase screen is potentially useful for the production of a pseudo-nondiffracting beam.

6.4 Discussion and conclusions

In this chapter we have described two types of structured surfaces that produce optical effects that planar unstructured surfaces cannot produce. The effects we have considered consist of the production of electromagnetic waves, which possess specified spatial or angular dependencies, by the transmission of volume electromagnetic waves through suitably structured surfaces.

Thus we have presented a random phase screen that, when illuminated by an electromagnetic field with a prescribed intensity distribution and a specified spectral degree of coherence, produces a transmitted beam with a different prescribed intensity distribution in the far field. We have also designed a two-dimensional circularly symmetric thin random phase screen that, when illuminated at normal incidence by a scalar plane wave, produces a transmitted field with a specified intensity distribution along the optical axis, and we have shown how this result can be used to produce a pseudo-nondiffracting beam.

The approaches we have used to transform a specified spatial or angular dependence of a transmitted field into the phase of a thin random phase screen that produces that dependence are based on the use of an expression for the transmitted field in the form of an integral that has been obtained with the use of the phase screen sought and contains the phase function in its integrand in a simple exponential form. The spatial variation of the phase function is then transformed into a spatial dependence of the surface profile function of the phase screen by a standard procedure. The latter dependence is then represented in terms of triangular facets with random slopes along two perpendicular directions for each facet in the case of a two-dimensional surface, or linear segments with random slopes in the case of a one-dimensional surface. Such representations lead to a Fredholm integral equation of the first kind for the joint probability density function or the probability density function of slopes defining the surface profile function of the phase screen, respectively. The inhomogeneous term in this equation is the desired intensity distribution of the transmitted field. The geometrical optics limit of this equation can be solved analytically to yield the corresponding probability density function. The analytic solutions of these equations are then tested by generating random surafces on the basis of these functions, and then solving the problem of wave transmission through them by the use of methods that are more rigorous than those used in their derivations.

There are many applications requiring optical elements that shape light in specified ways that have not been considered in this chapter. Many of them can be effected by the use of structured surfaces. Nevertheless, the examples presented here, and the approaches used in their realization, give an indication of how broadly distributed these applications, and how robust the results obtained by these approaches, can be.

Acknowledgments

The research of A.A.M. and T.A.L. was supported in part by AFRL Contract no. FA 9453-08-C-0230. The research of E.R.M. was supported in part by CONACyT grant no. 47712-F.

Appendix

```
----------------------
solid surface
facet normal 0.0 0.0 -1.0
   outer loop
      vertex 0.0   0.0    0.0
      vertex 1.0   0.0    0.0
      vertex 0.5   0.866 0.0
   endloop
endfacet
facet normal -0.171 -0.171 0.97
   outer loop
      vertex 1.0   0.0    0.176
      vertex 0.5   0.866 0.24
      vertex 0.0   0.0    0.0
   endloop
endfacet
facet normal 0.0 -1.0 0.0
   outer loop
      vertex 0.0   0.0    0.0
      vertex 1.0   0.0    0.0
      vertex 1.0   0.0    0.176
   endloop
endfacet
```

```
facet normal 0.5 0.866 0.0
  outer loop
    vertex 1.0   0.0    0.0
    vertex 0.5   0.866 0.0
    vertex 0.5   0.866 0.24
  endloop
endfacet
facet normal 0.5 0.866 0.0
  outer loop
    vertex 1.0   0.0    0.0
    vertex 0.5   0.866 0.24
    vertex 1.0   0.0    0.176
  endloop
endfacet
facet normal -0.5 -0.866 0.0
  outer loop
    vertex 0.0   0.0    0.0
    vertex 0.5   0.866 0.0
    vertex 0.5   0.866 0.24
  endloop
endfacet
endsolid
```

References

[1] B. R. Frieden, "Lossless conversion of a plane laser wave to a plane wave of uniform irradiance," *Appl. Opt.* **4**, 1400–1403 (1965).

[2] T. E. Horton and J. H. McDermit, "Design of a specular aspheric surface to uniformly radiate a flat surface using a nonuniform collimated radiation source," *J. Heat Trans. ASME C* **94**, 453–458 (1972).

[3] P. W. Rhodes and D. L. Shealy, "Refractive optical systems for irradiance redistribution of collimated radiation: their design and analysis," *Appl. Opt.* **19**, 3545–3553 (1980).

[4] Wai-Hon Lee, "Method for converting a Gaussian laser beam into a uniform beam," *Opt. Commun.* **36**, 469–471 (1981).

[5] M. Quintanilla and A. M. de Frutos, "Holographic filter that transforms a Gaussian into uniform beam," *Appl. Opt.* **20**, 879–880 (1981).

[6] W. B. Veldkamp, "Laser beam profile shaping with interlaced binary diffraction gratings," *Appl. Opt.* **21**, 3209–3212 (1982).

[7] N. Streibl, "Beam shaping with optical array generators," *J. Mod. Opt.* **36**, 1559–1573 (1989).

[8] C. C. Aleksoff, K. K. Ellis, and B. D. Neagle, "Holographic conversion of a Gaussian beam to a near-field uniform beam," *Opt. Eng.* **3**, 537–543 (1991).

[9] L. A. Romero and F. M. Dickey, "Lossless laser beam shaping," *J. Opt. Soc. Am. A* **13**, 751–760 (1995).

[10] X. Tan, B.-Y. Gu, G.-Z. Yang, and B.-Z. Dong, "Diffractive phase elements for beam shaping: a new design method," *Appl. Opt.* **34**, 1314–1320 (1995).

[11] Y. Lin, T. J. Kessler, and G. N. Lawrence, "Distributed phase plates for super-Gaussian focal-plane irradiance profiles," *Opt. Lett.* **20**, 764–766 (1995).

[12] F. X. Wagner, M. Scaggs, A. Koch, H. Endert, H.-M. Christen, L. A. Knauss, K. S. Harshavarardhan, and S. M. Green, "Epitaxial HTS thin films grown by PLD with a beam homogenizer," *Appl. Surf. Sci.* **127–129**, 477–480 (1998).

[13] J. A. Hoffnagle and C. M. Jefferson, "Design and performance of a refractive optical system that converts a gaussian to a flattop beam," *Appl. Opt.* **39**, 5488–5499 (2000).

[14] D. Shealy, "Geometrical methods," in *Laser Beam Shaping – Theory and Techniques*, eds. F. M. Dickey and S. C. Holswade (New York: Marcel Dekker, 2000), chap. 4.

[15] D. M. Brown, F. M. Dickey, and L. S. Weichman, "Multi-aperture beam integration systems," in *Laser Beam Shaping – Theory and Techniques*, eds. F. M. Dickey and S. C. Holswade (New York: Marcel Dekker, 2000), chap. 7.

[16] D. L. Shealy, "History of beam shaping," in *Laser Beam Shaping Applications*, eds. F. M. Dickey, S. C. Holswade, and D. L. Shealy (Boca Raton, FL: CRC Press, Taylor & Francis Group, 2006), chap. 9.

[17] J. Jia, C. Zhou, X. Sun, and L. Liu, "Superresolution laser beam shaping," *Appl. Opt.* **43**, 2112–2117 (2004).

[18] W. M. Lee, X.-C. Yuan, and W. C. Cheong, "Optical vortex beam shaping by use of highly efficient irregular spiral phase plates for optical micro-manipulation," *Opt. Lett.* **29**, 1796–1798 (2004).

[19] D. Varentsov, I. M. Tkachenko, and D. H. H. Hoffmann, "Statistical approach to beam shaping," *Phys. Rev. E* **71**, 066501 (1-7) (2005).

[20] R. El-Agmy, H. Bulte, A. H. Greenaway, and D. T. Reid, "Adaptive beam profile control using a simulated annealing algorithm," *Opt. Express* **13**, 6085–6091 (2005).

[21] J. Hahn, H. Kim, K. Choi, and B. Lee, "Real-time digital holographic beam – shaping system with a genetic feedback tuning loop," *Appl. Opt.* **45**, 915–924 (2006).

[22] T. Takaoka, N. Kawano, Y. Awatsuji, and T. Kuboto, "Design of a reflective aspherical surface of a compact beam-shaping device," *Opt. Rev.* **13**, 77–86 (2006).

[23] D. L. Shealy and J. A. Hoffnagle, "Laser beam shaping profiles and propagation," *Appl. Opt.* **45**, 5118–5131 (2006).

[24] E. Tefouet Kana, S. Bollanti, P. D. Lazzaro, D. Mirra, O. Bouba, and M. Boyomo Onana, "Laser beam homogenization: modeling and comparison with experimental results," *Opt. Commun.* **264**, 187–192 (2006).

[25] R. de Saint Denis, N. Passilly, M. Leroche, T. Mohammed-Brahim, and K. Ait-Ameur, "Beam-shaping longitudinal range of a binary diffractive optical element," *Appl. Opt.* **45**, 8136–8141 (2006).

[26] W. Chen and Q. Zhan, "Three-dimensional focus shaping with cylindrical vector beams," *Opt. Commun.* **265**, 411–417 (2006).

[27] J. S. Liu, A. J. Caley, and M. R. Taghizadeh, "Diffractive optical elements for beam shaping of monochromatic spatially incoherent light," *Appl. Opt.* **45**, 8440–8447 (2006).

[28] B. Hao and J. Leger, "Polarization beam shaping," *Appl. Opt.* **46**, 8211–8217 (2007).

[29] C. Dorrer and J. D. Zuegel, "Design and analysis of binary beam shapers using error diffusion," *J. Opt. Soc. Am. B* **24**, 1268–1275 (2007).

[30] D. Palima, C. A. Alonzo, P. J. Rodrigo, and J. Glückstad, "Generalized phase contrast matched to Gaussian illumination," *Opt. Express* **15**, 11971–11977 (2007).

[31] D. Palima and J. Glückstad, "Gaussian to uniform intensity shaper based on generalized phase contrast," *Opt. Express* **16**, 1507–1516 (2008).

[32] T. van Dijk, G. Gbur, and T. D. Visser, "Shaping the focal intensity distribution using spatial coherence," *J. Opt. Soc. Am. A* **25**, 575–581 (2008).

[33] J. Rubinstein and G. Wolansky, "A diffractive optical element for shaping arbitrary beams," *Opt. Rev.* **15**, 140–142 (2008).

[34] T. G. Jabbour and S. M. Kuebler, "Vectorial beam shaping," *Opt. Express* **16**, 7203–7213 (2008).

[35] C. Li and B. Lü, "Transformation and spatial shaping of partially coherent cosh-Gaussian beams through an astigmatic lens," *Optik* **120**, 374–378 (2009).

[36] M. Fratz, S. Sinzinger, and D. Giel, "Design and fabrication of polarization-holographic elements for laser beam shaping," *Appl. Opt.* **46**, 2669–2677 (2009).

[37] C. Zhang, N. R. Quick, and A. Kar, "Diffractive optical elements for pitchfork beam shaping," *Opt. Eng.* **48**, 078001(1-9) (2009).

[38] C. Dorrer, "High-damage-threshold beam shaping using binary phase plates," *Opt. Lett.* **34**, 2330–2332 (2009).

[39] R. N. Gaster, "Excimer laser photorefractive surgery of the cornea," *SPIE* **3343**, 212–220 (1998).

[40] W. D. Veldkamp and C. J. Kastner, "Beam profile shaping for laser radars that use detectors arrays," *Appl. Opt.* **21**, 345–356 (1982).

[41] N. Sanner, N. Huot, E. Audonard, C. Larat, and J.-P. Huignard, "Direct ultrafast laser microstructuring of materials using programmable beam shaping," *Opt. Lasers Eng.* **45**, 737–741 (2007).

[42] H. Koebner, ed., *Industrial Applications of Lasers* (New York: Wiley-Interscience, 1988).

[43] V. W. S. Chan, "Free-space optical communications," *J. Light. Technol.* **34**, 4750–4762 (2006).

[44] G. E. Marshall, ed., *Optical Scanning* (New York: Marcel Dekker Inc., 1991).

[45] F. M. Dickey, S. C. Holswade, and D. L. Shealy, eds., *Laser Beam Shaping Applications* (Boca Raton, FL: CRC Press, 2006).

[46] J. Durnin, "Exact solutions for nondiffracting beams," *J. Opt. Soc. Am. A* **4**, 651–654 (1987).

[47] Z. Bouchal, "Nondiffracting optical beams: physical properties, experiments, applications," *Czech. J. Phys.* **53**, 537–624 (2003).

[48] A. S. Ostrovsky, *Coherent Mode Representation in Optics* (Bellingham, WA: SPIE Press, 2006), chap. 3.

[49] J. Durnin, J. J. Micelli, Jr., and J. H. Eberly, "Diffraction-free beams," *Phys. Rev. Lett.* **58**, 1499–1501 (1987).

[50] F. Gori, G. Guattari, and C. Padovani, "Bessel–Gauss beams," *Opt. Commun.* **64**, 491–495 (1987).

[51] J. Turunen, A. Vasara, and A. T. Friberg, "Holographic generation of diffraction-free beams," *Appl. Opt.* **27**, 3959–3962 (1988).

[52] G. Häusler and W. Heckel, "Light sectioning with large depth and high resolution," *Appl. Opt.* **27**, 5165–5169 (1988).

[53] L. Vicari, "Truncation of nondiffracting beams," *Opt. Commun.* **70**, 263–266 (1989).

[54] M. Florjanczyk and R. Tremblay, "Guiding of atoms in traveling-wave laser trap formed by axicon," *Opt. Commun.* **73**, 448–450 (1989).

[55] G. Indebetouw, "Nondiffracting optical fields: some remarks on their analysis and synthesis," *J. Opt. Soc. Am. A* **6**, 150–152 (1989).

[56] A. Vasara, J. Turunen, and A. T. Friberg, "General diffraction-free beams produced by computer-generated holograms," *SPIE* **311**, 85–89 (1989).

[57] A. Vasara, J. Turunen, and A. T. Friberg, " Realization of general nondiffracting beams with computer-generated holograms," *J. Opt. Soc. Am. A* **6**, 1748–1754 (1989).

[58] K. Uehara and H. Kikuchi, "Generation of nearly diffraction-free laser beams," *Appl. Phys. B* **48**, 125–129 (1989).

[59] J. Turunen, A. Vasara, and A. T. Friberg, "Propagation invariance and self-imaging in variable-coherence optics," *J. Opt. Soc. Am. A* **8**, 282–289 (1991).

[60] T. Hidaka, "Generation of a diffraction-free laser beam using a specific Fresnel zone plate," *Jpn. J. Appl. Phys.* **30**, 1738–1739 (1991).

[61] R. M. Herman and T. A. Wiggins, "Production and uses of diffractionless beams," *J. Opt. Soc. Am. A* **8**, 932–942 (1991).

[62] N. Davidson, A. A. Friesem, and E. Hasman, "Efficient formation of nondiffracting beams with uniform intensity along the propagation direction," *Opt. Commun.* **88**, 326–330 (1992).

[63] A. J. Cox and J. D'Anna, "Constant-axial-intensity nondiffracting beam," *Opt. Lett.* **17**, 232–234 (1992).

[64] A. J. Cox and D. C. Dibble, "Nondiffracting beam from a spatially filtered Fabry–Perot resonator," *J. Opt. Soc. Am. A* **9**, 282–286 (1992).

[65] L. C. Laycock and S. C. Webster, "Bessel beams: their generation and application," *GEC J. Res.* **10**, 36–51 (1992).

[66] K. M. Iftekharunddin and M. A. Karim, "Heterodyne detection by using a diffraction-free beam: tilt and offset effects," *Appl. Opt.* **31**, 4853–4856 (1992).

[67] J. Sochacki, A. Kolodziejcryk, Z. Jareszewicx, and S. Bará, "Nonparaxial design of generalized axicons," *Appl. Opt.* **31**, 5326–5330 (1992).

[68] G. Scott and N. McArdle, "Efficient generation of nearly diffraction-free beams using an axicon," *Opt. Eng.* **31**, 2641–2643 (1992).

[69] R. P. McDonald, J. Chrostowski, S. A. Boothroyd, and B. A. Syrett, "Holographic formation of a dipole laser nondiffracting beam," *Appl. Opt.* **32**, 6470–6474 (1993).

[70] J. A. Davis, J. Guertin, and D. M. Cottrell, "Diffraction-free beams generated with programmable spatial light modulators," *Appl. Opt.* **32**, 6368–6370 (1993).

[71] J. Rosen, "Synthesis of nondiffracting beams in free-space," *Opt. Lett.* **19**, 369–371 (1994).

[72] C. Ozkul, S. Leroux, N. Anthore, M. K. Amara, and S. Rasset, "Optical amplification of diffraction-free beams by photorefractive two-wave mixing and its application to laser Doppler velocimetry," *Appl. Opt.* **34**, 5485–5491 (1995).

[73] J. Rosen, B. Salik, and A. Yariv, "Pseudo-nondiffracting beams generated by radial harmonic functions," *J. Opt. Soc. Am. A* **12**, 2446–2457 (1995).

[74] Z. Bouchal and M. Olivik, "Nondiffractive vector Bessel beams," *J. Mod. Opt.* **21**, 1555–1566 (1995).

[75] J. Rosen, B. Salik, A. Yariv, and H.-K. Liu, "Pseudo-nondiffracting slitlike beam and its analogy to the pseudonondispersing pulse," *Opt. Lett.* **20**, 423–425 (1995).

[76] R. P. McDonald, S. A. Boothroyd, T. Okamoto, J. Chrostowski, and B. A. Syrett, "Interboard optical data distribution by Bessel beam shadowing," *Opt. Commun.* **122**, 169–177 (1996).

[77] T. Aruga, "Generation of long-range nondiffracting narrow light beams," *Appl. Opt.* **36**, 3762–3768 (1997).

[78] L. Niggl, T. Lanzi, and M. Maier, "Properties of Bessel beams generated by periodic gratings of circular symmetry," *J. Opt. Soc. Am. A* **14**, 27–33 (1997).

[79] R. Piestun and J. Shamir, "Generalized propagation-invariant wave fields," *J. Opt. Soc. Am. A* **15**, 3039–3044 (1998).

[80] R. Liu, B. Z. Dong, G. Z. Yang, and B. Y. Gu, "Generation of pseudo-nondiffracting beams with use of diffractive phase elements designed by the conjugate-gradient method," *J. Opt. Soc. Am. A* **15**, 144–151 (1998).

[81] M. R. Wang, C. Yu, and A. J. Varela, "Efficient pseudo-nondiffracting beam shaping using a quasicontinuous-phase diffractive element," *Opt. Eng.* **40**, 517–524 (2001).

[82] A. S. Fedor, "Binary optics diffuser design," in *Micromachining and Microfabrication Process Technology VII*, eds. J. M. Karan and J. A. Yasaitis, Proc. SPIE **4557**, 378–385 (2001).

[83] J. N. Mait, "Understanding diffractive optic design in the scalar domain," *J. Opt. Soc. Am. A* **12**, 2145–2158 (1995).

[84] W. T. Welford, "Optical estimation of statistics of surface roughness from light scattering measurements," *Opt. Quant. Electron.* **9**, 269–287 (1977).

[85] W. T. Welford, "Laser speckle and surface roughness," *Contemp. Phys.* **21**, 401–412 (1980).

[86] Zu-Han Gu, H. M. Escamilla, E. R. Méndez, A. A. Maradudin, J. Q. Lu, T. Michel, and M. Nieto-Vesperinas, "Interaction of two optical beams at a symmetric random surface," *Appl. Opt.* **31** 5878–5889 (1992).

[87] E. R. Méndez and D. Macías, "Inverse problems in optical scattering," in *Light Scattering and Nanoscale Surface Roughness*, ed. A. A. Maradudin (New York: Springer, 2006), pp. 435–465.

[88] M. A. Golub, "Generalized conversion from the phase function to the blazed surface-relief profile of diffractive optical elements," *J. Opt. Soc. Am. A* **16**, 1194–1201 (1999).

[89] L. Mandel and E. Wolf, *Optical Coherence and Quantum Optics* (New York: Cambridge University Press, 1995), sect. 4.3.2.

[90] A. C. Schell, The multiple plate antenna. Ph.D. dissertation, Massachusetts Institute of Technology (1961), sect. 7.5.

[91] E. R. Méndez, T. A. Leskova, A. A. Maradudin, M. Leyva-Lucero, and J. Muñoz-Lopez, "The design of two-dimensional random surfaces with specified scattering properties," *J. Opt. A* **7**, S141–S151 (2005).

[92] W. H. Press, S. A. Teukolsky, W. T. Vetterling, and B. P. Flannery, *Numerical Recipes in Fortran*, 2nd edn (New York: Cambridge University Press, 1992), pp. 281–282.

[93] M. Burns, *Automated Fabrication: Improving Productivity in Manufacturing* (Englewood Cliffs, NJ: Prentice Hall, 1993), sect. 6.5.

[94] StereoLithography interface specification. http://www.Ennex.com/fabbers/StL.sht, Ennex Corporation, 1989.

[95] J. M. Stone, *Radiation and Optics* (New York: McGraw-Hill, 1963), sect. 12-6.

[96] C.-W. Tsai, Y.-C. Chang, Y. Sh. Shmavonyan, Y.-S. Su, and C.-F. Lin, "Extremely broad band superluminescent diodes/semiconductor optical amplifiers in optical communications band," *Proc. SPIE* **4989**, 69–77 (2003).

[97] W. T. Welford, "Use of annular aperture to increase focal depth," *J. Opt. Soc. Am.* **50**, 749–753 (1960).

[98] M. Born and E. Wolf, *Principles of Optics*, 7th edn (Cambridge: Cambridge University Press, 1999), sect. 8.8.

[99] J. C. Dainty, "The image of a point for an aberration free lens with a circular pupil," *Opt. Commun.* **1**, 176–178 (1969).

[100] A. A. Maradudin, T. A. Leskova, and E. R. Méndez, "Pseudo-nondiffracting beams from rough surface scattering," in *Wave Propagation, Scattering and Emission in Complex Media*, ed. Ya-Qiu Jin (Singapore: World Scientific, 2004), pp. 100–118.

[101] E. E. García-Guerrero, E. R. Méndez, H. M. Escamilla, T. A. Leskova, and A. A. Maradudin, "Design and fabrication of random phase diffusers for extending the depth of focus," *Opt. Express* **15**, 910–923 (2007).

[102] W. B. Lee, Benny C. F. Cheung, *Surface Generation in Ultra-precision Diamond Turning Modelling and Practices* (London: Professional Engineering Publishing, 2003).

[103] R. Leutz and A. Suzuki, *Nonimaging Fresnel Lenses: Design and Performance of Solar Concentrators* (New York: Springer, 2001).

7

Surface electromagnetic waves on structured perfectly conducting surfaces

A. I. FERNÁNDEZ-DOMÍNGUEZ, F. GARCÍA-VIDAL,
AND L. MARTÍN-MORENO

7.1 Introduction

The ability to localize electromagnetic energy below the diffraction limit of classical optics featured by surface plasmon polaritons (SPPs) (electromagnetic surface waves sustained at the interface between a conductor and a dielectric) is currently being exploited in numerous studies ranging from photonics, optoelectronics, and materials science to biological imaging and biomedicine [1]. While the basic physics of SPPs has been described in a number of seminal papers spanning the twentieth century [2, 3], the more recent emergence of powerful nanofabrication and characterization tools has catalyzed a vast interest in their study and exploitation. The dedicated field of plasmonics [4] brings together researchers and technologists from a variety of disciplines, with the common aim to take advantage of the subwavelength light confinement associated with the excitation of SPPs.

Most interest is focused on the optical regime, where SPPs are strongly confined to the respective metal/dielectric interface, i.e. where subwavelength mode localization is achieved in the direction perpendicular to the interface. These strongly confined SPPs occur at frequencies which are still an appreciable fraction of the intrinsic plasma frequency of the metal in question. In this regime, the motion of the conduction electrons at the interface is dephased with respect to the driving electromagnetic fields, leading to a reduction in both phase and group velocities of the SPP, and, therefore, to strong localization. A considerable fraction of the SPP field energy resides inside the conductor. This fraction increases with confinement of the SPP inside the dielectric, causing the well known tradeoff between confinement and propagation loss.

Most SPP research has thus far focused on the noble metals, such as Ag, Au, and Cu, which show plasma frequencies in the UV. Therefore, the aforementioned

Structured Surfaces as Optical Metamaterials, ed. A. A. Maradudin. Published by Cambridge University Press.
© Cambridge University Press 2011.

strong localization is only achieved for visible frequencies. As the frequency is lowered, the spread of the field in the direction normal to the interface increases, from being subwavelength to extending over many wavelengths. In this limit, SPPs acquire the character of a grazing-incidence light field, with phase velocities asymptoting the phase velocity of light in the dielectric. It is interesting to note that the first theoretical descriptions of SPPs considered this regime, namely the seminal publications by Sommerfeld [2] and Zenneck [5] on electromagnetic surface wave propagation at radio frequencies along cylindrical metal wires and planar metal interfaces. The link with more localized SPP excited via optical beams on diffraction gratings [6] or via electron impact [3] occurred decades later.

It would be highly desirable to achieve light localization of subwavelength dimensions at low frequencies such as microwave or terahertz (THz) frequencies. Since the middle of the twentieth century, it has been known that the addition of a subwavelength corrugation (arrays of holes, for example) to the metal surface produces an enhanced surface impedance and binds a surface mode to the interface, even in the limit of perfect conductivity [7, 8], and it is at the origin of the whole field of frequency selective surfaces [9]. This can be understood in the following way. The presence of a periodic array of small holes can be considered within a perturbative approach. The result is that the dispersion relation (band) of surface electromagnetic (EM) modes in the corrugated structure will closely follow the one for SPPs, except for values of the momentum lying close to the boundary of the Brillouin zone, where the SPP band bends in order to accommodate for band gaps. The band sector below the first gap represents a truly bound surface mode. As a result of the band bending caused by the periodic array of holes, the lowest band of the surface modes separates from the light line, therefore binding the EM field more strongly to the surface. This line of reasoning has been presented before [10].

In 2004, Pendry and co-workers [11] discovered that there is an additional mechanism for periodicity-induced binding of the EM fields to the surface. As will be extensively discussed in Section 7.3, these authors found that the effective surface layer can be described with a dielectric permittivity of the Drude form, with the plasma frequency given by the cutoff frequency of the hole waveguide. In this way, the surface EM modes supported by corrugated metal surfaces can be entirely controlled by geometry. In the perfect electrical conductor (PEC) limit, these designed surface EM modes are known as *spoof* surface plasmon polaritons.

In this chapter we focus our study on the recent developments in the field of surface EM waves on structured metal surfaces that have been reported since 2004. More specifically, we will show how the concept of spoof SPPs can be applied to very different geometries, such as planar (Section 7.3), cylindrical (Section 7.4), and even to more complex structures such as helically grooved wires, V-grooves, wedges, and domino structures (Section 7.5). In order to study the emergence of the corresponding spoof SPP modes, we will apply a common theoretical formalism, a coupled mode method, the details of which are introduced in Section 7.2.

7.2 Theoretical formalism: coupled mode method

We introduce the theoretical formalism that we employ to investigate the geometry dependence of the modal properties of the spoof SPPs supported by planar and cylindrical metal structures partially or fully perforated with periodic arrays of indentations. It is based on the modal expansion technique, a powerful theoretical framework which has been successfully applied to study different EM phenomena such as extraordinary transmission [12, 13] or negative index metamaterials [14]. Unlike other approaches used to analyze the emergence of spoof SPP modes [15–17], this theoretical method yields, under certain conditions, analytical expressions for the dispersion relation of the guided modes. Despite their approximate character, these simple dispersion relations allow us to reach a deeper understanding of the spoof SPP concept and its potential applications.

Our approach consists in the expansion of the EM fields into eigenmodes of Maxwell's equations within the various regions comprising the system under study. By imposing the appropriate continuity conditions at all the internal boundaries, EM fields can be constructed in all space. Although the so-called surface impedance boundary conditions [18] can be implemented into the formalism, in this chapter we treat metals as PECs, which is an excellent approximation at microwave or terahertz (THz) frequencies. Note that the scattering properties of a corrugated PEC remain invariant if all the lengths are scaled by the same factor, which allows the transfer of results from one frequency range to another.

Figure 7.1 shows a schematic of the expansion procedure for the case of a two-dimensional (2D) array of square holes with periods d_x and d_y. Here, we illustrate the construction of our theoretical framework in planar structures. In Section 7.4 we will introduce the modifications which allow us to treat cylindrical geometries. We denote by z the direction normal to the metal structure. The interfaces are placed at $z = 0$ and $z = h$. When treating fully pierced structures, the latter interface corresponds to the lower surface of the metal slab, whereas in the case of partial perforation, it yields the bottom of the blind indentations decorating the upper metal surface. Thus, the system is divided into three regions along the z-direction: the upper semi-space with dielectric constant ϵ_{I} (region I), the metal perforated with filled indentations with dielectric constant ϵ_{II} (region II), and the substrate, comprising a dielectric with permittivity ϵ_{III} (a PEC medium) for fully (partially) pierced geometries (region III). Note that the periodic character of the system

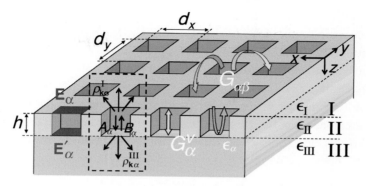

Figure 7.1. Schematic of the modal expansion procedure for the case of a periodic array of square holes. Both the expansion coefficients and the different terms in Eqs. (7.12) and (7.16) are rendered schematically.

enables us to apply Bloch's theorem and to expand the EM fields only within the unit cell of area $d_x \times d_y$ containing one single indentation.

A convenient notation, which simplifies the calculations presented in this chapter, is to use a Dirac nomenclature for the EM fields. In this way, we define the bi-vectors $\langle \mathbf{r} | \mathbf{E} \rangle = \mathbf{E}(\mathbf{r}) = (E_x(\mathbf{r}), E_y(\mathbf{r}))^T$ and $\langle \mathbf{r} | \mathbf{H} \rangle = \mathbf{H}(\mathbf{r}) = (H_x(\mathbf{r}), H_y(\mathbf{r}))^T$ (T standing for transposition). Note that $\mathbf{r}$ refers to the parallel components (x and y) of the spatial vector and that the z-components of the EM fields can be found using the Maxwell equations and the direction of propagation of the field.

Then, in region I, EM fields can be expanded into an infinite set of plane waves, $|\mathbf{k}_{mn}, \sigma \rangle$, which are characterized by their polarization σ (s or p) and parallel wave vector $\mathbf{k}_{mn} = \mathbf{k}_{\parallel} + \mathbf{K}_{mn}$, where $\mathbf{k}_{\parallel}$ is the in-plane wave vector of the spoof SPP mode and $\mathbf{K}_{mn} = m(2\pi/d_x)\mathbf{u}_x + n(2\pi/d_y)\mathbf{u}_y$ is a vector of the two-dimensional (2D) reciprocal lattice. The expressions for these plane waves in real space are:

$$\langle \mathbf{r} | \mathbf{k}p \rangle = (k_x, k_y)^T \exp(\imath \mathbf{k} \cdot \mathbf{r})/\sqrt{d_x d_y |k|^2},$$
$$\langle \mathbf{r} | \mathbf{k}s \rangle = (-k_y, k_x)^T \exp(\imath \mathbf{k} \cdot \mathbf{r})/\sqrt{d_x d_y |k|^2}. \tag{7.1}$$

These modes are orthonormal when integrated over a unit cell, i.e. $\langle \mathbf{k}\sigma | \mathbf{k}'\sigma' \rangle = \delta_{\mathbf{k},\mathbf{k}'}\delta_{\sigma,\sigma'}$, where δ is the Kronecker delta.

By introducing the unknown expansion coefficients $\rho^{\mathrm{I}}_{\mathbf{k}_{mn}\sigma}$, the parallel components of the electric and magnetic fields can be written as follows:

$$|\mathbf{E}^{\mathrm{I}}_{\mathrm{t}} \rangle = \sum_{m,n,\sigma} \rho^{\mathrm{I}}_{\mathbf{k}_{mn}\sigma} |\mathbf{k}_{mn}, \sigma \rangle e^{\kappa^{\mathrm{I}}_{mn}z}, \tag{7.2}$$

$$|-\mathbf{u}_z \times \mathbf{H}^{\mathrm{I}}_{\mathrm{t}} \rangle = \sum_{m,n,\sigma} Y^{\mathrm{I}}_{\mathbf{k}_{mn}\sigma} \rho^{\mathrm{I}}_{\mathbf{k}_{mn}\sigma} |\mathbf{k}_{mn}, \sigma \rangle e^{\kappa^{\mathrm{I}}_{mn}z}. \tag{7.3}$$

where the normal plane wave vector is $\kappa_{mn}^I = \sqrt{|\mathbf{k}_{mn}|^2 - \epsilon_I k_0^2}$, and $k_0 = 2\pi/\lambda = 2\pi f/c$ is the wave vector modulus in a vacuum. The electric and magnetic fields are related through the admittances $Y_{\mathbf{k}_{mn} s}^I = i\kappa_{mn}^I/k_0$ and $Y_{\mathbf{k}_{mn} p}^I = -i\epsilon_I k_0/\kappa_{mn}^I$. Note that we have assumed that all plane waves in region I are evanescent along the z-direction, i.e. $|\mathbf{k}_{mn}| > \sqrt{\epsilon_I} k_0$.

In region II, as we are modeling the metal response through the PEC approach, EM fields are nonzero only within the perforations. Therefore, parallel components of the fields can be expressed in terms of the corresponding waveguide modes, labeled with index α, having

$$|\mathbf{E}_t^{II}\rangle = \sum_\alpha [A_\alpha e^{iq_\alpha(z-h)} + B_\alpha e^{-iq_\alpha(z-h)}]|\alpha\rangle, \tag{7.4}$$

$$|-\mathbf{u}_z \times \mathbf{H}_t^{II}\rangle = \sum_\alpha Y_\alpha^{II}[A_\alpha e^{iq_\alpha(z-h)} - B_\alpha e^{-iq_\alpha(z-h)}]|\alpha\rangle, \tag{7.5}$$

where q_α is the propagation constant of mode $|\alpha\rangle$ and A_α and B_α are the unknown expansion coefficients. The mode admittances are given by $Y_\alpha^{II} = q_\alpha/k_0$ (for s-polarization) and $Y_\alpha^{II} = \epsilon_{II} k_0/q_\alpha$ (for p-polarization). Note that in this case we do not impose the propagating/evanescent character of the basis elements along the z-direction. In the case of blind indentations, the PEC boundary at the bottom of the perforations implies $\mathbf{E}_t|_{z=h} = 0$ and therefore $A_\alpha = -B_\alpha$ in Eqs. (7.4) and (7.5).

In region III, EM fields can be expressed again in terms of plane waves decaying along the z-direction as

$$|\mathbf{E}_t^{III}\rangle = \sum_{m,n,\sigma} \rho_{\mathbf{k}_{mn}\sigma}^{III} |\mathbf{k}_{mn}, \sigma\rangle e^{-\kappa_{mn}^{III} z}, \tag{7.6}$$

$$|-\mathbf{u}_z \times \mathbf{H}_t^{III}\rangle = -\sum_{m,n,\sigma} Y_{\mathbf{k}_{mn}\sigma}^{III} \rho_{\mathbf{k}_{mn}\sigma}^{III} |\mathbf{k}_{mn}, \sigma\rangle e^{-\kappa_{mn}^{III} z}, \tag{7.7}$$

where the definitions of all terms are the same as in Eqs. (7.2) and (7.3), substituting ϵ_I by ϵ_{III}. Importantly, when considering blind perforations, this region is filled with PEC metal, and, therefore, electric and magnetic fields must vanish within it, i.e. $\rho_{\mathbf{k}_{mn}\sigma}^{III} = 0$ for all $|\mathbf{k}_{mn}, \sigma\rangle$.

The unknowns $\rho_{\mathbf{k}_{mn}\sigma}^I$, A_α, B_α, and $\rho_{\mathbf{k}_{mn}\sigma}^{III}$ are calculated by imposing continuity of the parallel components of the EM field at $z = 0$ and $z = h$. Thus, we obtain four vectorial equations (one for each field component at each interface) which depend on the parallel spatial coordinates x and y:

$$\sum_{m,n,\sigma} \rho_{\mathbf{k}_{mn}\sigma}^I |\mathbf{k}_{mn}, \sigma\rangle = \sum_\alpha [A_\alpha e^{-iq_\alpha h} + B_\alpha e^{iq_\alpha h}]|\alpha\rangle, \tag{7.8}$$

$$\sum_{m,n,\sigma} Y_{\mathbf{k}_{mn}\sigma}^I \rho_{\mathbf{k}_{mn}\sigma}^I |\mathbf{k}_{mn}, \sigma\rangle = \sum_\alpha Y_\alpha^{II}[A_\alpha e^{-iq_\alpha h} - B_\alpha e^{iq_\alpha h}]|\alpha\rangle, \tag{7.9}$$

$$\sum_{\alpha}[A_\alpha + B_\alpha]|\alpha\rangle = \sum_{m,n,\sigma} \rho^{\mathrm{III}}_{\mathbf{k}_{mn}\sigma}|\mathbf{k}_{mn},\sigma\rangle e^{-\kappa^{\mathrm{III}}_{mn}h}, \tag{7.10}$$

$$\sum_{\alpha}Y^{\mathrm{II}}_\alpha[A_\alpha - B_\alpha]|\alpha\rangle = -\sum_{m,n,\sigma} Y^{\mathrm{III}}_{\mathbf{k}_{mn}\sigma}\rho^{\mathrm{III}}_{\mathbf{k}_{mn}\sigma}|\mathbf{k}_{mn},\sigma\rangle e^{-\kappa^{\mathrm{III}}_{mn}h}. \tag{7.11}$$

Note that in the case of blind indentations, Eqs. (7.10) and (7.11), which reflect field continuity at the boundary between regions II and III, are redundant. Let us stress that the PEC approximation leads to a relevant difference between equations for the parallel electric (Eqs. (7.8) and (7.10)) and magnetic fields (Eqs. (7.9) and (7.11)): whereas $\mathbf{E}_t$ must be continuous everywhere within the xy plane, $\mathbf{H}_t$ must be continuous only at the perforation openings. By projecting the electric (magnetic) continuity equations onto vacuum plane waves (indentation waveguide modes) we take into account this fact, as well as removing the dependence of the continuity equations on the spatial coordinates x and y. The analytical expressions for the overlap integrals between plane waves and waveguide modes of different aperture shapes can be found in the literature [19, 20].

We can combine the projected equations obtained from Eqs. (7.8)–(7.11) to express the continuity of the EM fields at the two interfaces of the system in the form of tight-binding-like equations:

$$\begin{aligned}
(G^{\mathrm{I}}_{\alpha\alpha} - \epsilon_\alpha)E_\alpha + \sum_{\alpha'\neq\alpha} G^{\mathrm{I}}_{\alpha\alpha'}E_{\alpha'} - G^V_\alpha E'_\alpha &= 0, \\
(G^{\mathrm{III}}_{\alpha\alpha} - \epsilon_\alpha)E'_\alpha + \sum_{\alpha'\neq\alpha} G^{\mathrm{III}}_{\alpha\alpha'}E'_{\alpha'} - G^V_\alpha E_\alpha &= 0,
\end{aligned} \tag{7.12}$$

whose unknowns are now the modal amplitudes of $\mathbf{E}_t$ at the openings of the perforations, which can be written in terms of the expansion coefficients as $E_\alpha = A_\alpha e^{-iq_\alpha h} + B_\alpha e^{iq_\alpha h}$ (at $z = 0$) and $E'_\alpha = -[A_\alpha + B_\alpha]$ (at $z = h$).

We can give a simple physical interpretation to all the magnitudes appearing in the matching equations. First, let us stress that the upper (lower) equation in system (7.12) is obtained from Eqs. (7.8) and (7.9) (Eqs. (7.10) and (7.11)) and therefore it can be associated with the field continuity at the I–II (II–III) interface of the system. Thus, the term

$$\epsilon_\alpha = Y^{\mathrm{II}}_\alpha \cot(q_\alpha h), \tag{7.13}$$

which takes into account how fields on one of the system interfaces affect the modal amplitudes at that interface, can be interpreted as resulting from the bouncing back and forth of the EM fields inside the indentations. On the other hand,

$$G^V_\alpha = Y^{\mathrm{II}}_\alpha \frac{1}{\sin(q_\alpha h)}, \tag{7.14}$$

which links the amplitudes at one interface with the fields at the other, is reflecting the EM coupling at the two sides of the metal slab through the perforations. Note that these two terms involve only one waveguide mode $|\alpha\rangle$. However, the terms

$$G_{\alpha\alpha'}^{\mathrm{I,III}} = i \sum_{m,n,\sigma} Y_{\mathbf{k}_{mn}\sigma}^{\mathrm{I,III}} \langle\alpha|\mathbf{k}_{mn}\sigma\rangle\langle\mathbf{k}_{mn}\sigma|\alpha'\rangle \qquad (7.15)$$

describe the interaction between modal amplitudes corresponding to different waveguide modes at a given interface. Note that $Y_{\mathbf{k}_{mn}p}^{\mathrm{I,III}}$, and therefore the G terms, diverge when the grazing condition $\kappa_{mn}^{I,III} = \sqrt{|\mathbf{k}_{mn}| - \epsilon_{\mathrm{I,III}}k_0^2} = 0$ is satisfied. In Fig. 7.1 the physical interpretations of the different terms in the set of homogeneous equations (7.12) are represented schematically.

Let us now describe how this result is modified for the case of blind holes, i.e. when a PEC substrate is considered. Then, Eqs. (7.8) and (7.9), together with the condition $A_\alpha = -B_\alpha$, express the appropriate continuity of the EM fields. Following the same matching procedure as before, we end up with a set of equations of the form

$$(G_{\alpha\alpha}^{\mathrm{I}} - \epsilon_\alpha)E_\alpha + \sum_{\alpha'\neq\alpha} G_{\alpha\alpha'}^{\mathrm{I}} E_{\alpha'} = 0, \qquad (7.16)$$

where the definition of all the terms remains the same as before and the unknown amplitudes are given by $E_\alpha = 2i B_\alpha \sin(q_\alpha h)$ (note that, as expected, $E'_\alpha = 0$). This result agrees with our interpretation of Eqs. (7.12). Now, only the interface between regions I and II is relevant, which removes the equations related to the II–III interface and the coupling between interfaces through the perforations (which are blind), yielding $G_\alpha^V = 0$.

Although the equations presented up to now are general and can be applied to both 1D and 2D geometries, it is worth commenting briefly how our approach is simplified when treating the 1D case with $k_y = 0$. In this case, it can be shown that light polarizations are decoupled, which permits the independent treatment of s- and p-polarized waves. Thus, when studying spoof SPPs in 1D indentations, we consider only p-polarization, as the appearance of nonzero solutions in Eqs. (7.12) and (7.16) is linked to the divergent behavior of $G_{\alpha\alpha'}$ when a p-polarized plane wave goes grazing. Therefore, we can restrict the expansion basis in our theoretical framework to only p-polarized modes when treating 1D systems, which simplifies considerably the calculations.

Spoof SPP modes supported by perforated metals are given by the nonzero solutions of Eqs. (7.12) and (7.16). Specifically, the dispersion relation of these bound modes can be calculated by finding the parallel wave vector, $\mathbf{k}_{||}$, and frequency, f, for which the determinant associated to the matching equations vanishes. This problem must be solved numerically in general, but analytical expressions can be obtained by using two approximations:

- only the fundamental waveguide mode (that we denote as $\alpha = 0$) is taken into account in the modal expansion inside the indentations, minimizing the size of the set of matching equations;
- only the zero-order diffracted (p-polarized) plane wave is considered in $G_{\alpha\alpha'}$, which provides us with simple expressions for this term.

The first approximation leads to accurate results for subwavelength indentations, failing when the cross section of the indentations is comparable to the size of the array unit cell. The second approximation, which is equivalent to considering the perforated structure as an homogenous metamaterial penetrated by the average EM fields, must be corrected close to the band edges, as it does not reflect the presence of band gaps due to diffraction effects.

Let us first consider the analytical results obtained from those two approximations for the case of periodic blind perforations. The condition for the existence of bound modes in these systems is given by

$$G - \epsilon = 0, \tag{7.17}$$

where we have made $G \equiv G_{00}^{I}$ (Eq. (7.15)) and $\epsilon \equiv \epsilon_0$ (Eq. (7.13)). Neglecting diffraction effects, the G term is given by $G = i Y_{\mathbf{k}_{||} p} |S|^2$, where $S = \langle \mathbf{k}_{||}, p | 0 \rangle$ is the overlap between the p-polarized zero-order diffracted mode and the lowest indentation waveguide mode. For apertures much smaller than the wavelength, the dependence of S on the parallel wave vector can be neglected. Imposing $\mathbf{k}_{||} = k_{||} \mathbf{u}_x$, we can write the dispersion relation of the spoof surface plasmon modes as

$$k_{||} = k_0 \sqrt{1 + \frac{|S|^4}{\epsilon^2}}. \tag{7.18}$$

Note that, as expected, $k_{||} > k_0$, which reflects the confined nature of the modes.

We now focus on metal slabs fully pierced by periodic indentations. The symmetric character of these structures with respect to the $z = h/2$ plane enables us to rewrite Eqs. (7.12) into a single approximate equation of the form

$$(G - \epsilon) \pm G^V = 0, \tag{7.19}$$

where $G^V \equiv G_0^V$. Importantly, the presence of the negative (positive) sign in Eq. (7.19) indicates the symmetric (antisymmetric) character of the modes with respect to the slab center. Following the same procedure as before, the analytical spoof surface plasmon bands can be now written as

$$k_{||} = k_0 \sqrt{1 + \frac{|S|^4}{(\epsilon \pm G^V)^2}}. \tag{7.20}$$

Again, we have $k_{||} > k_0$ (bound modes). Equation (7.19) indicates that the frequency shift between the spoof SPP dispersion relation for partially and fully perforated metals is governed by the ratio $G^V/\epsilon = 1/\cos(qh)$, which only depends on the propagation constant of the indentation waveguide mode, q, and the thickness of the metal slab, h.

7.3 Planar geometries

7.3.1 Textured surfaces

In this section we analyze the EM modes bound to textured PEC surfaces by considering the two simple geometries for which the spoof SPP concept was first developed: 1D arrays of grooves and 2D arrays of square dimples [11, 21]. We have shown that guided modes in these two structures are given by the nonvanishing solutions of the matching equations describing the continuity of the fields at the metal–vacuum interface.

First we consider the case of 1D arrays of grooves. The inset of Fig. 7.2(a) shows the geometrical parameters of the structure: the array period, d, and the grooves' width and depth, a and h, respectively. As mentioned before, the PEC approximation makes all lengths scalable. Thus, from now on, we take the system periodicity, d, as the unit of length. The approximate spoof SPP band obtained from Eq. (7.18) is given by

$$k_{||} = k_0\sqrt{1 + \left(\frac{a}{d}\right)^2 \tan^2(k_0 h)}, \qquad (7.21)$$

where we have used the result that $S = \sqrt{a/d}$ for very narrow 1D apertures and that the propagation constant for the lowest waveguide mode inside the grooves is $q = k_0$. This expression shows clearly the geometrical origin of the modes and allows us to predict the dependence of the spoof SPP bands on the width and depth of the indentations. Note that, according to Eq. (7.21), enlarging the groove depth translates into the deviation of the dispersion relation from the light line towards larger wave vectors, and that when $a = 0$ or $h = 0$ (flat surface), $k_{||} = k_0$, and no confined modes are supported by the structure.

Figure 7.2(a) displays the normalized frequency (d/λ) versus wave vector $(k_{||}d/2\pi)$ for the spoof SPP modes, in groove arrays of width $a = 0.2d$ and depths ranging from $h = 0.2d$ to $h = d$. As predicted by Eq. (7.21), the bands shift to lower frequencies when the depth of the grooves is increased. This result can be understood in terms of cavity resonances occurring inside the indentations. Note that the lowest waveguide mode supported by 1D apertures is always propagating. Thus, EM fields explore completely the groove depth, making them strongly

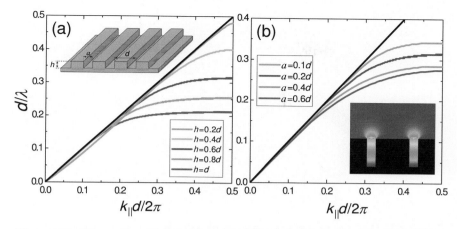

Figure 7.2. Dispersion relation of the spoof SPPs supported by periodic groove arrays. (a) Dependence on h for a fixed groove width $a = 0.2d$. The inset sketches the structure considered. (b) Bands for $h = 0.6d$ at different groove widths. The inset depicts the electric field amplitude for $h = 0.6d$ and $a = 0.2d$ evaluated at the band edge ($k_{\parallel} = \pi/d$). See color plates section.

dependent on h. This is reflected through the tangent function in Eq. (7.21), which diverges when the Fabry–Perot condition, $\sin(k_0 h) = 0$, is satisfied. We can interpret this behavior as resulting from the fact that spoof SPPs in 1D blind indentations have a hybrid nature with characteristics of both surface and cavity EM modes.

Figure 7.2(b) shows the dependence of the dispersion relation on the groove width (a) for $h = 0.6d$. Four groove sizes between $a = 0.1d$ and $a = 0.6d$ are considered. The mode frequency shifts to the red with larger a. This is also predicted by Eq. (7.21), where $k_{\parallel}$ grows linearly with a far from the light line. As the ratio a/d controls the overlap between the zero-order diffracted wave and the first waveguide mode, we can conclude that the EM coupling at the interface is larger for wider indentations, which increases the binding of the fields lowering the mode frequency. The inset depicts the electric field amplitude for a groove array with $a = 0.2d$ and $h = 0.6d$ evaluated at band edge ($k_{\parallel} = \pi/d$). It decays more rapidly into the vacuum than inside the grooves, which agrees with the interpretation of the spoof surface plasmons as hybrid modes between surface and cavity modes.

The previous calculations have been done within the PEC approximation, and, therefore, the propagation length of the corresponding spoof SPP modes is infinite. This length is reduced to finite values in a real metal due to absorption. Recently, Shen and co-workers [22] have studied the propagation length of geometrically modified SPP modes supported by 1D arrays of grooves in the THz regime. As expected, the loss is large at frequencies close to the asymptotic frequency.

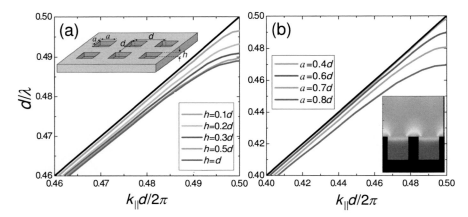

Figure 7.3. Dispersion relation of the spoof SPPs sustained by periodic dimple arrays. (a) Bands for $a = 0.2d$ and several dimple depths h. The inset shows a schematic of the system. (b) Bands for $h = 0.6d$ and different a. The inset displays the electric field amplitude at the band edge for the case $h = 0.6d$ and $a = 0.6d$. See color plates section.

Very recently, two different types of applications based on the propagation characteristics of spoof SPP modes in 1D arrays of grooves have been proposed. In the first one [23], the slowing down of THz waves based on 1D arrays of grooves with graded depths can be achieved by a proper design of the dispersion curves and asymptotic frequencies. Another proposal is that of active THz switches, controlled by an electro-optical material placed inside the grooves, which can be activated by a low-voltage control-signal [24].

Now we analyze the characteristics of the spoof SPP modes supported by 2D dimple arrays. For simplicity, we consider only the case of square arrays of square apertures; see the inset of Fig. 7.3(a). The geometry of the system is now given by the array period, d, taken as a reference length, and the side and depth of the dimples, a and h, respectively. The analytical expression for the dispersion relation of the modes has the form

$$k_{||} = k_0 \sqrt{1 + \left(\frac{2\sqrt{2}a}{\pi d}\right)^4 \frac{k_0^2}{(\pi/a)^2 - k_0^2} \tanh^2\left(\sqrt{(\pi/a)^2 - k_0^2}\, h\right)}, \quad (7.22)$$

where we have used the result that $S = (2\sqrt{2}/\pi)(a/d)$ in the limit of deep subwavelength indentations [21] and that the propagation constant for the lowest waveguide mode (TE_{11}) supported by 2D square apertures is $q = i\sqrt{(\pi/a)^2 - k_0^2}$. Note that, in the range of validity of our approximate approach, $\lambda >> 2a$, q for the TE_{11} mode is imaginary, and the fields decay evanescently within the perforations. This evanescent character of the fields is reflected in the hyperbolic tangent dependence on h that Eq. (7.22) yields for the spoof SPP dispersion relation. We can anticipate that the distinct character (propagating/evanescent) of the fundamental waveguide modes supported by 1D and 2D indentations leads to fundamental differences between the bound modes supported by these two geometries.

Figure 7.3(a) plots the exact spoof SPP bands for dimples of side $a = 0.6d$, as calculated by looking at zeroes of the determinant of the whole system of Eq. (7.16). The depth of the indentations is varied from $h = 0.1d$ to $h = d$. As in Fig. 7.2(a), the mode frequency is lowered when the depth of the perforations is increased. However, the displacement of the bands is much smaller than in the 1D case. This difference is related to the evanescent nature of the fields within the dimples and can be understood through Eq. (7.22), where the tanh function leads to a low sensitivity of the modes to variations in h.

In Fig. 7.3(b), the dependence of the spoof SPP bands on the dimple area is analyzed. Dimples of depth $h = 0.6d$ and sides between $a = 0.4d$ and $a = 0.8d$ are considered. As predicted by Eq. (7.22), the dispersion relation bends at lower frequencies when a is enlarged. Similar to groove arrays, this effect is due to an increase in the EM coupling of diffracted and dimple waveguide modes, which in our analytical approach is proportional to the ratio a/d. The inset depicts the electric field amplitude evaluated at the band edge for $a = 0.6d$ and $h = 0.5d$. As predicted by Eq. (7.22), the electric field is mostly located at the system interface ($z = 0$) and decays into both the indentations and the vacuum superstrate. Recently, the propagation and confinement of THz radiation in planar copper surfaces pierced by square arrays of square dimples have been experimentally analyzed [25], obtaining results in accordance with the analytical results described above.

7.3.2 Perforated slabs

In this subsection, we study the modal properties of the spoof SPPs supported by periodic arrays of 1D and 2D apertures drilled in PEC slabs of thickness h and compare them with those of textured surfaces. As in the case of blind indentations, it is possible to construct analytical expressions for the dispersion relation of the modes by recalling Eq. (7.20). Thus, for 1D arrays of slits of width a we have

$$k_{\parallel} = k_0 \sqrt{1 + \left(\frac{a}{d}\right)^2 \frac{\sin^2(k_0 h)}{[\cos(k_0 h) \pm 1]^2}}, \tag{7.23}$$

where the negative (positive) sign corresponds to bound modes whose parallel component of the electric field is symmetric (antisymmetric) with respect to the mid-plane of the film. Note that the overlap and the propagation constant remain the same as in Eq. (7.21).

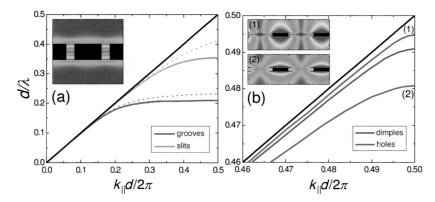

Figure 7.4. Dispersion relation of the spoof SPPs supported by fully perforated films. (a) Comparison between 1D slits and grooves of the same size ($a = 0.2d$ and $h = d$). Dotted lines show the analytical band obtained from Eqs. (7.21) and (7.23). The inset depicts the electric field amplitude at the band edge for the slit array. (b) Comparison between holes and dimples with $a = 0.6d$ and $h = 0.3d$. The insets depict the fields at the edges of the two spoof SPP bands for the hole array. See color plates section.

Figure 7.4(a) presents the dispersion relation of the spoof SPP modes supported by slits and grooves of the same dimensions. The width and depth of the indentations are $a = 0.2d$ and $h = d$, respectively. For this set of geometrical parameters, both structures sustain only one mode, which is tightly bound to the metal slab. The dispersion relation for the slit array is raised with respect to that for the grooves, which indicates that the modes are more weakly bound to the structure. This is a direct consequence of the propagating nature of the waveguide modes in 1D apertures, which leads to the bouncing of the fields inside the perforations. Whereas the bottom of the grooves acts as a mirror, slit openings allow the coupling to diffracted waves. This fact blueshifts the spectral position of the cavity resonances and permits spoof SPP modes to extend out of the structure. Dotted lines plot the analytical bands for both systems. The inset of Fig. 7.4(a) displays the electric field amplitude at the band edge for the fully perforated slab. Note that the electric field vanishes at the center of the slits, showing the odd parity of the mode with respect to the mid-plane of the film.

Similar results to those previously presented have been obtained by different groups [26–29]. Due to the similarity of these confined modes to those of waveguide modes supported by a dielectric slab, these works have been focused on obtaining the effective dielectric response of the 1D periodic metallic structure. It has been demonstrated that, in the metamaterial limit (wavelength much larger than the period of the array), a 1D array of slits behaves as an anisotropic dielectric medium characterized by $\epsilon_x = d/a$, $\epsilon_z = \infty$, and $\mu_y = a/d$ [27].

By introducing S and q for 2D apertures in Eq. (7.20), the dispersion relation of the spoof SPP modes supported by 2D holes fully piercing a metal slab can be calculated. For the simple case of square holes it is given by

$$k_{||} = k_0 \sqrt{1 + \left(\frac{2\sqrt{2}a}{\pi d}\right)^4 \frac{k_0^2}{(\pi/a)^2 - k_0^2} \frac{\sinh^2\left(\sqrt{(\pi/a)^2 - k_0^2}\, h\right)}{\left(\cosh\left(\sqrt{(\pi/a)^2 - k_0^2}\, h\right) \pm 1\right)^2}}. \quad (7.24)$$

Note that, similarly to dimple arrays, the evanescent character of the EM fields inside the apertures is reflected in the appearance of hyperbolic functions describing the dependence of the mode properties on h.

Similarly to blind indentations, the distinct behavior of EM fields within 1D and 2D apertures piercing the metal structure gives rise to different mode properties for these two systems. Whereas in 1D geometries the complete perforation of the metal slab blueshifts the mode frequency, in 2D perforations, this effect leads to the splitting of the spoof SPP band into two. This is clearly shown in Fig. 7.4(b), which plots the dispersion relation for dimples and holes of the same dimensions ($a = 0.6d$ and $h = 0.3d$). Note that, whereas the former support only one bound mode two different modes are sustained by the latter.

The origin of the two spoof SPP modes supported by hole arrays is clarified in the insets of Fig. 7.4(b). They depict the electric field amplitude at $k_{||} = \pi/d$ for the two spoof SPP bands. In both cases, EM fields are strongly localized at the film surfaces and decay into the holes. The two modes emerge from the interaction through the holes of the evanescent tails of the surface EM modes at each side of the film. The field patterns show that the lower (higher) band corresponds to bound modes having an even (odd) parity with respect to the symmetry plane of the perforated slab. Note that this phenomenology is similar to that observed in long- and short-range SPPs in thin metallic films [30].

It is worth analyzing the dispersion relation of these spoof SPP modes in the limit $h \to \infty$ (i.e. when the perforated PEC film is thick enough). In this limit, the coupling between the two sides of the PEC film vanishes and the two modes merge into one, whose dispersion relation reads

$$k_{||} = k_0 \sqrt{1 + \left(\frac{2\sqrt{2}a}{\pi d}\right)^4 \frac{k_0^2}{(\pi/a)^2 - \epsilon_{\text{hole}}k_0^2}}, \quad (7.25)$$

where ϵ_{hole} is the dielectric constant of the medium filling the holes. The effective parameters for the dielectric and magnetic response of a 2D array of holes perforating a semi-infinite PEC film can be obtained [11, 21] by:

- comparing the dispersion relation given by Eq. (7.25) with that of the canonical SPPs propagating on a *real* metal surface;
- forcing the EM fields to decay in the same way inside the corrugated metal and in the metamaterial.

This procedure yields the following expressions for ϵ_{eff} and μ_{eff}:

$$\epsilon_{\text{eff}}(\omega) = (\pi^2 d^2 \epsilon_{\text{hole}}/8a^2) \left(1 - \omega_p^2/\omega^2\right),$$
$$\mu_{\text{eff}} = 8a^2/\pi^2 d^2. \tag{7.26}$$

This functional form for $\epsilon_{\text{eff}}(\omega)$ is similar to Drude's expression for the dielectric constant of a metal. We can define an *effective plasma frequency*, $\omega_p = (c/\sqrt{\epsilon_{\text{hole}}})\pi/a$, which coincides with the cutoff frequency of the hole waveguide. Actually, the system is anisotropic, so care must be taken about the different components of the effective dielectric constant tensor. The anisotropy is also responsible for the fact that the flat region of the dispersion curve for the spoof SPP mode appears at $\epsilon_{\text{eff}} = 0$, whereas the dispersion relation for true SPPs bound to the interface between two *isotropic* media flattens at $\epsilon = -1$. The important point of a Drude formula for the dielectric response for a perforated PEC film is that the cutoff frequency of the hole waveguide marks the separation between positive and negative values for the effective dielectric function.

The simple dispersion relation of the spoof SPP modes supported by a 2D array of holes as written in Eq. (7.25) has been obtained within the two approximations described in Section 7.2. Namely, only the fundamental mode inside the holes is introduced in the modal expansion and only the zero-order *p*-polarized diffracted mode is considered. By including more modes in both regions (in the vacuum and inside the holes), different authors have demonstrated that the *exact* dispersion relation moves closer to the light line and strong confinement only occurs for frequencies much closer to ω_p than that which the effective parameter expression (7.26) predicts [15–17, 31].

The results of experiments in which the angular dependence of the transmission peaks in 2D arrays of holes infiltrated with wax was analyzed showed band bending associated with the cutoff frequency of the hole waveguide [32]. These experiments were carried out in the microwave regime of the EM spectrum. In a more recent development by the same group [33], the excitation of spoof SPP modes in 2D arrays of holes on perforated metals has been demonstrated at microwave frequencies using the classical method of prism-coupling. On the other hand, the Drude-like response of a 2D array of holes has been investigated in the THz range of the EM spectrum [34, 35].

Hole shapes other than squares have also been analyzed during the last years. For example, it has been demonstrated that 1D arrays of rectangular holes support the propagation of spoof SPP modes exhibiting a very low group velocity and that are confined in a deep subwavelength region [36, 37]. In 2009, surface EM modes supported by a Sievenpiper mushroom decorating a very thin metallic layer have been demonstrated experimentally [38]. As in the case of square holes, this surface EM mode is asymptotic to the effective surface plasma frequency defined by the fundamental resonance of the sample. Arrays of complementary split ring resonators [39] or annular holes [40] have been proposed as good candidates to enlarge the operative bandwidth of the spoof SPP modes.

One of the possible applications of the spoof SPP concept is to mold the flow of plasmons in 2D surfaces by means of a proper design of the geometry. Along this line, 1D light waveguiding could be achieved by gradually increasing the hole size as one moves away from the center of the 2D hole array, leading to a lowering of the effective index and, subsequently, to an evanescent decay of the mode in the transverse direction [41]. Finally, it has also been shown that arrays of holes presenting a parabolic graded-index distribution are able to focus, collimate, and waveguide spoof SPPs in the transverse direction [42].

7.4 Cylindrical geometries

Recently, there has been a resurgence of interest in SPP propagation along metal wires in the THz regime of the spectrum [43, 44], mostly in the context of bio-chemical sensing. However, the delocalized nature of the Sommerfeld waves sets constraints upon the achievable sensitivity, and leads to significant radiation loss at bends and surface imperfections. As in the case of planar interfaces, the field confinement decreases with increasing conductivity of the conductor and, in the PEC limit, metallic wires no longer sustain electromagnetic surface waves. The idea of increasing the binding of Sommerfeld waves through the wire corrugation was already explored in the 1950s in the context of telecommunications technology [7, 45]. These early works demonstrated that the guiding capabilities of corrugated transmission lines could be enhanced by tailoring their surface geometry. In this section, we reformulate the analysis of these two seminal works using the coupled mode method.

First we implement the theoretical formalism as explained in Section 7.2 to analyze the formation of spoof SPPs on corrugated PEC wires [46, 47]. We will show how the theoretical approach presented in Section 7.2 can be modified in order to treat these systems. Figure 7.5 shows the simplest structure supporting cylindrical spoof SPPs, a PEC wire milled with a periodic array of rings. The geometrical parameters of the system are: the wire radius, R; the array period, d; and the rings' width and depth, a and h, respectively. We restrict our analysis to azimuthally (θ) independent modes, which present the lowest frequency and for

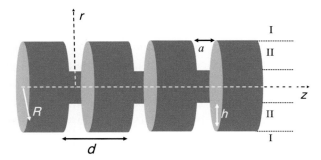

Figure 7.5. Schematic of the simplest structure supporting cylindrical spoof SPPs: a PEC wire of radius R perforated with an array of rings of period d, width a, and depth h.

which light polarizations are decoupled. This allows us to consider only p-polarized modes in our expansion basis.

As in planar geometries, we consider a unit cell of length d along the wire axis (z-direction) and divide the system into three regions: the vacuum space surrounding the wire (region I, $r \geq R$); the wire thickness occupied by the perforations (region II, $R > r \geq R - h$); and the wire core (region III, $r < R - h$). In region I, the relevant components of the EM fields (E_z and H_θ) can be expressed in terms of p-polarized plane waves as follows:

$$|E_z^I\rangle = \sum_n \rho_n K_0(\kappa_n r)|k_n\rangle, \tag{7.27}$$

$$|H_\theta^I\rangle = \sum_n Y_{k_n}^I \rho_n K_1(\kappa_n r)|k_n\rangle, \tag{7.28}$$

where $k_n = k_z + n(2\pi/d)$ and $\kappa_n = \sqrt{k_n^2 - k_0^2}$ are the wave vector components of the plane wave $|k_n\rangle$, and $Y_{k_n}^I = ik_0/\kappa_n$ is its admittance. Note that k_z denotes the propagation wave number of the guided modes. The radial dependence of the fields is given by the modified Bessel functions of the second kind K_0 and K_1 [48].

In region II, fields are only nonzero inside the rings. Thus, EM fields can be expanded as a sum over propagating and counter-propagating waveguide modes in the radial direction as follows:

$$|E_z^{II}\rangle = \sum_\alpha D_\alpha \Big(J_0(q_\alpha r) - \gamma_\alpha N_0(q_\alpha r) \Big)|\alpha\rangle, \tag{7.29}$$

$$|H_\theta^{II}\rangle = \sum_\alpha Y_\alpha^{II} D_\alpha \Big(J_1(q_\alpha r) - \gamma_\alpha N_1(q_\alpha r) \Big)|\alpha\rangle, \tag{7.30}$$

where $|\alpha\rangle$ are the ring waveguide modes and $Y_l^{II} = -ik_0/q_\alpha$ are their admittances. The radial dependence of the fields is now described by the Bessel and Neumann functions $J_{0,1}$ and $N_{0,1}$ [48]. Region III is filled with PEC material and therefore fields vanish within it. This provides us with a new condition on the fields in region II, which must satisfy $\mathbf{E}_t = 0$ at the ring bottom. Thus, from Eq. (7.29) we have $\gamma_\alpha = J_0[q_\alpha(R-h)]/N_0[q_\alpha(R-h)]$.

The matching of the EM fields at the I–II interface ($r = R$) is performed similarly to the case of planar geometries. The z-component of the electric field must be continuous everywhere on the interface, whereas the θ-component of the magnetic field is continuous only at the openings of the rings. Projecting the continuity equations for the electric (magnetic) field over plane waves (ring waveguide modes), we remove the dependence on z of the matching equations. Defining the quantities

$$E_\alpha = D_\alpha \Big(J_0(q_\alpha R) - \gamma_\alpha N_0(q_\alpha R) \Big), \qquad (7.31)$$

which correspond to the modal amplitudes of the z-component of the electric field at the openings of the rings, we can write the matching equations for the system in the same form as Eq. (7.16).

The physical interpretation of the various terms appearing in the matching equations are the same as for planar geometries, although their expression as a function of the modal expansion coefficients is different. Thus, the ϵ term is now given by

$$\epsilon_\alpha = Y_\alpha^{II} \frac{J_1(q_\alpha R) - \gamma_\alpha N_1(q_\alpha R)}{J_0(q_\alpha R) - \gamma_\alpha N_0(q_\alpha R)}. \qquad (7.32)$$

The G term is given by

$$G_{\alpha\alpha'} = \sum_n Y_{k_n}^I \frac{K_1(\kappa_n R)}{K_0(\kappa_n R)} \langle \alpha | k_n \rangle \langle k_n | \alpha' \rangle, \qquad (7.33)$$

where the overlap integrals are defined as

$$\langle k_n | \alpha \rangle = \int dz \langle z | k_n \rangle^* \langle z | \alpha \rangle. \qquad (7.34)$$

Following the same notation as for planar geometries, $\langle z | k_n \rangle$ and $\langle z | \alpha \rangle$ denote the real space wavefunctions for the p-polarized plane waves and the ring waveguide modes, respectively.

Once we have constructed the set of homogeneous matching equations, the spoof SPP modes correspond to the nonzero solutions for the modal amplitudes E_α. In Section 7.3 we have seen that for subwavelength perforations ($\lambda \gg a$), we can keep only the lowest (TM0) waveguide mode in the field expansion (note that this mode is always propagating, irrespective of the ratio a/λ). This allows us to reduce the

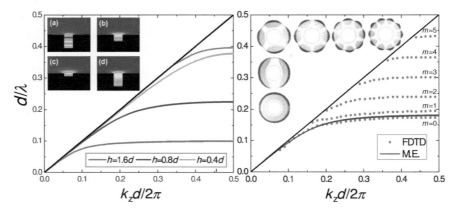

Figure 7.6. Spoof SPP dispersion relation for wires of radius $R = 2d$ perforated with rings of width $a = 0.2d$. Left panel: θ-independent bands for different ring depths. The inset shows the electric field amplitude $(r > R - h)$ at the edge of the bands. Right panel: FDTD bands for higher azimuthal orders (m) for $h = 0.5d$. The insets plot the electric field patterns at the band edge ordered with increasing m from the left bottom corner to the right top corner of the figure. See color plates section.

set of matching equations to a single one, $(G_{00} - \epsilon_0) = 0$, which, on introducing the expressions for the G and ϵ terms, reads as follows:

$$\sum_n \frac{k_0}{\kappa_n} \frac{K_1(\kappa_n R)}{K_0(\kappa_n R)} |\langle k_n|0\rangle|^2 = -\frac{J_1(k_0 R) - \gamma_0 N_1(k_0 R)}{J_0(k_0 R) - \gamma_0 N_0(k_0 R)}, \qquad (7.35)$$

where we have used the fact that $q_\alpha = k_0$ for the TM0 waveguide mode. The expressions for the overlap integrals $\langle k_n|0\rangle$ can be found elsewhere [46]. In the left panel of Fig. 7.6, the azimuthally independent spoof SPP bands obtained from Eq. (7.35) for three different ring arrays are plotted. We take d as the unit of length. Thus, the wire radius is $R = 2d$, and the ring width $a = 0.2d$. The three ring depths are: $h = 1.6d$, $h = 0.8d$, and $h = 0.4d$. The dispersion relations deviate farther from the light line when h is increased, in a similar way to what was observed in 1D groove arrays (see Fig. 7.2(a)).

At low frequencies $(\lambda \gg d, a)$, and for wires much thicker and rings much shallower than the array period $(R, R - h \gg d)$, we can neglect diffraction orders in the G term and obtain an analytical expression for the dispersion relation of the guided modes. By introducing the asymptotic expansions of the different Bessel functions involved in Eq. (7.35), we have

$$k_z = k_0 \sqrt{1 + \left(\frac{a}{d}\right)^2 \tan^2(k_0 h)}. \qquad (7.36)$$

Note that Eq. (7.36) coincides with Eq. (7.21), which gives the spoof SPP bands for 1D arrays of grooves of width a and depth h. This agrees with the fact that, as in the planar case, the key parameter governing the binding of the mode is the depth of the rings, h (see Fig. 7.6).

The right panel of Fig. 7.6 presents the dispersion relation of the spoof SPPs on a wire of radius $R = 2d$ perforated by periodic rings with $a = 0.2d$ and $h = d$. In this calculation, there is no restriction regarding the azimuthal dependence of the EM fields. The dispersion relations (dots) have been obtained by means of 3D finite difference time domain (FDTD) simulations. The different bands (labeled by the index m) correspond to the different azimuthal symmetries of the electric field amplitude shown in the insets of the figure. For the structure considered, m ranges from $m = 0$ (θ-independent modes) to $m = 5$ (see insets from left bottom corner to right top corner of the panel). The electric field associated to the mth azimuthal mode presents $2m$ nodes and maxima in θ. The solid line shows the $m = 0$ band calculated from Eq. (7.35). A very good agreement between FDTD and modal expansion (ME) results for the lowest spoof SPP band is observed.

7.5 Terahertz waveguides based on spoof SPPs

As shown in the preceding sections, spoof SPP modes allow the routing of EM fields along corrugated PEC surfaces. This fact opens the way to the design of THz or microwave waveguiding schemes based on the concept of spoof SPPs. Different theoretical works have been published exploring this idea in the microwave regime [39, 42]. Due to the recent technological interest raised by THz waves [49, 50], the guiding properties of spoof SPP modes in this range of the EM spectrum have been analyzed in different configurations [23, 24, 41]. In the following, we describe in detail four simple guiding schemes based on the spoof SPP concept that operate at THz frequencies: milled wires, corrugated channels and wedges, and domino structures.

7.5.1 Milled wires

The lack of lateral confinement of spoof SPPs in 1D and 2D indentations prevent their use as waveguides, aiming to transport EM energy within small transverse cross sections. Taking this into account, probably the most straightforward spoof SPP waveguide consists of a metallic wire periodically corrugated with an array of rings, as presented in Section 7.4. The tailoring of the wire geometry enables us to select the spectral range of operation of the structure. By choosing modulation sizes of the order of hundreds of microns, the guiding is optimized at the THz range.

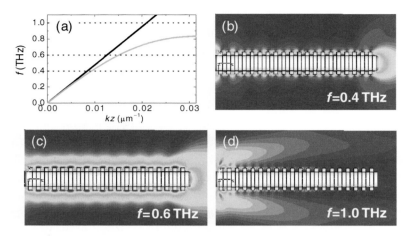

Figure 7.7. (a) Dispersion relation of the guided modes traveling along a corrugated wire of radius $R = 150\,\mu m$ perforated with an array of rings of period $d = 100\,\mu m$. The width and depth of the rings are both $50\,\mu m$. (b)–(d) Electric field amplitude for wires of length $20 \times d$ illuminated from the left by a radially polarized plane wave. Fields are evaluated at three different frequencies (0.4, 0.6, and 1.0 THz), indicated by dotted lines in (a). See color plates section.

Figure 7.7 shows the propagation of EM fields along a 2 mm long corrugated wire of 150 μm radius. The pitch of the corrugation is $d = 100\,\mu m$, and the width and depth of the rings are $a = h = 50\,\mu m$. The geometry of the ring array has been chosen so that the optimal frequencies for guiding are around 0.6–0.8 THz. Figure 7.7(a) shows the dispersion relation, $f(k_z)$, for the guided modes supported by an infinite wire with the same geometrical parameters. Figures 7.7(b), (c), and (d) depict the electric field amplitude (evaluated at three different frequencies) for the finite wire illuminated by a radially polarized broadband terahertz pulse from the left. The field patterns have been obtained through finite-integration-technique (FIT) simulations. At the lowest frequency considered, $f = 0.4$ THz, the band lies close to the light line, which leads to a weak binding of fields to the structure. At $f = 0.6$ THz, EM radiation is guided more efficiently as the modes are strongly confined to the wire surface. At $f = 1.0$ THz, as the system does not support the propagation of any guided mode, the incident radiation is scattered out from the wire.

Taking advantage of the strong dependence of the spoof SPP confinement on the wire geometry, it is feasible to design a structure able to concentrate EM energy at one of its ends [46, 51]. Here, we describe one candidate for this: a conical wire in which the external radius is gradually decreased along the direction of propagation, keeping the depth of the rings fixed. Figure 7.8(a) plots the guided mode bands for ring arrays of period 100 μm with $h = 30\,\mu m$ and $a = 50\,\mu m$. Four structures

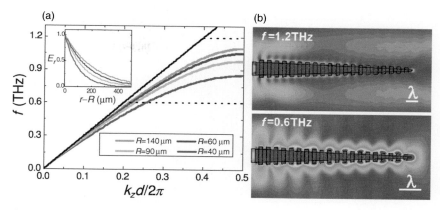

Figure 7.8. (a) Frequency versus propagation wave vector, k_z, for the guided modes supported by four corrugated wires of different radii. The inset plots the radial component of the electric field versus $r - R$ ($f = 0.6$ THz) for the four structures. (b) Electric field amplitude at 0.6 and 1.2 THz for a 2 mm corrugated cone whose radius is reduced from 140 to 40 µm. See color plates section.

with wire radii ranging between 140 µm and 40 µm are considered. Note that as R is decreased, the spoof surface plasmon bands deviate further from the light line. The inset renders the radial component of the electric field, E_r, as a function of the distance to the wire surface, $r - R$, for the four structures evaluated at 0.6 THz. We can see that the lowering of the bands leads to a stronger confinement of the modes with decreasing R.

Figure 7.8(b) shows the electric field pattern corresponding to a 2 mm long wire whose external radius is gradually reduced from 140 to 40 µm. The structures are milled with ring arrays with the same dimensions as in Fig. 7.8(a). At 0.6 THz, the guided modes are tightly bound to cylindrical wires, even for the smallest radius considered. However, at larger frequencies, 1.2 THz, modes are not supported by wires with $R \leq 140$ µm. This has the consequence that EM radiation is guided along the cone and focused at its tip at 0.6 THz. On the other hand, the absence of guided modes at 1.2 THz leads to EM waves being scattered out of the structure without reaching the end of the wire. Remarkably, the high confinement of EM fields featured by cylindrical spoof SPPs allows the concentration of THz waves into deep subwavelength volumes in conical geometries.

7.5.2 Helically grooved wires

In this subsection, we consider the case in which a metallic wire is periodically drilled with helical grooves. We present some experiments that verify the propagation of guided modes in this type of geometry. This experimental study is

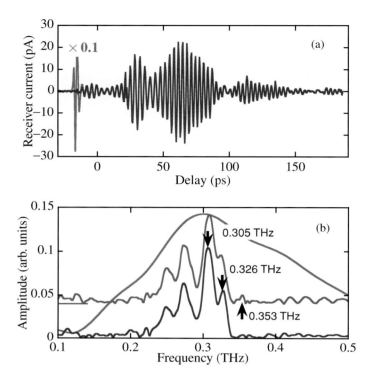

Figure 7.9. (a) Receiver current as a function of time delay for the smooth wire and the grooved structure. (b) Amplitude spectra of the time domain data in (a) together with the spectrum of another, nominally identical, helical sample (displaced for clarity). The arrows indicate the three azimuthal modes of the helical groove structure. The spectrum of the Sommerfeld wave on the smooth wire extends to ~1 THz. See color plates section.

accompanied by a theoretical analysis of the surface EM modes supported by this complex structure [52].

The experimental setup consists of a 150 mm long helically grooved wire, formed by tightly wrapping a steel wire (radius 200 μm) around a 200 μm radius core. For comparison, a bare copper wire of the same outer radius and length (600 μm and 150 mm, respectively) is also studied. Measurements are performed using time-domain THz spectroscopy. In order to discriminate the bound EM modes against unguided free space radiation, the wires are bent along the arc of a circle of radius 26 cm.

Figure 7.9(a) displays time-domain traces of the receiver current for the wires with smooth and helically grooved surfaces. It is clear that a single-cycle-like pulse, which can be associated with a Sommerfeld wave [2], propagates along the smooth wire. However, propagation on the helical wire exhibits significant

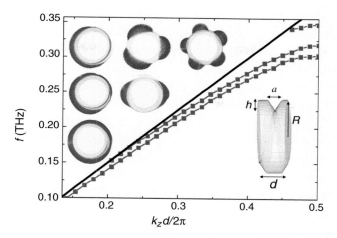

Figure 7.10. Dispersion relation of the guided modes supported by a PEC wire of radius $R = 600\,\mu m$ inscribed with a triangular-cross-section helical groove of pitch $d = 400\,\mu m$. The groove has width $a = 200\,\mu m$ and depth $h = 150\,\mu m$. The upper row of insets displays snapshots of the electric field at the three band edges, 0.305 THz (left), 0.320 THz (center), and 0.349 THz (right). The next lowest row correspond to the first mode at 0.280 THz (left) and the second mode at 0.180 THz (right). The pattern in the bottom row is for the first mode at 0.180 THz. See color plates section.

dispersion together with beating due to the presence of bound modes with different frequencies. Figure 7.9(b) plots the amplitude spectra of the traces in Fig. 7.9(a) together with the spectrum of a second, nominally identical sample of the helical structure, which shows the reproducibility of the experimental data in the presence of small variations in optical alignment. In Fig. 7.9(b), the frequency at the band edge ($k_z = \pi/d$) of the three lowest guided modes supported by the structure are indicated by vertical arrows. They are obtained by means of the theoretical FDTD calculations and correspond to the peaks in the amplitude spectra at 0.305 ± 0.002 THz, 0.326 ± 0.002 THz, and 0.353 ± 0.003 THz. We show below that the structure observed in the spectra at frequencies lower than 0.3 THz can be associated with the propagation of radiation along the wire at smaller wave vectors.

Figure 7.10 presents the spoof SPP bands for a helically grooved PEC wire calculated using the FDTD method. In accordance with the experimental parameters, the helix pitch is $d = 400\,\mu m$ and the wire radius $R = 600\,\mu m$. The EM fields are evaluated inside a unit cell along the direction parallel to the wire axis (z-direction). Due to design limitations of our computer code, the modeled groove has a triangular profile of width a and depth h (see lower inset). We find that $a = 200\,\mu m$ and $h = 150\,\mu m$ give a good match to the experimental results for the frequency at the edges of the bands. The main panel shows the dispersion relation for this set of geometrical parameters. The three theoretical values obtained, 0.305 THz, 0.320 THz, and 0.349 THz, are in excellent agreement with the spectral peaks found in the experiments.

The insets of Fig. 7.10 provide snapshots of the electric field amplitude of the spoof SPP modes supported by the wire at various frequencies. Note that, as expected, the lack of azimuthal symmetry of the metallic structure leads to nonsymmetrical field distributions. The upper row displays the electric field amplitude at the three band edges, increasing in frequency from left to right. The fields are confined within less than a wavelength of the wire surface. The field maps depicted in the lower two rows are evaluated at smaller k_z. The two on the left correspond to the first spoof surface plasmon mode at 0.280 THz and 0.180 THz, and that on the right of the second row corresponds to the second mode at 0.180 THz. At the band edge, the modes exhibit an odd number of azimuthal nodes (1, 3, and 5) whereas with decreasing k_z the number of nodes is gradually reduced by one and becomes even. Thus, the guided modes propagating along the helical structure at low k_z resemble the case of a ring array, where the number of nodes is even.

Now we analyze why the spoof SPPs on helically grooved wires exhibit such a k_z-dependent azimuthal symmetry. Any component of the EM fields bound to a helical structure [53] can be expanded in terms of diffracted waves as

$$F_m(r, \theta, z) = e^{ik_z z} e^{im\theta} \sum_n A_{n\,m-n}(r) e^{in(\frac{2\pi}{d}z - \theta)}, \tag{7.37}$$

where the modal amplitude $A_{n\,m-n}(r)$ contains the radial dependence of the nth-diffracted wave. Note that $F_m(r, \theta, z)$ is an eigenfunction of the helical translation operator, $S_{\phi\,\frac{d}{2\pi}\phi}$, satisfying

$$S_{\phi\,\frac{d}{2\pi}\phi} F_m(r, \theta, z) = F_m\left(r, \theta + \phi, z + \frac{d}{2\pi}\phi\right) = e^{i(m+k_z\frac{d}{2\pi})\phi} F_m(r, \theta, z), \tag{7.38}$$

where the index m controls the symmetry properties of the EM fields. We introduce the helical coordinate $\xi = z - d\theta/2\pi$, which is parallel to the cylindrical coordinate z, but measured from the surface $z = d\theta/2\pi$. Electromagnetic fields can be expressed in terms of ξ as

$$F_m(r, \theta, \xi) = f(r, \xi) e^{i(m+k_z\frac{d}{2\pi})\theta}. \tag{7.39}$$

It is now clear that this eigenfunction, evaluated along the helical surface ($\xi =$ constant), evolves in time as $\cos[(m + k_z(d/2\pi))\theta - 2\pi ft]$, where f and t are the mode frequency and time, respectively. Thus, snapshots of the EM fields with

$k_z = \pi/d$ show $2m + 1$ nodes along one helix pitch, whereas for $k_z = 0$ they show only $2m$ nodes. This result allows us to label the spoof SPP modes in Fig. 7.10 with the indices $m = 0$, $m = 1$, and $m = 2$.

7.5.3 *Corrugated channels*

We have shown that the presence of spoof SPP modes decorating the corrugated surface of free-standing PEC wires allows the transport of EM energy within subwavelength cross sections. However, this guiding scheme presents a major drawback: its nonplanar character makes it difficult to implement in a complex THz circuit. Here, we present a waveguide based on the spoof SPP concept that features subwavelength transverse confinement of EM fields at a planar surface. The design consists of corrugated V-grooves milled on a metal surface, and borrows ideas from the so-called channel plasmon polaritons (CPPs) [54] operating at visible and telecom frequencies [55]. The EM fields in uncorrugated V-grooves become more extended with increasing wavelength in such a way that, on PEC channels, they are not bound at all. Following the same strategy as in cylindrical geometries, we show how the texturing of the metal surface leads to the emergence of bound EM modes in corrugated channels, even in the PEC limit [56].

The upper inset of Fig 7.11 depicts the system under study: a V-channel of depth h and width w modulated with a periodic array of grooves of width a and depth h. Figure 7.11(a) shows the dispersion relation of the modes bound to the structure with $a = t = 0.5d$, $w = 0.76d$, and $h = 5d$. For this set of geometrical parameters, the angle of the channel is $20°$, similar to those considered in the telecom regime. FDTD calculations demonstrate the appearance of two guided modes (from now on termed spoof CPPs) for normalized frequencies $d/\lambda < 0.35$. Note that in corrugated V-channels of finite height, these modes possess a finite cutoff frequency, as in the case of conventional CPPs [57]. Note also the small frequency overlap between the two spoof CPP bands, which facilitates the monomode operation of the V-groove as a THz waveguide.

The longitudinal component of the electric field associated with the two spoof CPP modes is shown in the insets of Fig. 7.11(a). The fields are evaluated at the edges of the bands. Electric fields are plotted only inside the shallow part of the corrugated V-groove. The first mode has odd parity, as the longitudinal electric fields have two lobes of different sign at both sides of the channel, vanishing at the mid-plane. The second mode has even parity with respect to the symmetry plane. The tightly bound nature of the modes is clarified by introducing the modal size δ. It is defined as the transverse separation between the locations where the electric field amplitude has fallen to one-tenth of its maximum value; $\delta = 0.52\lambda$, and $\delta = 1.06\lambda$, respectively, for the two spoof CPPs at the band edge. Figures 7.11(b) render the

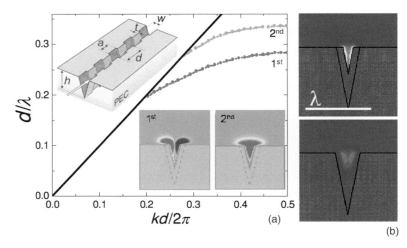

Figure 7.11. (a) Dispersion relation of the first two spoof CPP modes supported by a corrugated V-channel milled on a PEC surface. A schematic of the structure is shown in the upper inset. The lower insets depict the amplitude of the longitudinal component of the electric field evaluated at the band edge for both modes. (b) Electric field amplitude at the band edge for the lowest mode evaluated at the shallower (upper panel) and deeper (lower panel) sections of the channel. The horizontal white bar represents the wavelength of the mode in a vacuum. See color plates section.

electric field amplitude at the band edge for the lowest spoof CPP mode. The upper (lower) panel shows the field distribution within a transverse plane located at the shallower (deeper) part of the channel, displaced by $d/2$. Note that the electric field is confined into a subwavelength area, being strongly localized within the shallow section of the channel. Another interesting feature is that EM energy is not guided at the groove bottom but rather at the groove edges. We can anticipate that this is due to the strong hybridization of spoof CPPs with modes running on the edges of the groove, much in the same way as it occurs in conventional CPPs [57].

Once we have demonstrated that spoof CPPs are supported by infinitely long corrugated V-grooves, we analyze how these EM modes behave in waveguides of finite length. We choose the structure period $d = 200 \, \mu m$, keeping the relation between the rest of the geometrical parameters and d as in Fig. 7.11. Figure 7.12 shows the transmission spectra of THz waves through five different channels comprising 100 periods calculated through FIT simulations. Dashed arrows indicate the cutoff and band edge frequencies obtained from Fig. 7.11 for the lowest spoof CPP mode supported by the structure. Note that, as expected, the transmission of the straight waveguide approaches unity within the spectral region (0.3–0.42 THz) between these two frequencies.

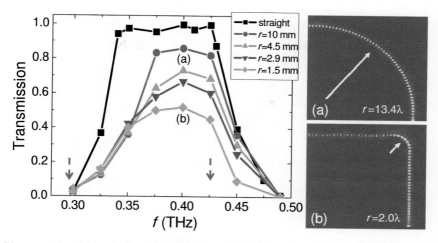

Figure 7.12. Transmission spectra for spoof CPP modes supported by a corrugated PEC V-channel of period $d = 200\,\mu m$ and total length 20 mm. Straight and four different 90° bent waveguides are shown. Dashed arrows indicate the spectral position of the cutoff and band edge frequencies in Fig. 7.11. The right panels depict the electric field amplitude at 0.40 THz evaluated at a height of 100 µm above the planar surface for the structures with (a) $r = 10$ mm and (b) $r = 1.5$ mm.

The possible use of spoof CPPs in corrugated channels for routing THz radiation requires a study of the bending losses suffered by these modes. The transmission of four 90° bends with radii of curvature, r, are also plotted in Fig. 7.12. In these structures, $d = 200\,\mu m$ in the straight part of the channel and is slightly adjusted at the bends in order to conform with the curved geometry. For the case of maximum r (10 mm), the transmission can be as large as 90%, but is reduced as r becomes smaller, reaching 50% in the case $r = 1.5$ mm (around twice the wavelength). Let us stress that these bending losses are much smaller than those reported for metallic wires at THz frequencies [43]. In the right-hand panels of Fig. 7.12, the electric field amplitude at 0.40 THz for $r = 10$ mm (a) and $r = 1.5$ mm (b) in a plane located 100 µm above the planar surface is depicted. It is clear how the bending losses in these structures stem from radiation into vacuum modes occurring just at the bend of the waveguide.

7.5.4 Corrugated wedges

The electric field distribution of spoof CPPs indicates that they are not guided at the bottom of corrugated channels but at its edges (see Fig. 7.11). In this subsection, we have related this field profile to the hybridization with modes traveling along

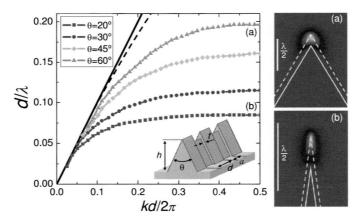

Figure 7.13. Dispersion relation of spoof WPPs traveling along corrugated wedges for different θ. Dashed line shows the dispersion band corresponding to the flat ($\theta = 180°$) case (groove array). The inset depicts the system geometry. Right panels show the electric field amplitude at the band edge for (a) $\theta = 60°$ and (b) $\theta = 20°$. In both panels, $\lambda/2$ is represented by white bars.

the edges of the structure. We analyze guided modes and present a new guiding scheme for THz waves based on them [58].

The system under study is depicted in the inset of Fig. 7.13: a PEC wedge milled with a periodic array of grooves. We will show that, as in the case of corrugated channels, the EM modes supported by such a geometry resemble wedge plasmon polaritons (WPPs) [59, 60] occurring at visible and telecom frequencies. The parameters defining the system supporting these EM guided modes (termed spoof WPPs from now on) are the height, h, and angle, θ. The grooves milled on the wedge have depth t and width a, and the period of the corrugation is d. In our analysis we fix the groove dimensions, $a = t = 0.5d$, and the wedge height, $h = 5d$.

Figure 7.13 presents the dispersion relation of the fundamental spoof WPPs supported by wedges with different θ. The dashed line plots the spoof SPP band for the limiting case of a flat ($\theta = 180°$) groove array. Reducing θ leads to the shift of the dispersion relation to lower frequencies, which implies the tighter confinement of the modes. Note that, as in the case of corrugated channels, the finite height of the structure provides the spoof WPP bands with a cutoff frequency, below which the modes lose their nonradiative nature. The right panels of Fig. 7.13 depict the electric field amplitude at the band edge for wedges with (a) $\theta = 60°$ and (b) $\theta = 20°$. The cross sections correspond to the deeper part of the corrugated wedge, where the EM fields are mainly localized. Remarkably, the field patterns resemble those corresponding to conventional WPPs. The half-wavelength ($\lambda/2$) is also represented by vertical white bars in both panels. The modal size for the

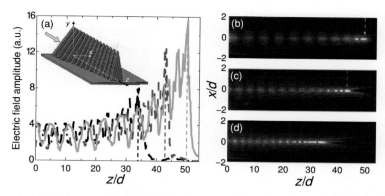

Figure 7.14. (a) Inset: Corrugated PEC wedge with θ varying smoothly along the z-direction from 60° to 20°. Plot shows the electric field amplitude versus z along the line located 100 μm above the structure apex. Three different frequencies are considered: 0.12 THz (solid line), 0.16 THz (dashed line), and 0.20 THz (dashed dotted line). (b)–(d) Electric field amplitude within the xz-plane located 100 μm above the apex for these three frequencies. Dashed arrows indicate the positions of the maxima of amplitude shown in (a).

60° wedge is $\delta = 0.78\lambda$, whereas for $\theta = 20°$ it is equal to 0.28λ. These results demonstrate the subwavelength transverse confinement featured by spoof WPPs.

As an illustrative example, we present a functional device exploiting the propagation capabilities of these spoof WPP modes. To make the design work at THz frequencies, the corrugation period, d, is set to 200 μm. Figure 7.13 provides a hint on how radiation can be focused and slowed down with the aid of spoof WPPs. The lowering of the dispersion bands for decreasing θ suggests that THz waves of a given frequency propagating in a wedge which is sharpened along its length (see inset of Fig. 7.14(a)) would be gradually concentrated within the transverse plane. Additionally, THz waves at frequencies above the band edge associated to a specific θ will never reach sections of the structure sharper than that angle, being slowed down as they approach it. In order to prevent back-reflection and scattering of EM fields out of the structure, impedance mismatches along the wedge can be minimized by performing the reduction in θ adiabatically.

The inset of Fig. 7.14(a) shows a diagram of the design proposed: a 10 mm long wedge with θ varying smoothly from 60° to 20° milled by 50 grooves disposed periodically, keeping the relation between d and the remaining geometrical parameters as in Fig. 7.13. The guiding properties of the structure are analyzed by means of FIT simulations under the PEC approximation. Figure 7.14(a) presents the electric field amplitude on a line parallel to the z-axis and 100 μm above the wedge apex. Fields are evaluated at three different frequencies within the spectral range spanned by the dispersion bands shown in Fig. 7.13. Waves at $d/\lambda = 0.08$

($f = 0.12$ THz) propagate until the sharpest end of the wedge, giving rise to a maximum in the electric field amplitude located at that position (solid line). At higher frequencies, radiation is slowed down and stopped before reaching the wedge end. For $d/\lambda = 0.1$ ($f = 0.16$ THz), a peak in $|E|$ is developed at the 42nd groove, for which $\theta = 26°$ (dashed line). At $d/\lambda = 0.13$ (0.20 THz) EM fields explore an even shorter section of the wedge and $|E|$ presents a maximum at the 34th groove, which corresponds to $\theta = 32°$ (dashed dotted line). Let us stress the excellent agreement of these results with the dispersion bands of Fig. 7.13.

Figures 7.14(b)–(d) depict the electric field amplitude within the xz-plane located 100 µm above the wedge apex, for the three frequencies considered in (a). Vertical dashed arrows indicate the position of the maxima in (a). These three contour plots show the reduction of the effective wavelength ($\lambda_{\text{eff}} = 2\pi/\text{Re}(k)$) experienced by the guided EM fields as they propagate along the structure. This indicates that THz waves slow down and stop at different frequency-dependent locations along the corrugated wedge. Figures 7.14(b)–(d) demonstrate that guided waves are not scattered out of the wedge as they travel in the z-direction. On the contrary, while propagating along the structure, EM fields are gradually concentrated, leading to frequency selective focusing of THz waves.

7.5.5 Domino structures

Perhaps the most promising route for THz waveguiding based on the concept of spoof SPPs is the domino structure (see inset of Fig. 7.15(a)), as first introduced in ref. [61]. This structure consists of a periodic array of metallic parallelepipeds standing on top of a metallic surface, resembling a chain of domino pieces. The properties of its guided modes, the so-called *domino plasmons* (DPs), are governed by the geometric parameters defining the dominoes: periodicity (d), height of the boxes (h), lateral width (L), and inter-domino spacing (a).

Here we analyze how the dispersion relation of DPs changes with the lateral width, L. To gain a physical insight, as in previous cases it is better to model the metal first as a PEC. Within the PEC approach we choose the periodicity d as the unit of length. The value of a is not critical for the properties of DPs and, in these simulations, is set equal to $a = 0.5d$. Domino plasmon bands present a typical plasmonic character, i.e. they approach the light line for low frequencies and reach a horizontal frequency limit at the edge ($k_{\text{edge}} = \pi/d$) of the first Brillouin zone (Fig. 7.15(a); note that only fundamental modes are plotted). While the limit frequency of SPPs for large k is related to the plasma frequency, the corresponding value for DPs is controlled by the geometry. In particular, the influence of the height h is clear: the band frequency rises for short dominoes ($h = 0.75d$) as compared with that of taller ones ($h = 1.5d$). The most striking characteristic

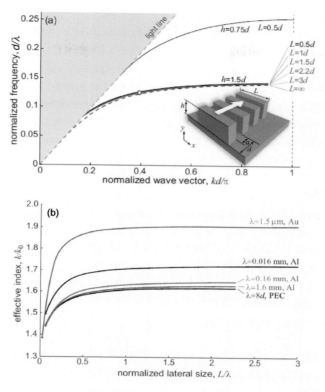

Figure 7.15. (a) Dispersion relation of DPs for various lateral widths L. The dashed line stands for infinitely wide dominoes ($L = \infty$). Inset: diagram of the domino structure and geometric parameters (the arrow depicts the mode propagation direction). (b) DP modal effective index as a function of lateral dimension L in units of wavelength. Various operating frequency regimes are considered: $\lambda = 1.6$ mm, $\lambda = 0.16$ mm, $\lambda = 0.016$ mm, and $\lambda = 1.5$ μm. A realistic description of the metals is used. As described in the main text, the periodicity d is different for the various operating frequencies, and $h = 1.5d$, $a = 0.5d$, $L = 0.5d, \ldots, 24d$.

of DPs is their behavior when the lateral width L is changed. All bands in the range $L = 0.5d, \ldots, 3d$ lie almost on top of each other (Fig. 7.15(a)). In other words, the modal effective index, $n_{\text{eff}} = k/k_0$, is rather insensitive to lateral width. Remarkably, the bands remain almost unchanged even for $L = 0.5d$, whose modal size is well inside the subwavelength regime. The described behavior is to be contrasted with that of conventional plasmonic modes in the optical regime for which subwavelength lateral confinement is not a trivial issue.

Now we study in more detail the role played by the lateral dimension L, considering realistic metals and paying attention to the spectral regime. The periodicity d is chosen to set the operating wavelength within the desired region of the EM spectrum, and L is varied in the range $L = 0.5d, \ldots, 24d$, while the remaining parameters are kept constant ($a = 0.5d$, $h = 1.5d$). Aluminum is selected for low frequencies, where metals behave almost like PECs. In order to work at $\lambda = 1.6$ mm, providing an operating angular frequency of the order of 1 THz, we first consider $d = 200\,\mu$m. The evolution of the modal effective index as a function of the lateral dimension normalized to the wavelength is plotted in Fig. 7.15(b). The curves corresponding to a PEC and aluminum at $\lambda = 1.6$ mm are, as expected, almost identical. We can now quantify the sensitivity of the effective index to L, its variation being only about 12% even when L goes from $L = \infty$ to $L = 0.5d = \lambda/16$, well inside the subwavelength regime. To investigate the performance of DPs at higher frequencies, the structures have been scaled down by factors $1/10$ and $1/100$. The fact that the curves corresponding to $\lambda = 0.16$ mm and $\lambda = 0.016$ mm do not lie on top of the previous ones is a signature of the departure of aluminum from the PEC behavior. Nevertheless, even at $\lambda = 0.016$ mm, the variation of the effective index is still smaller than 15%. When the operating frequency moves to the telecom regime ($\lambda = 1.5\,\mu$m) the variation of the effective index is much larger (about 38%).

The important message of Fig. 7.15(b) is that, although a variation of n_{eff} begins to be noticeable when the lateral dimension L goes below $\lambda/2$, DP bands are fairly insensitive to L in the range $L = 0.5d, \ldots, 24d$ when operating at low frequencies. Such a key property does not appear in spoof SPP modes in corrugated wedges, channels, or wires. Based on this remarkable characteristic (insensitivity of DPs to lateral dimensions), several THz devices based on domino plasmons such as tapers, power dividers, and directional couplers have recently been proposed [61].

7.6 Conclusions

In this chapter, we have described the fundamental physics behind the so-called spoof surface plasmon polaritons, i.e. surface EM modes propagating along structured perfectly conducting surfaces. By means of a theoretical formalism based on the modal expansion technique, we have studied spoof surface plasmon modes in textured planar and cylindrical surfaces. We have analyzed in detail the geometrical dependence of these modes in simple systems: planar surfaces decorated with arrays of fully and partially perforated indentations (slits and holes), and wires milled with periodic ring arrays. We have provided approximate analytical expressions for the dispersion relation of the spoof SPP modes in these geometries and for the effective electric and magnetic response of the structures supporting them.

We have also presented a number of guiding schemes for THz waves exploiting the modal properties of spoof SPPs: conical and helically grooved wires, corrugated channels and wedges, and domino structures. We have demonstrated that subwavelength routing of THz radiation in such geometries can be achieved by tailoring the parameters of the waveguiding design.

Throughout the chapter, we have linked the technological interest of spoof SPPs to their ability for transferring the capabilities of canonical SPPs in the optical regime to low-frequency domains such as the microwave or terahertz domains. However, the potential applications of these spoof SPP modes is not limited to that. Microwave and terahertz spoof SPP modes sometimes exhibit modal characteristics very different from conventional SPPs, which may lead to novel electromagnetic phenomena which do not have a counterpart in the optical regime. Moreover, the transfer of the spoof SPP concept to higher frequencies, where the hybridization of dielectric and geometrical effects may take place, promises to be a very exciting line of research for the future [62].

References

[1] W. L. Barnes, A. Dereux, and T. M. Ebbesen, "Surface plasmon subwavelength optics," *Nature* **424**, 824–830 (2003).

[2] A. Sommerfeld, "Uber die Fortpflanzung elektrodynamischer Wellen langs eines Drahtes," *Ann. Phys. Chemie* **303**, 233–290 (1899).

[3] R. H. Ritchie, "Plasma losses by fast electrons in thin films," *Phys. Rev.* **106**, 874–881 (1957).

[4] S. A. Maier, M. L. Brongersma, P. G. Kik, S. Meltzer, A. A. G. Requicha, and H. A. Atwater, "Plasmonics – a route to nanoscale optical devices," *Adv. Mater.* **13**, 1501–1505 (2001).

[5] J. Zenneck, "Ueber die Fortpflanzung ebener elektromagnetischer Wellen laengs einer ebenen Leiterflaeche und ihre Beziehung zur drahtlosen Telegraphie," *Ann. Phys.* **23**, 846–866 (1907).

[6] U. Fano, "The theory of anomalous diffraction gratings and of quasi-stationary waves on metallic surfaces (Sommerfelds waves)," *J. Opt. Soc. Am.* **31**, 213–222 (1941).

[7] G. Gobau, "Surface waves and their application to transmission lines," *J. Appl. Phys.* **21**, 1119–1128 (1950).

[8] D. L. Mills and A. A. Maradudin, "Surface corrugation and surface-polariton binding in the infrared frequency range," *Phys. Rev. B* **39**, 1569–1574 (1989).

[9] B. A. Munk, *Frequency Selective Surfaces: Theory and Design* (New York: Wiley, 2000).

[10] R. Ulrich and M. Tacke, "Submillimeter waveguiding on periodic metal structure," *Appl. Phys. Lett.* **22**, 251–253 (1973).

[11] J. B. Pendry, L. Martín-Moreno, and F. J. García-Vidal, "Mimicking surface plasmons with structured surfaces," *Science* **305**, 847–848 (2004).

[12] L. Martín-Moreno, F. J. García-Vidal, H. J. Lezec, K. M. Pellerin, T. Thio, J. B. Pendry, and T. W. Ebbesen, "Theory of extraordinary optical transmission through subwavelength hole arrays," *Phys. Rev. Lett.* **86**, 1114–1117 (2001).

[13] J. Bravo-Abad, F. J. García-Vidal, and L. Martín-Moreno, "Resonant transmission of light through finite chains of subwavelength holes in a metallic film," *Phys. Rev. Lett.* **93**, 227401(1-4) (2004).

[14] A. Mary, S. G. Rodrigo, F. J. García-Vidal, and L. Martín-Moreno, "Theory of negative-refractive-index response of double-fishnet structures," *Phys. Rev. Lett.* **101**, 103902(1-4) (2008).

[15] M. Qiu, "Photonic band structures for surface waves on structured metal surfaces," *Opt. Express* **13**, 7583–7588 (2005)

[16] F. J. García de Abajo and J. J. Saenz, "Electromagnetic surface modes in structured perfect-conductor surfaces," *Phys. Rev. Lett.* **95**, 233901(1-4) (2005).

[17] E. Hendry, A. P. Hibbins, and J. R. Sambles, "Importance of diffraction in determining the dispersion of designer surface plasmons," *Phys. Rev. B* **78**, 235426(1-10) (2008).

[18] J. D. Jackson, *Classical Electrodynamics*, 3rd edn (New York: John Wiley & Sons, 1999).

[19] P. M. Morse and H. Feshbach, *Methods of Theoretical Physics* (New York: McGraw-Hill, 1953).

[20] A. Roberts, "Electromagnetic theory of diffraction by a circular aperture in a thick, perfectly conducting screen," *J. Opt. Soc. Am. A* **4**, 1970–1983 (1987).

[21] F. J. García-Vidal, L. Martín-Moreno, and J. B. Pendry, "Surface with holes in them: new plasmonic metamaterials," *J. Opt. A: Pure Appl. Opt.* **7**, S97–S101 (2005).

[22] L. Shen, X. Chen, and T.-J. Yang, "Terahertz surface plasmon polaritons on periodically corrugated metal surfaces," *Opt. Express* **16**, 3326–3333 (2008).

[23] Q. Gan, Z. Fu, Y. J. Ding, and F. J. Bartoli, "Ultrawide-bandwidth slow-light system based on THz plasmonic graded metallic grating structures," *Phys. Rev. Lett.* **100**, 256803(1-4) (2008).

[24] K. Song and P. Mazumder, "Active terahertz spoof surface plasmon polariton switch comprising the perfect conductor metamaterial," *IEEE Trans. Electron. Dev.* **56**, 2792–2799 (2009).

[25] C. R. Williams, S. R. Andrews, S. A. Maier, A. I. Fernández-Domínguez, L. Martín-Moreno, and F. J. García-Vidal, "Highly confined guiding of terahertz surface plasmon polaritons on structured metal surfaces," *Nature Photon.* **2**, 175–179 (2008).

[26] J. T. Shen, P. B. Catrysse, and S. Fan, "Mechanism for designing metallic metamaterials with a high index of refraction," *Phys. Rev. Lett.* **94**, 197401(1-4) (2005).

[27] J. Shin, J. T. Shen, P. B. Catrysse, and S. Fan, "Cut-through metal slit array as an anisotropic metamaterial film," *IEEE J. Sel. Topics Quant. Electron.* **12**, 1116–1122 (2006).

[28] Y. M. Shin, J. K. So, J. H. Won, and G. S. Park, "Frequency-dependent refractive index of one-dimensionally structured thick metal film," *Appl. Phys. Lett.* **91**, 031102(1-3) (2007).

[29] X. F. Zhang, L. F. Shen, and L. X. Ran, "Low-frequency surface plasmon polaritons propagating along a metal film with periodic cut-through slits in symmetric and asymmetric environments," *J. Appl. Phys.* **105**, 013704(1-7) (2009).

[30] E. N. Economou, "Surface plasmons in thin films," *Phys. Rev.* **182**, 539–554 (1969).

[31] S. Collin, C. Sauvan, C. Billaudeau, F. Pardo, J. C. Rodier, J. L. Pelouard, and P. Lalanne, "Surface modes on nanostructured metallic surfaces," *Phys. Rev. B* **79**, 165405(1-7) (2009).

[32] A. P. Hibbins, B. R. Evans, and J. R. Sambles, "Experimental verification of designer surface plasmons," *Science* **308**, 670–672 (2005).

[33] A. P. Hibbins, E. Hendry, M. J. Lockyear, and J. R. Sambles, "Prism coupling to 'designer' surface plasmons," *Opt. Express* **16**, 20441–20447 (2008).

[34] A. Agrawal, Z. V. Vardeny, and A. Nahata, "Engineering the dielectric function of plasmonic lattices," *Opt. Express* **16**, 9601–9613 (2008).

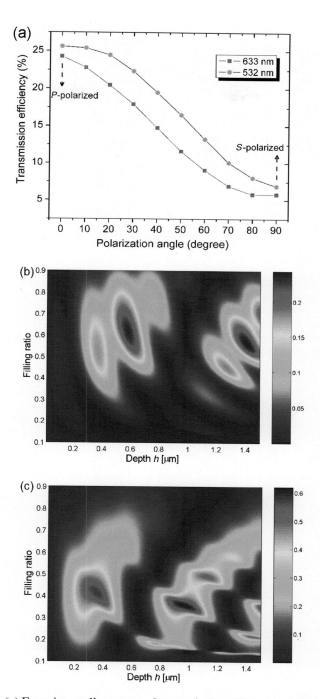

Figure 5.2. (a) Experimentally measured transmission efficiency of the beam negatively refracted through a BK7 grism with 2400 lines/mm grating. The electric vector is parallel to the groove for a 0° orientation (P polarization) and perpendicular for a 90° orientation of the polarizer. Calculated transmission efficiency in S polarization (b) and in P polarization (c) with an angle of incidence $\pi/4$ for $\lambda = 532$ nm through a lamellar grating on BK7 glass. The period of the lamellar grating has a density of 2400 lines/mm ($a_s = 416.7$ nm).

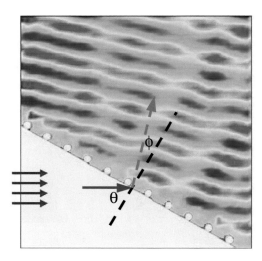

Figure 5.3. Microwave experiment demonstrating NR using a polystyrene grism with a surface grating period $a_s = 2$ cm and angle of incidence $\theta = \pi/3$ at 9 GHz. Plotted is the electric field (real part of the measured transmission coefficient S_{21}). The solid arrows on the left indicate the direction of the incident microwave beam. The dashed line is the surface normal, and the dashed arrow indicates the direction of propagation of the refracted beam.

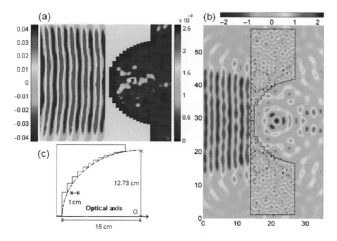

Figure 5.4. Demonstration of plano-concave grating lens focusing. (a) Composite figure of the microwave focusing experiment at 8.4 GHz using a plano-concave grating lens made of alumina with a grating on the curved surface. The electric field of the incident beam measured without the presence of the grating lens is plotted on the left. The intensity of the electric field is plotted on the right. In the middle is a photo of the lens. The grating lens behaves like a smooth plano-concave lens made of negative index material with $n_{\text{eff}} = -0.57$ at 8.4 GHz. (b) FDTD simulations at a plano-concave lens without aberration made with $n = 3$, $R = 15$ cm, and $a = 1$ cm at 8.5 GHz. Plotted is the electric field. The size of the system is measured in centimeters. (c) Details of the plano-concave lens (half of which is shown). The dashed curve is an ellipse with a semimajor axis of 15 cm and a semiminor axis of 12.73 cm. The horizontal length of the grooves is 1 cm.

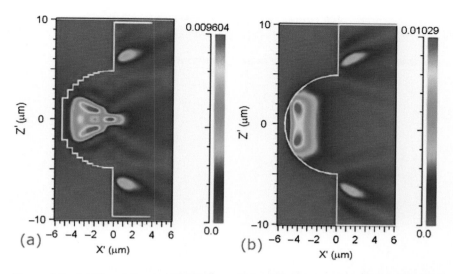

Figure 5.7. (a) Three-dimensional FDTD simulation of the plano-concave binary-staircase lens. (b) Three-dimensional FDTD simulation of the lens having the same geometrical dimensions as the binary-staircase one, but bearing no steps (or zones).

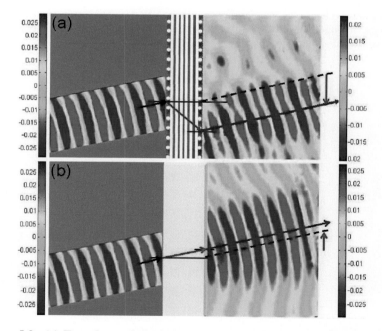

Figure 5.9. (a) Experimental demonstration of a negative lateral shift by a 1D PhC with a surface grating, at 6.96 GHz. A 5.6 cm negative lateral shift was observed. The 1D PhC is made of six layers of alumina bars with width $d = 0.5$ cm and spacing $a = 0.9$ cm. The surface grating was formed by rods of the same material, alumina, with diameter 0.63 cm and spacing $a_s = 1.8$ cm. The width of the incident beam is 10 cm and the angle of incidence is 13.5°. The incident and outgoing beams are plotted as the real parts of the measured transmission coefficient S_{21}. (b) A positive lateral shift for a microwave beam at 6.96 GHz by a slab of polystyrene with thickness 7.5 cm.

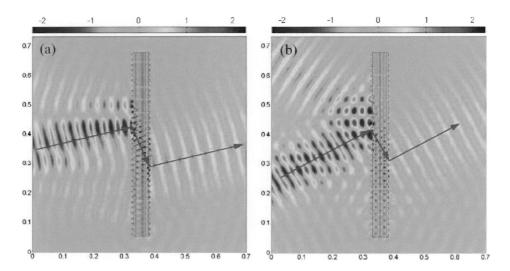

Figure 5.10. FDTD simulation of a negative lateral shift of microwave beams through a 1D PhC with surface gratings as specified in Fig. 5.8 at 6.96 GHz. (a) Microwave beam with an angle of incidence 13.5°. (b) Microwave beam with an angle of incidence 30°. The arrows indicate the energy flows of the incident and refracted beams. Lengths are measured in meters.

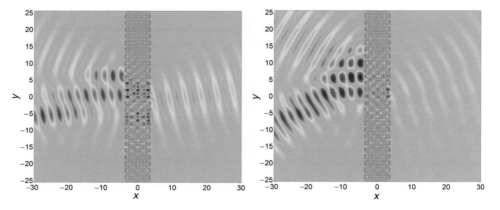

Figure 5.14. Negative lateral shift by a slab of PhC given in Fig. 5.13 for an incident Gaussian beam with angles of incidence 15° (left) and 30° (right) at $\omega = 0.219(2\pi c/a)$. The distance is measured in units of the lattice spacing a.

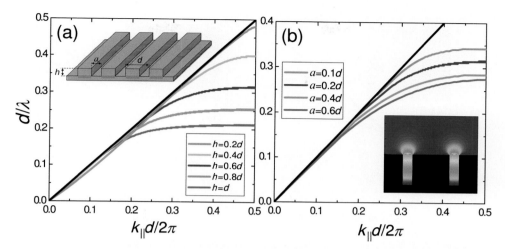

Figure 7.2. Dispersion relation of the spoof SPPs supported by periodic groove arrays. (a) Dependence on h for a fixed groove width $a = 0.2d$. The inset sketches the structure considered. (b) Bands for $h = 0.6d$ at different groove widths. The inset depicts the electric field amplitude for $h = 0.6d$ and $a = 0.2d$ evaluated at the band edge ($k_\parallel = \pi/d$).

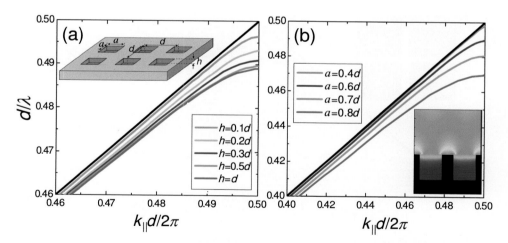

Figure 7.3. Dispersion relation of the spoof SPPs sustained by periodic dimple arrays. (a) Bands for $a = 0.2d$ and several dimple depths h. The inset shows a schematic of the system. (b) Bands for $h = 0.6d$ and different a. The inset displays the electric field amplitude at the band edge for the case $h = 0.6d$ and $a = 0.6d$.

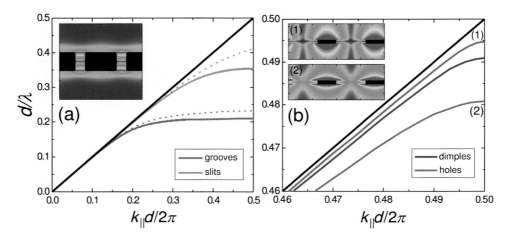

Figure 7.4. Dispersion relation of the spoof SPPs supported by fully perforated films. (a) Comparison between 1D slits and grooves of the same size ($a = 0.2d$ and $h = d$). Dotted lines show the analytical band obtained from Eqs. (7.21) and (7.23). The inset depicts the electric field amplitude at the band edge for the slit array. (b) Comparison between holes and dimples with $a = 0.6d$ and $h = 0.3d$. The insets depict the fields at the edges of the two spoof SPP bands for the hole array.

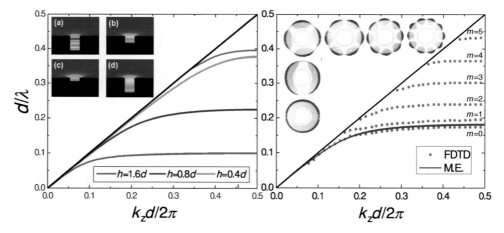

Figure 7.6. Spoof SPP dispersion relation for wires of radius $R = 2d$ perforated with rings of width $a = 0.2d$. Left panel: θ-independent bands for different ring depths. The inset shows the electric field amplitude ($r > R - h$) at the edge of the bands. Right panel: FDTD bands for higher azimuthal orders (m) for $h = 0.5d$. The insets plot the electric field patterns at the band edge ordered with increasing m from the left bottom corner to the right top corner of the figure.

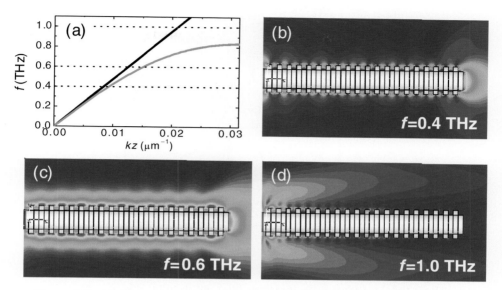

Figure 7.7. (a) Dispersion relation of the guided modes traveling along a corrugated wire of radius $R = 150\,\mu\text{m}$ perforated with an array of rings of period $d = 100\,\mu\text{m}$. The width and depth of the rings are both $50\,\mu\text{m}$. (b)–(d) Electric field amplitude for wires of length $20 \times d$ illuminated from the left by a radially polarized plane wave. Fields are evaluated at three different frequencies (0.4, 0.6, and 1.0 THz), indicated by dotted lines in (a).

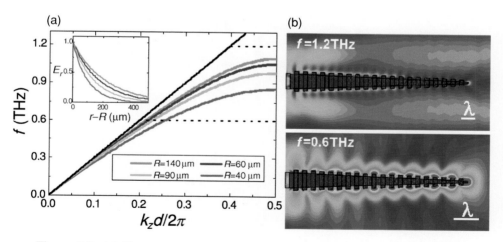

Figure 7.8. (a) Frequency versus propagation wave vector, k_z, for the guided modes supported by four corrugated wires of different radii. The inset plots the radial component of the electric field versus $r - R$ ($f = 0.6\,\text{THz}$) for the four structures. (b) Electric field amplitude at 0.6 and 1.2 THz for a 2 mm corrugated cone whose radius is reduced from 140 to $40\,\mu\text{m}$.

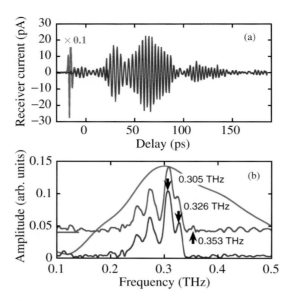

Figure 7.9. (a) Receiver current as a function of time delay for the smooth wire (red line) and the grooved structure (blue line). (b) Amplitude spectra of the time domain data in (a) together with the spectrum of another, nominally identical, helical sample (green curve, displaced for clarity). The arrows indicate the three azimuthal modes of the helical groove structure. The spectrum of the Sommerfeld wave (red curve) on the smooth wire extends to ~1 THz.

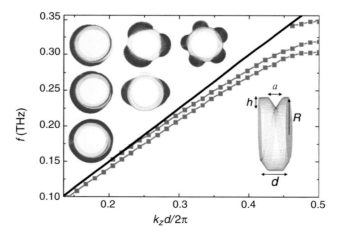

Figure 7.10. Dispersion relation of the guided modes supported by a PEC wire of radius $R = 600\,\mu m$ inscribed with a triangular-cross-section helical groove of pitch $d = 400\,\mu m$. The groove has width $a = 200\,\mu m$ and depth $h = 150\,\mu m$. The upper row of insets displays snapshots of the electric field at the three band edges, 0.305 THz (left), 0.320 THz (center), and 0.349 THz (right). The next lowest row correspond to the first mode at 0.280 THz (left) and the second mode at 0.180 THz (right). The pattern in the bottom row is for the first mode at 0.180 THz.

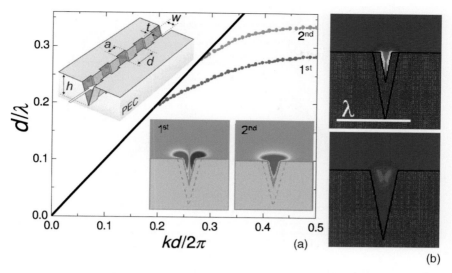

(a)

(b)

Figure 7.11. (a) Dispersion relation of the first two spoof CPP modes supported by a corrugated V-channel milled on a PEC surface. A schematic of the structure is shown in the upper inset. The lower insets depict the amplitude of the longitudinal component of the electric field evaluated at the band edge for both modes. (b) Electric field amplitude at the band edge for the lowest mode evaluated at the shallower (upper panel) and deeper (lower panel) sections of the channel. The horizontal white bar represents the wavelength of the mode in a vacuum.

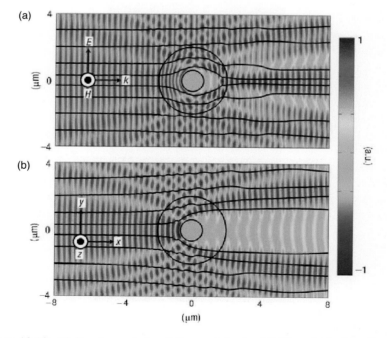

Figure 10.19. Finite element simulations of the invisibility cloak described by Cai *et al.* Cloak on (a); cloak off (b). The cloak is illuminated from the left with TM-waves at 632.8 nm. (From ref. [46], courtesy of Professor Vladimir Shalaev.)

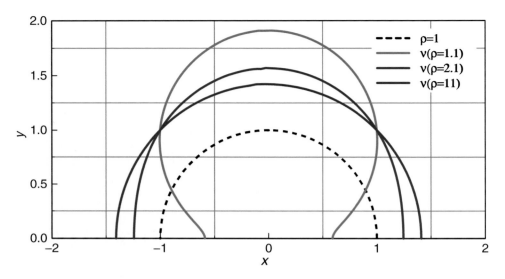

Figure 10.22. Refractive index variation corresponding to the mapping in Fig. 10.21.

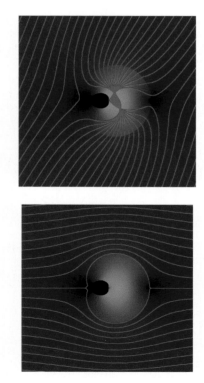

Figure 10.25. Ray paths in an invisibility device. (From refs. [3] and [48], courtesy of Professor Ulf Leonhardt.)

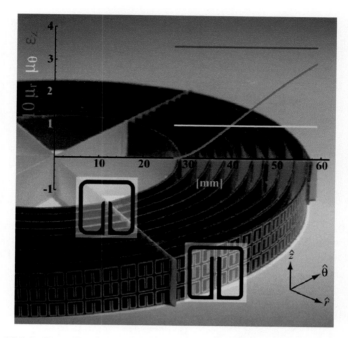

Figure 10.26. Two-dimensional microwave cloaking structure (background image) with a plot of the material parameters that are implemented. Note that: μ_r (red line) is multiplied by a factor of ten for clarity; μ_θ (green line) has the constant value 1; ε_z (blue line) has the constant value 3.423. The SRRs of cylinder 1 (inner) and cylinder 10 (outer) are shown in expanded schematic form (transparent square insets). (From ref. [58], courtesy of Professor David Smith.)

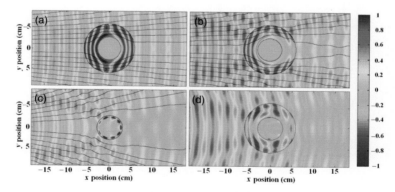

Figure 10.28. Snapshots of time-dependent, steady-state electric field patterns, with stream lines (black lines in (a)–(c)) indicating the direction of power flow (i.e. the Poynting vector). The cloak lies in the annular region between the black circles and surrounds a conducting Cu cylinder at the inner radius. The fields shown are (a) the simulation of the cloak with the exact material properties, (b) the simulation of the cloak with the reduced material properties, (c) the experimental measurement of the bare conducting cylinder, and (d) the experimental measurement of the cloaked conducting cylinder. Animations of the simulations and the measurements show details of the field propagation characteristics within the cloak that cannot be inferred from these static frames. The right-hand scale indicates the instantaneous value of the field. (From ref. [58], courtesy of Professor David Smith.)

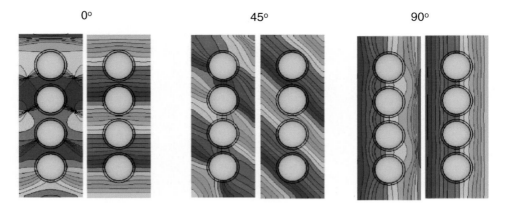

Figure 10.33. Phase of the total magnetic field distribution in the E plane for the case of four aligned spheres with and without plasmonic covers is shown for three different incidence angles. (From ref. [61], courtesy of Professor Nader Engheta.)

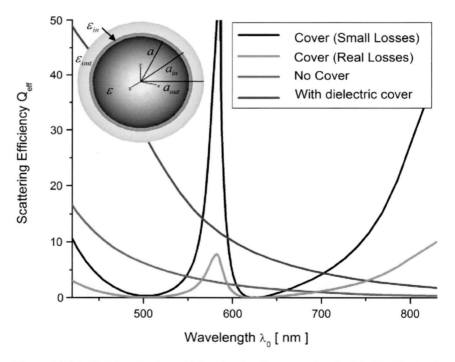

Figure 10.34. Total scattering efficiencies for the geometry depicted in the inset, i.e. a dielectric particle of radius $a = 100$ nm and permittivity $\varepsilon_r = 3$, cloaked by a two-layered shell designed for cloaking at $\lambda_0 = 500$ nm and $\lambda_0 = 625$ nm. The four curves refer to the following: a covered particle with small losses in the plasmonic materials (black), covered particle with reasonable losses (green), the original particle (red), and the same particle with a dielectric material replacing the shell region (blue). (From ref. [68], courtesy of Professor Nader Engheta.)

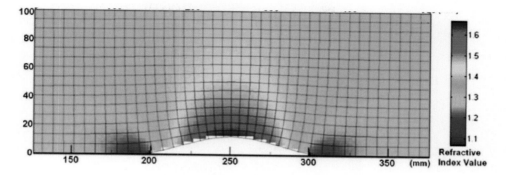

Figure 10.36. Metamaterial refractive index distribution in the ground plane cloak. The mesh lines indicate the quasi-conformal mapping. (From ref. [77], courtesy of Professor David Smith.)

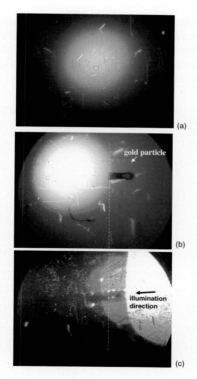

Figure 10.40. (a) Microscope image of the waveguide illuminated with white light from the top. The Newton rings are visible in the center of the field of view. (b) Microscope image of the waveguide with a gold particle placed inside and illuminated with white light from the top. (c) A long shadow has been cast by the gold particle upon coupling 515 nm laser light into the waveguide. The position of the particle edge is shown by the dashed line.

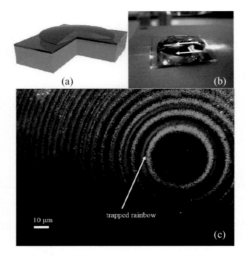

Figure 10.42. (a) Experimental geometry of the trapped rainbow experiment: a glass lens was coated on one side with a gold film. The lens was placed with the gold-coated side down on top of a flat glass slide also coated with a gold film. The air gap between these surfaces formed an adiabatically changing optical nano-waveguide. (b) Photo of the trapped rainbow experiment: He–Ne and Ar:ion laser light is coupled into the waveguide. (c) Optical microscope image of the trapped rainbow. (From ref. [89].)

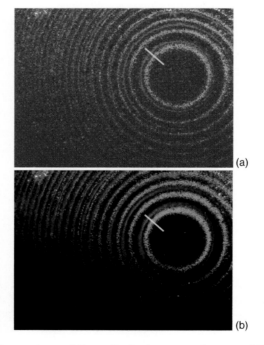

Figure 10.43. Comparison of the optical microscope images of the trapped rainbow effect from Fig. 10.42(c) (b) and the image (a) obtained when only two laser wavelengths (514 nm and 633 nm) are used for illumination. Individual spectral lines separated by only a few micrometers appear to be well resolved (see Fig. 10.44). (From ref. [89].)

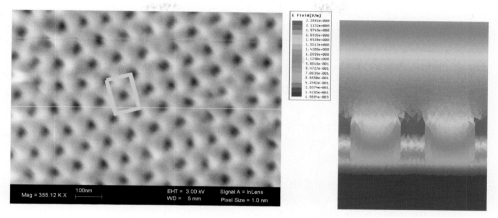

Figure 11.5. (a) Hole array fabricated in 50 nm oxide layer on top of aluminum (anodized aluminum oxide, or AAO). The yellow rectangle indicates the simulation cell. (b) Cross sectional view of electric field maps when the incident beam is launched at normal incidence. The refractive index for the oxide was 1.77. The electric field intensity values are: red is 2.25 V/m, green is 1.13 V/m and blue is 10^{-3} V/m. The incident beam was taken as 1 V/m [27]. Simulations were obtained with an Ansoft tool.

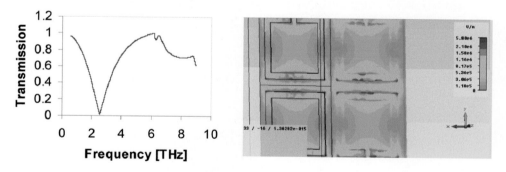

Figure 11.10. Transmission (left) showing the large absorption/reflection at resonance, and the field intensity plots (right). The polarization was set along the y-direction.

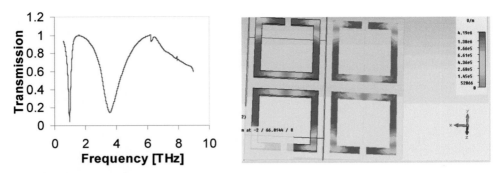

Figure 11.11. Transmission (left) and electric field thermal plots (right) when the polarization is along the x direction. The frequency of the peak has changed, and so has the field distribution.

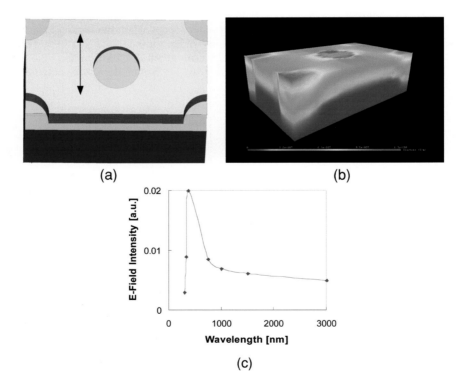

Figure 11.18. (a) The simulation cell: a 50 nm thick hexagonally perforated oxide (yellow) is lying on top of aluminum (blue). The hole radius is 40 nm and the refractive index of the oxide layer is 3.3. Graphene (gray) is deposited on top of the oxide layer. The arrow points to the direction of the TEM polarization state. (b) Linear electric field intensities map (red: 1.3×10^{-6} V/m; blue-green: 0.3×10^{-6} V/m; blue: 0). (c) Field intensity at the hole center as a function of incident optical wavelength.

[35] W. Zhu, A. Agrawal, and A. Nahata, "Planar plasmonic terahertz guided-wave devices," *Opt. Express* **16**, 6216–6226 (2008).

[36] Y. C. Lan and R. L. Chern, "Surface plasmon-like modes on structured perfectly conducting surfaces," *Opt. Express* **14**, 11339–11347 (2006).

[37] Z. C. Ruan and M. Qiu, "Slow electromagnetic wave guided in subwavelength regions along one-dimensional periodically structured metal surface," *Appl. Phys. Lett.* **90**, 201906(1-3) (2007).

[38] M. J. Lockyear, A. P. Hibbins, and J. R. Sambles, "Microwave surface-plasmon-like modes on thin metamaterials," *Phys. Rev. Lett.* **102**, 073901(1-4) (2009).

[39] M. Navarro-Cía, M. Beruete, S. Agrafiotis, F. Falcone, M. Sorolla, and S. A. Maier, "Broadband spoof plasmons and subwavelength electromagnetic energy confinement on ultrathin metafilms," *Opt. Express*, **17**, 18184–18195 (2009).

[40] C. R. Williams, M. Misra, S. R. Andrews, S. A. Maier, S. Carretero Palacios, S. G. Rodrigo, F. J. Garcia-Vidal, and L. Martin-Moreno, "Dual band terahertz waveguidng on a planar metal surface patterned with annular holes," *Appl. Phys. Lett.* **96**, 011101(1-3) (2010).

[41] S. A. Maier and S. R. Andrews, "Terahertz pulse propagation using plasmon-polariton-like surface modes on structured conductive surfaces," *Appl. Phys. Lett.* **88**, 251120(1-3) (2006).

[42] B. K. Juluri, S.-C. S. Lin, T. R. Walker, L. Jensen, and T. J. Huang, "Propagation of designer surface plasmons in structured conductor surfaces with parabolic gradient index," *Opt. Express* **17**, 2997–3006 (2009).

[43] K. Wang and D. M. Mittleman, "Metal wires for terahertz waveguiding," *Nature* **432**, 376–379 (2004).

[44] T.-I. Jeon, J. Zhang, and D. Grischkowsky, "THz Sommerfeld wave propagation on a single metal wire," *Appl. Phys. Lett.* **86**, 161904(1-3) (2005).

[45] G. Piefke, "The transmission characteristics of a corrugated wire," *IRE Trans. Antennas Propag.* **7**, 183–190 (1959).

[46] S. A. Maier, S. R. Andrews, L. Martín-Moreno, and F. J. García-Vidal, "Terahertz surface plasmon-polariton propagation and focusing on periodically corrugated metal wires," *Phys. Rev. Lett.* **97**, 176805(1-4) (2006).

[47] Y. Chen, Z. Song, Y. Li, M. Hu, Q. Xing, Z. Zhang, L. Chai, and C.-Y. Wang, "Effective surface plasmon polaritons on the metal wire with arrays of subwavelength grooves," *Opt. Express* **14**, 13021–13029 (2006).

[48] G. B. Arfken and H. J. Weber, *Mathematical Methods for Physicists*, 5th edn (London: Harcourt Academic Press, 2001).

[49] B. F. Ferguson and X.-C. Zhang, "Materials for terahertz science and technology," *Nature Mater.* **1**, 26–33 (2002).

[50] M. Tonouchi, "Cutting-edge terahertz technology," *Nature Photon.* **1**, 97–105 (2007).

[51] A. I. Fernández-Domínguez, L. Martín-Moreno, F. J. García-Vidal, S. R. Andrews, and S. A. Maier, "Spoof surface plasmon polariton modes propagating along periodically corrugated wires," *IEEE J. Sel. Top. Quant. Electron.* **14**, 1515–1521 (2008).

[52] A. I. Fernández-Domínguez, C. R. Williams, L. Martín-Moreno, F. J. García-Vidal, S. R. Andrews, and S. A. Maier, " Terahertz surface plasmon polaritons on a helically grooved wire," *Appl. Phys. Lett.* **93**, 141109(1-3) (2008).

[53] P. J. Crepeau and P. R. McIsaac, "Consequences of symmetry in periodic structures," *Proc. IEEE* **52**, 33–43 (1964).

[54] I. V. Novikov and A. A. Maradudin, "Channel polaritons," *Phys. Rev. B* **66**, 035403 (1-13) (2002).

[55] S. I. Bozhevolnyi, V. S. Volkov, E. Devaux, J.-Y. Laluet, and T. W. Ebbesen, "Channel plasmon subwavelength waveguide components including interferometers and ring resonators," *Nature* **440**, 508–511 (2006).

[56] A. I. Fernández-Domínguez, E. Moreno, L. Martín-Moreno, and F. J. García-Vidal, "Guiding terahertz waves along subwavelength channels," *Phys. Rev. B* **79**, 233104 (1-4) (2009).

[57] E. Moreno, F. J. García-Vidal, S. G. Rodrigo, L. Martín-Moreno, and S. I. Bozhevolnyi, "Channel plasmon-polaritons: modal shape, dispersion, and losses," *Opt. Lett.* **31**, 3447–3449 (2006).

[58] A. I. Fernández-Domínguez, E. Moreno, L. Martín-Moreno, and F. J. García-Vidal, "Terahertz wedge plasmon polaritons," *Opt. Lett.* **34**, 2063–2065 (2009).

[59] D. F. P. Pile and D. K. Gramotnev, "Channel plasmon-polariton in a triangular groove on a metal surface," *Opt. Lett.* **29**, 1069–1071 (2004).

[60] E. Moreno, S. G. Rodrigo, S. I. Bozhevolnyi, L. Martín-Moreno, and F. J. García-Vidal, "Guiding and focusing of electromagnetic fields with wedge plasmon polaritons," *Phys. Rev. Lett.* **100**, 023901(1-4) (2008).

[61] D. Martin-Cano, M. L. Nesterov, A. I. Fernández-Domínguez, F. J. García-Vidal, L. Martín-Moreno, and E. Moreno, "Domino plasmons for subwavelength terahertz circuitry," *Opt. Express* **18**, 754–764 (2010).

[62] M. L. Nesterov, D. Martin-Cano, A. I. Fernandez-Dominguez, E. Moreno, L. Martin-Moreno, and F. J. García-Vidal, "Geometrically-induced modification of surface plasmons in the optical and telecom regimes," *Opt. Lett.* **35**, 423–425 (2010).

8

Negative refraction using plasmonic structures that are atomically flat

PETER B. CATRYSSE, HOCHEOL SHIN, AND SHANHUI FAN

8.1 Introduction

All-angle negative refraction of electromagnetic waves [1, 2] has generated great interest because it provides the foundation for a wide range of new electromagnetic effects and applications, including subwavelength image formation [2] and a negative Doppler shift [1], as well as novel guiding, localization and nonlinear phenomena [3, 4]. There has been tremendous progress in achieving negative refraction in recent years using either dielectric photonic crystals [5–9] or metallic meta-materials [10–17]. For either approach, however, there is an underlying physical length scale that sets a fundamental limit [18]. Below such a length scale, the concept of an effective index no longer holds. For photonic crystals, it is the periodicity, which is smaller than but comparable to the operating wavelength of light [8]. For metallic meta-materials, it is the size of each individual resonant element. In the microwave wavelength range, constructing resonant elements that are far smaller than the operating wavelength is relatively straightforward. As one pushes towards shorter optical wavelengths, however, it becomes progressively more difficult to construct resonant elements at a deep subwavelength scale [15]. Moreover, in the optical wavelength range, the plasmonic effects of metals become prominent. The strong magnetic response of metallic structures, as observed in microwave and infrared wavelength ranges, may be fundamentally affected. It is therefore very desirable to accomplish all-angle negative refraction using structures that are flat at an atomic scale. Along these lines, a flat metal lens has been experimentally demonstrated using surface plasmons [19]. The structure does not operate, however, on the propagating components of a source [2]. Achieving a negative refractive index using nonmagnetic media has also been suggested [20, 21].

Structured Surfaces as Optical Metamaterials, ed. A. A. Maradudin. Published by Cambridge University Press.
© Cambridge University Press 2011.

It is not clear yet, however, what uniform physical medium would possess the required dielectric dispersion properties.

In this chapter, we review our work on the creation of all-angle negative refraction with atomically flat structures. We first introduce the plasmonic model for metals that we used to highlight the physics. We briefly review the types of modes that are supported by single metal–dielectric interfaces and metal–dielectric–metal structures. Next, we describe our approach for achieving all-angle negative refraction for surface plasmon waves with a hetero–metal–dielectric–metal structure. We demonstrate lens operation numerically for both a plasmonic metal model and real metal data. We then proceed to investigate the design of all-angle negative refraction and evanescent wave amplification with a metallo-dielectric photonic crystal in which each cell consists of a metal and a dielectric layer. We conclude with an overview of related work that has been done in this area of research.

8.2 Physics

To highlight the essential physics of the structures we are about to consider, we begin by describing the dielectric function of the metal with a Drude free-electron model:

$$\varepsilon_m(\omega) = 1 - \frac{\omega_p^2}{\omega(\omega - i\omega_\tau)}, \tag{8.1}$$

where ω_p and ω_τ are the bulk plasma frequency and the collision frequency of the metal, respectively. In this model, the dielectric function takes into account the contribution of free electrons only and, hence, displays plasma-like dispersion. We refer to Eq. (8.1) as the *plasmonic model* for metals; in particular, when ω_τ is set to zero, we call it the *lossless plasmonic model*. Despite its simplicity, the plasmonic model describes the main features of the dielectric function for real metals. Moreover, it has also provided valuable insights into their optical behavior [22]. In most metals, for example, ω_p is generally in the range of ultraviolet or even shorter wavelengths [23]. A metal exhibits therefore a negative dielectric constant at visible and infrared wavelengths, and bulk metal does not support propagating electromagnetic waves at these wavelengths. We use the plasmonic model to analyze the optical properties of metallic structures and to establish a sound theoretical background before proceeding to realistic designs based on a more complex model of real metals. The magnetic permeability μ is assumed to equal unity since metal is nonmagnetic at optical frequencies.

While metals do not allow bulk electromagnetic waves, a plasmonic metal can support a surface mode propagating along its interface with a dielectric with a positive dielectric constant. This mode comes in the form of a polariton and

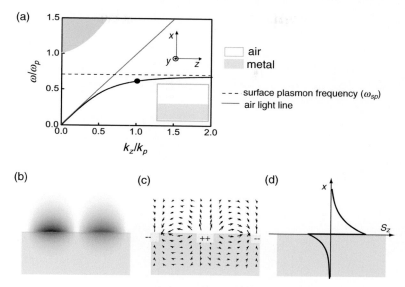

Figure 8.1. (a) Dispersion relation of a metal–dielectric interface. The inset shows the geometry. The frequency and the wave number are normalized with respect to ω_p, and $k_p = \omega_p/c$, where c is the speed of light in free space. The gray area respresents the continuum of modes that are extended in metals. For the surface plasmon polariton (SPP) mode indicated by the black dot, we show (b) a magnetic field distribution where the left gray spot, the white region, and the right gray spot correspond to positive, zero, and negative H_y, respectively, (c) a vector plot of the electric field with a schematic surface charge distribution superimposed, and (d) the time-averaged Poynting vector.

physically corresponds to the coupling of an electromagnetic wave and free electron charges. It is often referred to as a *surface plasmon polariton* (SPP). We briefly summarize the SPP properties using its dispersion relation, which relates frequency ω and wave number parallel to the interface k_z for the *eigenmodes* supported by the metal–dielectric interface (Fig. 8.1(a)) [22]. Note that SPP modes have transverse magnetic (TM) polarization with magnetic field perpendicular to the direction of propagation (Fig. 8.1(b)). Figure 8.1(c) shows the electric field with surface charge superimposed. The dispersion curve of this mode is located completely below the light line in the dielectric, which is the signature of a surface-bound mode. Hence, modal size is not subject to the diffraction limit [24] and can, in principle, be arbitrarily small. Unlike a conventional waveguide mode, the dispersion curve has an upper bound at $\omega_{sp} = \omega_p/\sqrt{1 + \varepsilon_d}$, which is commonly referred to as the surface plasmon frequency where the metal dielectric constant (ε_m) is equal in magnitude and opposite in sign to that of the dielectric (ε_d) [25]. Finally, the band has a positive group velocity ($v_g = d\omega/dk_z$) at all frequencies, and the eigenmode

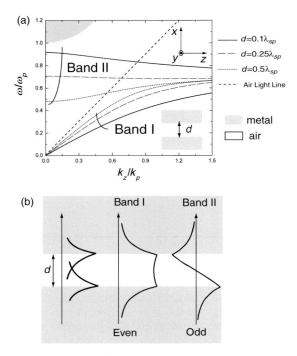

Figure 8.2. (a) Three sets of dispersion relations for MDM structures in which the separation $d = 0.1\lambda_p$, $d = 0.25\lambda_p$, and $d = 0.5\lambda_p$. The dashed line represents the light line in the dielectric (e.g. air). The gray area represents the continuum of extended modes in the metal. (b) Coupling through field overlap of the two otherwise independent surface plasmons (leftmost). As a result, the modes split into even (center) and odd (rightmost) modes. The fields shown here are the magnetic fields (H_y).

carries net power in the positive z direction. This is confirmed by the time-averaged Poynting vector S_z. Positive S_z means the power flows along with the phase front, whereas negative S_z means the power flows opposite to the phase front propagation direction. Figure 8.1(d) shows there is a sign change in the Poynting vector: such a unique power reversal phenomenon occurs due to the negative dielectric constant of the metal. Still, the SPP mode carries net power flow in the positive z direction because more power is carried in the air region than in the metal region.

We now consider a *metal–dielectric–metal* (MDM) structure, where we have two separate metal–dielectric interfaces in proximity to each other. In contrast to the metal–dielectric interface, the dispersion diagram features three bands of modes whose fields are guided in the dielectric region (Fig. 8.2(a)) [22]. We are interested in only the two lowest-frequency bands whose mode profiles exhibit the surface-bound nature. The third-order modes are similar to typical dielectric waveguide modes whose dispersion curves asymptotically approach the light line in the dielectric in the limit $k_z \to \infty$. The first two bands, on the other hand, approach $\omega_{sp}^{dielec} = \omega_p/\sqrt{1 + \varepsilon_d}$ in the limit $k_z \to \infty$. For the second band, the frequency $\omega_{II}(k_z = 0)$ depends on the dielectric thickness d. When $d < \lambda_p/4\sqrt{1 + \varepsilon_d}/\sqrt{\varepsilon_d}$, where $\lambda_p = 2\pi c/\omega_p$ and c is the speed of light in a vacuum, $\omega_{II}(k_z = 0)$ becomes greater than ω_{sp}^{dielec} and the band acquires a negative slope and a negative group velocity [26]. The aforementioned features in the dispersion diagram can be best explained with the coupling theory of two otherwise single-interface surface plasmons. If the coupling between the two interfaces were to be negligible, the dispersion diagram for the MDM structure would be identical to that of a single-interface surface plasmon. For finite d, the field overlap becomes zero as $k_z \to \infty$. Hence, the surface plasmon modes become degenerate at ω_{sp} in the limit $k_z \to \infty$. However, as the field overlap increases from zero with decreasing k_z, the degeneracy is lifted into an even (band I) mode and an odd (band II) mode with a frequency gap: the even (odd) mode is shifted up (down) (Fig. 8.2(b)).

In the MDM structure, the negative group velocity is related to the negative power flow of the mode. For any given eigenmode, the time-averaged Poynting vector always changes sign across a metal–dielectric interface, since the displacement fields normal to the interface are continuous, and the dielectric constant of the metal region is negative. As a result, the Poynting vector inside the metal is opposite to the phase velocity. For the MDM structure, the modes in the second band can have more power in the metal than in the dielectric, resulting in a net power flow that is opposite to the phase velocity [27]. Therefore, such a structure has an effective negative refractive index for the frequency range of band II.

8.3 All-angle negative refraction for surface plasmon waves

The second band for the MDM structure and the surface plasmon band for the metal–air interface overlap in frequency when $\varepsilon_d > 1$. With the proper choice of dielectric thickness, the MDM region can therefore function as a negative refraction lens for the propagating surface plasmon waves on the metal–air interface (Fig. 8.3). Since the MDM structure by itself is uniform in all directions parallel to the metal surface, the constant-frequency contour is exactly *circular* at all frequencies, which makes it a unique physical realization of a negative index medium. In addition, one can choose an operating frequency such that the wave numbers in the two regions are matched in magnitude (Fig. 8.3(c)). The phase-index matching ensures negative refraction at all angles of incidence, as well as aberration-free image formation, as can be shown using Fourier decomposition. In the design, as shown, there are reflections at the boundaries between the regions due to modal mismatch. The physical origin of such reflections can be seen by analyzing the modal profile for the

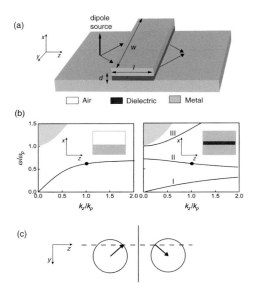

Figure 8.3. (a) Geometry of an imaging system consisting of a stripe of a metal–dielectric–metal (MDM) structure on a metal surface. The arrows on the metal surface indicate the lens operation. (b) Dispersion relations for modes of a metal–air interface (left panel), and for modes in an MDM structure with $d = 0.1\lambda_p$ and $\varepsilon_d = 4$ (right panel). The corresponding structures are shown in the insets of the respective panels. The gray areas represent the continuum of extended modes in the metal regions. The black dots in both panels indicate the frequency ω_0 at which the wave numbers match in magnitude. (c) The constant-frequency contour at ω_0 for both the metal surface region (left side) and the MDM region (right side). Arrows indicate group velocity direction. Dashed line represents conservation of parallel wave vector.

two structures, shown as insets in Fig. 8.3(b), at the index matching frequency. On the metal–air interface the mode intensity has a single maximum at the interface (Fig. 8.4(a), left panel), while the mode intensity in the MDM region has two maxima on the two metal–dielectric interfaces (Fig. 8.4(a), right panel). To optimize modal overlap, instead of using the *homo*-MDM structure, where both the top and the bottom metals are of the same kind, one can use a *hetero*-MDM structure, where the metals have different bulk plasma frequencies, to break the symmetry of the field profiles and enhance the fractional amplitude of the peak at the lower interface (Fig. 8.4(b)).

We now numerically demonstrate the lens operation of a hetero-MDM structure. We choose $\omega'_p = 0.6\omega_p$, where ω_p and ω'_p are the bulk plasma frequencies of the lower and upper metal sections, respectively. This ratio corresponds to the ratio between the plasma frequencies of Al (14.98 eV) and Ag (9.01 eV) in typical

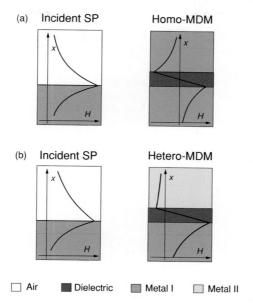

Figure 8.4. Magnetic field profiles. (a) A mode of the metal–air interface (left panel) and a mode in the second band of a homo-MDM structure (right panel). All metals have the same plasma frequency ω_p; the dielectric region has a thickness $d = 0.1\lambda_p$ and a dielectric constant $\varepsilon_d = 4$. Both modes have the same frequency, $\omega_0 = 0.62\omega_p$, and the same wave number, k_p. (b) A mode of the metal–air interface (left panel) and a mode in the second band of a hetero-MDM structure (right panel). The metal at the top has a plasma frequency, $\omega'_p = 0.6\omega_p$. The dielectric parameters are the same as in (a). Both modes have the same frequency, $\omega_0 = 0.54\omega_p$, and the same wave number, $0.69k_p$.

Drude models [23]. The MDM region has

$$d = 0.1\lambda_p < \frac{\lambda_{sp}}{4\sqrt{\varepsilon_d}}, \qquad w = 8.67\lambda_p, \qquad l = 2.5\lambda_p, \qquad (8.2)$$

where w and l are the width and the length of the MDM, respectively (Fig. 8.3(a)). The dielectric region has $\varepsilon_d = 4$. The simulations use the finite-difference time-domain (FDTD) method in three dimensions with a grid size of $\lambda_p/120$. The computational cell is surrounded by perfectly matched layer absorbing boundary conditions [28]. We assume a *near-lossless plasmonic model* Eq. (8.1) for metals with the collision frequencies set to one-thousandth of the metal plasma frequencies. We excite surface plasmon waves on the left region of the metal surface by placing a single dipole source at λ_p away from the edge of the MDM. The source has a frequency $\omega = 0.539\omega_p$ and is polarized perpendicular to the metal surface.

The steady state field distribution for the x component of the electric field is shown in Fig. 8.5(a). Two images are observed: one near the center in the

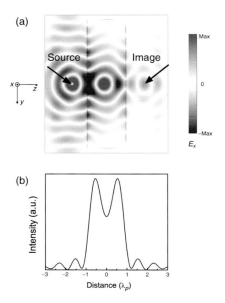

Figure 8.5.   The imaging process for surface plasmons with the hetero-MDM lens.
The computational setup is schematically shown in Fig. 8.3(a). Plotted is the steady
state E_x field distribution on a cross-section at $0.18\lambda_p$ above the bottom metal
surface. The image is formed clearly at large field values. (b) The time-average E_x
field intensity at the image plane for two point sources oscillating in phase. The
lens has a length of $1.7\lambda_p$. The two point sources are separated by $0.93\lambda_p$, and
are placed $0.5\lambda_p$ away from the edge of the lens. In comparison, at the operating
frequency $\omega_0 = 0.539\omega_p$, the surface plasmon wave at the metal–air interface has
a wavelength of $1.46\lambda_p$.

MDM region, and the other on the right region of the metal surface. We also
observe significant field enhancement at the boundaries of the MDM region, indicating excitation of edge states [29]. The actual field values at the boundaries are
underrepresented in Fig. 8.5(a), due to the shading scheme chosen, which satu-
rates at large field values in order to highlight the image. Our calculations show
that such a lens can resolve two sources, oscillating in phase, with a distance of
$0.93\lambda_p$ between them (Fig. 8.5(b)). In comparison, the surface plasmon wave at
the metal–air interface has a wavelength of $1.46\lambda_p$ at this operating frequency. We
do not observe perfect image recovery, which is consistent with previous FDTD
simulations on an ideal negative index lens that had both ε and μ simultaneously negative [29–31]. In addition to computational constraints due to the finite
widths [29] and numerical dispersion for large wave vector components [31], Smith
et al. have shown theoretically that to achieve perfect imaging for a lens with length

(a)

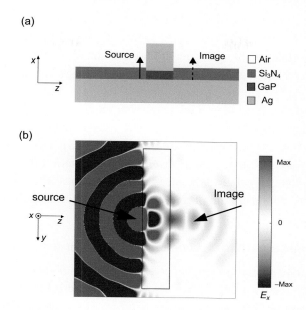

(b)

Figure 8.6. (a) Cross-sectional view of a simulated structure for the purpose of demonstrating the imaging process with realistic material parameters. The solid and dashed arrows indicate the positions of the source and image, respectively. (b) Steady state E_z field for the structure shown in (a), on a cross-section at 60 nm above the bottom Ag surface.

comparable to the wavelength of the incident wave requires parameter accuracy that is very difficult to realize in either simulations or experiments [18].

For practical metallic structures, material loss is a major issue. The effect of losses in the metal can potentially be mitigated by operating at low temperatures [32], or by introducing gain [33]. For the structure we have proposed, the lens can still function with proper design even in the presence of realistic material loss. As an MDM lens demonstration, we use a Ag–GaP–Ag structure with a length of 280 nm, a width of 1440 nm, and a dielectric thickness of 24 nm. The surface of Ag outside the MDM region is covered by a 28 nm thick Si_3N_4 film with a refractive index $n = 2$ (Fig. 8.6(a)). The use of GaP and Si_3N_4 pushes the operating free-space wavelength up to 479 nm, where Ag is less lossy. At this wavelength, however, there is absorption loss from GaP. The overall loss is lower compared with structures operating at shorter wavelengths, since the loss in Ag increases dramatically as wavelength decreases.

To simulate this structure with FDTD, we fit the experimentally determined dielectric constants of Ag and GaP with Lorentz–Drude models. Our models give $\varepsilon_{Ag} = -7.47 - i0.73$ and $\varepsilon_{GaP} = 13.69 - i0.045$ at the operating wavelength, which is in excellent agreement with experimental values for these materials [34].

The grid size is chosen to be 2 nm. The dipole source is placed on the Ag surface 40 nm away from the edge of the MDM region. Figure 8.6(b) shows the E_z field profile at steady state. Although the amplitude of the transmitted wave is strongly attenuated due to reflections and material losses, the image formation is still clearly visible.

With this work, we have identified a new route toward all-angle negative refraction. The simplicity of MDM structures should enable one to incorporate negative index materials into geometries that are more complex than a flat lens, which can lead to a wide range of unexplored novel electromagnetic effects [35]. The proposed structure also provides a new mechanism for controlling the propagation of surface plasmons, which are important for manipulating light at the nanoscale [36].

8.4 All-angle negative refraction and evanescent wave amplification

In general, negative refraction and super-lensing effects do not necessarily require a negative index medium [8]. It has been demonstrated, for example, that two- or three-dimensional photonic crystal structures can provide all-angle negative refraction in a positive refractive index medium, based on the *negative photonic mass*, where the photonic band dispersion $\omega(\mathbf{k})$ is a convex function [6]. Such a photonic crystal structure neither involves a negative refractive index for negative refraction nor does it require surface state excitations for evanescent wave amplification. Still, a lens based on such a photonic structure is shown to have super-resolution beyond the diffraction limit via an alternative mechanism. The resolution of such a photonic crystal lens is ultimately limited due to its surface periodicity at the interfaces [8].

In this section, we introduce a *metallo-dielectric photonic crystal* in which each cell consists of a metal and a dielectric layer (Fig. 8.7(a)). We show that, when the plasmonic properties of the metal become prominent, for example, in the visible or ultra-violet wavelength range, such a structure exhibits all-angle negative refraction at its interfaces. In contrast to the MDM negative refraction lens [37], it does not possess an effective negative refractive index. Instead, the all-angle negative refraction phenomenon is due to its convex dispersion band [38]. Moreover, its uniform surfaces parallel to the object plane distinguishes this lens from all previously demonstrated photonic crystal lenses [6–9]. The resolution is therefore not subject to the surface periodicity. With a proper design, such a structure provides a resolution beyond the diffraction limit. While such structures were initially studied by Bloemer and Scalora [39] as a transparent electrode, our work explicitly shows that all-angle negative refraction for propagating waves and recovery of evanescent components can be achieved simultaneously in such a system [40].

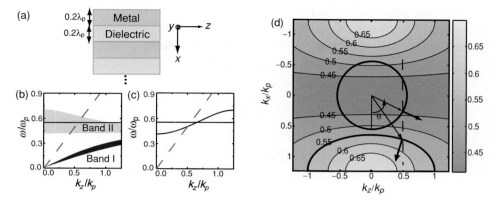

Figure 8.7. (a) Geometry of the metallo-dielectric photonic crystal structure.
(b) Projected band diagram $\omega(k_x, k_z)$ in the z direction and (c) band diagram
$\omega(k_x, k_z = 0)$ of the periodic structure in (a). The dashed lines are the light lines
in air and the horizontal lines indicate the operating frequency of $\omega = 0.55\omega_p$.
(d) The CFC representation of $\omega(k_x, k_z)$ in the frequency range of the second band,
with the contour at $\omega = 0.55\omega_p$ highlighted with a thick line. The CFC in air at
$\omega = 0.55\omega_p$ represented by a circle is overlaid. For light incident from air onto the
crystal at an angle of incidence θ, the thin arrows indicate the wave vectors in air
and in the crystal, whereas the thick arrows indicate the group velocity directions
in air and in the periodic layer region. These directions are determined by the
procedure represented by the dashed line, which arises from the conservation of
the parallel wave vectors.

As a starting point, for simplicity we use a *lossless plasmonic model*, Eq. (8.1),
for metal. For concreteness, we assume that both the metal and dielectric layers are
$0.2\lambda_p$ thick and choose a dielectric with a dielectric constant $\varepsilon_d = 4$ (Fig. 8.7(a)).
For the metallo-dielectric structure in Fig. 8.7(a), we numerically calculate the band
structure $\omega = \omega(k_x, k_z)$ in the two-dimensional wave vector (**k**) space, using the
transfer matrix formalism. All the modes we calculate here have TM polarization
with a magnetic field parallel to the layers. Since the structure is periodic in the
x direction and uniform in the z direction, the edges of the first Brillouin zone
along the k_x direction are located at $\pm\pi/0.4\lambda_p$, while k_z extends to $\pm\infty$. We show
the projected dispersion relation along the z direction in Fig. 8.7(b) [22], and the
dispersion relation $\omega(k_x, k_z = 0)$ in Fig. 8.7(c). The structure supports two bands
that both asymptotically approach the surface plasmon frequency of the metal–
dielectric interface at large k_z (Fig. 8.7(b)). A significant portion of band II lies
above the light line. Consequently, externally incident light from air can couple to
this band.

To study the refraction properties of light in band II, we plot the corresponding
constant-frequency contours (CFCs) in Fig. 8.7(d), in the frequency range from
$0.45\omega_p$ to $0.65\omega_p$, which frequencies are close to the lower and upper edges of

band II. Since the band maximum is located at $\mathbf{k}_m \equiv (k_x = 1.25k_p, k_z = 0)$, as can be inferred from Figs. 8.7(b) and (c), the CFCs in the frequency range close to $0.65\omega_p$ are of elliptical shapes centered around $\mathbf{k}_m$, with the size of the contour in the $\mathbf{k}$-space decreasing with increasing frequency. Such a CFC satisfies the condition for all-angle negative refraction of light externally incident from air [6]. As an illustration, we overlay in Fig. 8.7(d) a CFC of air at $\omega = 0.55\omega_p$ and plot the direction of the refracted beam for light incident from air at an incidence angle θ. The refraction direction, which clearly shows negative refraction behavior, is derived based upon conservation of the parallel wave vector k_z, and the fact that the gradient $\nabla_{\mathbf{k}}\omega$ determines the direction of electromagnetic energy propagation. Also at this frequency, since the radius of the air CFC is smaller than k_z^{max} (defined as the maximum $|k_z|$ allowed in the photonic crystal), negative refraction occurs for all angles of incidence. For this structure, all-angle negative refraction exists within the entire frequency range $0.447\omega_p < \omega < 0.615\omega_p$, where the metal dielectric constant is $-4.00 < \varepsilon_m < -1.64$. We have analyzed similar structures with other geometrical parameters. In general, all-angle negative refraction occurs when the dielectric thickness is smaller than $\lambda_p\sqrt{1 + \varepsilon_d}/4\sqrt{\varepsilon_d}$, which is closely related to the fact that with the same dielectric thickness a metal–dielectric–metal waveguide exhibits a negative group velocity [26] and the metal layer thickness has to be smaller than the skin depth to allow penetration of incident light into the structure.

In addition to focusing the propagating field components, our *metallo-dielectric* structure can provide recovery of the evanescent components of an object. For an object in air, the evanescent components have parallel wave vector components $|k_z| > k_0$, where $k_0 = \omega/c$. Using our structure, the evanescent components in the region $k_0 < |k_z| < k_z^{max}$ can be recovered. In this range of k_z, for a structure with an infinite number of periods, there exist propagating Bloch modes (Fig. 8.7). When the periodic structure is truncated with air, the Bloch modes undergo total internal reflection at the air interface. For a slab structure composed of a finite number of metal–dielectric layers, waveguide modes can therefore exist depending on the thickness of the slab structure. These waveguide modes provide the mechanism to amplify the optical near field. For our structure, which has a thickness of 4.5 periods (see the inset in Fig. 8.8(a)), we use the transfer matrix method to calculate the transmission coefficient through the multi-layer structure at different wave numbers k_z at $\omega = 0.55\omega_p$ (Fig. 8.8(a)). The transmission coefficient T is defined as the ratio of the complex amplitude of the transmitted wave to that of the incident wave. The transmission $|T|$ is fairly constant and smaller than unity in the propagating regime $k_z/k_0 \leq 1$, where negative refraction occurs. In the evanescent regime, where $k_z/k_0 > 1$, two sharp transmission peaks occur at $k_z = 1.23k_0$ and $k_z = 1.58k_0$, indicating the existence of waveguide modes. The width of the peaks is related to the loss in the metal, and can be made arbitrarily small when the loss in the metal

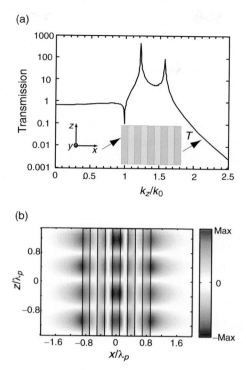

Figure 8.8. (a) Amplitude of the transmission coefficient through a multi-layer system as a function of k_z at $\omega = 0.55\omega_p$. The system consists of 4.5 metal–dielectric layers surrounded by air, as shown in the inset, where the dark and light gray layers represent the metal and the dielectric, respectively. (b) The steady state magnetic field H_y distribution with an incident evanescent wave of $k_z = 1.23k_0$ at $\omega = 0.55\omega_p$.

vanishes, in which case the amplitude of the peaks diverges. The height of the peaks greatly exceeds unity, indicating strong amplification of the optical near field at these vector components. Figure 8.8(b) shows the magnetic field H_y distribution on resonance at $k_z = 1.23k_0$, i.e. the signature of the waveguide mode where the field is highly concentrated in the structure. Such amplification, due to the presence of the waveguide mode, can be used to amplify the decaying evanescent fields from the source, leading to partial recovery of the evanescent components of the object [8]. Furthermore, it was noted by Luo *et al.* [8] that the use of the waveguide mode for subwavelength imaging has better tolerance to deviations from the ideal condition than a surface-plasmon-based mechanism.

The effects of negative refraction for propagating waves and amplification for evanescent waves, as discussed above using the lossless plasmonic model, also occur in real metal systems with realistic losses, which can be substantial at optical frequencies. As an example, we consider a structure composed of five layers of Ag and four layers of Si_3N_4. Both the Ag and Si_3N_4 layers are 40 nm thick. At $\lambda = 363.6$ nm, the dielectric constants for Ag and Si_3N_4 are $\varepsilon_{Ag} = -2.51 - i0.60$ and $\varepsilon_{Si_3N_4} = 4$ [34]. Using the transfer matrix method, we calculate the magnetic field H_y distribution both inside and outside the structure, in the presence of a point-like object located in air at 48 nm away from the slab surface. The field intensity of the object is Gaussian in the z direction, where the full width at half maximum (FWHM) is equal to one-tenth of the propagating wavelength. The object is formed by a Gaussian-weighted sum of plane waves with parallel wave vector components k_z ranging from $-11k_0$ to $+11k_0$. To calculate the image, we determine the amplitudes of the forward and backward waves for every parallel wave vector component in each layer, and superimpose all the plane wave components to obtain the full magnetic field distribution. Figure 8.9(a) shows the field distribution, which clearly shows an image formed on the opposite side of the structure in air. To assess the resolution performance of the slab lens, we plot the magnetic field intensity $|H_y|^2$ at the image plane ($x = 48$ nm) and compare the image against the diffraction-limited image (Fig. 8.9(b)). The diffraction-limited image is calculated by taking only propagating components from the Gaussian object field. The comparison in Fig. 8.9(b) clearly indicates recovery of the evanescent wave components from the original object field. While the FWHM of the solid curve is 0.44λ, the FWHM of the dashed curve is 0.31λ. The same design principle may also be applicable to longer wavelength ranges using dispersive materials such as polaritonic media.

We have demonstrated that both all-angle negative refraction and evanescent wave amplification can occur using a one-dimensional photonic crystal consisting of metal–dielectric multi-layers. Based on the theory, we designed a Ag–Si_3N_4 multi-layer structure that operates in the visible wavelength range, and we verified its subwavelength resolution. Our result not only provides an alternative approach to the practical realization of a super-lens at optical frequencies, but also opens a new perspective on light propagation in metals.

8.5 Related work

We now briefly describe a selection of works that are closely related to our work on negative refraction with atomic-scale media. The field of negative refraction is a very large one, and this selection is representative without being comprehensive.

Very recently, our approach towards all-angle negative refraction for surface plasmon waves with a hetero-MDM structure was experimentally demonstrated by the Atwater group at Caltech [40]. Lezec *et al.* demonstrate an experimental realization of a two-dimensional negative index material in the blue–green region of the visible spectrum. Their work features direct geometric visualization

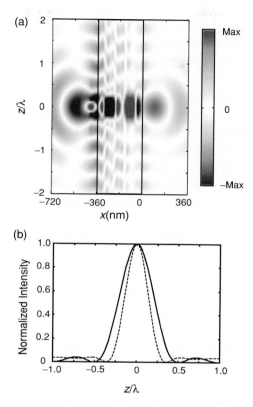

Figure 8.9. (a) Imaging process with a 4.5 period one-dimensional Ag–Si$_3$N$_4$ photonic crystal surrounded by air. Plotted is the magnetic field H_y distribution at the operating wavelength $\lambda = 363.6$ nm. (b) The field intensity at the image plane ($x = 48$ nm) with the multi-layer structure (dashed line) and the diffraction-limited field intensity (solid line).

of negative refraction. To achieve negative indices, they employed an ultrathin Au–Si$_3$N$_4$–Ag hetero-MDM structure, which, as we described, sustains a surface plasmon polariton mode with antiparallel group and phase velocities. All-angle negative refraction was observed at the interface between this hetero-MDM and a conventional Ag–Si$_3$N$_4$–Ag MDM structure. This experimental result is the first step in the development of practical negative index optical designs in the visible regime.

Bloemer *et al.* present a theoretical analysis of a broadband super-resolving lens with high transparency in the visible range [41]. The lens is based on one-dimensional metal–dielectric photonic band gap crystals composed of Ag/GaP multi-layers. The individual Ag layers are 22 nm thick and can be readily fabricated in conventional deposition systems. In this design the lens maintains a normal incidence transmittance of 50% for propagating waves over the super-resolving wavelength range of 500–650 nm.

Podolskiy and Narimanov developed an approach to build a material with negative refraction index in a similar geometry [21]. Their approach is also intrinsically nonmagnetic, but differs from ours in that it uses a material with a strong anisotropic dielectric constant to provide a left-handed behavior in a waveguide geometry.

Using the concepts of nanoscale circuit elements at optical frequencies, Alú and Engheta theoretically investigate optical nanotransmission lines that can be regarded as stacks of plasmonic and nonplasmonic planar slabs [42]. Such structures are similar to those described in Section 8.3 on achieving all-angle negative refraction for surface plasmon waves. The authors independently come to the conclusion that such structures may be designed to exhibit effectively the properties of planar metamaterials of "forward (right-handed) or backward (left-handed) operation." In particular, negative refraction and left-handed propagation are shown to be possible in these planar plasmonic guided-wave structures, providing possibilities for subwavelength focusing and imaging in planar optics and laterally confined waveguiding at IR and visible frequencies.

We have recently demonstrated that the entire dispersion behavior of coaxial plasmonic structures, including the number of modes at every frequency, the modal propagation constants, the propagation losses, and the cutoff frequencies of propagating modes, can be understood through a direct connection with the planar metal–dielectric–metal geometry [43]. This intuitive picture allows for a qualitative understanding of these technologically important structures and opens up the way to design meta-materials based on them as well. It is therefore not too surprising to find that structures based on closely packed deep-subwavelength plasmonic coaxial waveguides have been shown to act as negative index meta-materials [44].

References

[1] V. G. Veselago, "Electrodynamics of substances with simultaneously negative values of ϵ and μ," *Sov. Phys. Uspekhi* **10**, 509–514 (1968).

[2] J. B. Pendry, "Negative refraction makes a perfect lens," *Phys. Rev. Lett.* **85**, 3966–3969 (2000).

[3] G. D'Aguanno, N. Mattiucci, M. Scalora, and M. J. Bloemer, "Bright and dark gap solitons in a negative index Fabry-Perot etalon," *Phys. Rev. Lett.* **93**, 213902(1-4) (2004).

[4] I. V. Shadrivov, A. A. Sukhorukov, and Y. S. Kivshar, "Complete band gaps in one-dimensional left-handed periodic structures," *Phys. Rev. Lett.* **95**, 193903(1-4) (2005).

[5] M. Notomi, "Theory of light propagation in strongly modulated photonic crystals: refractionlike behavior in the vicinity of the photonic band gap," *Phys. Rev. B* **62**, 10696–10705 (2000).

[6] C. Luo, S. G. Johnson, J. D. Joannopoulos, and J. B. Pendry, "All-angle negative refraction without negative effective index," *Phys. Rev. B* **65**, 201104(1-4) (2002).

[7] E. Cubukcu, K. Aydin, E. Ozbay, S. Foteinopoulou, and C. M. Soukoulis, "Subwavelength resolution in a two-dimensional photonic-crystal-based superlens," *Phys. Rev. Lett.* **91**, 207401(1-4) (2003).

[8] C. Luo, S. G. Johnson, J. D. Joannopoulos, and J. B. Pendry, "Subwavelength imaging in photonic crystals," *Phys. Rev. B* **68**, 04511(1-8)(2003).

[9] A. Berrier, M. Mulot, M. Swillo, M. Qiu, L. Thylén, A. Talneau, and S. Anand, "Negative refraction at infrared wavelengths in a two-dimensional photonic crystal," *Phys. Rev. Lett.* **93**, 073901(1-4)(2004).

[10] J. B. Pendry, A. J. Holden, W. J. Stewart, and I. Youngs, "Extremely low frequency plasmons in metallic mesostructures," *Phys. Rev. Lett.* **76**, 4773–4776 (1996)

[11] J. B. Pendry, A. J. Holden, D. J. Robbins, and W. J. Stewart, "Magnetism from conductors and enhanced nonlinear phenomena," *IEEE Trans. Microwave Th. Tech.* **47**, 2075–2084 (1999).

[12] D. R. Smith, W. J. Padilla, D. C. Vier, S. C. Nemat-Nasser, and S. Schultz, "Composite medium with simultaneously negative permeability and permittivity," *Phys. Rev. Lett.* **84**, 4184–4187 (2000).

[13] R. A. Shelby, D. R. Smith, and S. Schultz, "Experimental verification of a negative index of refraction," *Science* **292**, 77–79 (2001).

[14] T. Koschny, M. Kafesaki, E. N. Economou, and C. M. Soukoulis, "Effective medium theory of left-handed materials," *Phys. Rev. Lett.* **93**, 107403(1-4) (2004).

[15] S. Linden, C. Enkrich, M. Wegener, J. Zhou, T. Koschny, and C. M. Soukoulis, "Magnetic response of metamaterials at 100 terahertz," *Science* **306**, 1351–1353 (2004).

[16] J. B. Pendry, "A chiral route to negative refraction," *Science* **306**, 1353–1355 (2004).

[17] T. J. Yen, W. J. Padilla, N. Fang, D. C. Vier, D. R. Smith, J. B. Pendry, D. N. Basov, and X. Zhang, "Terahertz magnetic response from artificial materials," *Science* **303**, 1494–1496 (2004).

[18] D. R. Smith, D. Schurig, M. Rosenbluth, S. Schultz, S. A. Ramakrishna, and J. B. Pendry, "Limitations on subdiffraction imaging with a negative refractive index slab," *Appl. Phys. Lett.* **82**, 1506–1508 (2003).

[19] N. Fang, H. Lee, C. Sun, and X. Zhang, "Sub-diffraction-limited optical imaging with a silver superlens," *Science* **308**, 534-537 (2005).

[20] Y.-F. Chen, P. Fischer, and F. W. Wise, "Negative refraction at optical frequencies in nonmagnetic two-component molecular media," *Phys. Rev. Lett.* **95**, 067401(1-4) (2005).

[21] V. A. Podolskiy and E. E. Narimanov, "Strongly anisotropic waveguide as a nonmagnetic left-handed system," *Phys. Rev. B* **71**, 201101(1-4) (2005).

[22] E. N. Economou, "Surface plasmons in thin films," *Phys. Rev.* **182**, 539–554 (1969).

[23] A. D. Rakic, A. B. Djurišic, J. M. Elazar, and M. L. Majewski, "Optical properties of metallic films for vertical-cavity optoelectronic devices," *Appl. Opt.* **37**, 5271–5283 (1998).

[24] M. Born and E. Wolf, *Principles of Optics: Electromagnetic Theory of Propagation, Interference and Diffraction of Light* (New York: Cambridge University Press, 1999).

[25] H. Raether, *Surface Plasmons on Smooth and Rough Surfaces and on Gratings* (New York: Springer-Verlag, 1988).

[26] H. Shin, M. F. Yanik, S. Fan, R. Zia, and M. L. Brongersma, "Omnidirectional resonance in a metal-dielectric-metal geometry," *Appl. Phys. Lett.* **84**, 4421–4423 (2004).

[27] P. Tournous and V. Laude, "Negative group velocities in metal-film optical wave-guides," *Opt. Commun.* **137**, 41–45 (1997).

[28] J. P. Berenger, "A perfectly matched layer for the absorption of electromagnetic-waves," *J. Comp. Phys.* **114**, 185–200 (1994).

[29] L. Chen, S. He, and L. Shen, "Finite-size effects of a left-handed material slab on the image quality," *Phys. Rev. Lett.* **92**, 107404(1-4) (2004).

[30] R. W. Ziolkowski and E. Heyman, "Wave propagation in media having negative permittivity and permeability," *Phys. Rev. E* **64**, 056701(1-15) (2001).

[31] S. A. Cummer, "Simulated causal subwavelength focusing by a negative refractive index slab," *Appl. Phys. Lett.* **82**, 1503–1505 (2003).

[32] A. Karalis, E. Lidorikis, M. Ibanescu, J. D. Joannopoulos, and M. Soljačić, "Surface-plasmon-assisted guiding of broadband slow and subwavelength light in air," *Phys. Rev. Lett.* **95**, 063901(1-4) (2005).

[33] S. A. Ramakrishna and J. B. Pendry "Removal of absorption and increase in resolution in a near-field lens via optical gain," *Phys. Rev. B* **67**, 201101(1-4) (2003).

[34] E. D. Palik and G. Ghosh, *Handbook of Optical Constants of Solids* (Orlando, FL: Academic Press, 1985).

[35] J. B. Pendry and S. A. Ramakrishna, "Near-field lenses in two dimensions," *J. Phys.* **14**, 8463–8479 (2002).

[36] W. L. Barnes, A. Dereux, and T. W. Ebbesen, "Surface plasmon subwavelength optics," *Nature* **424**, 824–830 (2003).

[37] H. Shin and S. H. Fan, "All-angle negative refraction for surface plasmon waves using a metal-dielectric-metal structure," *Phys. Rev. Lett.* **96**, 073907(1-4) (2006).

[38] H. Shin and S. H. Fan, "All-angle negative refraction and evanescent wave amplification using one-dimensional metallodielectric photonic crystals," *Appl. Phys. Lett.* **89**, 151101(1-3)(2006).

[39] M. J. Bloemer and M. Scalora "Transmissive properties of Ag/MgF_2 photonic band gaps," Appl. Phys. Lett. **72**, 1676–1678 (1998).

[40] H. J. Lezec, J. A. Dionne, and H. A. Atwater, "Negative refraction at visible frequencies," *Science* **316**, 430–432 (2007).

[41] M. Bloemer, G. D'Aguanno, N. Mattiucci, M. Scalora, and N. Akozbek, "Broadband super-resolving lens with high transparency in the visible range," *Appl. Phys. Lett.* **90**, 174113(1-3) (2007).

[42] A. Alú and N. Engheta "Optical nanotransmission lines: synthesis of planar left-handed metamaterials in the infrared and visible regimes," *J. Opt. Soc. Am. B* **23**, 571–583 (2006).

[43] P. B. Catrysse and S. H. Fan, "Understanding the dispersion of coaxial plasmonic structures through a connection with the planar metal-insulator-metal geometry," *Appl. Phys. Lett.* **94**, 231111(1-3) (2009).

[44] F. J. Rodríguez-Fortuño, C. García-Meca, R. Ortuño, J. Martí, and A. Martínez, "Coaxial plasmonic waveguide array as a negative-index metamaterial," *Opt. Lett.* **34**, 3325–3327 (2009).

9

Anomalous transmission in waveguides with correlated disorder in surface profiles

F. M. IZRAILEV AND N. M. MAKAROV

9.1 Introduction

In recent years, increasing attention has been paid to the so-called *correlated disorder* in low-dimensional disordered systems. Interest in this subject is mainly due to two reasons. First, it was found that specific correlations in a disordered potential can result in quite unexpected anomalous properties of scattering. Second, it was shown that such correlations can be relatively easily constructed experimentally, at least in the one-dimensional Anderson model and in Kronig–Penney models of various types. Therefore, it seems to be feasible to fabricate random structures with desired scattering properties, in particular when one needs to suppress or enhance the localization in given frequency windows for scattering electrons or electromagnetic waves. In addition, it was understood that, in many real systems, correlated disorder is an intrinsic property of the underlying structures. One of the most important examples is a DNA chain, for which strong correlations in the potential have been shown to manifest themselves in an anomalous conductance. Thus, the subject of correlated disorder is important both from the theoretical viewpoint, and for various applications in physics.

The key point of the theory of correlated disorder is that the localization length for eigenstates in one-dimensional models absorbs the main effect of correlations in disordered potentials. This fact has been known since the earliest analytical studies of transport in continuous random potentials. However, until recently the main interest was in delta-correlated potentials, or in potentials with a Gaussian-type of correlation. On the other hand, it was shown [1–7] that the most interesting effect is related to specific long-range correlations that can be fabricated in practice. In particular, it was demonstrated that in the standard one-dimensional Anderson model one can observe effective *mobility edges* in the energy spectrum, when the

Structured Surfaces as Optical Metamaterials, ed. A. A. Maradudin. Published by Cambridge University Press. © Cambridge University Press 2011.

pair correlator computed along the disorder is of a specific form. The important feature of the mobility edge ω_c in the frequency spectra of traveling waves is that it separates the region of strongly localized states from that of very extended states. This property is important for the construction of potentials with selective transmission or reflection. From the experimental viewpoint, many of the results obtained may have a strong impact on the creation of a new class of electron nanodevices, optical fibers, and acoustic and electromagnetic waveguides, with selective transport properties.

In spite of an asymptotic nature of the theoretical results obtained for infinitely large samples and weak disorder, the analytical predictions were found to work relatively well for a strong disorder. The first experimental study [8–10] of both the suppression and enhancement of localization due to correlated disorder, was performed on single-mode electromagnetic waveguides. It was shown that in the case of statistically correlated point-like surface scatterers one can create controlled frequency windows of enhanced transmission, or windows with a very strong reflection. The important point is that, in spite of many experimental imperfections, such as very strong absorbtion or a small number of scatterers with large amplitudes, selective transport was clearly observed in accordance with the theory.

For single-mode waveguides the problem of surface scattering can be reduced to that for one-dimensional disordered models. For this reason the methods and results obtained for the latter case can be directly applied for the waveguides [11–15]. The situation is fundamentally different for many-mode waveguides. The problem of wave propagation through such systems with corrugated surfaces has a long history, and remains a hot topic in the literature. This problem naturally arises in the analysis of spectral and transport properties of optical fibers, acoustic and radio waveguides, remote sensing, shallow water waves, multilayered systems and photonic lattices, etc. [16–21]. Similar problems emerge in quantum physics when describing the propagation of quasi-particles in thin metal films and semiconductor nanostructures, such as nanowires and strips, superlattices and quantum-well-systems [22–31]. Recently, new theoretical results [32, 33] have been obtained for one-dimensional disordered models describing photonic crystals, bi-layered metamaterials and electronic superlattices. Experiments performed on microwave guiding systems with intentionally randomized model parameters [34, 35] have confirmed the predictions of the theory.

As is well established, the scattering from corrugated surfaces results in diffusive transport [36–53], as well as in the effects of strong electron/wave localization [54–70]. Correspondingly, the eigenstates of periodic systems with corrugated surfaces turn out to have a chaotic structure [71–74]. Recent numerical studies of quasi-one-dimensional *surface-disordered* systems [67–70] have revealed a fundamental difference in their properties from those known in the standard models with *bulk* random potentials [29]. Specifically, it was found that transport properties of quasi-one-dimensional waveguides with rough surfaces essentially depend on many characteristic lengths. In comparison, for bulk scattering all transport characteristics depend on one parameter only, which is the ratio of the localization length to the size of a sample (the so-called *single-parameter scaling*).

The situation in many-mode waveguides in the presence of long-range correlations was found to be quite tricky [13, 75, 76]. It was shown that the long-range correlations, on the one hand, give rise to a suppression of the interaction between different propagating waveguide modes. On the other hand, the same correlations can provide a perfect transparency of each independent channel, similar to what happens in the one-dimensional geometry. The number of independent transparent modes is governed by the correlation length, and can be equal to the total number of propagating modes. Therefore, the transmission through such waveguides can be significantly enhanced in comparison with the case of uncorrelated surface roughness.

It should be stressed that the main results in the theory of surface scattering were obtained for random surfaces with rapidly decaying correlations along the structures. Therefore, it is of great importance to explore the role of specific long-range correlations in surface profiles, using the results found for one-dimensional systems with correlated disorder. Apart from its theoretical interest, this problem can be studied experimentally, since the existing experimental techniques allow for the construction of systems with sophisticated surfaces [77, 78].

We would like to note that in order to focus on the role of long-range correlations in surface profiles, in this chapter we do not discuss the so-called *square-gradient* mechanism of scattering. As was recently shown, this mechanism emerges due to a quite specific dependence of the scattering length on the second derivative of scattering profiles. The theoretical aspects of the square-gradient scattering and its possible experimental implications can be found in refs. [79–83].

9.2 Surface-corrugated waveguide

As a physically plausible and commonly used model to study multiple surface scattering, we consider an open planar waveguide of length L and average width d with perfectly conducting lateral walls. It is natural to require the waveguide length to be much greater than its width, $d \ll L$. Such a system is called quasi-one-dimensional. The x-axis is stretched along the structure and the z-axis is directed in the transverse direction. The lower boundary of the waveguide is assumed to have a rough (corrugated) profile $z = \xi(x)$, slightly deviated from its flat average $z = 0$.

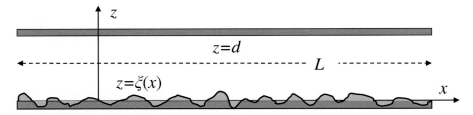

Figure 9.1. Planar waveguide with a lower corrugated edge.

The upper boundary is taken to be flat, $z = d$ (see Fig. 9.1). Thus, the surface-corrugated guiding system occupies the area defined by the following relations:

$$- L/2 < x < L/2, \quad \xi(x) \leqslant z \leqslant d. \tag{9.1}$$

The random function $\xi(x)$ describing the surface roughness is assumed to be statistically homogeneous with the following characteristics:

$$\langle \xi(x) \rangle = 0, \quad \langle \xi^2(x) \rangle = \sigma^2, \quad \langle \xi(x)\xi(x') \rangle = \sigma^2 \mathcal{W}(x - x'). \tag{9.2}$$

The angle brackets $\langle \ldots \rangle$ stand for ensemble averaging over the disorder, i.e. over different realizations of the random surface profile $\xi(x)$, or for a spatial average over the coordinate x of any prescribed realization. These two types of averaging are assumed to be equivalent due to ergodicity. The variance of $\xi(x)$ is denoted by σ^2, and, consequently, σ is the root-mean-square roughness height. The binary (two-point) correlator $\mathcal{W}(x)$ is normalized to its maximal value, $\mathcal{W}(0) = 1$, and is assumed to decrease with increasing $|x|$ on a characteristic scale termed the *correlation length*.

In what follows we consider weak surface scattering for which the corrugations are small, $\sigma \ll d$. This limitation is common in the surface scattering theories based on appropriate perturbative approaches [16]. As is known, for weak scattering all transport properties are entirely determined by the *roughness power spectrum* $W(k_x)$,

$$\mathcal{W}(x) = \int_{-\infty}^{\infty} \frac{dk_x}{2\pi} \exp(ik_x x) \, W(k_x), \tag{9.3a}$$

$$W(k_x) = \int_{-\infty}^{\infty} dx \exp(-ik_x x) \mathcal{W}(x). \tag{9.3b}$$

From Eq. (9.2) the pair correlator $\mathcal{W}(x)$ is seen to be a real and even function of the coordinate x. Its Fourier transform $W(k_x)$ is an even, real and non-negative function of the longitudinal wave number k_x. Note that the condition $\mathcal{W}(0) = 1$ is equivalent to the following normalization for $W(k_x)$:

$$\int_{-\infty}^{\infty} \frac{dk_x}{2\pi} W(k_x) = 1. \tag{9.4}$$

Since in the x-direction the system is open, at $x = \pm L/2$ we assume radiative boundary conditions. In the transverse z-direction the zero Dirichlet boundary conditions are applied at both lateral walls, $z = \xi(x)$ and $z = d$. Thus, the analysis of the surface scattering in our model is reduced to the study of the following two-dimensional boundary-value problem:

$$\left(\frac{\partial^2}{\partial x^2} + \frac{\partial^2}{\partial z^2} + k^2 \right) \Psi(x, z) = 0, \tag{9.5a}$$

$$\Psi(x, z = \xi(x)) = 0, \qquad \Psi(x, z = d) = 0. \tag{9.5b}$$

Here $k = \omega/c$ is the total wave number for an electromagnetic wave of frequency ω and TE polarization, propagating through a waveguide with perfectly conducting walls. Note that, in contrast with bulk scattering, here the wave equation does not contain any scattering potential since the source of the scattering is the roughness of the boundaries.

In an ideal waveguide with flat walls, $\xi(x) = 0$, the solution of the problem (9.5) has the canonical form of normal waveguide modes, i.e.

$$\Psi_{n,\pm}(x, z) = \frac{1}{\sqrt{\pi d}} \sin\left(\frac{\pi n z}{d} \right) \exp(\pm i k_n x). \tag{9.6}$$

Here the integer $n = 1, 2, 3, \ldots$ enumerates the normal modes (9.6) with the transverse wave number $k_z = \pi n/d$. The longitudinal wave number $k_x = \pm k_n$ for the nth mode is given by

$$k_n = \sqrt{k^2 - (\pi n/d)^2}. \tag{9.7}$$

Evidently, the transport properties depend only on normal modes that can propagate along the waveguide, i.e. have a real value of k_n. As follows from Eq. (9.7), such *propagating modes* have indices $n \leqslant N_d$, and their total number N_d is equal to the integer part $[\ldots]$ of the ratio kd/π,

$$N_d = [kd/\pi]. \tag{9.8}$$

The waveguide modes with indices $n > N_d$ have purely imaginary wave numbers k_n. These *evanescent modes* decay exponentially rapidly on a scale of the order of the wavelength. As one can see, the unperturbed (flat) waveguide is equivalent to a set of N_d one-dimensional noninteracting *conducting channels* occupied by the corresponding propagating modes.

9.3 Single-mode structure

Keeping in mind the relevance of wave scattering to the Anderson localization, let us first consider a single-mode waveguide. When the mode parameter kd/π is restricted by the relation $1 < kd/\pi < 2$, and the number of conducting channels equals one, $N_d = 1$. The transmission through such a waveguide depends on the longitudinal wave number k_1:

$$k_1 = \sqrt{k^2 - (\pi/d)^2}. \tag{9.9}$$

All other waveguide modes with $n \geqslant 2$ are evanescent and do not contribute to the transport properties. From the single-mode condition it follows that the wave number k_1 is confined within the interval

$$0 < k_1 d/\pi < \sqrt{3}. \tag{9.10}$$

Note that the weak surface-scattering condition $\sigma \ll d$ leads to the inequality $k_1\sigma \ll 1$.

As was shown in refs. [56–59], the transport problem (9.5) for the surface-disordered single-mode waveguide is equivalent to a one-dimensional disordered model,

$$\left[\frac{d^2}{dx^2} + k^2 - V(x)\right]\psi(x) = 0, \tag{9.11}$$

where k_1 replaces k. In such a description, the effective potential $V(x)$ has the form

$$V(x) = \frac{2}{\pi}\left(\frac{\pi}{d}\right)^3 \xi(x), \tag{9.12}$$

which is entirely determined by the rough surface profile $\xi(x)$.

As one can see, the surface scattering in one-mode waveguides is equivalent to the bulk scattering emerging in one-dimensional disordered systems. The latter problem can be solved with the use of well developed methods, such as the perturbative diagrammatic technique of Berezinski [23, 24], the invariant imbedding method [56–59, 84–86] or the two-scale approach [61–63, 87]. All these methods allow one to take adequately into account the effects of the coherent multiple scattering from the corrugated surface giving rise to Anderson localization.

The main theoretical result is that the average transmittance $\langle T \rangle$ as well as all of its moments $\langle T^s \rangle$, are described by the universal function

$$\langle T^s(L/L_{loc}) \rangle = \frac{1}{2\sqrt{\pi}} \left(\frac{L}{2L_{loc}} \right)^{-3/2} \exp \left(-\frac{L}{2L_{loc}} \right)$$

$$\times \int_0^\infty \frac{z \, dz}{\cosh^{2s-1} z} \exp \left(-z^2 \frac{L_{loc}}{2L} \right) \int_0^z dy \, \cosh^{2(s-1)} y \, ,$$

$$s = 0, \pm 1, \pm 2, \dots . \tag{9.13}$$

This function depends solely on the ratio L/L_{loc} between the waveguide length L and the *localization length* L_{loc} (see, for example, ref. [25]).

In one-dimensional disordered systems a wave can be scattered either forward or backward. However, it was shown that the transport properties are determined exclusively by the backscattering while the forward scattering has no influence. Therefore, the quantity L_{loc} is, in fact, the *backscattering length* emerging in an infinite one-dimensional structure. It is important that the inverse value L_{loc}^{-1} can be associated with the Lyapunov exponent appearing in various transfer matrix approaches [25]. In the latter description, the Lyapunov exponent gives the average rate of decrease of the wave function $\langle \psi(x) \rangle$ away from the center of its localization.

It should be stressed that the dependence of the transport properties on the ratio L/L_{loc} manifests a principal concept of *one-parameter scaling* that constitutes the phenomenon of one-dimensional Anderson localization. The nontrivial point of this concept is that in order to describe the transport properties of *finite* samples of size L, it is sufficient to know how the wave function is localized in an *infinite* sample with the same disorder.

From Eq. (9.13) one can derive relatively easily the expressions for low moments of the transmittance T. Specifically, for $s = 1$ one obtains the average transmittance $\langle T(L/L_{loc}) \rangle$:

$$\langle T(L/L_{loc}) \rangle = \frac{1}{2\sqrt{\pi}} \left(\frac{L}{2L_{loc}} \right)^{-3/2} \exp \left(-\frac{L}{2L_{loc}} \right)$$

$$\times \int_0^\infty \frac{z^2 \, dz}{\cosh z} \exp \left(-z^2 \frac{L_{loc}}{2L} \right) . \tag{9.14}$$

The second moment, $s = 2$, is important for obtaining the variance of the transmittance. It can be shown that, for $L_{loc} \lesssim L$, the variance is of the order of the squared average transmittance itself. This means that for strong localization the transmittance is not a self-averaging quantity. Hence, by changing the length L of the waveguide, or the disorder itself, one should expect very large fluctuations of the transmittance. Such fluctuations are known as the *mesoscopic fluctuations* that are characteristic of strong interference effects on a macroscopic scale.

In order to characterize properly the transport properties of one-dimensional structures for any degree of localization (ratio L/L_{loc}), one should refer to the self-averaging logarithm of the transmittance,

$$\langle \ln T(L/L_{loc}) \rangle = -2L/L_{loc}. \tag{9.15}$$

This result is consistent with an exponential decrease of the transmittance averaged over the so-called representative (nonresonant) realizations of the random disorder [25],

$$\langle T(L/L_{loc}) \rangle_{rep} = \exp(-2L/L_{loc}). \tag{9.16}$$

Note that Eq. (9.15) is quite often used as the definition of the localization length L_{loc} itself. It is highly nontrivial that by exploring the transmission properties of finite samples, one can extract the localization length that is defined for infinite samples. This fact is again the manifestation of one-parameter scaling.

In accordance with the scaling concept, in the one-dimensional geometry there are only two characteristic regimes, corresponding to (i) ballistic and (ii) localized transport.

(i) The *ballistic transport* occurs if the localization length L_{loc} is much larger than the system length L. In this case the one-dimensional structure is practically fully transparent, since its average transmittance is close to unity,

$$\langle T(L/L_{loc}) \rangle \approx 1 - 2L/L_{loc} \qquad \text{for} \qquad L_{loc} \gg L. \tag{9.17}$$

This asymptotic expression results from both Eq. (9.14) and Eq. (9.16).
(ii) Otherwise, the disordered structures exhibit *localized transport*, when the localization length L_{loc} is smaller than the sample length L. In this case the average transmittance, Eq. (9.14), is exponentially small,

$$\langle T(L/L_{loc}) \rangle \approx \frac{\pi^3}{16\sqrt{\pi}} (L/2L_{loc})^{-3/2} \exp(-L/2L_{loc}) \tag{9.18}$$

for $L_{loc} \ll L$. This means that in the localization regime disordered single-mode waveguides perfectly (with an exponential accuracy) reflect the incoming waves.

According to Eq. (9.18), the transmission exponentially decreases on the scale $L \approx 2L_{loc}$, with an additional power prefactor. In contrast, the transmittance, Eq. (9.16), has an exponential dependence with a much faster decrease on the scale $L \approx L_{loc}/2$. This fact can be explained as follows. The main contribution to the asymptotic form (9.18) for the average transmittance (9.14) is given by resonant realizations of the random potential $V(x)$. For these realizations the transmittance is almost equal to unity; however, they have an exponentially small probability. On the other hand, for representative realizations (most probable, but nonresonant) the transmittance is described by Eq. (9.16). This effect is peculiar to the mesoscopic nature of Anderson localization.

Equations (9.13)–(9.18) are universal and applicable for any one-dimensional system with a weak static disorder. As one can see, in order to describe transport properties of finite structures, one needs to know the localization length L_{loc}. According to different approaches [23–25, 84–87], the inverse localization length for any kind of weak disorder is determined by the $2k$-harmonic in the randomness power spectrum $S(k_x)$ of the scattering potential $V(x)$:

$$L_{loc}^{-1}(k) = S(2k)/8k^2, \tag{9.19a}$$

$$\langle V(x)V(x')\rangle = C(x - x'), \tag{9.19b}$$

$$S(k_x) = \int_{-\infty}^{\infty} dx \exp(-ik_x x) C(x). \tag{9.19c}$$

For elastic backward scattering, the wave vector $\mathbf{k}$ conserves its value k, changing its sign, $|\Delta \mathbf{k}| = 2k$. Accordingly, Eqs. (9.19) reflect the fact that the localization length is defined by the backscattering length only.

Equations (9.19) indicate that the global properties of the wave transmission through one-dimensional disordered media depend on the two-point correlations in the random scattering potential. Therefore, if the power spectrum $S(2k)$ is very small, or vanishes within some interval of the wave number k, then the localization length L_{loc} appears to be very large ($L_{loc} \gg L$), or even diverges. Evidently, the localization effects can be neglected in this case, and the structure, even of a large length, is fully transparent. This means that, in principle, by a proper choice of the disorder, one can design the disordered structures with selective (*anomalous*) ballistic transport within a prescribed range of k.

Taking into account the form (9.12) of the potential and its correlation properties (9.2), from Eqs. (9.19) one can readily derive the following explicit formula for the localization length in the single-mode waveguide [56–59]:

$$L_{loc}^{-1}(k_1) = \frac{2\sigma^2}{\pi^2} \left(\frac{\pi}{d}\right)^6 \frac{W(2k_1)}{(2k_1)^2}. \tag{9.20}$$

Since the potential, Eq. (9.12), is entirely determined by the rough surface profile $\xi(x)$, the localization length, Eq. (9.20), is specified by the roughness power spectrum $W(k_x)$. Therefore, by a proper fabrication of a random profile $\xi(x)$ with specific long-range correlations, one can arrange a desirable anomalous transport within a given window of $k = \omega/c$ inside the single-mode region given in Eq. (9.10).

Before we begin with a practical implementation of Eq. (9.20) for the inverse localization length, it is worthwhile clearing up some points. First, one should stress that this result, as well as Eqs. (9.19) obtained by various methods, is an asymptotic one. This means that the higher terms are non-controlled; however, they can be neglected in the limit $\sigma^2 \to 0$. Second, the main assumptions used in the derivation of Eq. (9.20) are based on the validity of averaging over different realizations of disorder (see, for instance, ref. [87]). The condition for such an average is that two lengths, L_{loc} and L, are much larger than two other characteristic lengths, the wavelength k_1^{-1} and the correlation length k_c^{-1} determining the maximal value of the power spectrum $W(2k_1)$. One should stress that, for any finite values of k_1 and k_c, this condition can always be satisfied due to the asymptotic nature of Eq. (9.20) (or, what is the same thing, due to the small value of σ^2).

9.4 Design of a random surface profile with predefined correlations: convolution method

From the preceding considerations it is seen that, in principle, by a proper choice of surface disorder one can artificially create systems with selective transparency or reflectivity. Thus, the important practical problem arises of how to construct a corrugated surface profile from a predefined roughness power spectrum $W(k_x)$. This problem can be solved by employing a widely used *convolution method* that was originally proposed in ref. [88]. The modern applications of this method for the generation of random structures with specific correlations, including long-range nonexponential correlations, can be found in refs. [77] and [89–98] and in other papers cited in this chapter.

The method consists of the following steps. First, having a desired form for the power spectrum $W(k_x)$, we derive the *modulation function* $\beta(x)$, whose Fourier transform is $W^{1/2}(k_x)$:

$$\beta(x) = \int_{-\infty}^{\infty} \frac{dk_x}{2\pi} \exp{(ik_x x)}\, W^{1/2}(k_x). \tag{9.21}$$

Then the random surface profile $\xi(x)$ is generated as a convolution of a white-noise $Z(x)$ with the modulation function $\beta(x)$:

$$\xi(x) = \sigma \int_{-\infty}^{\infty} dx'\, Z(x - x')\,\beta(x'). \tag{9.22}$$

The delta-correlated random process $Z(x)$ is determined by the standard relations

$$\langle Z(x)\rangle = 0, \qquad \langle Z(x)Z(x')\rangle = \delta(x - x'), \tag{9.23}$$

and can be numerically created with the use of random-number generators. Here, $\delta(x)$ is the Dirac delta-function.

Equations (9.22) and (9.23) give the solution of the inverse scattering problem of constructing random roughness from its power spectrum. Note that this method is valid in the case of a weak disorder only. That is why only the binary correlator is involved in the construction of $\xi(x)$, while the higher correlators do not contribute. Note also that the profile obtained by the proposed method is not unique. Indeed, there is an infinite number of realizations of delta-correlated noise $Z(x)$ that give rise to different profiles $\xi(x)$ having the same power spectrum $W(k_x)$.

The importance of this method is due to the possibility of obtaining profiles resulting in a sharp transition between ballistic and localized transport, when changing the wave number k. In this case the corresponding power spectrum $W(k_x)$ abruptly vanishes at prescribed values of k_c. This means that the binary correlator $\mathcal{W}(x)$ has to be a slowly decaying function of the distance $|x|$. In other words, the corresponding corrugated surface profiles $\xi(x)$ should be of a specific form, revealing long-range correlations along the structure. Because of the abrupt nature of transmission properties, the transition point k_c can be regarded as an effective *transparency edge*.

As was pointed out above, a statistical treatment is meaningful if the scale of decrease k_c^{-1} of the correlator $\mathcal{W}(x)$ is much smaller than both the sample length L and the localization length L_{loc}. In this connection, one should stress that the long-range correlations we speak about do not assume large values for the correlation length k_c^{-1}. Indeed, the simplest correlator, $\mathcal{W}(x) = \sin k_c x / k_c x$, has the finite scale k_c^{-1} of its decrease, and k_c^{-1} can be quite small; see the examples in Section 9.6. On the other hand, the effective width of the transparency edge is determined by the product $(L_{loc}k_c)^{-1}$, not by k_c^{-1}, and turns out to be very small (for details, see, e.g., ref. [87]). One can say that the sharpness of the transition is defined by the form of the pair correlator rather than by the value of its correlation length.

Note that systems with very complicated scattering potentials are not exotic. For example, bulk random potentials have been constructed in experiments [8–10, 78], while rough surfaces with a rectangular power spectrum have been fabricated in a study of the enhanced backscattering effect [77].

9.5 Gaussian correlations

As was mentioned above, the general expression, Eq. (9.20), for the localization length $L_{loc}(k_1)$ indicates that all features of the wave transmission through a surface-disordered single-mode waveguide depend on two-point correlations in the surface profile. In order to demonstrate how to realize the properties of long-range correlated disorder, let us first consider the surface roughness $\xi(x)$ with a widely used Gaussian correlator,

$$\mathcal{W}(x) = \exp\left(-k_c^2 x^2\right),$$ (9.24a)

$$W(k_x) = \sqrt{\pi}\, k_c^{-1} \exp\left(-k_x^2/4k_c^2\right).$$ (9.24b)

This correlator decreases exponentially on the scale of the correlation length k_c^{-1}.

Using the convolution method, Eqs. (9.21)–(9.23), one can obtain that the corrugated surface profile $\xi(x)$ with the correlation properties in Eqs. (9.24) is described by the function

$$\xi(x) = \frac{\sigma\sqrt{2k_c}}{\pi^{1/4}} \int_{-\infty}^{\infty} dx'\, Z(x - x') \exp\left(-2k_c^2 x'^2\right).$$ (9.25)

Correspondingly, the inverse localization length, Eq. (9.20), takes the following explicit form:

$$L_{loc}^{-1}(k_1) = \frac{2\sigma^2}{\pi\sqrt{\pi k_c}} \left(\frac{\pi}{d}\right)^6 \frac{\exp\left(-k_1^2/k_c^2\right)}{(2k_1)^2}.$$ (9.26)

One can see that, within the single-mode interval given in Eq. (9.10), the localization length increases exponentially with k_1 from zero at $k_1 = 0$ to a large value at $k_1 = \pi\sqrt{3}/d$. Clearly, in the vicinity of $k_1 = 0$ the localization length $L_{loc}(k_1)$ is much smaller than L and the waveguide is nontransparent. Thus, the localization regime, Eq. (9.18), occurs within the whole single-mode interval, provided by the condition $L_{loc}(\pi\sqrt{3}/d) \ll L$ at $k_1 = \pi\sqrt{3}/d$.

In contrast, when $L_{loc}(\pi\sqrt{3}/d) \gg L$, one can observe the crossover from the localized transport, Eq. (9.18), to the ballistic one, Eq. (9.17). For the Gaussian correlations (9.24), both the crossing point, where $L_{loc}(k_1) = L$, and the crossover width depend on the values of k_c^{-1} and L. Because of the smooth nature of the crossover, the crossing point cannot be regarded as the transparency edge. However, one can see that the longer the correlation length k_c^{-1}, the larger the localization length $L_{loc}(k_1)$. Hence, the larger is the ballistic region, the narrower is the crossover. One can conclude that, in general, *correlations suppress localization*.

The surface profile $\xi(x)$ with Gaussian correlations admits the uncorrelated roughness of the white-noise type. Indeed, since $\sigma^2 k_c^{-1} = \text{const}$, from Eqs. (9.24) for $k_c^{-1} \to 0$ one obtains the delta-like correlator and constant power spectrum:

$$\mathcal{W}_{wn}(x) = \sqrt{\pi}\, k_c^{-1} \delta(x), \qquad (9.27a)$$

$$W_{wn}(k_x) = \sqrt{\pi}\, k_c^{-1}. \qquad (9.27b)$$

The convolution method, Eqs. (9.21)–(9.23), results in the following expression for the surface profile $\xi(x)$:

$$\xi_{wn}(x) = \frac{\sigma \pi^{1/4}}{\sqrt{k_c}}\, Z(x). \qquad (9.28)$$

According to Eq. (9.20), the localization length is given by

$$\frac{1}{L_{loc}^{wn}(k_1)} = \frac{\sigma^2 k_c^{-1}}{2\pi \sqrt{\pi} k_1^2} \left(\frac{\pi}{d}\right)^6. \qquad (9.29)$$

Equations (9.27)–(9.29) can be considered as the asymptotic limits of the corresponding equations (9.24)–(9.26) when $(k_1/k_c)^2 \ll 1$.

A comparison of Eqs. (9.29) and (9.26) leads to the conclusion that the best way to observe localized transport is to employ an uncorrelated disordered surface. Indeed, the condition $L_{loc}(k_1) \ll L$ is stronger for the Gaussian correlations than for the case of white-noise profiles. On the other hand, for Gaussian correlations with a small value of k_c, the ballistic regime, Eq. (9.17), can be realized even for such lengths L and wave numbers k_1 for which strong localization, Eq. (9.18), takes place for delta-like correlations. Again, this fact confirms that correlations suppress localization.

9.6 Two complementary examples of selective transparency

As we discussed in Section 9.4, rough surfaces with prescribed two-point correlations can be constructed with the use of the convolution method. In the following we demonstrate the construction of surface-disordered structures with selective transparency by considering two examples of long-range correlations.

9.6.1 Example 1

Let us first consider a waveguide that is nontransparent when the wave number k_1 is smaller than some value k_c, and completely transparent for $k_1 > k_c$. Such a behavior can be observed if the transition point (transparency edge) $k_1 = k_c$ is located inside the allowed single-mode interval given by Eq. (9.10), i.e.

$$0 < k_c d/\pi < \sqrt{3}. \qquad (9.30)$$

For this case, one can obtain the following expressions for the binary correlator $\mathcal{W}(x)$ and power spectrum $W(k_x)$:

$$\mathcal{W}_a(x) = \frac{\sin(2k_c x)}{2k_c x}, \tag{9.31a}$$

$$W_a(k_x) = \frac{\pi}{2k_c} \Theta(2k_c - |k_x|). \tag{9.31b}$$

Here $\Theta(x)$ is the Heaviside unit-step function, $\Theta(x < 0) = 0$ and $\Theta(x > 0) = 1$, and k_c is the correlation parameter to be specified.

According to Eqs. (9.21)–(9.23), the surface profile with the properties given in Eqs. (9.31) is given by the expression

$$\xi_a(x) = \frac{\sigma}{\sqrt{2\pi k_c}} \int_{-\infty}^{\infty} dx' \, Z(x - x') \frac{\sin(2k_c x')}{x'}. \tag{9.32}$$

Correspondingly, the inverse localization length has the *step-down* form

$$L_{loc}^{-1}(k_1) = \frac{\sigma^2}{\pi k_c} \left(\frac{\pi}{d}\right)^6 \frac{\Theta(k_c - k_1)}{(2k_1)^2}. \tag{9.33}$$

In line with this expression, as k_1 increases, the localization length $L_{loc}(k_1)$ also smoothly increases and then diverges at $k_1 = k_c$. Thus, within the region $0 < k_1 < k_c$ the average transmittance $\langle T(L/L_{loc}) \rangle$ is expected to be exponentially small (see Eq. (9.18)) due to strong localization. The condition for strong localization to the left from $k_1 = k_c$ is as follows:

$$\frac{L}{L_{loc}(k_c - 0)} = \frac{\sigma^2 L}{4\pi k_c^3} \left(\frac{\pi}{d}\right)^6 \gg 1. \tag{9.34}$$

Otherwise, inside the interval $k_c < k_1 < \pi\sqrt{3}/d$ a ballistic regime occurs with perfect transparency, $\langle T(L/L_{loc}) \rangle = 1$.

9.6.2 Example 2

The second example refers to a complementary situation when, for $k_1 < k_c$, the waveguide is perfectly transparent and for $k_1 > k_c$ is nontransparent. The corresponding expressions for $\mathcal{W}(x)$ and $W(k_x)$ are given by

$$\mathcal{W}_b(x) = \pi\delta(2k_c x) - \frac{\sin(2k_c x)}{2k_c x}, \tag{9.35a}$$

$$W_b(k_x) = \frac{\pi}{2k_c} \Theta(|k_x| - 2k_c). \tag{9.35b}$$

In this case, the corrugated surface is described by a superposition of white noise and roughness of the first type:

$$\xi_b(x) = \frac{\sigma}{\sqrt{2\pi k_c}} \left[\pi Z(x) - \int_{-\infty}^{\infty} dx' \, Z(x - x') \frac{\sin(2k_c x')}{x'} \right]. \tag{9.36}$$

Correspondingly, the inverse localization length is expressed by the *step-up* function,

$$L_{loc}^{-1}(k_1) = \frac{\sigma^2}{\pi k_c} \left(\frac{\pi}{d}\right)^6 \frac{\Theta(k_1 - k_c)}{(2k_1)^2}. \tag{9.37}$$

As a consequence, in contrast with the first case, here the surface-scattering localization length $L_{loc}(k_1)$ diverges below the transparency edge $k_1 = k_c$. At this point $L_{loc}(k_1)$ sharply decreases to a finite value $L_{loc}(k_c + 0)$, and then smoothly increases with further increase of k_1. In order to observe localization within the whole region $k_c < k_1 < \pi\sqrt{3}/d$, one should assume that strong localization is retained at the upper point $k_1 = \pi\sqrt{3}/d$ of the single-mode region given by Eq. (9.10):

$$\frac{L}{L_{loc}(\pi\sqrt{3}/d)} = \frac{\sigma^2 L}{12\pi k_c} \left(\frac{\pi}{d}\right)^4 \gg 1. \tag{9.38}$$

Therefore, in this example, the ballistic transport is abruptly replaced by strong localization at the transparency edge, $k_1 = k_c$.

One should stress that the surface profiles given by Eqs. (9.32) and (9.36) with binary correlators and power spectra Eqs. (9.31) and (9.35), respectively, are substantially different from the delta-correlated white noise (9.27), (9.28), and from random processes, Eq. (9.25) with exponentially decaying Gaussian correlations (9.24). Specifically, here the profiles are characterized by the variation scale $(2k_c)^{-1}$ and have long power-decaying tails in the expressions for their two-point correlator. Such tails originate from the stepwise discontinuity at the points $k_x = \pm 2k_c$. The location of the transparency edge is defined by these discontinuity points, and does not depend on other parameters, in contrast to the case of Gaussian or delta-like correlations.

Now we demonstrate the above predictions by a direct numerical simulation. For this, the inverse localization length L_{loc}^{-1} was computed with the use of the Hamiltonian map approach developed in refs. [1–3]. First, the continuous scattering potential (9.12) was approximated by the sum of delta kicks, with the spacing δ chosen much smaller than any physical length scale in the model. Then, a discrete analog of the one-dimensional wave equation (9.11) was analyzed numerically, with the longitudinal wave number k_1 in place of k and with the surface scattering potential $V(x)$ given by Eq. (9.12). In this way the wave equation was expressed in the form of a two-dimensional Hamiltonian map describing the dynamics of a

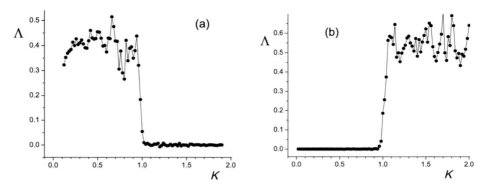

Figure 9.2. Selective dependence of the rescaled Lyapunov exponent on wave number for two realizations of a random surface with specific long-range correlations. To show the main effect of correlations, the complementary dependence of the Lyapunov exponent on K is shown: (a) eigenstates are localized for $K < 1$ and delocalized for $K > 1$; (b) complete delocalization for $K < 1$ alternates with strong localization for $K > 1$. (After ref. [7]).

classical linear oscillator under parametric noise determined by $V(x)$. As a result, the analysis of the localization length was reduced to the computation of the Lyapunov exponent L_{loc}^{-1} associated with this map (see details in refs. [1–3]).

Numerical data reported in Figure 9.2 represent the dependence of the dimensionless Lyapunov exponent, $\Lambda = c_0 L_{loc}^{-1}$, on the normalized wave number, $K = k_1/k_c$, in the range $0 < K < 2$, corresponding to the single-mode interval. The normalization coefficient c_0 was set to have $\Lambda = K^{-2}$ for the delta-correlated potential. Two surface profiles $\xi(x)$ were generated according to discrete versions of Eqs. (9.32) and (9.36), determining complementary stepwise dependencies of the localization length $L_{loc}(k_1)$ in accordance with Eqs. (9.33) and (9.37).

One can clearly see a nontrivial dependence of Λ on the wave vector K, which is due to the specific long-range correlations in $\xi(x)$. The data display sharp dependencies of Λ on K when crossing the point $K = 1$. Thus, by taking the size L of the scattering region according to the requirements in Eqs. (9.34) or (9.38), one can arrange anomalous transport in the single-mode guiding structure, as predicted by the analytical theory.

9.7 Random narrow-band reflector

Let us now consider one more example, namely a binary correlator that gives rise to a power spectrum of the following rectangular form:

$$\mathcal{W}(x) = \frac{\sin(2k_+ x) - \sin(2k_- x)}{2(k_+ - k_-)x}, \tag{9.39a}$$

$$W(k_x) = \frac{\pi}{2(k_+ - k_-)} \left[\Theta(2k_+ - |k_x|) - \Theta(2k_- - |k_x|)\right], \ 0 < k_- < k_+ < \pi\sqrt{3}/d. \tag{9.39b}$$

In these relations the factor $1/2(k_+ - k_-)$ provides the normalization requirement, Eq. (9.4), or equivalently, $\mathcal{W}(0) = 1$. Such a power spectrum has been employed to create specific rough surfaces in the experimental study of enhanced backscattering [77]. This spectrum was also used in the theoretical analysis of light scattering from amplifying media, as well as in the study of the localization of plasmon polaritons on random surfaces [99, 100].

In accordance with the convolution method, Eqs. (9.21)–(9.23), the random surface profile $\xi(x)$ with the correlation properties (9.39) can be obtained from the following expression:

$$\xi(x) = \frac{\sigma}{\sqrt{2\pi}} \int_{-\infty}^{\infty} dx' \, Z(x - x') \frac{\sin(2k_+ x') - \sin(2k_- x')}{(k_+ - k_-)^{1/2} x'}. \tag{9.40}$$

The peculiarity of such surfaces is that they have two characteristic scales, $(2k_+)^{-1}$ and $(2k_-)^{-1}$. Consequently, the binary correlator and its power spectrum, Eqs. (9.39), are specified by two correlation parameters.

From Eqs. (9.20) and (9.39), one can find the inverse localization length,

$$L_{loc}^{-1}(k_1) = \frac{\sigma^2}{\pi(k_+ - k_-)} \left(\frac{\pi}{d}\right)^6 \frac{\Theta(k_+ - k_1)\Theta(k_1 - k_-)}{(2k_1)^2}. \tag{9.41}$$

As one can see, there are two transparency edges, at the points $k_1 = k_-$ and $k_1 = k_+$. The localization length $L_{loc}(k_1)$ diverges below the first point, $k_1 = k_-$, and above the second one, $k_1 = k_+$. Between these points, for $k_- < k_1 < k_+$, the localization length has a finite value and smoothly increases with an increase of wave number k_1.

Let us now choose the parameters for which the regime of strong localization occurs at the upper transition point $k_1 = k_+$, where $L_{loc}(k_1)$ has its maximal value. This automatically provides strong localization within the whole interval $k_- < k_1 < k_+$. The condition for this situation is given by

$$\frac{L}{L_{loc}(k_+ - 0)} = \frac{\sigma^2 L}{4\pi k_+^2 (k_+ - k_-)} \left(\frac{\pi}{d}\right)^6 \gg 1. \tag{9.42}$$

As a result, there are two regions of perfect transparency for waveguides of finite length L with the surface profile specified above. Between these regions the average transmittance $\langle T \rangle$ is exponentially small according to Eq. (9.18). Due to this fact, the system exhibits localized transport within the interval $k_- < k_1 < k_+$, and the ballistic regime with $\langle T \rangle = 1$ outside this interval. In an experiment one can observe that, with an increase of the wave number k_1, the perfect transparency below $k_1 = k_-$ abruptly alternates with a complete reflection and recovers at $k_1 = k_+$. From Eqs. (9.41) and (9.42) one can see that the smaller the value $k_+ - k_-$ of the reflecting region, the smaller the surface-scattering localization length $L_{loc}(k_1)$ and, consequently, the stronger is the localization within this region. This remarkable fact may find important applications in creating a new class of random narrow-band filters or reflectors.

9.8 Multi-mode waveguide

Now we examine the correlated surface scattering in *multi-mode* waveguides, i.e. in waveguides with a large number, $N_d > 1$, of conducting channels; see Eq. (9.8). According to Landauer's concept [28], the *total average transmittance* $\langle T \rangle$ of any disordered quasi-one-dimensional guiding structure can be expressed as a sum of *partial average transmittances* $\langle T_n \rangle$ that describe the transport for every nth propagating normal mode, i.e.

$$\langle T \rangle = \sum_{n=1}^{N_d} \langle T_n \rangle. \tag{9.43}$$

When all conducting channels are open, i.e. all $T_n = 1$, the total transmittance attains its maximal value equal to the total number N_d of propagating modes. Therefore, our definition of the transmittance differs from the canonical one in which the maximal value of the total transmittance is equal to unity. Nevertheless, we shall use the definition in Eq. (9.43) in order to discriminate clearly the intervals of the mode parameter kd/π with different numbers of the conducting channels and more clearly display the role of correlated surface disorder. In order to pass to the definition used in wave theories, one should simply divide Eq. (9.43) by N_d.

From the general theory of quasi-one-dimensional scattering systems, it follows that the transmission properties of any nth conducting channel ($1 \leqslant n \leqslant N_d$) are determined by *two* attenuation lengths: the *forward scattering length* $L_n^{(f)}$ and the *backscattering length* $L_n^{(b)}$. For multi-mode quasi-one-dimensional waveguides with surface disorder, the inverse scattering lengths are given by

$$\frac{1}{L_n^{(f)}} = \sigma^2 \frac{(\pi n/d)^2}{2k_n d} \sum_{n'=1}^{N_d} \frac{(\pi n'/d)^2}{k_{n'} d} W(k_n - k_{n'}), \tag{9.44}$$

$$\frac{1}{L_n^{(b)}} = \sigma^2 \frac{(\pi n/d)^2}{2k_n d} \sum_{n'=1}^{N_d} \frac{(\pi n'/d)^2}{k_{n'} d} W(k_n + k_{n'}). \tag{9.45}$$

Here the longitudinal wave number k_n is defined by Eq. (9.7). Equations (9.44) and (9.45) can be obtained for the boundary-value problem in Eq. (9.5) by the diagrammatic Green's function method [16], as well as by the technique developed in ref. [54]. Also, these expressions can be derived by using the invariant imbedding method extended to quasi-one-dimensional structures [69, 70]. Note that in a single-mode waveguide with $N_d = 1$, the sum over n' contains only one term with

$n' = n = 1$. Therefore, in this case the backscattering length $L_1^{(b)}$ is exactly equal to the single-mode localization length given in Eq. (9.20), i.e. $L_1^{(b)} = L_{loc}(k_1)$; see Eq. (9.13).

The sums in Eqs. (9.44) and (9.45) show that, in general, the scattering of a given nth propagating mode into all other modes contributes to both attenuation lengths. This is the case when, for example, a rough surface profile is either a delta-correlated random process with a constant power spectrum, $W(k_x) = \text{const}$, or has a rapidly decreasing binary correlator $\mathcal{W}(x)$ and, correspondingly, a slowly decreasing roughness power spectrum $W(k_x)$.

Another feature is that Eqs. (9.44) and (9.45) display a rather strong dependence on the mode index n. Namely, the larger the number n, the smaller the corresponding mode scattering lengths and, as a consequence, the stronger is the scattering of this mode into the others. This strong dependence is due to the squared transverse wave number $k_z = \pi n/d$ in the numerator and to the longitudinal wave number k_n in the denominator of Eqs. (9.44) and (9.45). Evidently, with an increase of the mode index n the value of k_n decreases. An additional dependence appears because of the roughness power spectrum $W(k_n \mp k_{n'})$. Since the binary correlator $\mathcal{W}(x)$ of random surfaces is a decreasing function of $|x|$, its Fourier transform $W(k_n \mp k_{n'})$ increases with n (note that it is constant for the delta-correlated roughness only). Therefore, all the factors contribute in the same direction for the dependence of $L_n^{(f)}$ and $L_n^{(b)}$ on the mode index n. As a result, we arrive at the following hierarchy of mode scattering lengths:

$$L_{N_d}^{(f,b)} < L_{N_d-1}^{(f,b)} < \cdots < L_2^{(f,b)} < L_1^{(f,b)}. \tag{9.46}$$

The smallest mode attenuation lengths $L_{N_d}^{(f)}$ and $L_{N_d}^{(b)}$ belong to the highest channel with the mode index $n = N_d$, while the largest scattering lengths $L_1^{(f)}$ and $L_1^{(b)}$ correspond to the lowest channel with $n = 1$. Note that a similar hierarchy was also found in refs. [7], [14] and [15] in the model of quasi-one-dimensional systems with a stratified disorder.

As is known, the quasi-one-dimensional systems with isotropic volume disorder reveal three typical transport regimes: the regimes of *ballistic*, *diffusive* (metallic) and *localized* transport. In contrast to this conventional picture, in refs. [69] and [70] it was shown that, in the case of surface disorder, a very important phenomenon of the *coexistence* of ballistic, diffusive and localized transport emerges. This happens due to the hierarchy, Eq. (9.46), of scattering lengths, even in the absence of correlations in $\xi(x)$. Specifically, while the lowest modes can be in the ballistic regime, the intermediate and highest modes can exhibit diffusive and localized behavior, respectively. This effect seems to be generic for transport through waveguides with random surfaces.

One can see now that, unlike the single-mode case, the concept of one-parameter scaling is no longer valid for the transport in multi-mode surface-disordered systems. There are two points that should be stressed in this respect. On the one hand, the average partial transmittances $\langle T_n \rangle$ entering Eq. (9.43) are very different for different conducting channels. On the other hand, and what is even more important, all propagating modes turn out to be mixed due to inter-mode transitions. Therefore, the transmittance $\langle T_n \rangle$ of a given nth mode depends on the scattering into all modes, and the total average transmittance, Eq. (9.43), is determined by the whole set of attenuation lengths, Eqs. (9.44) and (9.45), with $1 \leqslant n \leqslant N_d$.

Summarizing our brief discussion, it becomes clear that, in quasi-one-dimensional guiding structures with delta-correlated or Gaussian correlations in surface disorder, the crossover from the ballistic to localized transport is realized through the successive localization of the highest propagating modes. Otherwise, if we start from the localized regime, the crossover to the ballistic transport is realized via the successive opening (delocalization) of the lowest conducting channels.

From this analysis one can conclude that for multi-mode structures with surface disorder the role of specific long-range correlations is much more sophisticated in comparison with single-mode waveguides. First, such correlations should result in the suppression of the interaction between different propagating modes. This nontrivial fact turns out to be crucial for the reduction of a system of mixed channels with quasi-one-dimensional transport to the subset of independent waveguide modes with a purely one-dimensional transport. Second, the same correlations can provide a complete transparency of each independent channel, similar to what happens in a strictly one-dimensional geometry.

To demonstrate these effects, let us take a random surface profile $\xi(x)$ of the form

$$\xi(x) = \frac{\sigma}{\sqrt{\pi k_c}} \int_{-\infty}^{\infty} dx' \, Z(x - x') \frac{\sin(k_c x')}{x'} , \qquad (9.47)$$

with a slowly decaying (on average) binary correlator. It results in the "window function" for the roughness power spectrum, given by

$$\mathcal{W}(x) = \frac{\sin(k_c x)}{k_c x} , \qquad (9.48a)$$

$$W(k_x) = \frac{\pi}{k_c} \Theta(k_c - |k_x|), \qquad k_c > 0. \qquad (9.48b)$$

From Eqs. (9.44) and (9.45) one can see that in the case of long-range correlations in a disordered surface, Eq. (9.47), the number of modes into which a given nth mode is scattered, i.e. the actual number of summands in the Eqs. (9.44) and (9.45), is entirely determined by the width k_c of the rectangular power spectrum, Eq. (9.48).

It is clear that if the distance $|k_n - k_{n\pm1}|$ between neighboring quantum values of k_n is larger than the correlation width k_c,

$$|k_n - k_{n\pm1}| > k_c, \tag{9.49}$$

then all inter-mode transitions (between different propagating modes) are forbidden. As a consequence, the sum over n' in Eq. (9.44) for the inverse forward scattering length contains only the diagonal term with $n' = n$ that describes a direct intra-mode scattering *inside* the channels. Moreover, each term in the sum of Eq. (9.45) for the inverse backscattering length is equal to zero. As a result, the following interesting phenomena arise.

(i) All *high* propagating modes with indices n that satisfy Eq. (9.49), turn out to be independent of the others, in spite of the interaction with the rough surface. Therefore, they form a subset of one-dimensional noninteracting conducting channels with a finite length of forward scattering $L_n^{(f)}$ and an infinite backscattering length $L_n^{(b)}$:

$$\frac{1}{L_n^{(f)}} = \frac{\pi \sigma^2}{k_c} \frac{(\pi n/d)^4}{(k_n d)^2}, \qquad \frac{1}{L_n^{(b)}} = 0. \tag{9.50}$$

(ii) As is well known from the standard theory of one-dimensional localization (see, e.g., refs. [23–26], [56–59], [61–63] and [87]), the transport through any one-dimensional disordered system is determined only by the backscattering length $L_n^{(b)}$, which in our consideration equals the localization length, and does not depend on the forward scattering length $L_n^{(f)}$. Since the former diverges for every independent channel in line with Eqs. (9.50), all of them are completely transparent because they exhibit ballistic transport with the partial average transmittance $\langle T_n \rangle = 1$. This means that, according to Landauer's formula, Eq. (9.43), the transmittance of the subset of such independent ballistic modes is simply equal to their total number.

(iii) As for *low* propagating modes with indices n contradicting Eq. (9.49), they stay mixed by surface scattering because the roughness power spectrum, Eq. (9.48b), is nonzero for them, $W(k_n - k_{n'}) = \pi/k_c$. These *mixed modes* have finite forward and backscattering lengths and, consequently, stay in the diffusive or localized transport regime for a large enough waveguide length L. As a result, they are nontransparent and do not contribute to the total transmittance $\langle T \rangle$. Therefore, the latter is equal to the number of independent ballistic modes.

Note that the distance $|k_n - k_{n\pm1}|$ between neighboring wave numbers k_n and $k_{n\pm1}$ increases as the mode index n increases. Therefore, the inequality (9.49) restricts the mode index n from below. That is why, in contrast with the conventional situation associated with the hierarchy of mode scattering lengths in Eq. (9.46), the low modes are mixed and nontransparent whereas the high propagating modes are independent and ballistic. Because of the sharp behavior of the roughness power spectrum, the transition from mixed to independent modes is also sharp.

More analytical results can be obtained for waveguides with large numbers of conducting channels N_d, if the quantum numbers n of independent ballistic modes are also large:

$$N_d = [kd/\pi] \approx kd/\pi \geqslant n \gg 1. \tag{9.51}$$

In this case the inequality (9.49) is reduced to the requirement $|\partial k_n/\partial n| > k_c$, which can be rewritten in the following explicit form:

$$n > N_{mix} = \left[\frac{(kd/\pi)}{\sqrt{1 + (k_c d/\pi)^{-2}}} \right]. \tag{9.52}$$

We recall that square brackets denote the integer part of the inner expression.

The condition (9.52) determines the total number N_{mix} of mixed nontransparent modes, the total number $N_{bal} = N_d - N_{mix}$ of independent ballistic modes, and the critical value of the mode index n that divides these two groups. All propagating modes with $n > N_{mix}$ are independent and fully transparent, otherwise they are mixed for $n \leqslant N_{mix}$ and characterized by finite scattering lengths $L_n^{(f)}$ and $L_n^{(b)}$. Therefore, the total average transmittance, Eq. (9.43), of the multi-mode structure is given by

$$\langle T \rangle = [kd/\pi] - [kd/\pi\alpha_c], \qquad \alpha_c = \sqrt{1 + (k_c d/\pi)^{-2}}. \tag{9.53}$$

The numbers N_{mix} and N_{bal} of mixed nontransparent and independent ballistic modes are governed by two parameters: the mode parameter kd/π and the dimensionless correlation parameter $k_c d/\pi$. In the case of "weak" correlations, when $k_c d/\pi \gg 1$, the number of mixed modes N_{mix} is of the order of N_d:

$$N_{mix} \approx \left[\left(\frac{kd}{\pi} \right) - \frac{1}{2} \left(\frac{kd}{\pi} \right) \left(\frac{k_c d}{\pi} \right)^{-2} \right] \qquad \text{for} \quad k_c d/\pi \gg 1. \tag{9.54}$$

Consequently, in this case the number of ballistic modes N_{bal} is small, or there are no such modes at all. If the parameter $k_c d/\pi$ tends to infinity, $k_c d/\pi \to \infty$, the rough surface profile becomes white-noise-like and, naturally, $N_{mix} \to N_d$.

The most appropriate case is when a random surface profile is strongly correlated so that the correlation parameter is small, $k_c d/\pi \ll 1$. Then the number of mixed nontransparent modes N_{mix} is much smaller than the total number of propagating

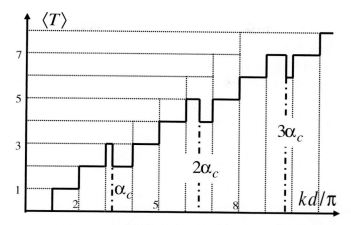

Figure 9.3. Stepwise transmittance, Eq. (9.53), of a surface-disordered guiding system versus the mode parameter kd/π. The value of the normalized correlation parameter $k_c d/\pi = 0.32$ (after ref. [7]).

modes N_d, i.e.

$$N_{mix} \approx \left[\left(\frac{kd}{\pi} \right) \left(\frac{k_c d}{\pi} \right) \right] \ll N_d \quad \text{for} \quad k_c d/\pi \ll 1. \tag{9.55}$$

Therefore, the number of independent modes N_{bal} is large. When the correlation parameter $k_c d/\pi$ decreases and becomes anomalously small, $k_c d/\pi < (kd/\pi)^{-1} \ll 1$, the number of mixed modes N_{mix} vanishes, and *all modes* become independent and perfectly transparent. Evidently, if the correlation parameter $k_c d/\pi$ vanishes, $k_c d/\pi \to 0$, the roughness power spectrum, Eq. (9.48b), becomes delta-function-like and, as a consequence, $N_{mix} = 0$. In this case the correlated disorder results in a perfect transmission of waves.

Finally, let us briefly discuss Eq. (9.53) for the total average transmittance. In Fig. 9.3 an unusual *nonmonotonic stepwise dependence* of $\langle T \rangle$ on the mode parameter kd/π is shown that is governed by the width k_c of the rectangular power spectrum, Eq. (9.48b).

Let us discuss Fig. 9.3. Within the region where $kd/\pi < \alpha_c$, the inter-mode transitions caused by specific surface correlations are forbidden for all conducting channels. Therefore, all propagating modes are independent and ballistic, and the second term (the number of mixed modes) in Eq. (9.53) for the total transmittance is equal to zero. Here the transmittance exhibits a ballistic *stepwise increase* with an increase of the parameter kd/π. Each step "up" arises for an integer value of the mode parameter kd/π, when a new conducting channel emerges. Such a stepwise increase of the total transmittance is similar to that known to occur in quasi-one-dimensional ballistic *nondisordered* structures (see, e.g., ref. [101]).

When $kd/\pi \geqslant \alpha_c$, in addition to the standard steps "up" originated from the first term in Eq. (9.53), there are also steps "down" associated with the second term. These steps "down" are provided by the correlated surface scattering and arise when a successive low mode abruptly becomes mixed and nontransparent. In other words, the position of the nth step "down" is at the transparency edge point $kd/\pi = n\alpha_c$, where the nth conducting channel closes. Specifically, the first step "down" occurs at the *total transparency edge* $kd/\pi = \alpha_c$, where the first mode is closed. This transparency edge separates the region of complete transparency from the region where lower modes are mixed and nontransparent. The second step "down" is due to the *particular transparency edge* $kd/\pi = 2\alpha_c$ of the second mode, etc. Since the values of the ratio $kd/\pi\alpha_c$ are determined by the correlation parameter k_c, the positions of steps "down," in general, do not coincide with those of steps "up." The situation may also occur when the steps "up" and "down" cancel each other within some interval of the mode parameter kd/π. The interplay between steps "up" and "down" results in a new kind of *stepwise nonmonotonic dependence* of the total quasi-one-dimensional transmittance. The experimental observation of this nonconventional dependence would be highly interesting.

Acknowledgments

F. M. I. acknowledges the support by CONACyT grant no. 80715.

References

[1] F. M. Izrailev and A. A. Krokhin, "Localization and the mobility edge in one-dimensional potentials with correlated disorder," *Phys. Rev. Lett.* **82**, 4062–4065 (1999).

[2] A. A. Krokhin and F. M. Izrailev, "Anderson localization and the mobility edge in 1D random potentials with long-range correlations," *Ann. Phys. (Leipzig)* **SI-8**, 153–156 (1999).

[3] F. M. Izrailev, A. A. Krokhin, and S. E. Ulloa, "Mobility edge in aperiodic Kronig-Penney potentials with correlated disorder: perturbative approach," *Phys. Rev. B* **63**, 041102(R)(1-4) (2001).

[4] F. A. B. F. de Moura and M. L. Lyra, "Delocalization in the 1D Anderson model with long-range correlated disorder," *Phys. Rev. Lett.* **81**, 3735–3738 (1998).

[5] F. A. B. F. de Moura and M. L. Lyra, "Reply," *Phys. Rev. Lett.* **84**, 199 (2000).

[6] F. A. B. F. de Moura and M. L. Lyra, "Correlations-induced metal-insulator transition in the one-dimensional Anderson model," *Physica A* **266**, 465–470 (1999).

[7] F. M. Izrailev and N. M. Makarov, "Anomalous transport in low-dimensional systems with correlated disorder," *J. Phys. A: Math. Gen.* **38**, 10613–10637 (2005).

[8] U. Kuhl, F. M. Izrailev, A. A. Krokhin, and H.-J. Stöckmann, "Experimental observation of the mobility edge in a waveguide with correlated disorder," *Appl. Phys. Lett.* **77**, 633–635 (2000).

[9]   A. Krokhin, F. Izrailev, U. Kuhl, H. J. Stöckmann, and S. E. Ulloa, "Random 1D structures as filters for electrical and optical signals," *Physica E* **13**, 695–698 (2002).

[10] U. Kuhl, F. M. Izrailev, and A. A. Krokhin, "Enhancement of localization in one-dimensional random potentials with long-range correlations," *Phys. Rev. Lett.* **100**, 126402(1-4) (2008).

[11] F. M. Izrailev, and N. M. Makarov, "Selective transparency of single-mode wave-guides with surface scattering," *Opt. Lett.* **26**, 1604–1606 (2001).

[12] F. M. Izrailev and N. M. Makarov, "Anomalous transmission in waveguides with correlated surface disorder," *Proc. Progress In Electromagnetic Research Symposium* (Cambridge, MA: The Electromagnetic Academy, 2002) p. 389.

[13] F. M. Izrailev and N. M. Makarov, "Onset of delocalization in quasi-one-dimensional waveguides with correlated surface disorder," *Phys. Rev. B* **67**, 113402(1-4) (2003).

[14] F. M. Izrailev and N. M. Makarov, "Selective transport and mobility edges in quasi-one-dimensional systems with a stratified correlated disorder," *Appl. Phys. Lett.* **84**, 5150–5152 (2004).

[15] N. M. Makarov and F. M. Izrailev, "Anomalous transmission in waveguides with correlated disorder," *Proc. 5th Intl. Kharkov Symposium on Physics and Engineering of Microwaves, Millimeter, and Submillimeter Waves, vol. 1* (Piscataway, NJ: IEEE, 2004), pp. 122–127.

[16] F. G. Bass and I. M. Fuks, *Wave Scattering from Statistically Rough Surfaces* (New York: Pergamon, 1979).

[17] J. A. DeSanto and G. S. Brown, "Analytical techniques for multiple scattering from rough surfaces," in *Progress in Optics*, vol. 23, ed. E. Wolf (Amsterdam: North-Holland, 1986), pp.1–62.

[18] J. A. Ogilvy, *Theory of Wave Scattering from Random Surfaces* (Bristol: Adam Hilger, 1991).

[19] V. Freilikher and S. Gredeskul, "Localization of waves in media with one-dimensional disorder," in *Progress in Optics, vol. 30*, ed. E. Wolf (Amsterdam: North-Holland, 1992), pp. 137–203.

[20] P. Sheng, *Introduction to Wave Scattering, Localization and Mesoscopic Phenomena* (San Diego, CA: Academic, 1995).

[21] I. Ohlidal, K. Navratil, and M. Ohlidal, "Scattering of light from multilayer systems with rough boundaries," in *Progress in Optics*, vol. 34, ed. E. Wolf (Amsterdam: North-Holland, 1995), pp. 249–331.

[22] K. L. Chopra, *Thin Film Phenomena* (New York: McGraw-Hill, 1969).

[23] V. L. Berezinski, "Kinetics of a quantum particle in a one-dimensional random potential," *Zh. Eksp. Teor. Fiz.* **65**, 1251–1266 (in Russian) (1973). Translated in *Sov. Phys. JETP* **38**, 620–627 (1974).

[24] A. A. Abrikosov and I. A. Ryzhkin, "Conductivity of quasi-one-dimensional metal systems," *Adv. Phys.* **27**, 147–230 (1978).

[25] I. M. Lifshits, S. A. Gredeskul, and L. A. Pastur, *Introduction to the Theory of Disordered Systems* (New York: Wiley, 1988).

[26] P. A. Mello and N. Kumar, *Quantum Transport in Mesoscopic Systems: Complexity and Statistical Fluctuations* (New York: Oxford University Press, 2004.)

[27] C. W. J. Beenakker and H. van Houten, "Semi-classical theory of magnetoresistance anomalies in ballistic multi-probe conductors," in *Electronic Properties of Multi-layers and Low-Dimensional Semiconductor Structures*, eds. J. M. Chamberlain, L. Eaves, and J. C. Portal, NATO Advance Study Institute Series, vol. 231 (London: Plenum, 1990), pp. 75–94.

[28] R. Landauer, "Conductance from transmission: common sense points," *Phys. Scr.* **T42**, 110–114 (1992).

[29] Y. V. Fyodorov and A. D. Mirlin, "Statistical properties of eigenfunctions of random quasi 1D one-particle Hamiltonians," *Int. J. Mod. Phys.* B **8**, 3795–3842 (1994).

[30] S. Datta, *Electronic Transport in Mesoscopic Systems* (Cambridge: Cambridge University Press, 1995).

[31] T. Dittrich, P. Hänggi, G.-L. Ingold, B. Kramer, G. Schön, and W. Zwerger, *Quantum Transport and Dissipation* (Weinheim: Wiley-VCH, 1998).

[32] F. M. Izrailev and N. M. Makarov, "Localization in correlated bilayer structures: from photonic crystals to metamaterials and semiconductor superlattices," *Phys. Rev. Lett.* **102**, 203901(1-4) (2009).

[33] F. M. Izrailev, N. M. Makarov, and E. J. Torres-Herrera, "Anderson localization in bi-layer array with compositional disorder: conventional photonic crystals versus metamaterials," *Physica B* **405**, 3022–3025 (2010).

[34] G. A. Luna-Acosta, F. M. Izrailev, N. M. Makarov, U. Kuhl, and H.-J. Stöckmann. "One-dimensional Kronig-Penney model with positional disorder: theory versus experiment," *Phys. Rev. B* **80**, 115112(1-8) (2009).

[35] G. A. Luna-Acosta and N. M. Makarov, "Effect of Fabry-Perot resonances in dis-ordered one-dimensional array of alternating dielectric bi-layers," *Annal. Phys.* **18**, 887–890 (2009).

[36] A. V. Chaplik and M. V. Entin, "Energy spectrum and electron mobility in a thin film with non-ideal boundary," *Zh. Eksp. Teor. Fiz.* **55**, 990–998 (1968) (in Russian). Translated in *Sov. Phys. JETP* **28** 514–517 (1969).

[37] Z. Tešanović, M. Jarić, and S. Maekawa, "Quantum transport and surface scattering," *Phys. Rev. Lett.* **57**, 2760–2763 (1986).

[38] N. Trivedi and N. W. Ashcroft, "Quantum size effects in transport properties of metallic films," *Phys. Rev. B* **38**, 12298–12309 (1988).

[39] G. Fishman and D. Calecki, "Surface-induced resistivity of ultrathin metallic films: a limit law," *Phys. Rev. Lett.* **62**, 1302–1305 (1989).

[40] G. Fishman and D. Calecki, "Influence of surface roughness on the conductivity of metallic and semicondicting quasi-two-dimensional structures," *Phys. Rev. B* **43**, 11581–11585 (1991).

[41] A. A. Krokhin, N. M. Makarov, and V. A. Yampol'skii, "Theory of the surface scattering of electrons in metals with gently sloping surface irregularities," *Zh. Eksp. Teor. Fiz.* **99**, 520–529 (1991) (in Russian). Translated in *Sov. Phys. JETP* **72**, 289–294 (1991).

[42] A. A. Krokhin, N. M. Makarov, and V. A. Yampol'skii, "Microscopic theory of conduction electron scattering from a random metal surface with mildly sloping asperities," *J. Phys.: Condens. Matter* **3**, 4621–4632 (1991).

[43] N. M. Makarov, A. V. Moroz, and V. A. Yampol'skii, "Classical and quantum size effects in electron conductivity of films with rough boundaries," *Phys. Rev. B* **52**, 6087–6101 (1995).

[44] X.-G. Zhang and W. H. Butler, "Conductivity of metallic films and multilayers," *Phys. Rev. B* **51**, 10085–10103 (1995).

[45] L. Sheng, D. Y. Xing, and Z. D. Wang, "Transport theory in metallic films: crossover from the classical to the quantum regime," *Phys. Rev. B* **51**, 7325–7328 (1995).

[46] A. E. Meyerovich and S. Stepaniants, "Transport phenomena at rough boundaries," *Phys. Rev. Lett.* **73**, 316–319 (1994).

[47] A. E. Meyerovich and S. Stepaniants, "Transport in channels and films with rough surfaces," *Phys. Rev. B* **51**, 17116–17130 (1995).

[48] A. E. Meyerovich and S. Stepaniants, "Ballistic transport in ultra-thin films with random rough walls," *J. Phys.: Condens. Matter* **9**, 4157–4174 (1997).

[49] A. E. Meyerovich and S. Stepaniants, "Transport equation and diffusion in ultrathin channels and films," *Phys. Rev. B* **58**, 13242–13263 (1998).

[50] A. E. Meyerovich and S. Stepaniants, "Quantized systems with randomly corrugated walls and interfaces," *Phys. Rev. B* **60**, 9129–9144 (1999).

[51] A. E. Meyerovich and S. Stepaniants, "Interference of bulk and boundary scattering in films with quantum size effect," *J. Phys.: Condens. Matter* **12**, 5575–5598 (2000).

[52] M. A. Meyerovich and I. V. Ponomarev, "A new type of size effect in the conductivity of quantized metal films," *J. Phys.: Condens. Matter* **14**, 4287–4295 (2002).

[53] M. A. Meyerovich and I. V. Ponomarev, "Quantum size effect in conductivity of multilayer metal films," *Phys. Rev. B* **67**, 165411(1-10) (2003).

[54] A. R. McGurn and A. A. Maradudin, "Localization of electrons in thin films with rough surfaces," *Phys. Rev. B* **30**, 3136–3140 (1984).

[55] G. Brown, V. Celli, M. Haller, A. A. Maradudin, and A. Marvin, "Resonant light scattering from a randomly rough surface," *Phys. Rev. B* **31**, 4993–5005 (1985).

[56] N. M. Makarov and I. V. Yurkevich, "Localization of 2D electrons by coherent surface scattering," *Zh. Eksp. Teor. Fiz.* **96**, 1106–1108 (1989) (in Russian). Translated in *Sov. Phys. JETP* **69**, 628–629 (1989).

[57] N. M. Makarov and I. V. Yurkevich, "Localization caused by surface scattering," in *Proceedings of the 6th International School on Microwave Physics and Technique, Varna, Bulgaria, 1989*, eds. A. I. Spasov and M. A. Tsankov (Singapore: World Scientific Publishing Co., 1990), pp. 566–570.

[58] V. D. Freilikher, N. M. Makarov, and I. V. Yurkevich, "Strong one-dimensional localization in systems with statistically rough boundaries," *Phys. Rev. B* **41**, 8033–8036 (1990).

[59] A. A. Krokhin, N. M. Makarov, V. A. Yampol'skii, and I. V. Yurkevich, "Problem of localization and scattering of the conduction electrons at the rough metal surface," *Physica B* **165**, **166**, 855–856 (1990).

[60] Y. Takagaki and D. K. Ferry, "Conductance of quantum waveguides with a rough boundary," *J. Phys.: Condens. Matter* **4**, 10421–10432 (1992).

[61] N. M. Makarov and Yu. V. Tarasov, "Conductance of a single-mode electron waveguide with statistically identical rough boundaries," *J. Phys.: Condens. Matter* **10**, 1523–1537 (1998).

[62] N. M. Makarov and Yu. V. Tarasov, "Electron localization in narrow surface-corrugated conducting channels: manifestation of competing scattering mechanisms," *Phys. Rev. B* **64**, 235306(1-14) (2001).

[63] N. M. Makarov and Yu. V. Tarasov, "Electron localization in narrow rough-bounded wires: evidence of different surface scattering mechanisms," *Superficies y Vacío* **13**, 120–125 (2001).

[64] S. A. Bulgakov and M. Nieto-Vesperinas, "Competition of different scattering mechanisms in a one-dimensional random photonic lattice," *J. Opt. Soc. Am. A* **13**, 500–508 (1996).

[65] A. García-Martín, J. A. Torres, J. J. Sáenz, and M. Nieto-Vesperinas, "Transition from diffusive to localized regimes in surface corrugated optical waveguides," *Appl. Phys. Lett.* **71**, 1912–1914 (1997).

[66] A. García-Martín, J. J. Sáenz, and M. Nieto-Vesperinas, "Spatial field distributions in the transition from ballistic to diffusive transport in randomly corrugated waveguides," *Phys. Rev. Lett.* **84**, 3578(1-4) (2000).

[67] A. García-Martín, J. A. Torres, J. J. Sáenz, and M. Nieto-Vesperinas, "Intensity distribution of modes in surface corrugated waveguides," *Phys. Rev. Lett.* **80**, 4165–4168 (1998).

[68] M. Leadbeater, V. I. Falko, and C. J. Lambert, "Lévy flights in quantum transport in quasiballistic wires," *Phys. Rev. Lett.* **81**, 1274–1277 (1998).

[69] J. A. Sánchez-Gil, V. Freilikher, I. V. Yurkevich, and A. A. Maradudin, "Coexistence of ballistic transport, diffusion, and localization in surface disordered waveguides," *Phys. Rev. Lett.* **80**, 948–951(1998).

[70] J. A. Sánchez-Gil, V. Freilikher, A. A. Maradudin, and I. V. Yurkevich, "Reflection and transmission of waves in surface-disordered waveguides," *Phys. Rev. B* **59**, 5915–5925 (1999).

[71] G. A. Luna-Acosta, Kyungsun Na, L. E. Reichl, and A. Krokhin, "Band structure and quantum Poincaré sections of a classically chaotic quantum rippled channel," *Phys. Rev. E* **53**, 3271–3283 (1996).

[72] G. A. Luna-Acosta, J. A. Méndez-Bermúdez, and F. M. Izrailev, "Quantum-classical correspondence for local density of states and eigenfunctions of a chaotic periodic billiard," *Phys. Lett. A* **274**, 192–199 (2000).

[73] G. A. Luna-Acosta, J. A. Méndez-Bermúdez, and F. M. Izrailev, "Periodic chaotic billiards: quantum-classical correspondence in energy space," *Phys. Rev. E* **64**, 036206(1–14) (2001).

[74] F. M. Izrailev, G. A. Luna-Acosta, J. A. Méndez-Bermúdez, and M. Rendón, "Amplitude and gradient scattering in waveguides with corrugated surfaces," *phys. stat. sol. (c)* **0**, 3032–3036 (2003).

[75] F. M. Izrailev and N. M. Makarov, "Controlled transparency of many-mode waveguides with rough surface," phys. stat. sol. (c), 3037–3041 (2003).

[76] F. M. Izrailev and N. M. Makarov, "Anomalous selective transparency of many-mode surface-corrugated waveguides," in *Extended Papers of Progress In Electromagnetic Research Symposium, Pisa, Italy, 2004* (Cambridge MA: The Electromagnetic Academy, 2004), pp. 277–280.

[77] C. S. West and K. A. O'Donnell, "Observations of backscattering enhancement from polaritons on a rough metal surface," *J. Opt. Soc. Am. A* **12**, 390–397 (1995).

[78] U. Kuhl and H.-J. Stöckmann, "Microwave realization of the Hofstadter butterfly," *Phys. Rev. Lett.* **80**, 3232–3235 (1998).

[79] F. M. Izrailev, N. M. Makarov, and M. Rendón, "Rough surface scattering in many-mode conducting channels: gradient versus amplitude scattering," *phys. stat. sol. (b)* **242**, 1224–1228 (2005).

[80] F. M. Izrailev, N. M. Makarov, and M. Rendón, "Gradient and amplitude scattering in surface-corrugated waveguides," *Phys. Rev. B* **72**, 041403(R) (2005).

[81] F. M. Izrailev, N. M. Makarov, and M. Rendón, "Square-gradient scattering mechanism in surface-corrugated waveguides," *Braz. J. Phys.* **36**, 971–974 (2006).

[82] F. M. Izrailev, N. M. Makarov, and M. Rendón, "Manifestation of the roughness-square-gradient scattering in surface-corrugated waveguides," *Phys. Rev. B* **73**, 155421(1-12) (2006).

[83] M. Rendón, F. M. Izrailev, and N. M. Makarov, "Square-gradient mechanism of surface scattering in quasi-1D rough waveguides," *Phys. Rev. B* **75**, 205404(1-18) (2007).

[84] R. Bellman and G. M. Wing, *An Introduction to Invariant Imbedding* (New York: Wiley, 1975).

[85] V. I. Klyatskin, *The Invariant Imbedding Method in a Theory of Wave Propagation* (Moscow: Nauka, 1986) (in Russian).

[86] V. I. Klyatskin, *Dynamics of Stochastic Systems* (Amsterdam: Elsevier, 2005).

[87] N. M. Makarov, Lectures on *Spectral and Transport Properties of One-Dimensional Disordered Conductors* (1999). http://www.ifuap.buap.mx/virtual/page_vir.html

[88] S. O. Rice, "Mathematical analysis of random noise," in *Selected Papers on Noise and Stochastic Processes*, ed. N. Wax (New York: Dover, 1954), pp. 133–294.

[89] D. Saupe, "Algorithms for random fractals," in *The Science of Fractal Images,* eds. H.-O. Peitgen and D. Saupe (New York: Springer, 1988), pp. 71–136.

[90] J. Feder, *Fractals* (New York: Plenum Press, 1988).

[91] A. Czirok, R. N. Mantegna, S. Havlin, and H. E. Stanley, "Correlations in binary sequences and a generalized Zipf analysis," *Phys. Rev. E* **52**, 446–452 (1995).

[92] H. A., Makse, S. Havlin, M. Schwartz, and H. E. Stanley, "Method for generating long-range correlations for large systems," *Phys. Rev. E* **53**, 5445–5449 (1996).

[93] A.-L. Barabási and H. E. Stanley, *Fractal Concepts in Surface Growth* (Cambridge: Cambridge University Press, 1995).

[94] A. Romero and J. Sancho, "Generation of short and long range temporal correlated noises," *J. Comput. Phys.* **156**, 1–11 (1999).

[95] J. García-Ojalvo and J. Sancho, *Noise in Spatially Extended Systems* (New York: Springer, 1999).

[96] R. Cakir, P. Grigolini, and A. A. Krokhin, "Dynamical origin of memory and renewal," *Phys. Rev. E* **74**, 021108(1–6) (2006).

[97] F. M. Izrailev, A. A. Krokhin, N. M. Makarov, and O. V. Usatenko, "Generation of correlated binary sequences from white-noise," *Phys. Rev. E* **76**, 027701(1-4) (2007).

[98] S. S. Apostolov, F. M. Izrailev, N. M. Makarov, Z. A. Mayzelis, S. S. Melnyk, and O. V. Usatenko, "The signum function method for the generation of correlated dichotomic chains," *J. Phys. A: Math. Th.* **41**, 175101(1-23) (2008).

[99] I. Simonsen, T. A. Leskova, and A. A. Maradudin, "Light scattering from an amplifying medium bounded by a randomly rough surface: a numerical study," *Phys. Rev. B* **64**, 035425(1–7) (2001).

[100] A. A. Maradudin, I. Simonsen, T. A. Leskova, and E. R. Méndez, "Localization of surface plasmon polaritons on a random surface," *Physica B* **296**, 85–97 (2001).

[101] B. J. van Wees, H. van Houten, C. W. J. Beenakker, J. G. Williamson, L. P. Kouwenhoven, D. van der Marel, and C. T. Foxon, "Quantized conductance of point contacts in a two-dimensional electron gas," *Phys. Rev. Lett.* **60**, 848–850 (1988).

10

Cloaking

CHRISTOPHER C. DAVIS AND IGOR I. SMOLYANINOV

10.1 Introduction, general background, and history

Cloaking is the ability to make a region of space, and everything in it, invisible to an external observer. It has been the dream of fantasy writers for decades. In 2009, John Mullan [1] of *The Guardian* newspaper summarized the ten most important works that use the theme: *The Invisible Man* by H. G. Wells, *The Republic* by Plato, *The Lord of the Rings* by J. R. R. Tolkien, the Harry Potter books by J. K. Rowling, *Theogony* by Hesiod, *Dr Faustus* by Christopher Marlowe, *The Tempest* by William Shakespeare, *The Voyage of the Dawn Treader* by C. S. Lewis, *The Emperor's New Clothes* by Hans Christian Andersen, and *The Hitchhiker's Guide to the Galaxy* by Douglas Adams. A true cloak allows the clear observation of the space behind the cloaked region, and the cloaked region casts no shadow and produces no wavefront changes in the light that has passed through the cloaked region. It is not possible to build a perfect invisibility cloak, as was perceptively observed in the *Star Trek* series in which cloaked Romulan and Klingon spaceships could be detected by the subtle disturbances of space that the cloak produced.

Interest in making real cloaking devices can be traced to two seminal articles, one by John Pendry and his co-workers [2], and the other by Ulf Leonhardt [3]. Their approach can be called the *transformational optics* approach to cloaking, which will be discussed in more detail later. Briefly, they propose a distortion of space around a cloaked region by a spatial distribution of electric permittivity and magnetic permeability that folds light around the cloaked region. In a simplistic sense they produce a continuous distribution of refracting or mirror-like boundaries around the cloaked region. A simple mirror geometry that cloaks a region in a limited way is shown in Fig. 10.1. The cloaked region is bounded by four off-axis cylindrical parabolic mirrors. Parallel light from behind the cloaked region

Structured Surfaces as Optical Metamaterials, ed. A. A. Maradudin. Published by Cambridge University Press.
© Cambridge University Press 2011.

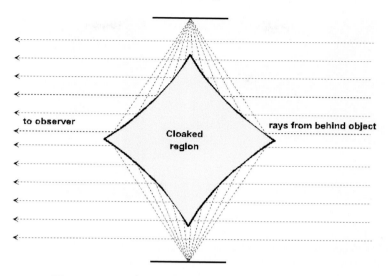

Figure 10.1. Simple geometrical optics mirror cloak.

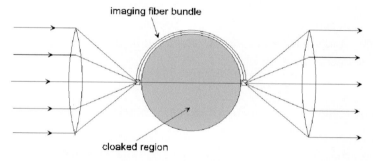

Figure 10.2. Cloaking by guiding light around the cloaked region.

reflects from these parabolic mirrors and auxiliary flat mirrors placed at the foci to provide unobstructed viewing of distant objects placed behind the cloaked region. This cloak works for all colors, but only for viewing within a narrow range of angles. It also delays different rays by different times, as will be seen later to be a common problem with more advanced cloaks. Another cloaking scheme is shown in Fig. 10.2. Parallel light from an object is focused onto an imaging fiber bundle, which diverts light around the cloaked region.

A type of cloaking can also be achieved if light that is absorbed or scattered by an object can be re-emitted either by the object itself or by one or more external scatterers so as to cancel out the effects of the scattering, and perhaps the absorption as well. We refer to this approach as *cloaking by scattering cancellation*. When

(a)

(b)

Figure 10.3. F-117 Nighthawk Stealth Ground Attack Aircraft. (a) From http://www.carzi.com/wp-content/uploads/f-117-Night-hawk.jpg. (b) From http://aircraft-list.com/db/images/F-117-Nighthawk/7145/.

a region of space is "cloaked," the surrounding region or objects that render the cloaked region invisible is called the *cloak*. There are also distinctions that arise between macroscopic cloaks, where geometrical optics can be used to describe the ray paths and microscopic cloaks, where a full wave description is required to describe the operation of the cloak.

10.2 The difference between "cloaking," "blackness," and "camouflage"

From a practical standpoint, cloaking renders objects within a cloaked region invisible to an external observer and the cloak itself must be invisible. The cloaked region appears transparent to an external observer. However, there are many instances where making an object undetectable does not require cloaking, and this distinction is worth pointing out. In the radar world objects are detected by the signals that backscatter to a receiver, which is generally co-located with the transmitter of the microwave pulses. An object does not need to be cloaked to produce no backscatter. If the object itself is shaped so as to have no surfaces that produce specular reflection back to the irradiating source, then the backscatter cross section is drastically reduced. This is the approach used in "stealth" technology, where aircraft are designed with angled surfaces to reduce their backscatter, as shown in Fig. 10.3. Radiation reflects from the aircraft, but away from the direction back to the transmitter.

An important part of stealth technology is the concept of "blackness." A "black" surface absorbs all the radiation that falls on it. A black object is invisible in reflection, but will still obscure objects behind it and cast a shadow. Blackness can be achieved in three principal ways: impedance matching, high intrinsic absorption, and geometrical loss.

10.2.1 Impedance matching

When a plane electromagnetic wave strikes a boundary there is intrinsic reflection because of impedance mismatch. The impedance of an homogeneous medium is given by

$$Z = \sqrt{\frac{\mu_r \mu_0}{\varepsilon_r \varepsilon_0}}. \tag{10.1}$$

In free space with $\mu_r = 1$, $\varepsilon_r = 1$, $Z = Z_0 = 376.7 \ \Omega$. If a plane wave strikes the boundary of a medium with nonunity μ_r, ε_r, then, depending on the angle of incidence θ_1 at the boundary and the polarization state of the incident wave, the wave both reflects and is transmitted into the second medium with reflection and transmission coefficients ρ and τ, respectively, where

$$\rho = \frac{Z_2' - Z_1'}{Z_2' + Z_1'}, \qquad \tau = \frac{2Z_2'}{Z_2' + Z_1'}. \tag{10.2}$$

The primed impedances are the effective impedances taking into account the angle of incidence θ_1 and the angle of refraction θ_2. For P-polarized (TM) waves $Z_1' = Z_1 \cos \theta_1$ and $Z_2' = Z_2 \cos \theta_2$, whereas for S-polarized (TE waves) $Z_1' = Z_1 / \cos \theta_1$ and $Z_2' = Z_2 / \cos \theta_2$. If the boundary between free space and the medium has an anti-reflection (AR) layer, then all the incident radiation is transmitted into the second medium and there is no reflection. Anti-reflection layers are simple to produce for a particular angle of incidence and wavelength. The effective impedance of the AR layer must be $Z_{AR} = \sqrt{Z_2' Z_1'}$ and its effective thickness is one-quarter of the wavelength of the radiation in the AR layer. Specifically, for an angle of refraction θ_{AR} into the AR layer, the thickness of the AR layer should be d, where $d = \lambda_2 / (4 \cos \theta_{AR})$.

Perfect impedance matching is difficult to accomplish for the surface of macroscopic objects for all angles of incidence. As we will see later, an ideal cloak should have a refractive index at its boundary that is the same as that of free space, which is then graded smoothly and adiabatically as light rays enter the cloak so that light rays bend within the cloak and follow light ray geodesics that divert them around the cloaked region.

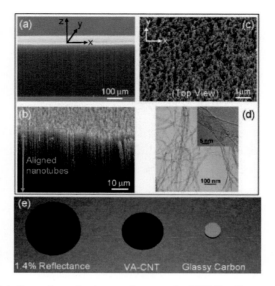

Figure 10.4. (a) Scanning electron micrograph (SEM) of a vertically aligned carbon nanotube (VA-CNT) sample. (b) Side-view SEM image of the same sample at a higher magnification. The nanotubes are vertically aligned, forming a highly porous nanostructure. (c) Top-view SEM image of the sample. The nanotubes are entangled with each other, forming a loosely connected random surface. The surface corrugation is on the order of 100–1000 nm. (d) Transmission electron micrograph of the sample, indicating that most of the nanotubes are multiwalled with a diameter $d \simeq 10$ nm. (e) Photograph of a 1.4% NIST reflectance standard, a VA-CNT sample, and a piece of glassy carbon, taken under flash light illumination. (From ref. [5], courtesy of Professor Shawn-Yu Lin.)

10.2.2 Highly absorbing and nonreflective surfaces

Reflections and backscatter from surfaces can also be reduced by tessellating the surface with crevices, which allow radiation to enter and be dissipated by multiple lossy reflections before the radiation can escape. Simple examples in the optical region of the spectrum include Wood's horns, hollow nonimaging devices, and the gaps between sharp wedges, as in a "razor blade" light absorber [4]. Conventional "black" paint can reduce scattering and backscatter to a few percent, and recently developed carbon nanotube black can reduce this to 0.045% [5]. Figure 10.4 shows a comparison between the "blackness" of the new carbon nanotube material, compared to a NIST reflection standard, and a sample of glassy carbon. In the radiofrequency (RF) region, surface absorber materials typically use large conical structures of polymer foam impregnated with carbon. The vertically aligned carbon nanotubes in Figs. 10.4 (a), (b), and (c) look similar, except for their scale, to the absorbing structures used in an RF absorber. Unfortunately, most of these

Figure 10.5. Adaptive camouflage. (From the Tachi Laboratory, Keio University and The University of Tokyo, courtesy of Professor Susumu Tachi.)

approaches to "blackness" produce surfaces that are fragile, and possess poor aerodynamic properties for application to aircraft flight surfaces.

10.2.3 Camouflage

Camouflage is an ability that has evolved in many animals to allow them to blend in with their surroundings, either for protection from predators, or to allow a predator to be less detectable by its prey. Everybody is familiar with the chameleon, which can change its color to blend into its background. Many insects have evolved colors or shapes to make them either less visible in a natural background or to look like inanimate objects in the background. Stick insects look like twigs, and butterflies even use photonic crystal-like structures to modify their reflectance and color, even beyond the normal visible part of the spectrum. Camouflage has been used for centuries in military applications to reduce the visibility of humans and machines. A more recent development is the use of re-projection to mask an object. Cameras aimed at the scene behind an object to be rendered less visible activate projectors that reproduce an image of this background onto a semi-transparent screen in front of the object, or onto a retroreflective "cloak" to give the illusion that the object is not there. An imaginary system of this kind was used to render James Bond's car "invisible" in the movie *Die Another Day*. There is continuing work on this form of "adaptive camouflage." An example is shown in Fig. 10.5.

The synthetic invisibility cloak in Fig. 10.5 relies on a camera to capture the background and requires the projection of the imagery onto a white raincoat-like wearable screen. The rapid development of flexible displays ("electronic paper") will soon likely allow for a comfortable full display suit. The micro-cams needed to capture the background imagery are already plentiful and relatively inexpensive. All of the elements of a self-contained synthetic optical invisibility cloak are either available off-the-shelf or soon will be, and these technologies will continue to improve.

10.3 Transformational optics and optical metamaterials

Metamaterials are artificial composite structures patterned on a subwavelength scale. They have enabled new ways to control and manipulate electromagnetic waves. The electromagnetic response of natural materials is no longer constrained by their chemical composition. Instead, the shape and size of the structural units of the metamaterial can be tailored, or their composition and morphology tuned, to provide new functionality.

The innovative field of *transformation optics*, which is enabled by metamaterials, has inspired a fresh look at the very foundations of optics, and has helped to create a new paradigm for the science of light. Transformation optics has shown that space can be *designed and engineered*, opening the fascinating possibility of controlling the flow of light with nanometer spatial precision.

The properties of many metamaterials critically depend on *plasmons* [6–8]. A specific important type of plasmon is a surface plasmon polariton (SPP) – a surface electromagnetic wave propagating along the interface between two media possessing permittivities with opposite signs, such as a metal–dielectric interface [9]. The ability to generate SPPs in various geometries and to control their 2-D propagation has given rise to the term *plasmonics*. Plasmonics has shown that it is possible to control a refractive index over a wide range, from high to low, and even into negative values, and promises new devices with capabilities that supplement photonics or electronics. Transformational cloaking uses metamaterials to control the flow of electromagnetic waves around a cloaked region.

10.4 Dielectric constants, relative permeabilities, and refractive indices

Central to our discussion of approaches to cloaking will be the consideration of the electromagnetic properties of the materials of the cloaks themselves. The fields around the cloaked region must obey Maxwell's equations, and will depend on the spatial variation of both the dielectric and relative permeability tensors.

The relative permittivity (dielectric constant) ε_r and relative permeability μ_r characterize a medium in terms of its difference from a vacuum. Both these dimensionless quantities are frequency-dependent. The refractive index is given by $n = \sqrt{\mu_r \varepsilon_r}$. Most traditional optical materials are not strongly magnetic, and it is generally legitimate to assume that for such materials $\mu_r = 1$. All passive materials are lossy to some extent, and the dielectric constant can be represented by a complex number. We write $\varepsilon_r = \varepsilon' - j\varepsilon''$, where $j = \sqrt{-1}$. The choice of the negative sign in defining ε_r is a matter of convention. If a wave propagates through an absorbing medium its field amplitudes vary as $\exp[j(\omega t - \mathbf{k} \cdot \mathbf{r})]$, where $k = |\mathbf{k}| = \omega\sqrt{\mu_r\mu_0\varepsilon_r\varepsilon_0}$. The attenuation constant of the wave is $\alpha = -Im(k)$. For a low loss material, $\varepsilon'' \ll \varepsilon'$ and $\alpha = \omega\sqrt{\mu_r\mu_0\varepsilon'\varepsilon_0}(\varepsilon''/2\varepsilon')$, where $\varepsilon''/\varepsilon'$ is often called the *loss tangent*. In any real material the refractive index is also complex and we write it as $n - jK$, where again the choice of the negative sign is a matter of convention. The definitions of ε_r and n must be consistent with attenuation of a wave in the direction of energy flow. For an amplifying medium, which represents a nonpassive situation where there is energy input, the attenuation coefficient becomes negative, and it is common practice to call $-\alpha$ the *gain coefficient*. In a metal the dielectric properties are dominated by the free electrons and are frequently described by the Drude model [10–12]. In this model the real and imaginary parts of the dielectric constant are given by

$$\varepsilon'(\omega) = 1 - \frac{\omega_p^2}{\omega^2 + \Gamma^2} \quad \text{and} \quad \varepsilon''(\omega) = \omega_p^2 \frac{\Gamma/\omega}{\omega^2 + \Gamma^2}, \tag{10.3}$$

where $\omega_p = \sqrt{Ne^2/(\varepsilon_0 m_e)}$ is the plasma frequency, and Γ is a damping coefficient. Consequently, for frequencies sufficiently below the plasma frequency, the real part of the dielectric constant becomes negative. For gold, the wavelength corresponding to the plasma frequency is about 138 nm and in silver it is 311 nm. At a wavelength of 633 nm for gold, the refractive index is $0.122 - j3.968$ and $\varepsilon_r = -15.73 - j0.968$. For silver, at 496 nm, the refractive index is $0.24 - j3.09$ and $\varepsilon_r = -9.49 - j1.483$. At 354 nm the real part of ε_r is -1. Consequently, in a mixed metal–dielectric structure it is possible to play off the negative real dielectric constant of the metal against the positive real dielectric constant of the dielectric. For this to occur, a wave passing through such a *metamaterial* must have a wavelength greater than the nanoscale structure of the material so that the wave perceives an intermediate average bulk dielectric constant. We acknowledge that in *anisotropic*, as distinct from *inhomogenous*, materials the dielectric constant and relative permeability are in general tensors, so that we would write $\mathbf{D} = \varepsilon_0\tilde{\varepsilon}_r\mathbf{E}$ and $\mathbf{B} = \mu_0\tilde{\mu}_r\mathbf{H}$, where in a Cartesian principal coordinate system, for example,

$$D = \varepsilon_0 \begin{pmatrix} \varepsilon_x & 0 & 0 \\ 0 & \varepsilon_y & 0 \\ 0 & 0 & \varepsilon_z \end{pmatrix} E. \tag{10.4}$$

In a lossless nonmagnetic biaxial material the three principal refractive indices are $n_x = \sqrt{\varepsilon_x}$, $n_y = \sqrt{\varepsilon_y}$, and $n_z = \sqrt{\varepsilon_z}$. In a uniaxial material it is common practice to write $n_x = n_y = n_o$ and $n_z = n_e$, where n_o and n_e and the ordinary and extraordinary refractive indices, respectively.

10.5 Negative refractive index materials

It is very interesting to examine the electromagnetic properties of materials for which both the real part of the dielectric constant and the relative permeability are negative. The electromagnetic properties of such materials were first examined by Veselago [13], but it is only recently that metamaterials have offered the possibility of making such materials a reality.

For so-called double-negative materials, both ε and μ have negative real parts. We write

$$\mu_r = \mu' - j\mu''. \tag{10.5}$$

In this case, the generalized propagation constant can be written as

$$\gamma = jk_0\sqrt{(\mu' - j\mu'')(\varepsilon' - j\varepsilon'')}, \tag{10.6}$$

where k_0 is the free space propagation constant given by $k_0 = \omega\sqrt{\mu_0\varepsilon_0}$, or alternatively as

$$\gamma = jk_0\sqrt{\mu'\varepsilon'\left(1 - j\frac{\mu''}{\mu'}\right)\left(1 - j\frac{\varepsilon''}{\varepsilon'}\right)}. \tag{10.7}$$

If both μ' and ε' are negative, this equation takes the following form:

$$\gamma = jk_0\sqrt{\mu'\varepsilon'(1 - jX)(1 - jY)}, \tag{10.8}$$

where both X and Y are absolutely positive for nonamplifying materials. The location of γ in the complex plane depends on the choice of the sign of the square root $\sqrt{\mu'\varepsilon'}$. For nonamplifying materials γ can only lie in the second or third quadrants of the complex plane, which requires the negative sign of the square root. Thus, for such materials the real part of the refractive index is $n = -\left|\sqrt{\mu'\varepsilon'}\right|$, which is negative. For a negative refractive index material the real part of the impedance $Z = Z_0/(n - jK)$ is also negative. Materials with a negative refractive index are often called left-handed media. If a plane wave strikes the boundary between two media with positive and negative refractive indices n_1 and n_2, respectively, at an angle of incidence θ_1, then the angle of refraction θ_2 obeys Snell's law, $n_1\sin\theta_1 = n_2\sin\theta_2$. So the refracted ray bends in the opposite direction from the usual one, as shown in Fig. 10.6. Energy flow goes towards the boundary in medium 1 and away from the boundary in medium 2, as shown in Fig. 10.6; otherwise there would be stored energy buildup at the boundary. Phase velocity, in contrast, is towards the boundary in both media. At the boundary between two anisotropic materials with positive principal refractive indices the wavevector direction (**k**) of a wave obeys Snell's law, but the ray direction deviates from the wavevector

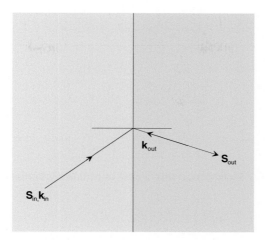

Figure 10.6. Negative refraction.

direction. At the boundary between two sufficiently birefringent materials it is possible for the ray direction to undergo an apparent negative refraction.

In the optical region, metals sufficiently below their plasma frequency provide a negative ε', but achieving a negative value for μ' is a challenge since most optical materials have $\mu' = \mu_r = 1$. It is easier to build metamaterials with controllable permeability at longer wavelengths, where resonant metallic structures can be incorporated that provide inductive and/or capacitive properties. For an inductor L the impedance is $Z = \omega L$ and the admittance is $Y = -j/(\omega L)$. For a capacitor the corresponding quantities are $Z = -j/(\omega C)$ and $Y = j\omega C$. To understand how inductive/capacitive structures allow manipulation of the propagating electromagnetic waves, it is instructive to examine the properties of generalized transmission lines, which provide a lumped circuit analog of various types of propagating electromagnetic field waves.

10.6 Generalized transmission lines and backward wave systems

If a space can be represented by a distributed series impedance (Ω/m)/shunt admittance (S/m) combination as shown in Fig. 10.7, then the generalized propagation constant is $\gamma = \sqrt{ZY}$. The simplest case is where $Z = j\omega L$ and $Y = j\omega C$, and $\gamma = j\omega\sqrt{LC}$, which corresponds to TEM wave propagation in a space with effective permittivity C (F/m) and permeability L (H/m). The wave vector magnitude is $\beta = \omega\sqrt{LC}$ and the phase and group velocities are $v_p = \omega/\beta = 1/\sqrt{LC}$ and $v_g = d\omega/d\beta = 1/\sqrt{LC}$, respectively. The dispersion ($\omega - \beta$) diagram is a straight line.

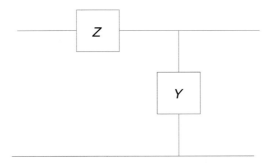

Figure 10.7. Generalized distributed impedance/admittance system.

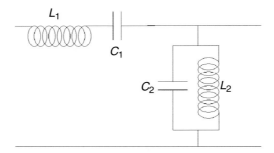

Figure 10.8. Generalized system with two pass bands and a stop band.

For the generalized system shown in Fig. 10.8, there are two characteristic cut-off frequencies $\omega_{c1} = 1/\sqrt{L_1 C_1}$ and $\omega_{c2} = 1/\sqrt{L_2 C_2}$. The generalized propagation constant is given by

$$\gamma = \sqrt{\frac{1 - \omega^2 L_1 C_1}{j\omega L_1} \frac{1 - \omega^2 L_2 C_2}{j\omega L_1}}. \tag{10.9}$$

For the case where $\omega_{c1} < \omega_{c2}$, the dispersion diagram in the pass band $0 \le \omega \le \omega_{c1}$ is shown in Fig. 10.9. There is a stop band (energy gap) for $\omega_{c1} < \omega < \omega_{c2}$, and a high-frequency pass band for $\omega \ge \omega_{c2}$. In the frequency range $0 \le \omega \le \omega_{c1}$, the phase velocity and group velocities have opposite signs. This is a *backward wave system*. It can be interpreted as a wave whose group velocity is positive when its phase velocity is negative, or vice versa. It can be viewed as a system where the phase refractive index is negative and the group refractive index is positive. This is an analog of a negative refractive index system. Examples of other types of dispersion diagram for distributed LC systems are given by Ramo *et al.* [14].

Generalized LC systems can be fabricated using distributed inductive and capacitive elements so as to provide spatially controlled variations in dielectric constant

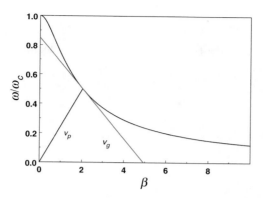

Figure 10.9. Dispersion diagram in the low-frequency pass band for the backward wave system shown in Fig. 10.8.

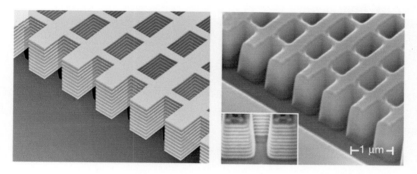

Figure 10.10. (a) Schematic structure of a "fishnet" metamaterial: 21 alternating layers of silver and magnesium fluoride. (From ref. [15], courtesy of Professor Xiang Zhang.) (b) Scanning electron microscope image of the fabricated fishnet structure, developed by UC Berkeley researchers. The alternating layers form small circuits that can bend light backwards. (From ref. [15], courtesy of Professor Xiang Zhang.)

and relative permeability. For example, composite structures of metal and dielectric incorporating nanowires, metal grids, and metal loops can be used to provide distributed inductance and capacitance. Metamaterials with a negative refractive index that incorporate these ideas of distributed inductive and capacitive elements have been fabricated using hybrid structures of metal and dielectric. Figure 10.10 shows the "fishnet" structure developed by Zhang and his co-workers [15]. Such metamaterial structures, which have many other applications beyond their use in cloaking, are very complex and require sophisticated 3-D lithography techniques. It is beyond the scope of this chapter to discuss the optical properties of metamaterials in detail. Their overall performance can be summarized as depending on

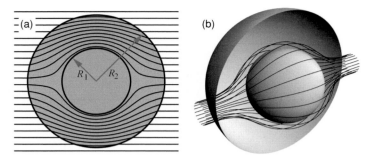

Figure 10.11.   Illustration of the ray paths inside a spherical cloak of inner radius R_1, outer radius R_2, that cloaks an inner spherical region of radius R_1. A bundle of parallel rays enters the cloak and its path is deviated by the spatial refractive index distribution inside the cloak. A ray-tracing program was used to calculate ray trajectories in the cloak, assuming that $R_2 \gg \lambda$. The rays essentially follow the Poynting vector. (a) Two-dimensional (2-D) cross section of rays entering the cloak and being diverted within the annulus of the cloaking material contained within $R_1 < r < R_2$ to emerge on the far side undeviated from their original course. (b) A 3-D view of the same process. (From ref. [2], courtesy of Professor Sir John Pendry.)

their isofrequency surfaces – the dependence of the components of the wavevector of a propagating wave on its direction relative to symmetry directions in the metamaterial [16].

10.7 Transformation optics and the ray optics of cloaks

As discussed previously, in the geometrical optics approximation a region can be cloaked if rays of light from behind an object flow around the object and then continue on the same trajectory that they would have had if the cloaked region were absent. This can be accomplished if the space around the object to be cloaked – the cloak itself – is a region of graded refractive index that bends light rays around the cloaked region, as shown in Fig. 10.11. It is clear from Fig. 10.11 that for rays to enter a cloak and avoid a cloaked region they must take longer geometric paths around the cloaked region, which requires that the refractive index inside the cloak over at least some fraction of each path must be smaller than unity. It is not necessary for the refractive index in the cloak to be negative for ray paths to exist that avoid the cloaked region. If the cloak excludes external rays from entering the cloaked region, it must also prevent any radiation from inside the cloaked region from escaping.

    In a uniform space, rays of light or wavevectors travel along straight lines, but a cloak must make light travel along curved trajectories. One way of looking

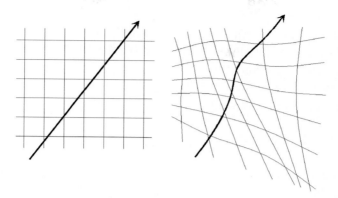

Figure 10.12. A mapping from one space to another transforms a straight line in one space to a curved line in the new space. The mapping shown here is not conformal.

at this is to imagine that space inside the cloak has been transformed so that, in the transformed space, the paths of rays of light have become straight again, even though the dielectric constant and relative permeability vary from point to point. The transformation between coordinates in the original (Cartesian) space and coordinates in the new space is called a mapping. For example, as shown in Fig. 10.12, if the original space is a uniform grid and the transformed space is a distorted grid, then a straight line in the original space becomes distorted in the new space. Each point in the original space corresponds to a single point in the new space. If the gridlines defining the coordinate systems in the original and transformed space are locally orthogonal, then the mapping is called *conformal*.

In a transformation from Cartesian coordinates (x, y, z) to a set of curvilinear coordinates (q_1, q_2, q_3), each new coordinate $q_1 = q_1(x, y, z), q_2 = q_2(x, y, z)$, and $q_3 = q_3(x, y, z)$ represents a surface such as $q_1 = const.$, etc. The space curves formed by the intersection of these surfaces in pairs are called the coordinate lines. The location of a point $P(x, y, z)$ in the new coordinate system is $P(q_1, q_2, q_3)$, and it lies at the intersection of the q_1, q_2, and q_3 surfaces. The most useful coordinate systems are ones in which locally the three surfaces q_1, q_2, and q_3 intersect at right angles. So a conformal mapping is a transformation from one set of orthogonal curvilinear coordinates to another. It is interesting to understand how Maxwell's equations change as we move between coordinate systems. Of most importance in this context are the curl equations:

$$\text{curl } \mathbf{E} = -\frac{\partial \mathbf{B}}{\partial t} \quad \text{and} \quad \text{curl } \mathbf{H} = \mathbf{j} + \frac{\partial \mathbf{D}}{\partial t}, \quad (10.10)$$

which for time-harmonic fields can be written as $\text{curl } \mathbf{E} = -j\omega\mu_r\mu_0\mathbf{H}$ and $\text{curl } \mathbf{H} = j\omega\varepsilon_r\varepsilon_0\mathbf{E}$, where, for conductive media, the current density has been incorporated into a lossy dielectric constant such that $\varepsilon_r = \varepsilon' - j(\sigma/\omega\varepsilon_0)$, where σ includes both ohmic conductivity and dielectric loss. For materials that have magnetic loss, we can introduce a "magnetic conductivity" σ^*, and the modified relative permeability becomes $\mu_r = \mu' - j(\sigma^*/\omega\varepsilon_0)$. If we perform a coordinate transformation to a general coordinate system (q_1, q_2, q_3), Maxwell's equations retain their form, and we can write

$$\text{curl}\,_q\hat{\mathbf{E}} = -j\omega\hat{\mu}_r\mu_0\hat{\mathbf{H}} \quad \text{and} \quad \text{curl}\,_q\hat{\mathbf{H}} = j\omega\hat{\varepsilon}_r\varepsilon_0\hat{\mathbf{E}}, \quad (10.11)$$

where in general $\hat{\mu}_r$ and $\hat{\varepsilon}_r$ are tensors and $\hat{\mathbf{E}}$ and $\hat{\mathbf{H}}$ are renormalized electric and magnetic fields [17]. We are still solving Maxwell's equations, but with new definitions of the dielectric constant and relative permeability. Transformations of vector operators and tensors in coordinate transformations have been dealt with in general by Margenau and Murphy [18], but the specific results that are relevant to a discussion of spaces for cloaking have been given by Pendry *et al.* [2] and in more general detail by Schurig *et al.* [19] and Rahm *et al.* [20]. Interesting and complementary approaches have been described by Ma *et al.* [21] and Qiu *et al.* [22]. If the new coordinates $q_1(x, y, z), q_2(x, y, z), q_3(x, y, z)$ are orthogonal curvilinear coordinates, the renormalized values of the dielectric constant and relative permeability are

$$\varepsilon^{i'i'} = \varepsilon^{ii}\frac{Q_1Q_2Q_3}{Q_i^2} \quad \text{and} \quad \mu^{i'i'} = \mu^{ii}\frac{Q_1Q_2Q_3}{Q_i^2}, \quad (10.12)$$

where $i = 1, 2, 3$ and

$$Q_i^2 = \left(\frac{\partial x}{\partial q_i}\right)^2 + \left(\frac{\partial y}{\partial q_i}\right)^2 + \left(\frac{\partial z}{\partial q_i}\right)^2. \quad (10.13)$$

The transformed fields are $\hat{E}_i = Q_i E_i$ and $\hat{H}_i = Q_i H_i$, where E_i and H_i are the original field components along the original orthogonal axes.

The Jacobian for the transformation is given by

$$\frac{\partial(q_1, q_2, q_3)}{\partial(x, y, z)} = \Lambda_{xyz}^{q_1q_2q_3} = \begin{pmatrix} \frac{\partial x}{\partial q_1} & \frac{\partial x}{\partial q_2} & \frac{\partial x}{\partial q_3} \\ \frac{\partial y}{\partial q_1} & \frac{\partial y}{\partial q_2} & \frac{\partial y}{\partial q_3} \\ \frac{\partial z}{\partial q_1} & \frac{\partial z}{\partial q_2} & \frac{\partial z}{\partial q_3} \end{pmatrix}, \quad (10.14)$$

where we have introduced the notation used by Schurig *et al.* [19]. It is instructive to examine the meaning of the transformation of ε_r and μ_r in going from one coordinate system to another in an undistorted space. For a transformation from spherical to Cartesian, coordinates, the Jacobian is given by

$$\Lambda_{r\theta\phi}^{xyz} = \begin{pmatrix} \sin\theta\cos\phi & r\cos\theta\cos\phi & -r\sin\theta\sin\phi \\ \sin\theta\sin\phi & r\cos\theta\sin\phi & r\sin\theta\cos\phi \\ \cos\theta & -r\sin\theta & 0 \end{pmatrix}, \quad (10.15)$$

with determinant $\det(\Lambda_{xyz}^{r\theta\phi}) = r^2\sin\theta$. In this transformation, $Q_r^2 = 1$, $Q_\theta^2 = r^2$, $Q_\phi^2 = r^2\sin^2\theta$, so the dielectric constant transformation is

$$\varepsilon' = \begin{pmatrix} \varepsilon_0 r^2\sin\theta & 0 & 0 \\ 0 & \varepsilon_0\sin\theta & 0 \\ 0 & 0 & \frac{\varepsilon_0}{\sin\theta} \end{pmatrix}. \quad (10.16)$$

The metric tensor for this transformation is

$$g_{ij} = \begin{pmatrix} 1 & 0 & 0 \\ 0 & r^2 & 0 \\ 0 & 0 & r^2\sin^2\theta \end{pmatrix} \quad \text{and} \quad g^{ij} = \begin{pmatrix} 1 & 0 & 0 \\ 0 & r^{-2} & 0 \\ 0 & 0 & (r^2\sin^2\theta)^{-1} \end{pmatrix}. \quad (10.17)$$

Note that

$$\left|\det(g_{ij})\right|^{-1/2} g_{ij}\varepsilon' = \begin{pmatrix} 1 & 0 & 0 \\ 0 & 1 & 0 \\ 0 & 0 & 1 \end{pmatrix}\varepsilon_0, \quad (10.18)$$

so, as pointed out by Rahm *et al.* [20], Eq. (10.16) still represents an isotropic medium.

Before discussing specific cloak designs, it is instructive to review the overall general approach to describing ray paths in transformed optical situations. We follow the approach described by Schurig *et al.* [19]. If the Cartesian coordinates in the original and transformed spaces are x^i and $x^{i'}$, where i, i' take the values 1, 2, 3, respectively, corresponding, for example, for i, to the familiar x, y, z coordinates in the original space. We are using the standard superscript notation for the components of a contravariant vector. The Jacobian transformation matrix can be written as

$$\Lambda_i^{i'} = \frac{\partial x^i}{\partial x^{i'}}. \quad (10.19)$$

In a situation where the transformation is time-invariant and the dielectric and magnetic properties are not directly dependent on each other, then the dielectric constant and relative permeabilities are tensors that change during the transformation according to Post [23]:

$$\varepsilon^{i'j'} = \frac{\Lambda_i^{i'} \Lambda_j^{j'} \varepsilon^{ij}}{\left|\det(\Lambda_i^{i'})\right|}, \qquad \mu^{i'j'} = \frac{\Lambda_i^{i'} \Lambda_j^{j'} \mu^{ij}}{\left|\det(\Lambda_i^{i'})\right|}. \qquad (10.20)$$

If the original medium is isotropic, then Eq. (10.20) can also be written in terms of the metric as

$$\varepsilon^{i'j'} = \frac{g^{i'j'} \varepsilon_r}{\sqrt{\left|\det(g^{i'j'})\right|}}, \qquad \mu^{i'j'} = \frac{g_j^{i'j'} \mu_r}{\sqrt{\left|\det(g^{i'j'})\right|}}, \qquad (10.21)$$

where ε_r and μ_r are the original isotropic dielectric constant and relative permeability, respectively. The metric is defined as $g^{i'j'} = \Lambda_k^{i'} \Lambda_l^{j'} \delta^{kl}$. For example, for a Cartesian to spherical coordinate transformation,

$$g^{i'j'} = \begin{pmatrix} 1 & 0 & 0 \\ 0 & \frac{1}{r^2} & 0 \\ 0 & 0 & \frac{1}{r^2 \sin^2 \theta} \end{pmatrix}. \qquad (10.22)$$

So, for example, $\varepsilon^{rr} = \varepsilon_r r^2 \sin\theta$, where ε_r is the original isotropic dielectric constant, as we saw previously.

To make a cloak, we perform a transformation that distorts space in the sense that rays of light take curved paths in the cloak with regard to the original coordinate system, but with the transformed distribution of dielectric constant and relative permeability making the rays of light "think" that they are taking straight line paths. If the space distortion involves a compression, then along the direction that is compressed both ε and μ are decreased by the compression factor; in the orthogonal directions, ε and μ are increased by the inverse of the compression factor. There are two ways of looking at this situation: the space is transformed by a coordinate transformation or mapping and the dielectric properties at a specific point don't change – *the topological interpretation*, or the material properties within the space are transformed, but the coordinate systems in the two versions of the space remain the same – *the materials interpretation*. These different viewpoints are illustrated in Fig. 10.13. If a ray of light travels in a straight line in an original coordinate system, then it will take a curved path in the transformed space. For example, if the surface of the spherical earth is mapped to a plane, then a straight line in the planar map will be curved in three dimensions on the spherical surface. If the ray of light takes the shortest time path between two points P_1 and P_2, then in the topological interpretation it follows the geodesics of the space, while in the materials interpretation it follows the path s along which $\int_{P_1}^{P_2} n(\mathbf{r}, \omega) ds$ is minimized.

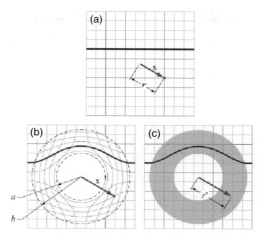

Figure 10.13. The thick line shows the path of the same ray in (a) the original Cartesian space, and under two different interpretations of the electromagnetic equations: (b) the topological interpretation and (c) the materials interpretation. The position vector x is shown in both the original and transformed spaces, and the length of the vector where the transformed components are interpreted as Cartesian components is shown in (c). (From ref. [19], courtesy of Professor David Smith.)

10.7.1 Spherical cloak

To design a spherical cloak we consider a spherically symmetric coordinate transformation and follow the details of the procedure described by Pendry *et al.* [2]. We compress all the space inside a sphere of radius b into a spherical shell of inner radius a and outer radius b. A mapping that takes all the points in the sphere of radius b and compresses this space into the shell $a \leq r' \leq b$ is:

$$r' = a + r\frac{b-a}{b}; \ \theta' = \theta; \ \phi' = \phi ; \text{note that } r = (r' - a)\frac{b}{b-a}. \quad (10.23)$$

The Jacobian for this transformation has elements $\partial r/\partial r'$, written out in full as

$$\Lambda^{i'}_i = \begin{pmatrix} \frac{b}{b-a} & 0 & 0 \\ 0 & 1 & 0 \\ 0 & 0 & 1 \end{pmatrix}, \quad (10.24)$$

so the dielectric tensor for the transformed medium is given by

$$\varepsilon' = \begin{pmatrix} \frac{b}{b-a}(r' - a)^2 \sin\theta & 0 & 0 \\ 0 & \varepsilon_0 \left(\frac{b}{b-a}\right) \sin\theta & 0 \\ 0 & 0 & \frac{\varepsilon_0}{\sin\theta}\left(\frac{b}{b-a}\right) \end{pmatrix}. \quad (10.25)$$

Finally, renormalizing with the metric

$$g_{i'j'} = \begin{pmatrix} 1 & 0 & 0 \\ 0 & r'^2 & 0 \\ 0 & 0 & r'^2 \sin^2 \theta \end{pmatrix} \tag{10.26}$$

yields

$$\varepsilon' = \begin{pmatrix} \frac{b}{b-a}\frac{(r'-a)^2}{r'^2} & 0 & 0 \\ 0 & \frac{b}{b-a} & 0 \\ 0 & 0 & \frac{b}{b-a} \end{pmatrix}. \tag{10.27}$$

It is worthwhile examining the cloak prescription given by Eq. (10.27) in a direct fundamental way as described by Schurig *et al.* [19]. Consider a position vector **x** in the original coordinate system, as shown in Fig. 10.13, with components x^i. In the transformed coordinate system it has components $x^{i'}$. The magnitude r of this position vector is independent of coordinate system, so

$$r = \left(x^i x^j \delta_{ij}\right)^{1/2} = \left(x^{i'} x^{j'} g_{i'j'}\right)^{1/2}. \tag{10.28}$$

In the materials interpretation of the new space we consider the components $x^{i'}$ to be those of a Cartesian vector whose magnitude is

$$r' = x^{i'} x^{j'} \delta_{i'j'}. \tag{10.29}$$

Because the transformation is radially symmetric the unit vectors in the materials interpretation and in the original space must be equal, so

$$\frac{x^{i'}}{r'} = \frac{x^i}{r}\delta_i^{i'}. \tag{10.30}$$

Therefore, from Eqs. (10.23) and (10.30),

$$x^{i'} = \frac{b-a}{b}x^i\delta_i^{i'} + a\frac{x^i}{r}\delta_i^{i'}, \tag{10.31}$$

which gives the Jacobian

$$\Lambda^{i'}_j = \frac{r'}{r}\delta^{i'}_j - \frac{a x^i x^k \delta^{i'}_i \delta_{kj}}{r^3}. \tag{10.32}$$

Written out in full, we obtain

$$\Lambda^{i'}_j = \begin{pmatrix} \frac{r'}{r} - \frac{ax^2}{r^3} & -\frac{axy}{r^3} & -\frac{axz}{r^3} \\ -\frac{ayx}{r^3} & \frac{r'}{r} - \frac{ay^2}{r^3} & -\frac{ayz}{r^3} \\ -\frac{azx}{r^3} & -\frac{azy}{r^3} & \frac{r'}{r} - \frac{az^2}{r^3} \end{pmatrix}. \tag{10.33}$$

If we rotate into a coordinate system where $(x^i) = (r, 0, 0)$, then

$$\det\left(\Lambda^{i'}_j\right) = \frac{r' - a}{r}\left(\frac{r'}{r}\right)^2.$$

(10.34)

If our original medium is free space, then $\varepsilon_r = \mu_r = 1$. Using Eqs. (10.20) and (10.34) yields

$$\varepsilon^{i'j'} = \mu^{i'j'} = \frac{b}{b-a}\left(\delta^{i'j'} - \frac{2ar' - a^2}{r'^4}x^{i'}x^{j'}\right).$$

(10.35)

At this point we can simplify the notation by dropping the primes, since we just need the spatial variation of the material properties in the original geometrical space. The corresponding transformed $\hat{\varepsilon}$ and $\hat{\mu}$ are given by[1]

$$\varepsilon^{r'r'} = \varepsilon_r = \mu^{r'r'} = \mu_r = \frac{b}{b-a}\frac{(r-a)^2}{r^2},$$

(10.36)

where we have simplified the notation. We have

$$\varepsilon^{\theta'\theta'} = \varepsilon_\theta = \mu^{\theta'\theta'} = \mu_\theta = \frac{b}{b-a},$$

(10.37)

$$\varepsilon^{\phi'\phi'} = \varepsilon_\phi = \mu^{\phi'\phi'} = \mu_\phi = \frac{b}{b-a}.$$

(10.38)

The corresponding components of the refractive index are $n_i = \sqrt{\varepsilon_i\mu_i} = \varepsilon_i$. The radial refractive index has the value

$$n_{r'} = \frac{b-a}{b}\frac{(r-a)^2}{r^2},$$

(10.39)

and the tangential refractive index is given by

$$n_{t'} = \frac{b}{b-a}.$$

(10.40)

Locally, the cloak is uniaxial, with an isofrequency surface (indicatrix) that is an ellipsoid of revolution.

Clearly, inside the cloak the radial component of the refractive index ellipsoid take values from zero to unity. The tangential index can take large values if the cloak is thin, $b - a \ll b$. Outside the region, $r > b$, $\varepsilon_r = \mu_r = \varepsilon_\theta = \mu_\theta = \varepsilon_\phi = \mu_\phi = 1$, so, at the boundary of the cloak, $r = b$, $\varepsilon_\theta = \varepsilon_\phi = 1/\varepsilon_r$, and $\mu_\theta = \mu_\phi = 1/\mu_r$. The impedance components at the boundary of the cloak are all $Z_i = \sqrt{\mu_i\mu_0/\varepsilon_i\varepsilon_0} = \sqrt{\mu_0/\varepsilon_0}$, so there is perfect impedance matching. We note that these are not strictly

[1] Note that there is a typographical error in eq. (7) of ref. [2].

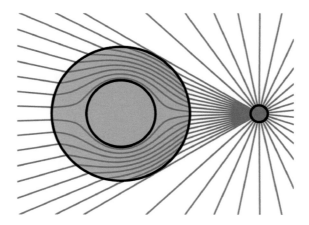

Figure 10.14. Point charge located near a cloaked sphere. It is assumed that $b \gg \lambda$, the near-field limit, and the electric displacement field lines are plotted. The field is excluded from the cloaked region, but emerges from the cloaking sphere undisturbed. The field lines are plotted closer together near the sphere to emphasize the screening effect. (From ref. [2], courtesy of Professor Sir John Pendry.)

the boundary conditions for a perfectly matched layer [24], which requires lossy $\hat{\varepsilon}_r$ and $\hat{\mu}_r$ such that $\sigma_{r'}/\varepsilon_0 = \sigma_{r'}^*/\mu_0$; $\sigma_{\theta'}/\varepsilon_0 = \sigma_{\theta'}^*/\mu_0$; $\sigma_{\phi'}/\varepsilon_0 = \sigma_{\phi'}^*/\mu_0$ [25], where σ and σ^* are the electrical and magnetic conductivities of the medium. These conductivities are related to the lossy parts of the dielectric constant and relative permeability for time-harmonic waves by $\sigma = \omega\varepsilon''\varepsilon_0$ and $\sigma^* = \omega\mu''\mu_0$. However, it is consistent to point out that there is no reflection from the outer surface of the cloak in the geometrical optics approximation because the ray does not enter the cloak and is therefore not reflected. On the other hand, if the refractive index distribution within the cloak is viewed as a dielectric light guide for waves being diverted around the cloaked region, then, since the ray directions are curved, there is also some light leakage from the cloak, as well as scattering from the material of the cloak.

The deviation of rays of light around the cloaked region, as shown in Fig. 10.12, also applies for any light source or object placed close to the cloak, as shown in Fig. 10.14.

10.7.2 Cylindrical cloak

The transformation that will cloak a cylindrical region $0 < r < b$ by compressing a cylindrical region into an annular shell with inner radius a and outer radius b is given by

$$r' = \left(\frac{b-a}{b}\right)r + a; \qquad \theta' = \theta; \qquad z' = z, \qquad (10.41)$$

where θ and z are the angular and axial coordinates in the original coordinate system, and θ' and z' are the angular and axial coordinates in the transformed system. The corresponding transformed $\hat{\varepsilon}$ and $\hat{\mu}$ in the original coordinate system are given by

$$\varepsilon_r = \mu_r = \frac{r-a}{r}; \quad \varepsilon_\theta = \mu_\theta = \frac{r}{r-a}; \quad \varepsilon_z = \mu_z = \left(\frac{b}{b-a}\right)^2 \frac{r-a}{r}. \qquad (10.42)$$

One interpretation of the action of the cloak is that it compresses the cloaked region to a line, but there is a problematic singularity at the cloak boundary where $r = a$.

10.7.3 Homogeneous isotropic cloak

As we have seen so far, the transformational optics approach to cloak design requires the use of anisotropic materials with spatially controlled dielectric and magnetic properties. However, Sun et al. [26] have described an approach to a cylindrical cloak that uses layers of isotropic material to fold light rays around a cloaked region. They use an inner diverging lens with an isotropic refractive index below unity with a Luneburg lens around the exterior. The Luneburg lens has its external focal point at infinity. This design requires no magnetic materials or anisotropy, only a radially dependent refractive index.

10.7.4 Cloaks of arbitrary shape

The validity of cloaking designs based on coordinate transformations has been confirmed by solving the ray trajectories in a geometrical optics model [2, 19] and confirmed in a full electromagnetic wave numerical simulation of a cylindrical cloak by Cummer et al. [27]. It is also possible from a theoretical standpoint to design a cloak for an arbitrarily shaped object by performing coordinate transformations that shrink the cloaked region to a point or a line. The theoretical validity of arbitrary cloak designs has been verified in simulations for square cloaks [20], elliptical cylindrical cloaks [28–30], conical cloaks [31], and arbitrarily shaped polygonal 2-D cloaks made up of connected triangular regions [32]. Qiu et al. [33] describe a spherical cloak made of concentric layers, where each layer is isotropic. Such cloaks do not perform as well as a continuously varying cloak, but are in principle easier to fabricate. In practice, of course, a spherical cloak can make an object

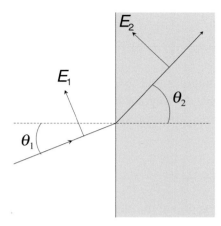

Figure 10.15. P-wave electric field vectors at a cloak boundary.

of any shape invisible provided the object fits inside the cloaked region $r < a$. The advantage of spherical cloaks for this purpose is that only the radial dielectric constant and relative permittivity need to vary, which, in principle, allows the fabrication of a metamaterial cloak of minimal complexity.

10.7.5 Cloak boundary conditions

At the boundary of the cloak, if all impedance components $Z_i = \sqrt{\mu_i \mu_0 / \varepsilon_i \varepsilon_0} = \sqrt{\mu_0 / \varepsilon_0}$, there is perfect impedance matching and no reflection. It is instructive to examine how refraction occurs for a light ray entering such a cloak, as shown in Fig. 10.15. This figure shows the locally plane boundary of the cloak at a point that would correspond to a ray entering above the axis in the meridional plane of either Fig. 10.10 or Fig. 10.14. The ray is shown refracting such that the angle of refraction is greater than the angle of incidence because the ray is being deviated around the cloaked region. There is no reflected ray because of perfect impedance matching. If both the region outside the cloak and the cloak itself are lossless, there is no mechanism for charge transport up to the cloak boundary, and for the P-wave (TM-wave) shown, the continuity of the normal component of $\mathbf{D}$ yields

$$E_1 \sin \theta_1 = \varepsilon_{rc}(\theta_2) E_2 \sin \theta_2 , \qquad (10.43)$$

where $\varepsilon_{rc}(\theta_2)$ is the effective dielectric constant of the cloak at angle θ_2, E_1 and E_2 are the magnitudes of the electric field vector in the two media, and the region outside the cloak has $\varepsilon_r = \mu_r = 1$. Continuity of the tangential components of the electric fields yields

$$E_1 \cos \theta_1 = E_2 \cos \theta_2 . \tag{10.44}$$

Combination of Eqs. (10.43) and (10.44) yields

$$\tan \theta_1 = \varepsilon_{rc} \tan \theta_2 = n_c^2(\theta_2) \tan \theta_2 . \tag{10.45}$$

In the transformed medium this is equivalent to every ray at any angle of incidence being at Brewster's angle. For a cloak that is very large in relation to the wavelength, so that the boundary can be viewed as locally plane, the phase velocities of the waves must match on both sides of the boundary so that the effective refractive index, n_{eff}, seen by the refracting component parallel to the boundary, satisfies Snell's law:

$$n_{eff} = \frac{\sin \theta_1}{\sin \theta_2} , \tag{10.46}$$

which yields

$$n_{eff}^2 = n_c^4(\theta_2) + [1 - n_c(\theta_2)]^2 \sin^2 \theta_1 . \tag{10.47}$$

It is important to note that there is no birefringence in these structures of anisotropic $\hat{\varepsilon}$ and $\hat{\mu}$. Because both the dielectric and magnetic properties are equally anisotropic, there is no angular deviation between the ray direction and the wavevector direction in these transformed structures [34]. A ray in the original space cannot become two rays in the transformed structure: there is no equivalent of an ordinary and an extraordinary ray.

10.7.6 Ray dynamics in cloaks

It is worth examining some general characteristics of the refractive index distribution inside the cloak before going into a detailed analysis of the actual required profiles and a discussion of how these could be fabricated in practice. In the geometric optics model of ray propagation, the path of a ray from a point P_1 to a second point P_2 follows Fermat's principle. In its simplest form, Fermat's principle says that a ray follows the path of least transit time from P_1 to P_2, which can be stated as follows [35]:

$$\delta \int_{P_1}^{P_2} \frac{n(\mathbf{r}, \omega)}{c_0} \, ds = 0 , \tag{10.48}$$

where $n(\mathbf{r}, \omega)$ is the refractive index at vector position $\mathbf{r}$ measured from an origin **O** along a path s, where the element of length along the ray trajectory is ds. In an

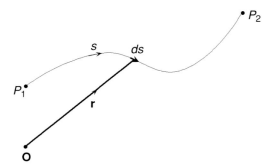

Figure 10.16. Schematic ray path through an inhomogeneous medium.

anisotropic medium the polarization state of the ray must be stated explicitly, since ordinary and extraordinary rays can take different paths through such a medium. The geometry in question is shown in Fig. 10.16. For the geometrical optics approach to be valid, the refractive index should not change much on spatial scales on the order of the wavelength. In addition, if the medium is not only inhomogeneous but also anisotropic, then the refractive index is related to a tensor dielectric constant and relative permeability.

In a general sense, the variation of the path integral in Eq. (10.48) is an extremum, although treating it as a minimum is satisfactory for most situations. It should be pointed out that if P_1 and P_2 are conjugate points in an imaging system, there is more than one path that satisfies Eq. (10.48). From Fermat's principle we can derive the equation of light rays as follows:

$$\frac{d}{ds}\left[n(\mathbf{r}, \omega)\frac{d\mathbf{r}}{ds}\right] = \frac{1}{2n}\mathrm{grad}(n^2) = \mathrm{grad}[n(\mathbf{r}, \omega)],\qquad(10.49)$$

where $d\mathbf{r}/ds$ is a unit vector in the direction of the trajectory and $|d\mathbf{r}/ds| = 1$. Along the trajectory of a ray, $ds = c\,dt = (c_0/n)dt$, where c_0 is the velocity of light in a vacuum. Making this substitution in Eq. (10.49) gives an alternative form of the equation of light rays:

$$\frac{n^2}{c_0}\frac{d}{dt}\left(\frac{n^2}{c_0}\frac{d\mathbf{r}}{dt}\right) = \frac{\mathrm{grad}(n^2)}{2}.\qquad(10.50)$$

10.7.7 The Hamiltonian optics of rays

A powerful approach to determining the trajectory of a light ray in an inhomogeneous medium is to use Hamilton's approach, in which the trajectory of the ray can be regarded as equivalent to the motion of a particle in a field of force [36–39]. It is worth noting that Eq. (10.48) is analogous to Maupertuis's principle of least action in particle dynamics, where the equation takes the form

$$\delta \int_{P_1}^{P_2} mv \, ds = \delta \int_{P_1}^{P_2} \sqrt{2m(E - U)} \, ds = 0 \,, \tag{10.51}$$

where m and v are the particle mass and velocity, respectively, and E and U are the total energy and potential energy, respectively. Using this mechanical analogy between particle motion and light rays, the light ray equivalent of mv is n/c_0, so clearly the effective mass in the light ray equation is unity. Equation (10.49) can be written for a particle as follows:

$$\frac{d}{ds} \left(v \frac{d\mathbf{r}}{ds} \right) = \text{grad } v \,. \tag{10.52}$$

Note that $d\mathbf{r}/ds$ is a unit vector tangential to the trajectory, so grad $v = dv/ds$. We introduce an effective "time" τ for the light ray defined from

$$\frac{c_0}{n^2} \, dt = d\tau \,. \tag{10.53}$$

With this definition, Eq. (10.50) becomes

$$\frac{d^2\mathbf{r}}{d\tau^2} = \frac{\text{grad } n^2}{2} \,, \tag{10.54}$$

which is equivalent to Newton's second law of motion, where the Newtonian "force" is given by

$$\mathbf{F} = \frac{\text{grad } n^2}{2} \,. \tag{10.55}$$

This force can be regarded as the negative gradient of an overall effective potential function,

$$U = -\frac{n^2}{2} \,. \tag{10.56}$$

Note that

$$n = \frac{c_0 \, dt}{n \, d\tau} = \left| \frac{d\mathbf{r}}{dt} \right| \frac{dt}{d\tau} = \left| \frac{d\mathbf{r}}{d\tau} \right| \,; \tag{10.57}$$

$d\mathbf{r}/d\tau$ is the equivalent of the velocity, in this view, of the light ray propagation as a light "particle" of mass $m = 1$. The "kinetic energy" of the light particle is

$$T = \frac{1}{2} \left| \frac{d\mathbf{r}}{d\tau} \right|^2$$

and its total energy is given by

$$E = T + U = \frac{1}{2}\left|\frac{d\mathbf{r}}{d\tau}\right|^2 - \frac{n^2}{2} = 0; \qquad (10.58)$$

the "momentum" of the light particle in this formalism is

$$p = \left|\frac{d\mathbf{r}}{d\tau}\right| = n. \qquad (10.59)$$

So, the mechanical analog of the motion of a light "particle" is that of a particle with zero total energy and zero frequency [37]. It is important to note that if the medium is inhomogeneous and anisotropic then the optical path now depends on the polarization state of the wave and a different approach is required. This is best accomplished through the use of the *eikonal* [35] and the use of a Hamiltonian approach, in which the components of the wavevector are associated with generalized momenta [40]. This approach will be described in detail later. Although the Hamiltonian is generally associated with the total energy of a particle, this is not always the case when coordinate transformations or mappings are used. The Hamiltonian of a light ray is written in many different ways, for example as follows [41]:

$$H = -\sqrt{n^2 - p^2}. \qquad (10.60)$$

So, in the mechanical analog (since $p = n$), $H = 0$, and the mechanical analog suggests $\omega = 0$. If the motion of the light ray is equivalent to the motion of a particle of zero total energy in a spherically symmetric refractive index profile, then, for a potentially curved path, the ray has linear and angular momentum, and we can write, in polar coordinates, the following:

$$\left(\frac{dr}{d\tau}\right)^2 + r^2\left(\frac{d\phi}{d\tau}\right)^2 - n^2(r) = 0, \qquad (10.61)$$

so the trajectory of the ray can be determined in much the same way that it would be determined for a particle in an orbit. If the refractive index is independent of position, the trajectory is a straight line, and Eq. (10.61) yields

$$\frac{dr}{d\tau} = n\frac{c_0}{n^2} = \frac{c_0}{n}, \qquad (10.62)$$

as expected.

Some care must be exercised in not carrying the zero energy particle model too far because the photon momentum in a material medium is still a matter of debate [42]. This involves the Minkowski–Abraham controversy as to whether the photon momentum is

$$\hbar k = \frac{nh}{\lambda_0} = nh\upsilon \text{ (Minkowski) or } m\upsilon = \frac{E}{c_0^2}\frac{c_0}{n} = \frac{h\upsilon}{nc_0} \text{ (Abraham).} \quad (10.63)$$

Although recent experimental evidence [43] appears to support the Abraham interpretation, Barnett [44] has indicated that there is no ambiguity. He asserts that both Minkowski and Abraham are correct: the Minkowski momentum is the canonical momentum for a photon viewed as a wave, while the Abraham momentum is the kinetic momentum where a photon is viewed as a particle.

10.7.8 Ray and wave paths in inhomogeneous and anisotropic materials

Most 3-D cloaking designs to date have required the use of anisotropic media in which the ray direction can depend on polarization state. Because the dielectric and magnetic properties of the material are tensors, we can no longer use a scalar refractive index that varies from point to point to describe the behavior of rays in a cloak. We concentrate here on the ray dynamics in the nonmagnetic TM cloak design described by Jacob and Narimanov [45]. A more general discussion has been given by Schurig *et al.* [19].

We assume that time-harmonic fields propagate with a spatial field variation of the form $U \sim \exp[-jk_0\psi(r)]$, where ψ is the eikonal and $k_0 = \omega/c$ is the propagation constant in a vacuum. For TM-wave propagation in a locally uniaxial nonmagnetic medium the eikonal obeys the following equation [40]:

$$\frac{1}{\varepsilon_{22}h_3^2}\left(\frac{\partial\psi_{TM}}{dq_3}\right)^2 - \frac{2}{\varepsilon_{23}h_2h_3}\left(\frac{\partial\psi_{TM}}{\partial q_2}\right)\left(\frac{\partial\psi_{TM}}{\partial q_3}\right)$$

$$+ \frac{1}{\varepsilon_{33}h_2^2}\left(\frac{\partial\psi_{TM}}{dq_2}\right)^2 = \mu_0\varepsilon_0, \quad (10.64)$$

where q_1, q_2, q_3 are generalized coordinates and h_1, h_2, h_3 are the associated scale factors. If cylindrical symmetry is assumed, with a dielectric constant tensor of the form

$$\hat{\varepsilon}_r = \begin{pmatrix} \varepsilon_r & 0 \\ 0 & \varepsilon_\theta \end{pmatrix}, \quad (10.65)$$

then, for a principal coordinate system oriented locally in the r, θ directions, Eq. (10.64) gives a Hamiltonian representation of the ray trajectory as

$$H = \sqrt{\frac{p_r^2}{\varepsilon_\theta} + \frac{p_\theta^2}{r^2\varepsilon_r}}, \quad (10.66)$$

or (written as a dispersion equation)

$$\frac{k_r^2}{\varepsilon_\theta} + \frac{k_\theta^2}{r^2 \varepsilon_r} = \frac{\omega^2}{c_0^2}, \tag{10.67}$$

where we have used the fact that the canonical momenta satisfy $p_i = \partial\psi/\partial q_i$. In this formulation, angular frequency can be regarded as the Hamiltonian, and the wavevector **k** as the canonical momentum. Hamilton's equations in this context can be written as follows:

$$\frac{d\mathbf{r}}{dt} = \frac{\partial\omega}{\partial\mathbf{k}}, \quad \frac{d\mathbf{k}}{dt} = -\frac{\partial\omega}{\partial\mathbf{r}}. \tag{10.68}$$

In a vacuum the Hamiltonian

$$H_{vacuum} = \sqrt{p_r^2 + \frac{p_\theta^2}{r^2}} \tag{10.69}$$

corresponds to photon motion in a straight line, where angular momentum is conserved. The angular momentum of a photon, $p_\theta = \rho\hbar k = L$, is conserved, where ρ is the impact parameter of the ray defined with respect to the origin. This corresponds to the following ray trajectory:

$$r \sin\theta = \rho. \tag{10.70}$$

As pointed out by Jacob *et al.* [45], one angular momentum-conserving trajectory that avoids a cylindrical region of radius a is

$$(r - a)\sin\theta = \rho. \tag{10.71}$$

They assert that a cloak Hamiltonian that accomplishes this is

$$H_{cloak} = \sqrt{p_r^2 + \frac{p_\theta^2}{(r - a)^2}}. \tag{10.72}$$

The simplest dielectric constants that reduce Eq. (10.66) to this form are

$$\varepsilon_r = C\left(\frac{r - a}{r}\right), \quad \varepsilon_\theta = C, \tag{10.73}$$

where C is a constant. It is important that these parameters be independent of angle so that the cloak works for rays incident from all directions in any circular cross section. Jacob *et al.* [45] show that the choice of the constant $C = b^2/(b - a)^2$ provides a finite size cloak with inner radius a and outer radius b. At the outer boundary of the cloak, rays refract into the birefringent medium and then take trajectories that conserve angular momentum and avoid the cloaked region. The ray trajectory inside the cloak is given by

$$r(\theta) = a + \frac{b \sin \theta}{\sqrt{C} \sin(\theta - \theta_0)},$$ (10.74)

where θ_0 is a constant that makes sure that angular momentum is conserved as the ray enters the cloak:

$$\theta_0 = \theta_1 - \arcsin \left(\frac{b \sin \theta_1}{b - a} \right).$$ (10.75)

The cloak parameters given by (Eq. 10.73) are exactly the same as those obtained by a transformational optics approach, which shows the equivalence between the transformed space and the materials interpretations of cloaking. Later in this chapter we will describe how this quasi-classical cloak may be emulated using tapered waveguides.

10.7.9 Nonmagnetic cloak for visible light

In the visible region of the spectrum, where the notion of invisibility has the most emotional impact and potentially greatest utility, the fabrication of metamaterials with spatially controlled permeability is challenging. If the cloak is limited to P-wave (TM-wave) illumination in a cylindrical or 2-D circular geometry, then the only material parameters that must satisfy Eq. (10.42), are μ_z, ε_r, and ε_θ. The ray paths inside the cloak are the same as predicted by Eq. (10.42) provided the values of $\mu_z \varepsilon_r$ and $\mu_z \varepsilon_\theta$ are maintained at the values required by Eq. (10.42). With these constraints, a reduced cloak specification is the one described above [46]:

$$\mu_z = 1, \quad \varepsilon_\theta = \left(\frac{b}{b - a} \right)^2, \quad \varepsilon_r = \left(\frac{b}{b - a} \right)^2 \left(\frac{r - a}{r} \right)^2.$$ (10.76)

The penalty incurred in using this reduced cloak design is that there is no longer perfect impedance matching at the boundary of the cloak. For waves incident normally on the cloak, the effective impedance at the boundary of the cloak is given by

$$Z = \sqrt{\frac{\mu_r \mu_0}{\varepsilon_\theta \varepsilon_0}} = \left(\frac{b}{b - a} \right) Z_0,$$ (10.77)

so there is significant reflection. For grazing incidence on the cloak, the reflection goes to zero. This cloak design does not require the incorporation of any optical magnetic response; however, it does require a radial dielectric constant that rises from zero at the boundary of the cloaked region to a value of unity at the boundary of the cloak, as shown in Fig. 10.17.

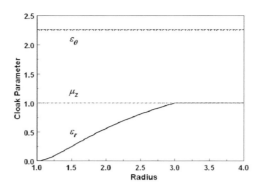

Figure 10.17. Cloak parametric variation for $a = 1$, $b = 3$.

Cai *et al.* [46] have carried out simulations of their nonmagnetic metamaterial cloak in which the radial variation in dielectric constant was accomplished by embedding nanowires of subwavelength diameter in the radial direction. If loss in the metal is neglected, then the dielectric constant of the metal sufficiently below the plasma frequency is negative, and, for the embedded wires inside a dielectric, the effective dielectric constant in the radial direction can be adjusted between zero and unity from the inside of the cloak to its outside. The azimuthal permittivity inside the cloak is essentially the same as the dielectric because of the minimal dielectric response of the nanowires in the direction normal to their length. The wires do not need to be periodically spaced, and can be random, but their average fill factor should not vary much over the scale of the wavelength inside the dielectric. The effective dielectric constant for a metal particle–dielectric composite is

$$\varepsilon_{\mathit{eff}} = \frac{1}{2\kappa} \left(\bar{\varepsilon} \pm \sqrt{\bar{\varepsilon}^2 + 4\kappa\varepsilon_m\varepsilon_d} \right) , \tag{10.78}$$

where ε_m and ε_d are the dielectric constants of the metal and dielectric, respectively, κ is a screening factor, and

$$\bar{\varepsilon} = [(\kappa + 1)f - 1]\varepsilon_m + [\kappa - (\kappa + 1)f]\varepsilon_d . \tag{10.79}$$

The idealized structure is shown in Fig. 10.18, and its cloaking performance is shown in Fig. 10.19. The cloaking is not perfect, as expected, but visibility of the cloaked cylinder is markedly reduced, as is its shadow.

In the last few years there has been considerable work on theoretical cloaking structures that have potential advantages over the spherical and cylindrical structures described so far. Jiang *et al.* [28] have described a cloak based on a coordinate transformation in elliptical–cylindrical coordinates. In their approach, the cloak still effectively hides a cylindrical region by compressing it, but not to a

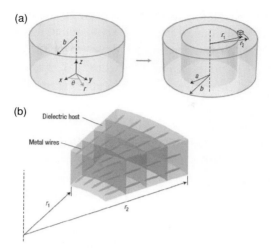

Figure 10.18. Coordinate transformation and structure of the nonmagnetic optical cloak. (a) The coordinate transformation that compresses a cylindrical region r, b into a concentric cylindrical shell a, r, b. There is no variation along the z direction; r_1 and r_2 define the internal and external radius of a fraction of the cylindrical cloak. (b) A small fraction of the cylindrical cloak. The wires are all perpendicular to the cylinder's inner and outer interfaces, but their spatial positions do not have to be periodic, and can be random. Also, for large cloaks, the wires can be broken into smaller pieces that are smaller in size than the wavelength.

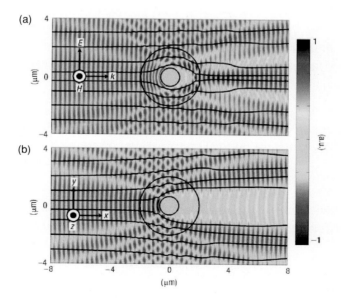

Figure 10.19. Finite element simulations of the invisibility cloak described by Cai *et al.* Cloak on (a); cloak off (b). The cloak is illuminated from the left with TM-waves at 632.8 nm. (From ref. [46], courtesy of Professor Vladimir Shalaev.) See color plates section.

point in planes perpendicular to the cylinder axis, but to a line, which avoids the singularity at the inside radius of the cloak. Unfortunately, this cloak design still requires spatial control of relative permeability as well as dielectric constant, which is challenging in the optical region of the spectrum. Cai *et al.* [47] have extended their work on nonmagnetic cloaks to include nonlinear coordinate transformations, which allow the impedance mismatch at the cloak boundary to be avoided. They consider a general transformation $r' = f(r)$, of which Eq. (10.41) represents a linear example. They are still compressing the region $r < b$ into a cylindrical shell for which $a < r' < b$. For any cylindrically symmetric transformation of this kind,

$$\varepsilon_{r'} = \mu_{r'} = \frac{r}{r'} \frac{\partial f(r)}{\partial r}, \quad \varepsilon_{\theta'} = \mu_{\theta'} = \frac{1}{\varepsilon_{r'}}, \quad \varepsilon_{z'} = \mu_{z'} = \frac{r}{r'} \left[\frac{\partial f(r)}{\partial r} \right]^{-1}, \quad (10.80)$$

which reduces to Eq. (10.42) for the linear transformation of Eq. (10.41). For an incident P-wave only, μ_z, ε_r, and ε_θ enter into Maxwell's equations. A reduced set of nonmagnetic cloak parameters can be obtained by multiplying ε_r and ε_θ by μ_z to give

$$\varepsilon_{r'} = \left(\frac{r}{r'}\right)^2, \quad \varepsilon_{\theta'} = \left[\frac{\partial f(r)}{\partial r}\right]^{-2}, \quad \mu_{z'} = 1. \quad (10.81)$$

The normalized impedance at the outer boundary of the cloak is given by

$$\zeta_{r'=b} = \frac{Z_{r'=b}}{Z_0} = \left(\sqrt{\frac{\mu_{z'}}{\varepsilon_{z'}}}\right)_{r'=b} = \frac{\partial f(r)}{\partial r}, \quad (10.82)$$

which by an appropriate choice of $f(r)$ can be set equal to unity to provide a perfectly matched nonmagnetic cloak. For example, with the nonlinear transformation

$$r' = f(r) = \left[1 - \frac{a}{b} + p(r - b)\right] r + a, \quad (10.83)$$

the boundary conditions $f(0) = a$ and $f(b) = b$ are satisfied. To achieve $\zeta_{r'=b} = 1$ requires $p = a/b^2$, and the optimal transformation becomes

$$r' = f(r) = \left[\frac{a}{b}\left(\frac{r}{b} - 2\right) + 1\right] r + a. \quad (10.84)$$

The nonmagnetic material properties can then be determined from Eq. (10.79). To keep the transformation monotonic requires the shape factor $a/b < 1/2$.

10.8 Conformal mapping for cloaking

A parallel approach to the general space transformations pioneered by John Pendry and his colleagues is the elegant conformal mapping procedure described by Leonhardt [42, 48]. Conformal mapping is an elegant and well established technique for

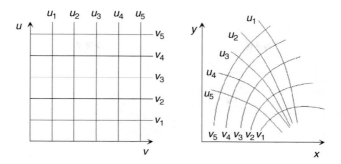

Figure 10.20. Mapping of coordinate lines in the W plane to the Z plane.

the solution of two-dimensional electrostatics and wave problems [14]. The fields must be independent of the third dimension. Conformal mapping can transform an electromagnetic wave problem with complicated boundaries into a mapping in the complex plane that makes the boundaries simpler. The original 2-D plane is the Z plane, where the coordinates of a point are represented by the complex number $Z = x + jy$. The conformal mapping is to the W plane, where the coordinates of points are represented by $W(Z) = u + jv$, where $W(Z)$ is an analytic function. In the W plane lines of constant u and v intersect at right angles and form an orthogonal grid. In the Z plane lines of constant u and v also create a grid, but the $u = const.$ and $v = const.$ lines are curved, as shown schematically in Fig. 10.20.

A Smith chart is a fine example of a conformal transformation between the Z (normalized complex impedance) plane and the W (normalized complex reflection) plane with $W(Z) = (Z - 1)/(Z + 1)$.

The Helmholtz equation for time-harmonic waves in two-dimensional (x, y) space is

$$\frac{\partial^2 \psi}{\partial x^2} + \frac{\partial^2 \psi}{\partial y^2} + k^2 \psi = 0,$$ (10.85)

where ψ is any field or potential associated with the wave. A transformation from the Z plane to the W plane transforms the Helmholtz equation to [14]

$$\frac{\partial^2 \psi}{\partial x^2} + \frac{\partial^2 \psi}{\partial y^2} + k^2 \psi = \left| \frac{dW}{dZ} \right|^2 \left(\frac{\partial^2 \psi}{\partial u^2} + \frac{\partial^2 \psi}{\partial v^2} \right) + k^2 \psi = 0,$$ (10.86)

so

$$\frac{\partial^2 \psi}{\partial u^2} + \frac{\partial^2 \psi}{\partial v^2} + k^2 \left| \frac{dZ}{dW} \right|^2 \psi = 0.$$ (10.87)

Note that $|dZ/dW|$ is a scale factor that relates an incremental length $|dW|$ in the W plane to the corresponding incremental length $|dZ|$ in the Z plane. In the

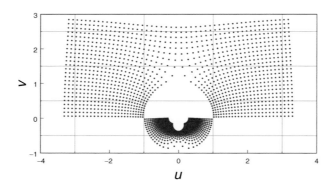

Figure 10.21. Mapping of $W(Z) = (1/2)\left(Z \pm \sqrt{Z^2 - 4}\right)$. There are two branch cuts at $Z = \pm 2$, where the mapping flips between the upper and lower half planes.

Z plane, $k \propto n$, so, in the transformation to the W plane, $n|dZ/dW| = n'$ can be regarded as a transformed refractive index profile.

Because of Fermat's principle, light rays do not know the difference between a long path in a low refractive index medium and a short path in a higher index medium. In a similar way, if light takes a straight line path in (u, v) space, then the corresponding paths in (x, y) space will in general be curved. The transformed refractive index profile causes curved lines in the Z plane to become straight lines in the W plane. A well known conformal transformation is

$$W(Z) = \frac{1}{2}\left(Z \pm \sqrt{Z^2 - 4}\right), \qquad Z = W + \frac{1}{W}, \tag{10.88}$$

which maps the upper half of the Z plane to the region in the upper half of the W plane that is outside the unit circle $|W| = 1$ and to the region of the lower half plane that is inside the unit circle $|W| = 1$, as shown in Fig. 10.21.

If the mapping includes the lower half of the Z plane, then the combined mapping is as shown in Fig. 10.21. Some of the geodesic curves for this mapping are shown. In the context of the equivalent refractive index,

$$n' = \left|\frac{dZ}{dW}\right| = \left|1 - \frac{1}{W^2}\right|, \tag{10.89}$$

this is equivalent to a refractive index distribution $n'(z) = |1 - 1/Z^2|$ that will exclude some rays from the region of unit radius, as shown in Fig. 10.21. Unfortunately, this is not a good cloaking device because the refractive index distribution is azimuthally nonuniform. In real space the refractive index in polar

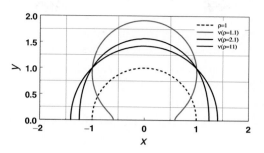

Figure 10.22. Refractive index variation corresponding to the mapping in Fig. 10.21. See color plates section.

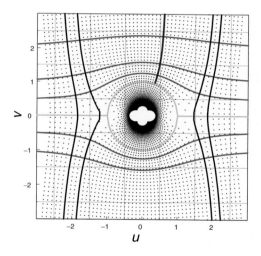

Figure 10.23. Geodesics of the mapping $W(Z) = (1/2)(Z \pm \sqrt{Z^2 - 4})$. The ray enters the cloak and is lost in the singularity at the origin.

coordinates is

$$n(r, \theta) = \sqrt{2}\sqrt{1 - \frac{\cos(2\theta)}{r^2}}, \qquad (10.90)$$

as shown in Fig. 10.22.

Rays of light that follow the geodesic curves in Fig. 10.23 approaching the "cloak" horizontally are diverted around the unit circle, but rays of light approaching the cloak vertically can enter the unit circle and be trapped by the singularity at $r = 0$, where the refractive index goes to infinity. This would be the fate of the ray shown in Fig. 10.23. Although a singularity of the refractive index of this kind could never occur in the real world, if it did, rays that reached the singularity would slow down to zero velocity and never re-emerge. This is analogous to the behavior of a black hole. A "cloak" designed in this way would cast a large shadow for rays approaching the cloak in the vertical direction shown in Fig. 10.23.

Leonhardt and his collaborators describe the behavior of rays in conformal maps in topological terms [3, 48]. Rays that enter the unit circle enter a new Riemannian surface that is physically separated from the original Riemannian surface, which can be identified as the real two-dimensional space where a spatially controlled refractive index diverts rays around a cloaked region. In their approach to cloak design they find conformal maps that consist of two or more Riemannian surfaces, where one surface corresponds to the exterior of the cloak. Light rays that enter the cloaked region perform closed orbits and then re-emerge onto the original surface as if they had never left it. The invisible part of the cloaked region is inaccessible to rays. To accomplish this, the potential under which light "particles" move in the cloaked region must correspond to a potential that provides closed orbits around one of the branch points, which in real space correspond to the points in w space that separate two Riemannian sheets, for example the upper and lower half planes in Fig. 10.21. Two cloak designs described by Leonhardt and his colleagues are related to Luneburg lenses [49] and Eaton lenses [50]. The Luneburg lens has a spherically symmetric refractive index of the form

$$n(r) = \sqrt{2 - \left(\frac{r}{a}\right)^2},\tag{10.91}$$

where a is the radius of the spherical lens. This lens is not immediately a cloaking device, although it does possess the attribute that it does not reflect light since its exterior index is unity, the same as surrounding free space. Parallel light rays entering a Luneburg lens are brought to a focus on the diametrically opposite surface, and then continue to propagate as diverging rays coming from that point. Inside the lens, ray paths are arcs of ellipses [51]. There is no cloaked region inside the lens.

The Eaton lens has a spherically symmetric refractive index profile given by

$$n(r) = \sqrt{\frac{2a}{r} - 1},\tag{10.92}$$

where a is the radius of this spherical lens. An Eaton lens works like a perfect "cats-eye" retroreflector. Parallel light rays entering the lens are translated laterally by the lens and returned parallel to their original paths [52]. The Luneburg lens posseses a refractive index singularity at $r = 0$, but its design can be modified to

(a) (b)

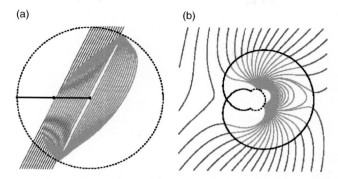

Figure 10.24. Light guiding using a harmonic oscillator profile (similar to a Luneburg lens) with the refractive index profile Eq. (10.91) in w space. The device guides light that has entered its interior layer back to the exterior, represented in (a) using two Riemann sheets that correspond to the two layers, seen from above. (b) Corresponding ray propagation in physical space with the optical conformal map Eq. (10.98) and $r_1 = 8r_0$. At the branch cut in (a), the thick black line between the two points, the branch points, light passes from the exterior to the interior sheet. Here light is refracted according to Snell's law. On the lower sheet, the refractive index profile Eq. (10.91) guides the rays to the exterior sheet in elliptic orbits with one branch point in the center. Finally, the rays are refracted back to their original directions and leave on the exterior sheet as if nothing had happened. The dotted circle in (a) indicates the maximal elongations of the ellipses. This circle limits the region in the interior of the device that light does not enter. The outside of the circle and the other Riemann sheets of the map correspond to the inside of the device in physical space, as shown in (a). Anything inside this area is invisible. (From ref. [3], courtesy of Professor Ulf Leonhardt.)

avoid this [53]. In the conformal mapping approach, the Luneburg lens has the transformed harmonic oscillator refractive index profile given by

$$n' = \sqrt{1 - \frac{|w - w_1|^2}{r_1^2}}, \tag{10.93}$$

which in w space corresponds to a central potential around the branch point w_1 of the form

$$U(|w - w_1|) = -\frac{n'^2}{2} = -\frac{1}{2} + \frac{|w - w_1|^2}{2r_1^2}. \tag{10.94}$$

Note that r_1 is a maximum radius in w space beyond which the refractive index becomes imaginary, so the region beyond this circle corresponds to the region of real space where an object can be hidden. The behavior of rays in this mapping is shown in Fig. 10.24.

For the profile based on an Eaton lens, the refractive index profile in w space is

$$n' = \sqrt{\frac{r_2}{|w - w_1|} - 1},\tag{10.95}$$

for which the central potential is

$$U(|w - w_1|) = -\frac{r_2}{2|w - w_1|} - \frac{1}{2},\tag{10.96}$$

which corresponds to a Kepler potential. For both the harmonic oscillator and Kepler potentials all orbits are closed, so they provide the necessary properties for closed orbits on the interior Riemannian sheet that then return to the exterior sheet.

Another interesting example is provided by the refractive index profile corresponding to a Maxwell fish eye lens, where

$$n' = \frac{n_0}{1 + (|w - w_1|/r_3)^2}.\tag{10.97}$$

In real space the ray trajectories in a Maxwell fish eye lens are circles. Figure 10.25 shows the ray trajectories in real space for this conformal invisibility device using the map

$$w = 4r_0 J\left(-\frac{\ln(432z/r_0)}{2j\pi}\right) - \frac{31r_0}{19},\tag{10.98}$$

where $J(z)$ is the Klein invariant, $r_3 = 4r_0$, and $n_0 = 2$. Note that rays traveling both horizontally and vertically avoid the cloaked region.

Leonhardt and his colleagues have shown that for these conformal cloaks, where rays entering the interior Riemannian sheet make closed orbits and then re-emerge on the exterior sheet, all of these rays experience the same time delay, so in principle do not suffer from the phase distortions experienced by rays passing through cloaks based on transformation optics. Leonhardt and his coworkers [3, 48, 54, 55] have described a more general approach to cloaking using non-Euclidean transformations, but a fundamental problem remains: how to fabricate the complex spatial distributions of ε_r and μ_r required to turn these concepts into reality. They point out a problem with cloaks based on transformation optics in that cloaking a finite region requires expanding a point to a circle (in a simple 2-D case). In real space, light takes zero time to pass a point, but when the point is expanded to a circle the light in the transformed space must cover a finite path in zero time. This requires the light to travel around the inner lining of the cloak at infinite phase velocity. In the materials interpretation of the transformation described by Schurig *et al.* [19], it is not clear that this is a practical issue for geometrical optics, since a specific path through the cloak could be mimicked with a series of mirrors. There

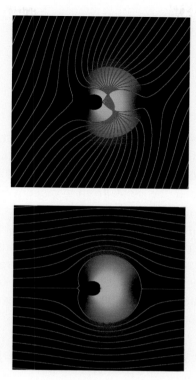

Figure 10.25. Ray paths in an invisibility device. (From refs. [3] and [48], courtesy of Professor Ulf Leonhardt.) See color plates section.

will be a path-dependent delay for a light pulse passing through the cloak, but no infinite group velocity is required.

It is important to note, however, that these cloaks based on 2-D conformal mapping cannot cloak a 3-D region as can the transformation cloaks described by Pendry and his colleagues, although they provide important guidance towards the design of all dielectric cloaks without loss. For example, in the approach described by Jacob and Narimanov [45], the ray trajectories in the cloak are generalized to an anisotropic cloak where $n(r)$ is not a simple scalar function of position but is a local 2-D uniaxial tensor of the form

$$\varepsilon_r = \begin{pmatrix} \varepsilon_r & 0 \\ 0 & \varepsilon_\theta \end{pmatrix}. \tag{10.99}$$

In this medium the dispersion relation is given by [56]

$$\frac{k_r^2}{\varepsilon_\theta} + \frac{k_\theta^2}{r^2 \varepsilon_r} = \frac{\omega^2}{c_0^2}. \tag{10.100}$$

If the light trajectory is described in polar coordinates, then the Hamiltonian is

$$H = \omega = c_0 \sqrt{\frac{k_r^2}{\varepsilon_\theta} + \frac{k_\theta^2}{r^2 \varepsilon_r}}. \tag{10.101}$$

10.9 Ray dynamics entering a dielectric cloak

As shown by Jacob and Narimanov [45], the modeling of a cloak for light particles where angular momentum is conserved leads to ray trajectories that avoid the cloaked region. For a nonmagnetic cylindrical cloak of finite radius b and TM geometry, a ray must enter the cloak from a vacuum before it can execute its trajectory around the cloaked region. Because the cloak is birefringent in this case, in cylindrical symmetry the ray in the cloak propagates as an extraordinary ray. Consequently, the ray angle after a ray enters the cloak must be calculated from the ray angle relative to the optical axis for a wave that refracts according to Snell's law for the extraordinary wave vector. At the boundary of the cloak the angle of incidence is θ_1. The wave refracts according to Snell's law at an angle of refraction θ_2, where θ_2 is the solution of

$$\theta_2 = \arctan\left(\frac{n_e(b)\sin\theta_1}{\sqrt{n_o^2 n_e^2(b) - n_o^2 \sin^2\theta_1}}\right), \tag{10.102}$$

and we have written $n_o = \sqrt{\varepsilon_\theta}$ and $n_e(b) = \sqrt{\varepsilon_r(b)}$, where ε_θ, $\varepsilon_r(b)$ are the tangential and radial components of the local diagonal dielectric tensor at the cloak boundary $r = b$; θ_2 is the angle the wavevector makes with the optic axis. The ray angle θ_s with respect to the optic axis (radial direction) then satisfies the following equation [57]:

$$\theta_s = \arctan\left(\frac{\varepsilon_\theta}{\varepsilon_r(b)}\tan\theta_2\right). \tag{10.103}$$

It is easy to show that the ray bends away from the normal on entering the cloak, although the wavevector direction does not. Because the nonmagnetic cloak described by Jacob *et al.* is locally negative uniaxial, the ray bends further away from the optic axis than the wavevector.

10.10 Practical cloaking experiments

10.10.1 Microwave cloak

The idealized cloaks described previously require 3-D spatial control of both the dielectric and magnetic properties of the cloaking medium. In particular, the refractive index tensor of the cloaking medium must take values below unity, but not necessarily negative values. It is difficult to fabricate cloaking structures with the

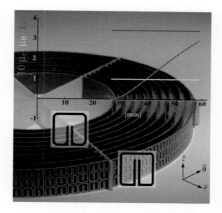

Figure 10.26. Two-dimensional microwave cloaking structure (background image) with a plot of the material parameters that are implemented. Note that: μ_r is multiplied by a factor of ten for clarity; μ_θ has the constant value 1; ε_z has the constant value 3.423. The SRRs of cylinder 1 (inner) and cylinder 10 (outer) are shown in expanded schematic form (transparent square insets). (From ref. [58], courtesy of Professor David Smith.) See color plates section.

full spatial variations given, for example, by Eqs. (10.36)–(10.38) and (10.42), but for specific geometries there are simplified structures. In a cylindrical structure with the incident wave polarized along the cylinder axis (an S wave (TE-wave)), the only tensor components that enter are ε_z, μ_r, and μ_θ, so a simplified cloaking structure has

$$\varepsilon_z = \left(\frac{b}{b-a}\right)^2, \quad \mu_r = \left(\frac{r-a}{r}\right)^2, \quad \text{and} \quad \mu_\theta = 1. \quad (10.104)$$

This structure will divert rays around the cloaked cylinder, but with a penalty: there is no longer perfect impedance matching at the boundary of the cloak so there is some reflection of incident waves. Schurig *et al.* [58] have built a microwave metamaterial structure incorporating split ring resonators (SRRs) arranged in concentric rings around a central region to mimic the average properties described by Eq. (10.104). The split ring resonators provide *LC* elements that provide an effective negative index, as in a backward wave structure, to compensate for the positive refractive index between these structures to provide the radially dependent variation of μ_r and the required value of ε_z. Figure 10.26 shows a picture of the cloaking structure developed by Schurig *et al.* [58]. The structure consists of ten concentric rings of SRRs, with the SRRs arranged three high with alternating alignment along the z axis. The structure operates at 8.5 GHz, where the free space wavelength is 35 mm, which is significantly larger than the discrete nature of the fabricated cloak, so effective values of the cloak parameters result from averaging over the electromagnetic properties of the rings and the air space in between. The

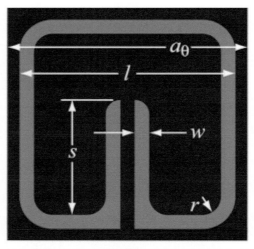

cyl.	r	s	μ_r
1	0.260	1.654	0.003
2	0.254	1.677	0.023
3	0.245	1.718	0.052
4	0.230	1.771	0.085
5	0.208	1.825	0.120
6	0.190	1.886	0.154
7	0.173	1.951	0.188
8	0.148	2.027	0.220
9	0.129	2.110	0.250
10	0.116	2.199	0.279

Figure 10.27. Spilt ring resonator design of the microwave cloaking structure. The in-plane lattice parameters are $a_\theta = a_z = 10/3$ mm. The ring is square, with edge length $l = 3$ mm and tracewidth $w = 0.2$ mm. The substrate is 381-μm-thick Duroid 5870 ($\varepsilon = 2.33$, $t_d = 0.0012$ at 10 GHz, where t_d is the loss tangent). The Cu film, from which the SRRs are patterned, is 17 μm thick. The parameters r and s are given in the table together with the associated value of μ_r. The extractions gave roughly constant values for the imaginary parts of the material parameters, yielding 0.002 and 0.006 for the imaginary part of ε_z and μ_r, respectively. The inner cylinder (cyl.) is 1 and the outer cylinder is 10. (From ref. [58], courtesy of Professor David Smith.)

actual design of the SRRs is shown in Fig. 10.27. This microwave cloaking structure was tested in a parallel plane waveguide and was shown to reduce scattering, although not perfectly, and the losses in the structure caused the cloak to cast a shadow. However, the general theoretical predictions about the behavior of the cloak were verified experimentally. Figure 10.28 shows the flow of electromagnetic fields around the cloak.

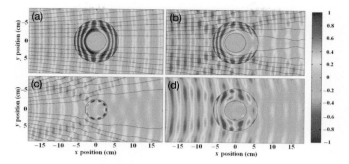

Figure 10.28. Snapshots of time-dependent, steady-state electric field patterns, with stream lines (black lines in (a)–(c)) indicating the direction of power flow (i.e. the Poynting vector). The cloak lies in the annular region between the black circles and surrounds a conducting Cu cylinder at the inner radius. The fields shown are (a) the simulation of the cloak with the exact material properties, (b) the simulation of the cloak with the reduced material properties, (c) the experimental measurement of the bare conducting cylinder, and (d) the experimental measurement of the cloaked conducting cylinder. Animations of the simulations and the measurements show details of the field propagation characteristics within the cloak that cannot be inferred from these static frames. The right-hand scale indicates the instantaneous value of the field. (From ref. [58], courtesy of Professor David Smith.) See color plates section.

10.10.2 Visible light cloak

An electromagnetic cloak based on coordinate transformations cannot be easily implemented in the visible frequency range. The main reason is the need to vary the magnetic permeability of the metamaterials in a nontrivial way, which is difficult to implement at optical frequencies.

In the nonmagnetic optical cloak, which has been suggested by Cai *et al.* [46, 47], the abovementioned difficulties have been alleviated for a specific polarization state of the illuminating light. In this approach the electromagnetic field may be treated as a scalar field, and the only metamaterial parameters it is necessary to control are the radial and tangential components of the dielectric permittivity as given in Eq. (10.76). Experimental realization of the radial component of the dielectric permittivity distribution approximately described by Eq. (10.76) in the frequency range around 500 nm is possible in a two-dimensional geometry with surface plasmon polaritons. The two-dimensional optics of surface plasmon polaritons offers viable ways to demonstrate various theoretical designs of electromagnetic cloaks that have been suggested in the literature [2, 3, 46, 47, 59–61]. Even though electromagnetic cloaking cannot be perfect, and may only be achieved in a narrow frequency range, there may be practical benefits of reduced visibility in various

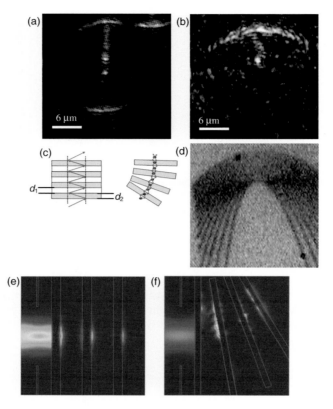

Figure 10.29. (a) Plasmon ray propagation in a "magnifying superlens" concentric ring structure from ref. [65]. (b) Bending of the plasmon ray by the slanted array of PMMA stripes in a parabolic lens structure shown in (d). (c) Ray optics simulation of beam bending by a stack of slanted negative index layers. The refractive index of gray stripes is assumed to be $n_2 = -1$. (e), (f) Numerical simulations of the same effect performed using COMSOL Multiphysics 3.3a.

applications related to surface plasmon polaritons, such as sensing, which explains high current research interest in this topic.

Our approach is based on the plasmonic metamaterials described in detail by Smolyaninov and co-workers [62–65], which are ideally suited for experimental realization of the scalar electromagnetic cloak described in refs. [46] and [47], since the surface plasmon polariton (SPP) field has only one polarization state [66].

The plasmonic metamaterials used are based on layers of polymethylmethacrylate (PMMA) deposited on a gold film surface. In the 500 nm frequency region, PMMA exhibits effective negative refraction as perceived by surface plasmons (the group velocity is opposite to the phase velocity Fig. 10.29(a)) [62]. We have fabricated various surface patterns consisting of stripes of PMMA separated by

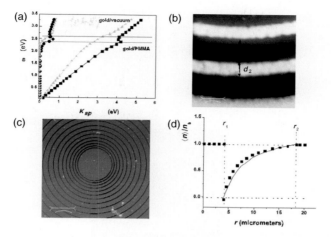

Figure 10.30. (a) Real and imaginary parts of the wavevector of the symmetric SPP mode propagating along a 50-nm-thick gold film in the PMMA/gold/glass and vacuum/gold/glass geometries as a function of frequency. In the frequency range marked by the box, PMMA areas have effective negative refractive index as perceived by SPPs, while the gold/vacuum interface looks like a medium with positive refractive index. The antisymmetric plasmon mode exhibits very high propagation losses and is not shown. (b), (c) AFM images of the central area of the 2-D cloak at different magnifications. (d) Distribution of n_{av} in the fabricated 2-D model reduced visibility device compared to the theoretical distribution given by Eq. (10.76).

uncoated regions containing gold/air interfaces. The local orientation of PMMA stripes may be either parallel, as shown in Figs. 10.29(c) and (e), or slanted, as shown in Figs. 10.29(d) and (f). The width of the PMMA stripes d_2, the width d_1 of the gold/air portions of the interface, and the relative angle of the stripes may be chosen freely. Figures 10.29(a) and (b) demonstrate that the developed metamaterial allows a high degree of control of SPP propagation: narrow plasmon rays may be formed as a result of focusing and repeated self-imaging of the focal spot by the negative index stripes, as reported in ref. [62].

The successive stripes of effective positive and negative refractive index may continuously redirect the plasmon ray propagation along some curvilinear path, as shown in Figs. 10.29(b), (c), and (f). The ability to bend the light path around some given area constitutes a necessary condition for a successful cloaking experiment.

The internal structure of the 2-D plasmonic model reduced visibility structure [65] is shown in Figs. 10.30(b) and (c) and Fig. 10.31(a). It consists of concentric PMMA rings deposited on a gold film surface, in which the width of the PMMA rings d_2 and the width d_1 of the gold/vacuum portions of the interface were varied during the e-beam fabrication process. Since the gold/air portions of the interface

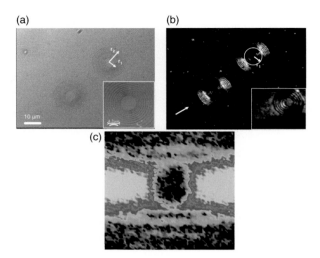

Figure 10.31. (a) Two plasmonic reduced visibility devices are observed using an optical microscope with white light illumination. The inset shows an AFM image of the central area of the device. (b) Optical image of surface plasmon polariton propagation through these structures at 532 nm. The area inside the circle of radius r_1 is cloaked, except for a very small fraction of plasmon rays, which propagate exactly through the center of the cloak. The illumination direction is shown by the arrow. The inset demonstrates plasmon scattering by a typical concentric ring structure, which is not optimized for cloaking at 532 nm. The plasmons are strongly scattered by the edge of the structure and by the circular area in the middle. (c) Measured plasmon field scattering around the central area of the device. The flow of energy around the "cloaked" region is visualized. (From ref. [65].)

have an effective group index $n_1 > 0$, by changing $(d_1 + d_2)$ and the d_1/d_2 ratio the average group refractive index of the multilayer material may be continuously varied locally from large effective negative to large positive values [62]. The distribution of average group refractive index in the fabricated 2-D device is shown in Fig. 10.30(d). It is reasonably close to the theoretical distribution given by Eq. (10.76). Numerical simulations performed using COMSOL Multiphysics 3.3a validate this approach (Fig. 10.32). The effective refractive index of the variable-diameter negative index ring structure is made reasonably close to the distribution given by Eq. (10.76), which produces satisfactory reduced visibility performance. The tangential component of the dielectric constant in this design also follows Eq. (10.76).

In our experiments the plasmonic metamaterial device was illuminated by an external laser operating at 532 nm, at an illumination angle that provides phase-matched excitation of SPPs at the left and right outer rims of the structure. Since the periodicity of the structure changes away from the outer rim, SPPs are not excited anywhere else, and the picture of light scattering presented in Figs. 10.31(b) and (c)

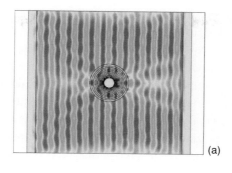

(a)

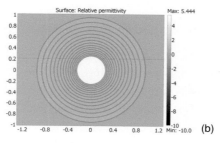

(b)

Figure 10.32. (a) Numerical validation of the 2-D cloaking design based on the variable-diameter negative index rings performed using COMSOL Multiphysics 3.3a. Distribution of the magnetic field is shown upon the illumination of the cloaking structure from the left. (b) Refractive index distribution within the 2-D cloaking structure shown in (a). (From ref. [65].)

corresponds to SPP propagation inside the structure. Figure 10.31(b) demonstrates that most of the plasmon energy cannot penetrate beyond the internal radius r_1 of the device, which corresponds to the $n_1 d_1 = -n_2 d_2$ (or $\varepsilon_r = 0$) boundary. However, a very small fraction of the plasmon rays, which propagate exactly through the center of the device, do reach the "cloaked" circular PMMA region in the middle of the structure. This practical difficulty has been mentioned in ref. [2] as an unavoidable singularity, since these rays do not know whether to deviate left or right. On the other hand, compared to the circular PMMA region shown in the inset, which is surrounded by a periodic concentric ring structure, the amount of scattered energy has been reduced considerably. Figure 10.31(c) demonstrates the flow of plasmon energy around the "cloaked" region. Visualization is achieved due to weak scattering of plasmons into photons by the edges of the PMMA rings, which is observed using a regular optical microscope. Since scattering efficiency depends on the orientation of these edges, the optical signal appears to be considerably weaker in the top and bottom of the image (the image in Fig. 10.31(c) was obtained by considerable overexposure, compared to the image in Fig. 10.31(b)).

Thus, some basic properties of the electromagnetic cloak, i.e. (i) considerable isolation of the cloaked region and (ii) the flow of energy around the cloak boundary, appear to match theoretical predictions. However, this structure may only be considered as an approximation to an ideal cloak, since surface waves experience losses and considerable scattering into "normal" 3-D photons. The effect of scattering is similar to the effect of losses in electromagnetic metamaterials. In general, losses lead to nonideal cloaking. Since some energy of the electromagnetic wave is lost while the wave is traveling around the cloaked area, the cloak appears as somewhat "gray" (instead of "clear") when observed in transmission.

10.11 "Cloaking" by scattering compensation (plasmonic cloaks)

Another approach to a type of cloaking is based on scattering reduction by appropriately chosen plasmonic nanoparticles [60, 61, 67–69]. In principle, a careful design of a "plasmonic cover" allows one to reduce the scattering from an isolated conducting, plasmonic or insulating sphere through scattering cancellation. This cloaking phenomenon is based on the fact that the multipolar radiation from a given object may be canceled term by term (for the significant scattering orders) with a judicious design of the plasmonic cover. This results from the local negative polarizability of a plasmonic layer. The relevant mathematics can be found in refs. [60], [61], [67], [68], and the references therein. In one example, the scattering reduction was carried out for the electric and magnetic dipole radiation from an impenetrable sphere. This implies that, even for an observer located very close to an object, its presence becomes hardly detectable after the cover is used. Theoretically, there exists a possibility of cloaking multiple wavelength-sized objects placed in close proximity of each other, as shown in Fig. 10.33, or even joined together to form a single object of large electrical size. The total scattering cross section of the object has been shown theoretically to be lowered by more than 99% with respect to the uncovered case. The impinging wave, both in the case of plane wave excitation and more complex forms of excitation, was shown to be "re-routed" through the plasmonic cover without any substantial reflection or perturbation of its wavefronts. This cloaking phenomenon, being "nonresonant," is considerably robust to frequency variations near the design frequency, to changes in the shape of the cloaked object, and/or to variations of geometrical and electromagnetic design parameters. By covering a small object with layers of plasmonic shells, the scattering reduction can be accomplished over a broad frequency band [68]. An example of the scattering reduction from multilayer objects is given in Figs. 10.34 and 10.35. Baumeier *et al.* [70] have shown that in two dimensions two concentric rings of point scatterers on a metal surface can significantly reduce the scattering of surface plasmon polaritons (2-D waves) by an object "cloaked" within this ring

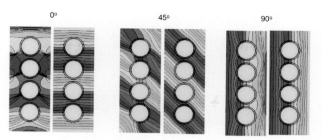

Figure 10.33. Phase of the total magnetic field distribution in the E plane for the case of four aligned spheres with and without plasmonic covers is shown for three different incidence angles. (From ref. [61], courtesy of Professor Nader Engheta.) See color plates section.

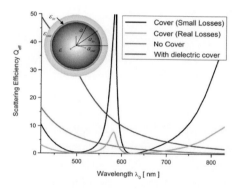

Figure 10.34. Total scattering efficiencies for the geometry depicted in the inset, i.e. a dielectric particle of radius $a = 100$ nm and permittivity $\varepsilon_r = 3$, cloaked by a two-layered shell designed for cloaking at $\lambda_0 = 500$ nm and $\lambda_0 = 625$ nm. The four curves refer to the following: a covered particle with small losses in the plasmonic materials (solid line), covered particle with reasonable losses (dashed line), the original particle (dotted line), and the same particle with a dielectric material replacing the shell region (dash-dot). (From ref. [68], courtesy of Professor Nader Engheta.) See color plates section.

structure. Alú [71] has shown that similar scattering cancellation from small objects can be accomplished by patterning the surface of an object on a sub-wavelength scale. There has been experimental verification that a reduction in scattering cross section of approximately 50% can be obtained at microwave frequencies by utilizing the local negative polarizability of metamaterials [72]. Unfortunately, although scattering reduction does reduce the detectability of small objects whose size is on the order of the wavelength, it cannot realistically be applied to larger objects, as is the case in principle at least to ray-optics cloaks where Poynting vectors are folded around a cloaked object and then continue undisturbed. Wavelength-sized

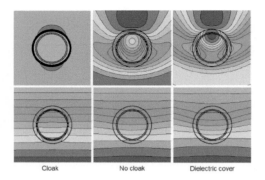

Figure 10.35. Distribution of the total electric field orthogonal to the H plane in the three cases corresponding to a covered particle with losses, the original particle, and a particle covered by the dielectric of Fig. 10.34 for $\lambda_0 = 625$ nm. Brighter regions correspond to higher values of the field. (From ref. [68].)

objects are generally visible as a result of their scattering and diffraction patterns, so, unlike large-sized objects, do not cast a geometric shadow, whose removal is a requirement of a macroscopic cloak.

In related work, Milton and his colleagues have analyzed a cloaking concept called exterior cloaking, in which the cloaked region is outside the cloaking device [73–75]. The principle is similar to that used in sound cancelation. The fields from the cloaking device(s) cancel out the fields scattered by the cloaked object(s). This active approach to cloaking requires a controlled phase relationship between the cloaking devices and the illumination fields, which is reasonable in the microwave region, but which is problematic in the visible region of the spectrum because of the short coherence time of optical sources.

10.12 Carpet cloaks

A cloak design that conceals a perturbation on a flat conducting plane, under which an object can be hidden, appears to be much simpler than a "stand alone" cloak [76]. The metamaterial parameters required for its experimental realization do not have any singularities, as shown in Fig. 10.36.

As a result, such a cloak was realized almost immediately after its theoretical conception [77]. The internal structure of the cloak and the comparison between theoretical and measured field distributions are shown in Fig. 10.37. To match the complex spatial distribution of the required constitutive parameters, Liu *et al.* constructed a metamaterial consisting of thousands of elements (Fig. 10.37). The geometry of each element was determined by an automated design process. The

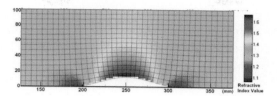

Figure 10.36. Metamaterial refractive index distribution in the ground plane cloak. The mesh lines indicate the quasi-conformal mapping. (From ref. [77], courtesy of Professor David Smith.) See color plates section.

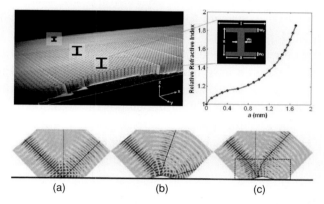

(a)   (b)   (c)

Figure 10.37. Design of the nonresonant elements of the ground plane cloak and the relation between the unit cell geometry and the effective refractive index. The dimensions of the metamaterial unit cells are $l = 2$ mm, $w_1 = 0.3$ mm, $w_2 = 0.2$ mm, and a varying from 0 to 1.7 mm. The plots show measured field mapping (E field) of the ground, perturbation, and ground plane cloaked perturbation. The rays display the wave propagation direction, and the dashed line indicates the normal of the ground in the case of free space and that of the ground plane cloak in the case of the transformed space. (a) Collimated beam incident on the ground plane at 14 GHz. (b) Collimated beam incident on the perturbation at 14 GHz (control). (c) Collimated beam incident on the ground plane cloaked perturbation at 14 GHz.

ground plane cloak has been realized with the use of nonresonant metamaterial elements, resulting in a structure having a broad operational bandwidth (covering the range of 13–16 GHz) and exhibiting extremely low loss. This approach indicates that the carpet cloak is easily scalable for any wavelength range of electromagnetic radiation. Unfortunately, its applications are limited to situations in which a ground plane is indeed present near the object to be hidden. Xu *et al.* [78] have proposed a carpet cloak design using only anisotropic all-dielectric materials. Using silicon grating structures they have achieved relatively broadband cloaking in the near infrared (1372–2000 nm). Kallos *et al.* [79] have proposed a simple ground plane cloak based on relatively simple surface structures made from isotropic all-dielectric materials. While it is often the case in RF applications that the Earth's surface may behave as such a ground plane, carpet cloaks have limited applications in the infrared and visible spectral ranges.

10.13 Metamaterial emulation using tapered waveguides

While current interest in electromagnetic metamaterials has been motivated by a solid body of theoretical work on cloaking and transformation optics [2, 3, 46, 47, 59–61], it appears to be difficult to develop metamaterials with low-loss, broadband performance. The difficulties are especially severe in the visible frequency range, where good magnetic performance is limited. While interesting metamaterial devices have been suggested based on nonmagnetic designs [45–47], the development of anisotropic magnetic metamaterials for the visible range would be highly desirable. Other limitations of "traditional" metamaterials in any portion of the electromagnetic spectrum are high losses and narrowband performance. These limitations may once again be illustrated using cloaking theoretical effort, as there exist only a few experimental demonstrations performed in rather narrow frequency ranges. The first experimental realization of an electromagnetic cloak in the microwave frequency range was reported in a two-dimensional cylindrical waveguide geometry [58]. In addition, a plasmonic metamaterial structure exhibiting reduced visibility at 500 nm has been demonstrated in ref. [65]. In both experimental demonstrations, the dimensions of the "cloaked" area were comparable with the wavelength of the incident electromagnetic radiation, meaning that the shadow produced by an uncloaked object of the same size would not be pronounced in these cases anyway.

On the other hand, it appears that many metamaterial devices requiring anisotropic dielectric permittivity and magnetic permeability could be emulated by specially designed tapered waveguides. This approach leads to low-loss, broadband performance in the visible frequency range, which is difficult to achieve by other means. This technique has been recently applied to electromagnetic cloaking. Broadband, two-dimensional, electromagnetic cloaking in the visible frequency range on a scale roughly 100 times larger than that of the incident wavelength has been demonstrated.

As a starting point, let us show that the transformation optics approach allows us to map a planar region of space (a waveguide) filled with inhomogeneous, anisotropic metamaterial into an equivalent region of empty space with curvilinear boundaries. This mapped region could be a planar wedge waveguide or a circular waveguide, for example. We begin by considering the following formal (material) notation for Maxwell's curl equations [17, 80]:

$$\text{curl}_q \hat{\mathbf{E}} = -j\omega\hat{\mu}_r\mu_0\hat{\mathbf{H}}, \quad \text{curl}_q \hat{\mathbf{H}} = j\omega\hat{\varepsilon}_r\varepsilon_0\hat{\mathbf{E}}, \tag{10.105}$$

for vector fields $\hat{\mathbf{E}} = \sum \hat{e}_i\hat{\mathbf{x}}_i$ and $\hat{\mathbf{H}} = \sum \hat{h}_i\hat{\mathbf{x}}_i$ in an orthogonal curvilinear system with the unit vectors $\hat{\mathbf{x}}_i$. The components of the actual physical fields, $\mathbf{E} = \sum e_i\hat{\mathbf{x}}_i$ and $\mathbf{H} = \sum h_i\hat{\mathbf{x}}_i$, are connected with the vectors in the material coordinates, $\hat{\mathbf{E}}$ and $\hat{\mathbf{H}}$, through $e_i = \hat{e}_i/g_i$ and $h_i = \hat{h}_i/g_i$, using the metric coefficients g_1, g_2, and g_3. The tensors $\hat{\varepsilon}_r$ and $\hat{\mu}_r$ are given by $\mathbf{t}\varepsilon_r$ and $\mathbf{t}\mu_r$, with

$$\mathbf{t} = g_1 g_2 g_3 \begin{pmatrix} g_1^{-2} & 0 & 0 \\ 0 & g_2^{-2} & 0 \\ 0 & 0 & g_3^{-2} \end{pmatrix}. \tag{10.106}$$

This description formally links the curvilinear components e_i and h_i in the domain characterized by anisotropic material diagonal tensors ε_r and μ_r to the components $\hat{e}_i$ and $\hat{h}_i$ in a formal, Cartesian domain characterized by nonuniform, anisotropic $\hat{\varepsilon}_r$ and $\hat{\mu}_r$. However, a Cartesian domain is not ideal for comparing confined, rotationally symmetric material systems encapsulated within axisymmetric coordinate surfaces. For the comparison of axisymmetric cloaking and imaging systems, it is more practical to match a given axisymmetric material domain to an equivalent inhomogeneous axisymmetric material distribution between two planes in a classical circular cylinder coordinate system, as shown in Fig. 10.38(a).

The transformation optics approach allows us to achieve this in a similar manner. The metric coefficients are $g_\rho = g_z$, $g_\phi = \rho$. Therefore, $\rho e_\phi = \hat{e}_\phi$, $\rho h_\phi = \hat{h}_\phi^2$, and $e_i = \hat{e}_i$, $h_i = \hat{h}_i$ for $i = \rho, z$. Thus, any rotational coordinate system $(\hat{\rho}, \hat{\phi}, \hat{z})$ converted into the cylindrical format

$$\text{curl}_q \mathbf{H} = -j\omega \begin{pmatrix} \hat{\rho} & 0 & 0 \\ 0 & \hat{\rho}^{-1} & 0 \\ 0 & 0 & \hat{\rho} \end{pmatrix} \varepsilon_0\hat{\varepsilon}_r\hat{\mathbf{E}}, \tag{10.107a}$$

$$\text{curl}_q \mathbf{E} = j\omega \begin{pmatrix} \hat{\rho} & 0 & 0 \\ 0 & \hat{\rho}^{-1} & 0 \\ 0 & 0 & \hat{\rho} \end{pmatrix} \mu_0\hat{\mu}_r\hat{\mathbf{H}}, \tag{10.107b}$$

is equivalent to a classical circular cylinder coordinate system with material tensors $\hat{\varepsilon}_r$ and $\hat{\mu}_r$.

Let us now consider an interesting application of the above formalism, which will lead us to another experimental demonstration of electromagnetic cloaking. We map an axisymmetric space between two spherical surfaces onto a space between

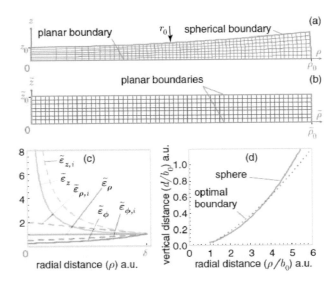

Figure 10.38. Space between a spherical surface and a planar surface (a) mapped onto a layer with planar boundaries (b). (c) Distribution of the radial, azimuthal, and axial (or vertical) diagonal components of permittivity in the equivalent planar region (solid lines). Dashed lines show the same components in the equivalent planar region obtained for a tapered waveguide with the radius-dependent refractive index for an ideal cloak. (d) Normalized profile of the optimal waveguide shape plotted for a cloak radius of $b_0 = 172$ μm. The shape of the optimal waveguide may be approximated by a spherical surface placed on top of a flat surface, as shown by the dotted line. (From ref. [81].)

two planes. The parametric description,

$$s^2 = (z - z_0)^2 + \rho^2, \quad \rho = \frac{s\sqrt{(2s - \hat{\rho})\hat{\rho}}}{s - \hat{\rho} + \sqrt{s^2 + \hat{z}^2}}, \quad \rho = \frac{s\hat{z}}{s - \hat{\rho} + \sqrt{s^2 + \hat{z}^2}}$$

(10.108)

provides this 3-D mapping (see Fig. 10.38). Both coordinate systems share the same azimuthal parametrization $\phi = \hat{\phi}$. The tensors $\hat{\varepsilon}_r$ and $\hat{\mu}_r$ are now given by $t\hat{\varepsilon}_r$ and $t\hat{\mu}_r$, with t explicitly written as

$$t = g_\rho g_\phi g_z \begin{pmatrix} \rho^{-1}g_\rho^{-2} & 0 & 0 \\ 0 & \rho g_\phi^{-2} & 0 \\ 0 & 0 & \rho^{-1}g_z^{-2} \end{pmatrix}.$$

(10.109)

The transformation optics technique yields the following diagonal components:

$$\hat{\varepsilon}_\rho = \frac{s(2s - \hat{\rho})}{a(a + s - \hat{\rho})}, \quad \hat{\varepsilon}_\phi = \frac{s^3}{a(a + s - \hat{\rho})(2s - \hat{\rho})}, \quad \hat{\varepsilon}_z = \frac{as}{\hat{\rho}(a + s - \hat{\rho})}$$

(10.110)

of the inhomogeneous anisotropic dielectric constant $\hat{\varepsilon}_r$, as well as the anisotropic relative permeability tensor $\hat{\mu}_r = \hat{\varepsilon}_r$, distributed in an equivalent layer between two planes (where $a = \sqrt{s^2 + z^2}$). Analysis of Eqs. (10.110) and Fig. 10.38(c) indicates that Eqs. (10.110) can be approximated by

$$\hat{\varepsilon}_\rho = \hat{\mu}_\rho \approx 1, \quad \hat{\varepsilon}_\phi = \hat{\mu}_\phi \approx \frac{s^2}{(2s - \hat{\rho})^2}, \quad \hat{\varepsilon}_z = \hat{\mu}_z = \frac{s^2}{\hat{\rho}(2s - \hat{\rho})}. \quad (10.111)$$

Note that the requirement for an ideal cloak in the effective material coordinate system $(\hat{\rho}, \hat{\phi}, \hat{z})$ should be written as follows [80]:

$$\hat{\varepsilon}_\rho = \hat{\mu}_\rho = \frac{\rho\hat{\rho}'}{\hat{\rho}}, \quad \hat{\varepsilon}_\phi = \hat{\mu}_\phi = \frac{1}{\hat{\varepsilon}_\rho}, \quad \hat{\varepsilon}_z = \hat{\mu}_z = \frac{\rho}{\hat{\rho}\hat{\rho}'}, \quad (10.112)$$

where $\hat{\rho} = \hat{\rho}(\rho)$ is a radial mapping function and $\hat{\rho}' = d\hat{\rho}/d\rho$. In general, $\hat{\varepsilon}_\phi\hat{\varepsilon}_z = (\rho/\hat{\rho})^2$. These requirements can be met if the refractive index $n = \sqrt{\varepsilon_r}$ inside the gap between the sphere and the plane is chosen to be a simple radius-dependent function $n = \sqrt{(2s - \hat{\rho})}/s$. In such a case we obtain

$$\hat{\varepsilon}_\rho \approx \frac{2s - \hat{\rho}}{s}, \quad \hat{\varepsilon}_\phi = \frac{s}{2s - \hat{\rho}}, \quad \hat{\varepsilon}_z = \frac{s}{\hat{\rho}}. \quad (10.113)$$

Since in general $\hat{\varepsilon}_\phi\hat{\varepsilon}_z = (\rho/\hat{\rho})^2$, we can recover the mapping function from Eq. (10.113) as $\rho(\hat{\rho}) = s\sqrt{\hat{\rho}/(2s - \hat{\rho})}$, which is consistent with the approximated version of the initial mapping shown in Eqs. (10.108). Note that the scale s is chosen to avoid singularities: $s > \max(\hat{\rho})$. Also note that Eqs. (10.113) represent the invisible body, i.e. a self-cloaking arrangement. It is also important that the filling substance has an isotropic effective refractive index ranging from 2 to 1 for $\rho = [0, s]$.

It is important to note that filling an initial domain between rotationally symmetric curvilinear boundaries, for example with an anisotropic dielectric, *allows for independent control over the effective magnetic and electric properties in the equivalent right-cylinder domain.* Thus, while the shape is controlling the general mapping and provides the identical transformations for the effective magnetic and electric components, further independent adjustment of material tensors $\hat{\varepsilon}_r$ and $\hat{\mu}_r$ could be achieved through either anisotropic magnetic or (what is more realistic for optics) anisotropic electric filling.

It is also interesting that in the semi-classical ray-optics approximation, the cloaking geometry may be simplified even further for a family of rays with similar parameters. This simplification lets us clearly demonstrate the basic physics involved. Our starting point is the semi-classical 2-D cloaking Hamiltonian (dispersion law), Eq. (10.69), introduced by Jacob *et al.* [45], which can be written as

$$\frac{\omega^2}{c_0^2} = k_r^2 + \frac{k_\phi^2}{(r-b)^2} = k_r^2 + \frac{k_\phi^2}{r^2} + k_\phi^2 \frac{b(2r-b)}{(r-b)^2 r^2}. \tag{10.114}$$

Jacob and Narimanov demonstrated that, for such a cylindrically symmetric Hamiltonian, the rays of light would flow smoothly without scattering around a cylindrical cloaked region of radius b. Our aim is to produce this cloaking Hamiltonian, Eq. (10.114), in an optical waveguide (Fig. 10.38). Let us allow the thickness d of the waveguide in the z direction to change adiabatically with radius r. The top and bottom surfaces of the waveguide are coated with metal. The dispersion law (Hamiltonian) of light in such a waveguide is given by

$$\frac{\omega^2}{c_0^2} = k_r^2 + \frac{k_\phi^2}{r^2} + \frac{\pi^2 m^2}{d^2(r)}, \tag{10.115}$$

where m is the transverse mode number. Note that a photon launched into the mth mode of the waveguide stays in this mode as long as d changes adiabatically [82]. In addition, since the angular momentum of the photon, $k_\phi = \rho k = L$, is conserved (where ρ is the impact parameter defined with respect to the origin), for each combination of m and L the cloaking Hamiltonian, Eq. (10.114), can be emulated precisely by an adiabatically changing $d(r)$. A comparison of Eqs. (10.114) and (10.115) produces the following desired radial dependence of the waveguide thickness:

$$d = \frac{m\pi r^{3/2}\left(1 - \frac{b_{mL}}{r}\right)}{L\left[2b_{mL}\left(1 - \frac{b_{mL}}{r}\right)\right]^{1/2}}, \tag{10.116}$$

where b_{mL} is the radius of the region that is "cloaked" for the photon launched into the (m, L) mode of the waveguide. The shape of such a waveguide is presented in Fig. 10.38(b), where we have chosen $b_{mL} = b_0 = 50\,\mu m$. Thus, an electromagnetic cloaking experiment in a waveguide may be performed in a geometry that is identical to the classic geometry of Newton's rings [83], as shown in Fig. 10.38(a).

An aspherical lens shaped according to Eq. (10.116) has to be used for the single (m, L) mode cloak to be ideal. It appears also that the shape of the "ideal" waveguide may be reasonably approximated by a spherical surface placed on top of a flat surface, as shown by the dotted line in Fig. 10.38(d). Moreover, the waveguide geometry for a single cloaking mode may be further improved to allow cloaking in a multimode waveguide geometry. It is clear from Eq. (10.116) that the choice of $b_{mL} = b_0 m^2/L^2$ leads to the same desired shape of the waveguide for all the

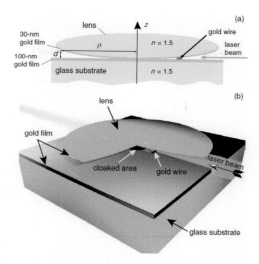

Figure 10.39. Tapered waveguide acting as an optical cloak (not to scale). (a) Cross section sketch for the waveguide experiment. (b) Three-dimensional rendition of the experimental setup sketched in (a); one-quarter of the gold-coated lens is removed to show the interior details. (From ref. [81].)

(m, L) modes of a multimode waveguide in the leading order of b_{mL}/r:

$$d = \pi \sqrt{\frac{r^3}{2b_0}}.$$ (10.117)

Equation (10.117) describes the best-shaped aspherical lens for the electromagnetic cloaking observation in the Newton's rings geometry with an air gap.

In the experiments, a 4.5-mm-diameter double convex glass lens (lens focus 6 mm) was coated on one side with a 30 nm gold film. The lens was placed with the gold-coated side down on top of a flat glass slide coated with a 70 nm gold film, as shown in Fig. 10.39. The air gap between these surfaces has been used as an adiabatically changing waveguide. The point of contact between two gold-coated surfaces is clearly visible in Fig. 10.40(a). The Newton's rings appear around the point of contact upon illumination of the waveguide with white light from the top. The radius of the mth ring is given by the expression $r_m = [((1/2) + m)\, R\lambda]^{1/2}$, where R is the lens radius. The central area around the point of contact appears to be bright, since light reflected from the two gold-coated surfaces has the same phase. Laser light from an argon ion laser was coupled to the waveguide formed between two gold-coated surfaces via side illumination. Light propagation through the waveguide was imaged from the top using an optical microscope. Figures 10.40(b) and (c) show microscope images of the light propagation through the waveguide in an experiment in which a gold particle cut from a 50 μm diameter gold wire is placed

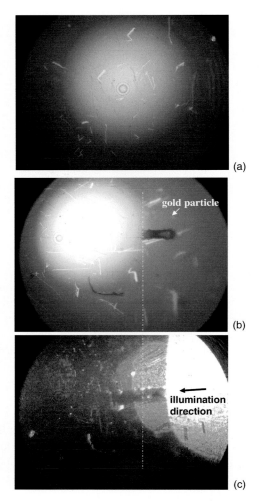

Figure 10.40. (a) Microscope image of the waveguide illuminated with white light from the top. The Newton rings are visible in the center of the field of view. (b) Microscope image of the waveguide with a gold particle placed inside and illuminated with white light from the top. (c) A long shadow has been cast by the gold particle upon coupling 515 nm laser light into the waveguide. The position of the particle edge is shown by the dashed line. See color plates section.

inside the waveguide. A very pronounced long shadow is cast by the particle inside the waveguide (Fig. 10.40(c)). This result is quite natural since the gold particle size is approximately equal to 100 μm (note that the first dark Newton ring visible in Figs. 10.40(a) and (b) has approximately the same size as the diameter of the gold wire). Since the gold particle is located 400 μm from the point of contact between the walls of the waveguide, the effective Hamiltonian around the gold particle

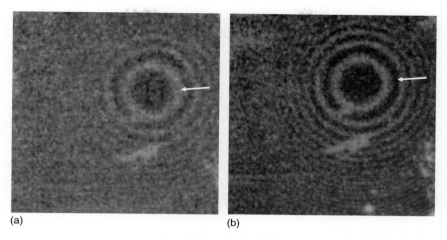

(a)   (b)

Figure 10.41. Magnified images of the rings for (a) 488 nm and (b) 515 nm laser illumination. In both cases, no wire is placed in the waveguide. (From ref. [81].)

differs strongly from the cloaking Hamiltonian of Eq. (10.114). Figure 10.40 represents the results of our best effort to insert a 150-μm-long, 50-μm-diameter gold particle inside the waveguide and orient it along the illumination direction. A few scratches visible in Fig. 10.40 resulted from achieving this nontrivial experimental task.

While the gold particle casts a long and pronounced shadow, it appears that the area around the point of contact between the two gold-coated surfaces casts no shadow at all (Fig. 10.41). This is an observation which would be extremely surprising in the absence of the theoretical description presented above. For the mth mode of the waveguide shown in Fig. 10.39(a), the cut-off radius is given by the same expression as that of the radius of the mth Newton ring: $r_m = [((1/2) + m) R\lambda]^{1/2}$, which means that no photon launched into the waveguide can reach an area within the radius $r_0 = \sqrt{R\lambda/2}$, or approximately 30 μm from the point of contact between the two gold-coated surfaces. This is consistent with the fact that the area around the point of contact appears dark in Fig. 10.41. Even though some photons may couple to surface plasmon polaritons [66] near the cut-off point of the waveguide, the propagation length of the surface plasmons at 515 nm is only a few micrometers. Thus, the area around the point of contact about 50 μm in diameter is about as opaque for guided photons as the 50 μm gold particle from Figs. 10.40(b) and (c), which casts a pronounced long shadow. Nevertheless, there appears to be no shadow behind the cut-off area of the waveguide (see the images in Figs. 10.41(a) and (b), which were taken at the 488 nm and 515 nm laser lines of the argon ion laser). The observed cloaking behavior appears to be broadband, which is consistent with the theory presented above. It is also interesting to note that the geometry of our cloaking experiment is similar to the geometry recently proposed by Leonhardt and Tyc [55].

10.14 Trapped rainbow

The concept of a "trapped rainbow" is closely related to electromagnetic cloaking, and has attracted considerable recent attention. According to various theoretical models, a specially designed metamaterial [84] or plasmonic [85, 86] waveguide has the ability to slow down and stop light of different wavelengths at different spatial locations along the waveguide, which is extremely attractive for such applications as spectroscopy on a chip. In addition, being a special case of the slow light phenomenon [87], the trapped rainbow effect may be used in applications such as optical signal processing and enhanced light–matter interactions [88]. On the other hand, unlike typical slow light schemes, the proposed theoretical trapped rainbow arrangements are extremely broadband, and can trap a true rainbow ranging from violet to red in the visible spectrum. Due to the necessity of complicated nanofabrication and the difficulty of producing broadband metamaterials, the trapped rainbow schemes remained in the theoretical domain only for quite a while. Very recently, Smolyaninova *et al.* [89] demonstrated an experimental realization of the broadband trapped rainbow effect which spans the 457–633 nm range of the visible spectrum. Similar to the recent demonstration of broadband cloaking [81], the metamaterial properties necessary for device fabrication were emulated using an adiabatically tapered optical nano-waveguide geometry. A 4.5 mm diameter double convex glass lens was coated on one side with a 30 nm gold film. The lens was placed with the gold-coated side down on top of a flat glass slide coated with a 70 nm gold film (Fig. 10.42(a)). The air gap between these surfaces has been used as an adiabatically changing optical nano-waveguide. The dispersion law of light in such a waveguide is given by

$$\frac{\omega^2}{c_0^2} = k_r^2 + \frac{k_\phi^2}{r^2} + \frac{\pi^2 l^2}{d^2(r)}, \tag{10.118}$$

where $l = 1, 2, 3 \ldots$ is the transverse mode number and $d(r)$ is the air gap, which is a function of radial coordinate r. Light from a multi-wavelength argon ion laser (operating at $\lambda = 457$ nm, 465 nm, 476 nm, 488 nm, and 514 nm) and 633 nm light from a He–Ne laser were coupled to the waveguide via side illumination. This multi-line illumination produced the appearance of white light illuminating the waveguide (Fig. 10.42(b)). Light propagation through the nano-waveguide was imaged from the top using an optical microscope (Fig. 10.42(c)).

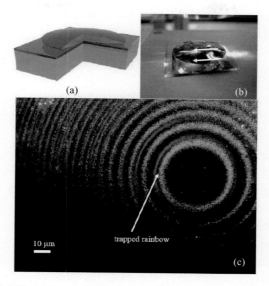

(a)

(b)

10 μm

trapped rainbow

(c)

Figure 10.42. (a) Experimental geometry of the trapped rainbow experiment: a glass lens was coated on one side with a gold film. The lens was placed with the gold-coated side down on top of a flat glass slide also coated with a gold film. The air gap between these surfaces formed an adiabatically changing optical nano-waveguide. (b) Photo of the trapped rainbow experiment: He–Ne and Ar:ion laser light is coupled into the waveguide. (c) Optical microscope image of the trapped rainbow. (From ref. [89].) See color plates section.

Since the waveguide width at the entrance point is large, the air gap waveguide starts as a multimode waveguide. Note that a photon launched into the lth mode of the waveguide stays in this mode as long as d changes adiabatically [82]. In addition, the angular momentum of the photons, $k_\phi = \rho k = L$, is conserved (where ρ is the impact parameter defined with respect to the origin). Gradual tapering of the waveguide leads to mode number reduction: only $L = 0$ modes may reach the vicinity of the point of contact between the gold-coated spherical and planar surfaces, and the group velocity of these modes,

$$v_g = c_0 \sqrt{1 - \left(\frac{l\lambda}{2d}\right)^2},$$

$$(10.119)$$

tends to zero as d is reduced: the rings around the central circular dark area in Fig. 10.42(c) each represent a location where the group velocity of the lth waveguide mode becomes zero. These locations are defined by

$$r_n = \sqrt{\left(l + \frac{1}{2}\right) R\lambda},$$

$$(10.120)$$

where R is the lens radius [81]. Finally, the light in the waveguide is completely stopped at a distance

$$r = \sqrt{R\lambda/2} \qquad\qquad (10.121)$$

from the point of contact between the gold-coated surfaces, where the optical nano-waveguide width reaches the $d = \lambda/2 \simeq 200$ nm range. The group velocity of the only remaining waveguide mode at this point is zero. This is consistent with the fact that the area around the point of contact appears dark in Fig. 10.42(c). In this area the waveguide width falls below 200 nm down to zero. Since the stop radius depends on the light wavelength, different light colors stop at different locations inside the waveguide, which is quite obvious from Fig. 10.42(c). Thus, the visible light rainbow has been stopped and "trapped." In principle, the same technique can be applied to any spectral range of interest.

The described experimental arrangement may be used in such important applications as spectroscopy on a chip. Figure 10.43 presents a comparison of the optical microscope images of the trapped rainbow effect from Fig. 10.42(c) and the image obtained when only two laser wavelengths (514 nm and 633 nm) are used for illumination (shown at the top of Fig. 10.42). Individual spectral lines separated by only a few micrometers appear to be well resolved in the latter image, which is evident from the cross section analysis presented in Fig. 10.44. Based on the image cross section analysis, spectral resolution of the order of 40 nm has been obtained. Further improvement of spectral resolution may be achieved by using a gold-coated spherical surface with a smaller radius of curvature.

10.15 The limitations of real cloaks

When the wave nature of light is taken into account, perfect cloaking of an object is impossible. This is well known for isotropic media [90, 91]. Cloaks based on spatial variations in refractive index, as described by Pendry *et al.* [2], can, however, be extremely good in principle, although remaining subtle effects will result in partial visibility of the cloaked region – perhaps a hazy outline of the cloaked object. Scattering from within the material of the cloak is inevitable, and there is an ambiguity for rays of light that enter along an axis of symmetry of the cloak – do they deviate around the cloaked region in the up or down direction? Unless there is perfect impedance matching for waves at all angles at the boundary of the cloak, there will be some reflection. In addition, although rays directed around the cloaked region each individually take a path of least time relative to their entry point into the cloak, different rays experience different time delays in passing through the cloak, so wavefronts that emerge are distorted. In a conventional imaging application

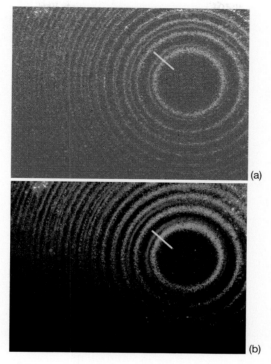

(a)

(b)

Figure 10.43. Comparison of the optical microscope images of the trapped rain-bow effect from Fig. 10.42(c) (b) and the image (a) obtained when only two laser wavelengths (514 nm and 633 nm) are used for illumination. Individual spectral lines separated by only a few micrometers appear to be well resolved (see Fig. 10.44). (From ref. [89].) See color plates section.

not involving coherent light, such wavefront distortion would not be noticeable. There are also a host of practical issues that make real cloaks less than perfect. The dielectric and magnetic properties of the cloak are frequency-dependent, so the mapping of space that produces the cloak will only work at a single wavelength. This can be a small effect provided the materials of the cloak are being used in their transparency region, where losses are very small and the dispersion of the refractive index $dn/d\lambda$ is small. Unfortunately, current metamaterials that use metal/dielectric composite structures are very lossy, which might have little impact on the cloaking of a microscopic object, but would be unacceptable for cloaking macroscopic objects. For example, for an effective average dielectric constant inside a spherical cloak of $1 - j0.5$, which is approximately what might be expected if silver is the metal used in the composite cloak, the absorption length for light at 632.8 nm is about 2.6 μm. Consequently, only objects on the scale of the wavelength can be cloaked. In summary, although cloak designs work, in principle, even for

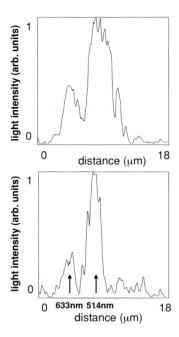

Figure 10.44. Cross sections of the optical microscope images along the yellow lines shown in Fig. 10.43. Individual spectral lines are clearly resolved in the bottom plot obtained using 514 nm and 633 nm illumination. Multiple spectral lines are visible in the top cross section.

macroscopic objects, there is a serious problem with the properties of currently available metamaterials. What is needed is more work on all-dielectric cloaks, where losses can be negligible, but unfortunately lossless cloak materials with refractive index smaller than unity are not yet available.

10.16 Prospects for the future

Research on electromagnetic cloaking has to date been largely a theoretical effort because of the difficulty in fabricating the metamaterials needed to provide the requisite spatially varying material properties. Given the ongoing activity in this active field, our overview here cannot be complete and exhaustive. New interesting theoretical ideas on cloaking and transformation optics are being put forward almost daily. Even though progress may slow because of numerous practical difficulties, we may expect considerable progress in experimental demonstration of these ideas. For example, we may expect that the problem of losses, which currently present the biggest challenge to experimental realizations, may be to some degree alleviated by the introduction of gain media into the metamaterial designs. In addition, some progress may be expected in the bandwidth extension of metamaterial devices. Geometrical ideas similar to the ones described in Section 10.13 may be used to alleviate the bandwidth problem. At the same time, electromagnetic cloaking ideas are being extended to such fields as acoustics, shoreline protection, and even quantum mechanics. We firmly believe that this new and exciting field of research will bring about many interesting and surprising discoveries.

Acknowledgments

In this review chapter we have drawn freely from the many papers published by our colleagues who are also interested in the cloaking of material objects. We thank them for their excellent work and the stimulation we have received in compiling this review of the field.

References

[1] J. Mullan, *The Guardian*, September 9 (2009).
[2] J. B. Pendry, D. Schurig, and D. R. Smith, "Controlling electromagnetic fields," *Science* **312**, 1780–1782 (2006).
[3] U. Leonhardt, "Optical conformal mapping," *Science* **312**, 1777–1780 (2006).
[4] J. H. Moore, C. C. Davis, and M. A. Coplan, *Building Scientific Apparatus,* 4th edn. (Cambridge: Cambridge University Press, 2009).
[5] Z. Yang, L. Ci, J. A. Bur, S. Y. Lin, and P. M. Ajayan, "Experimental observation of an extremely dark material made by a low-density nanotube array," *Nano Lett.* **8**, 446–451 (2008).
[6] R. H. Ritchie, "Surface plasmons in solids," *Surf. Sci.* **34**, 1–19 (1973).
[7] M. Fleischmann, P. J. Hendra, and A. J. McQuillan, "Raman spectra of pyridine adsorbed at a silver electrode," *Chem. Phys. Lett.* **26**, 163–166 (1974).
[8] M. Moskovits, "Surface-enhanced spectroscopy," *Rev. Mod. Phys.* **57**, 783–826 (1985).
[9] H. Raether, *Surface Plasmons on Smooth and Rough Surfaces and on Gratings* (Berlin: Springer-Verlag, 1988).
[10] P. Drude, "Zur ionen-theorie der metalle," *Phys. Z.* **1**, 161–165 (1900).
[11] P. Drude, "Zur elektronentheorie der metalle; II. Teil. galvanomagnetische und ther-momagnetische effecte," *Ann. Phys. Lpz.* **308** 369–402 (1900).
[12] P. Drude, "Zur elektronentheorie der metalle," *Ann. Phys. Lpz.* **306**, 566–613 (1900).
[13] V. G. Veselago, "The electrodynamics of substances with simultaneously negative values of ε and μ." *Sov. Phys. Uspekhi* **10**, 509–514 (1968).
[14] S. Ramo, J. R. Whinnery, and T. van Duzer, *Fields and Waves in Communication Electronics,* 1st edn (New York: Wiley, 1967).
[15] J. Valentine, S. Zhuang, T. Zentgraf, E. Ulin-Avila, D. A. Denov, G. Bartal, and X. Zhang, "Three-dimensional optical metamaterial with a negative refractive index," *Nature* **455**, 376–379 (2008).
[16] T. Paul, C. Rockstuhl, C. Menzel, and F. Lederer, "Anomalous refraction, diffraction, and imaging in metamaterials," *Phys. Rev. B* **79**, 115430(1-11) (2009).

[17] A. J. Ward and J. B. Pendry, "Refraction and geometry in Maxwell's equations," *J. Mod. Opt.* **43**, 773–793 (1996).

[18] H. Margenau and G. M. Murphy, *The Mathematics of Physics and Chemistry,* 2nd edn (Princeton, NJ: Van Nostrand, 1956).

[19] D. Schurig, J. B. Pendry, and D. R. Smith, "Calculation of material properties and ray tracing in transformation media," *Opt. Express* **14**, 9794–9804 (2006).

[20] M. Rahm, D. Schurig, D. A. Roberts, S. A. Cummer, D. R. Smith, and J. B. Pendry, "Design of electromagnetic cloaks and concentrators using form-invariant coordinate transformations of Maxwell's equations," *Photon. Nanostruct.* **6**, 87–95 (2008).

[21] J-J. Ma, X-Y. Cao, K-M. Yu, and T. Liu, "Determination the material parameters for arbitrary cloak based on Poisson's equation," *Prog. Electromag. Res.* **9**, 177–184 (2009).

[22] C.-W. Qiu, A. Novitsky, and L. Gao, "Inverse design mechanism of cylindrical cloaks without knowledge of the required coordinate transformation," *J. Opt. Soc. Am. A* **27**, 1079–1082 (2010).

[23] E. J. Post, *Formal Structure of Electromagnetics* (New York: Wiley, 1962).

[24] J.-P. Bérenger, "A perfectly matched layer for the absorption of electromagnetic waves," *J. Comput. Phys.* **114**, 185–200 (1994).

[25] R. Mittra and Ü. Pekel, "A new look at the perfectly matched layer (PML) concept for the reflectionless absorption of electromagnetic waves," *IEEE Microwave Guided Wave Lett.* **5**, 84–86 (1995).

[26] J. Sun, J. Zhou, and L. Kang, "Homogeneous isotropic invisible cloak based on geometrical optics," *Opt. Express* **16**, 17768–17773 (2008).

[27] S. A. Cummer, B-I. Popa, D. Schurig, D. R. Smith, and J. Pendry, "Full-wave simulations of electromagnetic cloaking structures," *Phys Rev. E* **74**, 036621(1-5) (2006).

[28] W. Xiang, T. J. Cui, X. M. Yang, Q. Cheng, R. Liu, and D. R. Smith, "Invisibility cloak without singularity," *Appl. Phys. Lett.* **93**, 194102(1-3) (2008).

[29] H. Ma, S. Qu, Z. Xu, J. Zhang, B. Chen, and J. Wang, "Material parameter equation for elliptical cylindrical cloaks," *Phys. Rev. A* **77**, 013825(1-3) (2008).

[30] D-H. Kwon and D. H. Werner, "Two-dimensional eccentric elliptic electromagnetic cloaks," *Appl. Phys. Lett.* **82**, 013505(1-3) (2008).

[31] Y. Luo. J. Zhang, B.-I. Wu, and H. Chen, "Interaction of an electromagnetic wave with a cone-shaped invisibility cloak and polarization rotator," *Phys. Rev. B* **78**, 125108(1-9) (2008).

[32] J. Zhang, Y. Luo, H. Chen, and B.-I. Wu, "Cloak of arbitrary shape," *J. Opt. Soc. Am.* **25**, 1776–1779 (2009).

[33] C.-W. Qiu, Li Hu, B. Zhang, B.-I. Wu, S. G. Johnson, and J. D. Joannopoulos, "Spherical cloaking using nonlinear transformations for improved segmentation into concentric isotropic coating," *Opt. Express* **17** 13467–13478 (2009).

[34] H. C. Chen, *Theory of Electromagnetic Fields: A Coordinate Free Approach* (New York: McGraw-Hill, 1985).

[35] M. Born and E. Wolf, *Principles of Optics,* 7th edn (Cambridge: Cambridge University Press, 1999).

[36] W. R. Hamilton, "On a general method of expressing the paths of light, and of the planets, by the coefficients of a characteristic function," *Dublin Univ. Rev. Quart. Mag.* **1**, 795–826 (1833).

[37] J. Evans and M. Rosenquist, "F=ma optics," *Am. J. Phys.* **54**, 876–883 (1986).

[38] T. Sekiguchi and K. B. Wolf, "The Hamiltonian formulation of optics," *Am. J. Phys.* **55**, 830–835 (1987).

[39] J. Evans, "The ray form of Newton's law of motion," *Am. J. Phys.* **61**, 347–350 (1993).

[40] J. A. Reyes, "Ray propagation in anisotropic inhomogeneous media," *J. Phys. A: Math. Gen.* **32**, 3409–3419 (1999).

[41] A. J. Dragt, "Lie algebraic theory of geometrical optics and optical aberrations," *J. Opt. Soc. Am.* **72**, 372–379 (1982).

[42] U. Leonhardt, "Momentum in an uncertain light," *Nature* **444**, 823–824 (2006).

[43] W. She, J. Yu, and R. Feng, "Observation of a push force on the end of a nanometer silica filament exerted by outgoing light," *Phys. Rev. Lett.* **101**, 234601(1-4) (2008).

[44] S. M. Barnett "Resolution of the Abraham–Minkowski dilemma," *Phys. Rev. Lett.* **104**, 070401(1-4)(2010).

[45] Z. Jacob and E. E. Narimanov, "Semiclassical description of non magnetic cloaking," *Opt. Express* **16**, 4597–4604 (2008).

[46] W. Cai, U. K. Chettiar, A. V. Kildishev, and V. M. Shalaev, "Optical cloaking with metamaterials," *Nature Photon.* **1**, 224–227 (2007).

[47] W. Cai, U. K. Chettiar, A. V. Kildishev, V. M. Shalaev, and G. W. Milton, "Nonmagnetic cloak with minimized scattering," *Appl. Phys. Lett.* **91**, 111105(1-3) (2007).

[48] U. Leonhardt, "Notes on conformal invisibility devices," *New J. Phys.* **8**, 118(1-16) (2006).

[49] R. K. Luneburg (1944). *Mathematical Theory of Optics* (Providence, RI: Brown University, 1944).

[50] J. E. Eaton, "On spherically symmetric lenses," *IRE Trans. Antennas Propag.* **AP-4**, 66–71 (1952).

[51] J. A. Lock, "Scattering of an electromagnetic plane wave by a Luneburg lens. I. Ray theory," *J. Opt. Soc. Am. A* **25**, 2971–2979 (2008).

[52] J. H. Hannay and T. M. Haeusser, "Retroreflection by refraction," *J. Mod. Opt.* **40**, 1437–1442 (1993).

[53] T. Tyc and U. Leonhardt, "Transmutation of singularities in optical instruments," *New J. Phys.* **10**, 115038(1-8) (2008).

[54] T. Ochial, U. Leonhardt, and J. N. Nacher, "A novel design of dielectric invisibility devices," *J. Math. Phys.* **49**, 032903(1-13)(2008).

[55] U. Leonhardt and T. Tyc, "Broadband invisibility by non-Euclidean cloaking," *Science* **323**, 110–112 (2009).

[56] Z. Jacob, L. V. Alekseyev, and E. Narimanov, "Semiclassical theory of the hyperlens," *J. Opt. Soc. Am. A* **24**, 52–59 (2007).

[57] L. D. Landau, E. M. Lifshitz, and L. P. Pitaevskii, *Electrodynamics of Continuous Media,* Landau and Lifshitz Course of Theoretical Physics, vol 8, 2nd edn (Amsterdam: Elsevier, 1984).

[58] D. Schurig, J. J. Mock, B. J. Justice, S. A. Cummer, J. B. Pendry, A. F. Starr, and D. R. Smith, "Metamaterial electromagnetic cloak at microwave frequencies," *Science* **314**, 977–980 (2006).

[59] A. Greenleaf, M. Lassas, and G. Uhlmann, "The Calderon problem for conormal potentials – I: Global uniqueness and reconstruction," *Commun. Pure Appl. Math.* **56**, 328–352 (2003).

[60] A. Alú and N. Engheta, "Plasmonic materials in transparency and cloaking problems: mechanism, robustness, and physical insights," *Opt. Express* **15**, 3318–3331 (2007).

[61] A. Alú and N. Engheta, "Cloaking and transparency for collections of particles with metamaterial and plasmonic covers," *Opt. Express* **15**, 7578–7590 (2007).

[62] I. I. Smolyaninov, Y. J. Hung, and C. C. Davis, "Magnifying superlens in the visible frequency range," *Science* **315**, 1699–1701 (2007).

[63] I. I. Smolyaninov, Y. J. Hung, and C. C. Davis, "Imaging and focusing properties of plasmonic metamaterial devices," *Phys. Rev. B* **76**, 205424(1-7) (2007).
[64] I. I. Smolyaninov, "Two-dimensional plasmonic metamaterials," *Proc. SPIE* **6638**, 663803(1-12) (2007).
[65] I. I. Smolyaninov, Y. J. Hung, and C. C. Davis, "Two-dimensional metamaterial structure exhibiting reduced visibility at 500 nm," *Opt. Lett.* **33**, 1342–1344 (2008).
[66] A. V. Zayats, I. I. Smolyaninov, and A. A. Maradudin, "Nano-optics of surface plasmon-polaritons," *Phys. Rep.* **408**, 131–314 (2005).
[67] A. Alú and N. Engheta, "Achieving transparency with plasmonic and metamaterial coatings," *Phys. Rev E.* **72**, 016623(1-9) (2005).
[68] A. Alú and N. Engeta "Multifrequency optical invisibility cloak with layered plasmonic shells," *Phys. Rev. Lett.* **100**, 113901(1-4) (2008).
[69] M. G. Silveirinha, A. Alú and N. Engheta, "Parallel-plate metamaterials for cloaking structures," *Phys. Rev. E* **75**, 036603(1-16)(2007).
[70] B. Baumeier, T. A. Leskova, and A. A. Maradudin, "Cloaking from surface plasmon polaritons by a circular array of point scatterers," *Phys. Rev. Lett.* **103**, 24603(1-4) (2009).
[71] A. Alú, "Mantle cloak: invisibility induced by a surface," *Phys. Rev. B* **80**, 245115(1-5) (2009).
[72] B. Edwards, A. Alú, M. G. Silveirinha, and N. Engheta, "Experimental verification of plasmonic cloaking at microwave frequencies with metamaterials," *Phys. Rev. Lett.* **103**, 153901(1-4) (2009).
[73] G. W. Milton and N.-A. P. Nicorovici, "On the cloaking effects associated with anomalous localized resonance," *Proc. Roy. Soc. Lon. Ser. A. Math. Phys. Sci.* **462**, 3027–3059 (2006).
[74] G. W. Milton, N.-A. P. Nicorovici, R. C. McPhedran, K. Cherednichenko, and Z. Jacob, "Solutions in folded geometries, and associated cloaking due to anomalous resonance," *New J. Phys.* **10**, 115(1-21) (2008).
[75] N.-A. P. Nicorovici, G. W. Milton, R. C. McPhedran, and L. C. Botten, "Quasistatic cloaking of two-dimensional polarizable discrete systems by anomalous resonance," *Opt. Express* **15**, 6314–6323 (2007).
[76] J. Li and J. B. Pendry, "Hiding under the carpet: a new strategy for cloaking," *Phys. Rev. Lett.* **101**, 203901(1-4) (2008).
[77] R. Liu, C. Ji, J. J. Mock, J. Y. Chin, T. J. Cui, and D. R. Smith, "Broadband ground-plane cloak," *Science* **323**, 366–369 (2009).
[78] X. Xu, Y. Feng, Y. Hao, J. Zhao, and T. Jiang, "Infrared carpet cloak designed with uniform silicon grating structure." *Appl. Phys. Lett.* **95**, 184102(1-3) (2009).
[79] E. Kallos, C. Argyropoulos, and Y. Hao, "Ground-plane quasicloaking for free space," *Phys. Rev. A* **79**, 063825(1-5) (2009).
[80] A.V. Kildishev, W. Cai, U. K. Chettiar, and V. M. Shalaev, "Transformation optics: approaching broadband electromagnetic cloaking," *New J. Phys.* **10**, 115029(1-14) (2008).
[81] I. I. Smolyaninov, V. N. Smolyaninova, A. V. Kildishev, and V. M. Shalaev, "Anisotropic metamaterials emulated by tapered waveguides: application to optical cloaking," *Phys. Rev. Lett.* **102**, 213901(1-4) (2009).
[82] L. D. Landau and E. M Lifshitz, *Quantum Mechanics* (Oxford: Reed, 1988).
[83] I. Newton, "A letter of Mr. Isaac Newton, Professor of the Mathematicks in the University of Cambridge; containing his new theory about light and colors," *Phil. Trans. Royal Soc.* **80**, 3075–3087 (1671).

[84] K. L. Tsakmakidis, A. D. Boardman, and O. Hess, "Trapped rainbow storage of light in metamaterials," *Nature* **450**, 397–401 (2007).

[85] M. I. Stockman, "Nanofocusing of optical energy in tapered plasmonic waveguides," *Phys. Rev. Lett.* **93**, 137404(1-4) (2004).

[86] Q. Gan, Y. J. Ding, and F. J. Bartoli, "Rainbow trapping and releasing at telecommunication wavelengths," *Phys. Rev. Lett.* **102**, 056801(1-4) (2009).

[87] L. V. Hau, S. E. Harris, Z. Dutton, and C. H. Behroozi, "Light speed reduction to 17 metres per second in an ultracold atomic gas," *Nature* **397**, 594–598 (1999).

[88] Y. A. Vlasov, M. O'Boyle, H. F. Hamann, and S. J. McNab, "Active control of slow light on a chip with photonic crystal waveguides," *Nature* **438**, 65–69 (2005).

[89] V. N. Smolyaninova, I. I. Smolyaninov, A. V. Kildishev, and V. M. Shalaev, *Appl. Phys. Lett.* **96**, 211121(1-3) (2010).

[90] A. I. Nachman, "Reconstructions from boundary measurements," *Ann. Math.* **128**, 531–576 (1988).

[91] E. Wolf and T. Habashy, "Invisible bodies and uniqueness of the inverse scattering problem," *J. Mod. Opt.* **40**, 785–792 (1993).

11

Linear and nonlinear phenomena with resonating surface polariton waves and their applications

HAIM GREBEL

11.1 Introduction

Surface plasmon polariton (SPP) modes have attracted much interest in recent years. Although known and studied for over 100 years [1–3], the dream of confining light to dimensions smaller than its propagating wavelength has led the way towards technological possibilities not previously addressed, such as optical circuitry within ultra small computer processors [4, 5], or small biochemical sensors [6, 7]. Confinement of light to sub-wavelength dimensions is also a possibility when one considers the field aspects of the electromagnetic waves near surfaces (near-field phenomena). Add to this the interest in materials and structures exhibiting a negative refractive index for the purpose of increasing the resolution of optical microscopy [8], and it is no wonder that the area of electromagnetic (EM) propagation in sub-wavelength structures is enjoying a renewed interest. Whether the far-field aspects of periodic resonating metallo-dielectric structures are the true manifestations of a negative refractive index or simply a unique, but already known, near-field dispersion phenomenon may be debated [9]. Nonetheless, the near-field aspects of periodic sub-wavelength metallo-dielectric structures, and especially recent advances in nano-fabrication of structures at dimensions smaller than optical wavelengths, deserve a closer look.

Artificial dielectrics (ADs) constitute a class of man-made materials: the effective permittivity and permeability of a given dielectric material may be altered by imbedding metallic or semiconductive structures on scales smaller than the propagating wavelength. For example, one may alter the equivalent capacitance and inductance of microwave waveguides by the addition of a pattern of fine metallic features along the waveguide axis. These features, with spacing much smaller than the propagating electromagnetic wavelength, affect the propagation constant

Structured Surfaces as Optical Metamaterials, ed. A. A. Maradudin. Published by Cambridge University Press.

of the wave within the guide. Moreover, by making these patterns light sensitive, one is able to shift the phase of the propagating mode using a controlling light beam [10]. Imbedding metallic structures within materials, or patterning metallo-dielctric composites, may be described by an effective permittivity (the capacitive portion of the material) and an effective permeability (the inductive component) [11, 12]. These may be determined self-consistently by the electrodynamic potential equations. Typically, the spatial perturbation of the material is considered small on a scale of a wavelength, and therefore the material or the structured film may be considered as homogeneous. Often, these scatterers are not at resonance with the propagating wave, so their effect is limited to their dimensional cross-section. Such is the case of traditional ADs, where the changes in the index of refraction are typically small, less than 10% [12]; here, the imbedded spheres, disks, or simply wires are considered as dipoles, added to the uniform dielectric medium. The imbedding of other "impurities," such as, chiral structures, e.g. springs, or chiral molecules, e.g. sugar, has also been known [13–15]. However, when the material is imbedded with small resonators, or when the structure itself is made to resonate, local scatterings may no longer be treated as a perturbation, and higher orders of diffraction need to be included in a self-consistent manner. Since resonators have a strong frequency dependence, dispersion and inevitably loss need to be considered. Putting it differently, the constitutive relations, the relationships between the fields in a vacuum and the fields impacted by the material polarization, may no longer be valid under these circumstances [16].

Periodic structures have been a subject of interest since the early days of optics and helped establish its wave formulation. For example, Young has shown that light interferes with itself when transmitted through two slits at close proximity. The experiment may be repeated with many slits, or with a periodically altered refractive index (gratings) [17]. Talbot has shown that periodic structures have wavelength-dependent self-imaging properties [18], an effect that was later explained by Lord Rayleigh [19]. Typically, interference between two optical beams requires some degree of coherence, namely their relative phases should be correlated or, in the simplest case, fixed. Otherwise, these uncorrelated phases would rapidly average to zero within a few wave cycles. Three-(3-D) and two-dimensional (2-D) distributed structures take advantage of such coherence properties: the scattering from each individual feature is added coherently to the scatterings from other features. Strong scattering ensues because the positions of all the scattering features are also correlated. Simply put, the scattering features have a crystallographic symmetry. One may realize 3-D structures by an assembly of metal spheres (Fig. 11.1). This structure exhibits an effect of self-focusing because all diffraction orders in the lateral direction (left and right; top and bottom of the figure) are summed coherently along the structure axis (pointing towards the page).

(a)   (b)

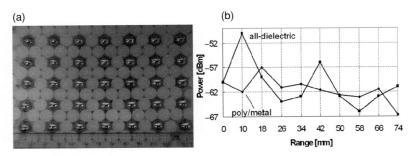

Figure 11.1. (a) A periodic structure made of interlaced metallic and nonmetallic (polypropylene) spheres. (b) Comparison of transmission through an all-dielectric polypropylene fcc crystal and poly/metal as a function of range along the propagation direction. The microwave frequency was 10 GHz and the sphere had a diameter of 1 cm [17].

Individual metallic or dielectric spheres display a resonance at some frequency owing to their size, geometry, their own dielectric constant, and its ratio with the dielectric constant of their surrounding. In essence they act like cavities, which are affected by their boundary conditions. If the sphere is small enough with respect to the radiation wavelength, it may be treated as a dipole antenna (Hertzian dipole). That does not mean that scattering from it is trivial. On the contrary; in some frequency range, a large scattering occurs if certain metals having negative dielectric constants are surrounded by dielectrics with positive dielectric constants. Furthermore, if the distance between these antennas is well defined (namely, the antennas are placed on a periodic structure or a lattice), the overall transmission of such phased-array antennas is dictated by the relative phase between elements, similarly to transmission of light through multiple slits. In retrospect, the linear properties of dielectric-filled periodic arrays of holes in a metal (metallo-dielectric screens) have been extensively investigated as spectral selective structures (optical filters), mostly in the infrared (IR) spectral region, long before they acquired fancy names [3]. Nonetheless, the ability to fabricate periodic and aperiodic structures at the nanoscale enabled us to investigate optical scatterings in the visible spectral region, and to assess related nonlinear properties of these, man-made, unique materials.

The aim of this chapter is to follow the traditional path of dielectric waveguides and interfaces in order to provide an introduction to those who are unfamiliar with the subject. It is also meant for those who wish to reflect upon some delicate derivation points along the way.

The chapter follows the formation of surface polariton modes from a waveguide point of view. As the waveguide thickness approaches zero, an interface is formed. Such is the interface between a metal and a dielectric (say, air) which sustains surface plasmon (SP) waves. A waveguide may be understood by the use of ray optics. A short description of reflection from an interface using ray optics terminology is provided. Ray optics, though, has its limitations when the interface becomes more complex. If the waveguide thickness is made extremely thin but finite, and is composed of another dielectric (say, oxide), the electric field intensity of the propagating surface modes maximizes at the discontinuity between the two dielectrics (in our example, between air and the oxide layer). If we further pattern the oxide layer with a periodic structure (say, an array of holes), a standing surface wave may be formed. Such a standing wave is responsible for an enhanced interaction between a molecule placed on the surface and the electric field carried by the surface mode. The enhanced interaction is useful for Raman spectroscopy, infrared spectroscopy, and the development of SP lasers, as shown below.

11.2 Two-dimensional surface polariton modes (a straightforward analysis)

Polaritons are formed through the interaction between electromagnetic (EM) radiation and collective wave-like entities, such as free-carrier charged particles, or a collection of free spins. Surface plasmon polariton (SPP) waves are formed through the interaction between EM radiation and charge carriers. These waves are confined to the interface between a lossy medium, typically a metal, and a dielectric. A similar phenomenon might be conjured to take place between a dielectric and a lossy medium of another type, which portrays negative permeability or surface magnetic polariton (SMP) waves, carried by free spins. While electrons in the metal do carry spin, the magnetic effect in the interaction between the EM radiation and the metal is typically weaker than the electric effect.

In the following we will concentrate on a simple slab guide because, in the end, we are interested in a thin oxide layer lying on a metallic substrate. A slab guide comprises three layers, and the structure is schematically described in Fig. 11.2. A surface guide may be deduced from a slab guide by extending the top layer to infinity and reducing the thickness of the guide layer to zero, forming one interface between two regions. Let us approach it systematically.

11.2.1 Homogeneous waveguides and interfaces

A slab waveguide is made up of three layers. We assume three dielectric layers (Fig. 11.2). The interfaces between the dielectric layers confine the electromagnetic field to the middle core layer. The wave is reflected back and forth between the interfaces while propagating along the opened waveguide axis. In the lateral direction (in our simple example, along the x direction) such counter reflections lead to

Haim Grebel

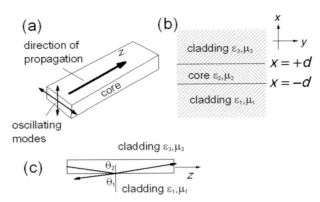

Figure 11.2. (a) A dielectric waveguide is a dielectric surrounded by another dielectric. In general, there are two independent linear oscillation modes (two linear polarization states). These are confined by the guide boundaries and propagate along the waveguide axis, along the z direction in this case. In the simplest case the three layers are placed on top of each other, as described in (b). (c) The refraction at the interface 2,1 when $Re n_2 < 0$ is shown just before the point of total internal reflection.

interference and formation of a stationary field pattern. The wave in the y direction is unbounded and is uniform in both $\pm y$ directions. For a positive dielectric (namely both real values of ε and μ are positive) the waveguide region should have a larger dielectric constant than its cladding. This is deduced from Snell's law. So, for example, between layers 2 and 1 we have $n_2 \sin \theta_2 = n_1 \sin \theta_1$, where n is the index of refraction $n = \pm \sqrt{\varepsilon \mu}$ of the corresponding layer and θ_1 is the refracted angle in layer 1 and measured from the surface normal. For a positive dielectric layer 2, $n_2 > 0$, and with a negative dielectric layer 1, $n_1 < 0$, then θ_1 is negative (as shown). At the point of total internal reflection, θ_1 becomes imaginary and the wave is totally reflected within the core layer. The angle of reflection equals the angle of incidence θ_1, regardless of whether the cladding is made of negative or positive cladding material. This is because of the momentum conservation assumption that leads to Snell's law: $\beta_{2z} = \beta_{1z} = \beta_{3z}$, where β_{jz} is the propagation constant in each region (see the following).

Metals have complex permittivity values $\varepsilon = \varepsilon_R + j\varepsilon_I$, albeit that their permeability values are close to the value for a vacuum, $\mu \approx \mu_0$. Certain metals, such as gold, silver, aluminum, and copper, exhibit large negative dielectric constants, which translates to a large imaginary refractive index with small real values. This separates ordinary metals from metamaterials, where both the real part of the permittivity and the permeability constants are negative, ε, $\mu < 0$. In that case, the refractive index may be real and negative, as was noted by Veselago [20]. To simplify the description, we will treat the dielectrics (ε, $\mu > 0$), metals ($\varepsilon < 0$, $\mu > 0$), and metamaterials (ε, $\mu < 0$) as having only real values of the permittivity and permeability constants. The treatment, though, is general, and one may replace these constants by their complex values.

Metal assisting surface guides may be understood as dielectric guides in the limiting case where the waveguide support is made of metal and the waveguide thickness approaches zero. Differently put, take layer 1 with $\varepsilon_1 < 0$ and $\mu_1 \sim \mu_0$, make the thickness of layer 2 extremely thin, and put $\varepsilon_3 > 0$ and $\mu_3 > 0$.

The standard approach to dealing with waveguides is to solve the wave equation for the electric or magnetic fields. While reiterating this extended description, it provides an opportunity to reflect on some points and re-examine their validity.

One starts with Maxwell's equations for the electric, **E**, magnetic, **H**, electric induction, **D**, and magnetic induction, **B**, fields (bold letters denote vectors):

$$\mathbf{curl\,E} = -\frac{\partial \mathbf{B}}{\partial t}; \ \mathbf{curl\,H} = \mathbf{J} + \frac{\partial \mathbf{D}}{\partial t}; \ \mathbf{div\,D} = \rho; \ \mathbf{div\,B} = 0, \quad (11.1)$$

where **J** is the current density and ρ is the charge density. We make use of the constitutive relations $\mathbf{B} = \mu\mathbf{H}$ and $\mathbf{D} = \varepsilon\mathbf{E}$, assuming source-free media (**J**, $\rho = 0$), and analyze the solution one frequency at a time by separating the spatial and the temporal contributions, $(\mathbf{E}(x, y, z), \mathbf{H}(x, y, z)) \exp(\pm j\omega t)$:

$$\nabla^2\mathbf{E} + \varepsilon\mu\frac{\omega^2}{c^2}\mathbf{E} + \frac{1}{\mu}(\nabla\mu) \times (\mathbf{curl\,E}) + \nabla\left[\mathbf{E} \cdot \frac{1}{\varepsilon}(\nabla\varepsilon)\right] = 0, \quad (11.2a)$$

$$\nabla^2\mathbf{H} + \varepsilon\mu\frac{\omega^2}{c^2}\mathbf{H} + \frac{1}{\varepsilon}(\nabla\varepsilon) \times (\mathbf{curl\,H}) + \nabla\left[\mathbf{H} \cdot \frac{1}{\mu}(\nabla\mu)\right] = 0. \quad (11.2b)$$

Here, c is the speed of light in vacuum. Since the field is a real quantity, and the time averaged energy of each photon has to be equal to a positive quantity $\hbar\omega$, in the end we will take $Re[(\mathbf{E}(x, y, z), \mathbf{H}(x, y, z)) \exp(-j\omega t)]$; radiation (propagation) is characterized by a phase $-(\omega t - \mathbf{k} \cdot \mathbf{r})$. We will further comment on this in the context of loss and dispersion in Section 11.2.6 .

The configuration of a slab waveguide simplifies the wave equations, Eqs. (11.2). According to Fig. 11.2 we can separate the modes of oscillation into two categories: one for which the **E** field oscillates parallel to the interfaces, and one in which it oscillates (or has a nonzero component) perpendicular to them. The first mode is known as the TE mode or s polarization and the other is known as the TM mode or p polarization. We are focusing on the TM mode; the TE mode may be deduced accordingly for metamaterials [21, 22]. Through Maxwell's equations, the TM mode may be described by the **H** field, whose oscillations are parallel to the slab interface. In our case, $\mathbf{H} = \mathbf{a_y}H_y$, where $\mathbf{a_y}$ is a unit vector along the y direction. Since the wave is oscillating in a direction transverse to the direction of propagation (in our case, the direction of propagation is along the z direction), there is no other **H** field component for this mode of oscillation. The corresponding **E** field has two components, E_x, E_z, which are derived from Maxwell's equations, Eqs. (11.1). We will use Eq. (11.2b) for convenience: it involves only one oscillation mode, along the y direction, and therefore the equation becomes a scalar equation. The last term in Eq. (11.2b) is zero because the interfaces are oriented perpendicularly to the x direction (or, to put it differently, their surface normal is parallel to the x direction). This means that **grad** μ is along the x direction, which is positioned perpendicularly to the chosen direction of oscillation, H_y, and the dot product between the two is zero.

If a solution is found, its dependence on the y coordinate must be similar at every point (up to a phase difference) because every segment of the guide along this direction resembles the others. We conveniently set $\partial/\partial y(\partial H_y/\partial y) = 0$. Equation (11.2b) is, therefore, written as follows:

$$\left[\frac{\partial^2}{\partial x^2} + \frac{\partial^2}{\partial z^2} + \varepsilon(x)\mu(x)\frac{\omega^2}{c^2} - \frac{1}{\varepsilon(x)}\frac{\partial\varepsilon(x)}{\partial x}\frac{\partial}{\partial x}\right]H_y = 0. \qquad (11.3a)$$

For a slab comprising three homogeneous films, the last term is zero; one simply solves Eq. (11.3a) separately in the three regions 1, 2, and 3 and matches the solution(s) at the interfaces.

When solving second order differential equations, boundary conditions assert the continuity of the field and its first derivative across the guide boundaries. Yet, because the boundaries are defined by two different materials, the (**E**, **H**) fields are coupled, and the derivatives of one field are related to the other and vice versa, one may use Maxwell's equations instead. These boundary conditions for the problem at hand are as follows:

(1) The magnetic field component, H_y, parallel to the interfaces is continuous across the boundaries.
(2) The electric field component, E_z, parallel to the interfaces and along the direction of propagation is continuous across the boundaries. The latter is deduced from the H_y solution using $E_z = (1/j\omega\varepsilon)\partial H_y/\partial x$.
(3) The electric field component, E_x, perpendicular to the interfaces obeys the relation $\varepsilon_1(x)E_{x1} = \varepsilon_2(x)E_{x2}$. The latter is deduced from the H_y solution using $E_x = -(1/j\omega\varepsilon)\partial H_y/\partial z$.

One is searching for a solution that is bound to the waveguide core (layer 2) and which propagates in unison in all waveguide regions at a single speed. This means that we are searching for a solution in the form of $H_y(x, y, z) = H_y(x, y)\exp(j\beta_z z)$. Inserting the proposed solution into Eq. (11.3a), we obtain

$$\left[\frac{\partial^2}{\partial x^2} + \left(\varepsilon_i\mu_i\frac{\omega^2}{c^2} - \beta_z^2\right)\right]H_y = 0, \qquad i = 1, 2, 3. \qquad (11.3b)$$

We now define the propagation constants (some of which may be complex):

$$\kappa_1 = \left(\beta_z^2 - n_1^2\frac{\omega^2}{c^2}\right)^{1/2}; \; \kappa_2 = \left(\beta_z^2 - n_2^2\frac{\omega^2}{c^2}\right)^{1/2}; \; \kappa_3 = \left(\beta_z^2 - n_3^2\frac{\omega^2}{c^2}\right)^{1/2}.$$

(11.4a)

Here, $n_i^2 = \varepsilon_i\mu_i$ is the refractive index of layer i. It is convenient to assume the solution to be as follows:

$$H_y(x) = \begin{cases} A\exp(-\kappa_1|x|) & x < -d \\ [B\exp(+\kappa_2 x) + C\exp(-\kappa_2 x)] & -d < x < +d \\ D\exp(-\kappa_3|x|) & x > +d. \end{cases}$$

(11.5)

If this is a solution to Eq. (11.3a), obeying the boundary conditions outlined before, then this is the only solution. By matching the boundary conditions (specifically, by using (1) and (2) from the recipe above), we arrive to a relationship between the parameters κ_1, κ_2, κ_2, and β_z for given ε_1, ε_2, and ε_3.

The condition that ascertains the existence of a nonzero solution is written as, and is known as, the characteristic equation for the waveguide, namely

$$\tanh(2\kappa_2 d) = -\frac{\kappa_2}{\varepsilon_2}\frac{(\kappa_1/\varepsilon_1) + (\kappa_3/\varepsilon_3)}{(\kappa_2/\varepsilon_2)^2 + (\kappa_1/\varepsilon_1)(\kappa_3/\varepsilon_3)}.$$

(11.6)

We may rearrange Eqs. (11.4a) as follows:

$$\kappa_2^2 d^2 + \kappa_1^2 d^2 = (-n_2^2 - n_1^2)\frac{\omega^2}{c^2}d^2;$$

$$\kappa_2^2 d^2 + \kappa_3^2 d^2 = (-n_2^2 - n_3^2)\frac{\omega^2}{c^2}d^2.$$

(11.4b)

Comment: the negative sign on the right-hand side of Eqs. (11.6) and (11.4b) is simply due to the definition of κ_2 – the wavenumber along the x direction within the waveguide core. It becomes purely imaginary for a positive dielectric layer, and it may be real for surface metamaterial guides.

Now consider the case where the waveguide thickness is approaching zero, $d \sim 0$; the left-hand side of Eq. (11.6) becomes very small, and the numerator on the right-hand side of the equation is set to zero, i.e. $(\kappa_1/\varepsilon_1 + \kappa_3/\varepsilon_3) = 0$. This is the condition for the propagation of a surface wave. The dielectric constant of plasmonic metals is frequency dependent and is typically taken as $\varepsilon_1 = \varepsilon_m = \varepsilon_0(1 - \omega_p^2/\omega^2)$, with ω_p being the plasma frequency. The dielectric constant of a dielectric is of order unity and positive. Since both κ_1 and κ_3 are positive numbers, namely the wave decays exponentially away from the surface, a solution to this waveguide construction may be found.

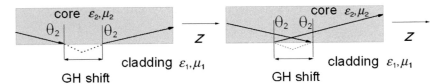

Figure 11.3. Goos–Hänchen (GH) shift may take positive (left) and negative (right) signs.

11.2.2 Goos–Hänchen shift in optical waveguides (ray optics approach)

In the ray optics approximation under the total reflection condition, the wave zigzags inside the guide between the two interfaces. It acquires a phase shift, as if it is partially penetrating the waveguide wall and refracts back. The shift for a *p*-polarized (TM) mode, where the **H**-field oscillations are parallel to the surface, is shown schematically in Fig. 11.3: it may take positive and negative values depending on the relative signs between the core and cladding of the interface. The phase shift amounts to a displacement of the beam under total internal reflection, as shown in the figure.

This actual beam displacement is given as follows (see ref. [23], where the standard Goos–Hänchen displacement, D, was replaced by the GH shift$= S_{21} = D/\cos\theta_2$, as shown in Fig. 11.3):

$$S_{21}^{(TM)} = \frac{2}{kn_2}\left[\frac{\varepsilon_{12}\cos\theta_2\sin\theta_2}{\varepsilon_{12}^2\cos^2\theta_2 + \sin^2\theta_2 - n_{12}^2}\right]\left[\frac{1}{\sqrt{\sin^2\theta_2 - n_{12}^2}}\right],$$

where $\varepsilon_{12} = \varepsilon_1/\varepsilon_2$ and $n_{12} = n_1/n_2$. This translation may be written as a phase delay. In a slab waveguide, there are two such delays from each of the waveguide interfaces. Therefore

$$2kn_2 d\cos\theta - 2\phi_{21} - 2\phi_{23} = 2\pi m \qquad (m = 1, \pm 1, \pm 2\ldots),$$

where θ_2 is the local angle of reflection between the ray and the normal to the interface and m is the mode number (see also Fig. 11.2(c)). For a TM mode, the phase shift at each interface is given by

$$\tan\phi_{12} = \frac{\kappa_1\varepsilon_2}{|\kappa_2|\varepsilon_1},$$

$$\tan\phi_{23} = \frac{\kappa_3\varepsilon_2}{|\kappa_2|\varepsilon_3},$$

$$2dkn_2\cos\theta = 2d|\kappa_2|.$$

At grazing angles, $\theta \sim \pi/2$ and the acquired phase shift approaches $\pi/2$. Therefore, for the lowest order mode, propagating between very close interfaces, we write $|\kappa_2|d = 0$, and $-2\phi_{21} - 2\phi_{23} = +2\pi$.

For a surface guide, only the lowest mode may propagate and, therefore, $m = 0$. As a result, the total acquired phase shift is $-2\phi_{21} - 2\phi_{23} = 0$. Due to the signs of the permittivity values, obviously $\phi_{21} < 0$ and $\phi_{23} > 0$, and these inequalities are automatically fulfilled. The above is true only for metals with $\varepsilon < 0$ and $\mu > 0$, or conversely for materials with $\varepsilon > 0$ and $\mu < 0$. The requirements for metamaterials are similar: while the refracted angle in Snell's law, $n_2 \sin \theta_2 = n_1 \sin \theta_1$, reverses itself for $n_1 < 0$ and $n_2 > 0$, from a critical angle point of view the angle of reflection will be always positive to the incident angle regardless of the definition of the refractive index n. As was pointed out by Veselago [20], the energy flux (the Poynting vector) always forms a right-hand system with the electric and magnetic field vectors. Therefore, the intensity energy flow in the waveguides with a dielectric core which is bounded by a metamaterial cladding is always positive. The situation with a metamaterial waveguide (namely, a left-handed core) encapsulated with a right-handed cladding is a bit different: since both ϕ_{21} and ϕ_{23} are positive, n_2 becomes negative. A large Goos–Hänchen shift has been calculated when a metamaterial slab is backed by a metal [24]. In that case, the slab with the metal acts as a resonator.

11.2.3 Surface modes

Standard surface plasmon case: metal–dielectric interface

The ε_1 region is a plasmonic metal, namely $Re\,\varepsilon_1 \ll 0$, while, the ε_3 region is a dielectric, $\varepsilon_3 > 0$. While, in general, ε_1 is described by a complex number, we will treat metals as if they do not exhibit a major loss component (in other words, $Im\,\varepsilon \sim 0$), unless otherwise noted. Both regions have $\mu_{1,3} > 0$. Equation (11.6) is simplified and becomes, $0 = \kappa_1/\varepsilon_1 + \kappa_3/\varepsilon_3$. The equality holds since both κ_1 and κ_3 are taken to be positive (otherwise the field envelope would not decay exponentially along the x direction). On substituting into this equation the expressions for κ_1 and κ_3 given by Eqs. (11.4), we obtain for the wavenumber β_z the result

$$\beta_z^2 = \frac{\omega^2}{c^2} \frac{\varepsilon_3^2 n_1^2 - \varepsilon_1^2 n_3^2}{\varepsilon_3^2 - \varepsilon_1^2}. \tag{11.7}$$

Here we used, $n^2 = \varepsilon\mu$. The condition for propagation is $\beta_z^2 > 0$. Remember that region 1 is the metal ($n_1^2 < 0$) whereas region 3 is the dielectric ($n_3^2 > 0$). The numerator is therefore negative, which dictates that the denominator ought to be negative too. This will occur if $|\varepsilon_1^2| > |\varepsilon_3^2|$ and it is true for silver, gold, copper, and aluminum. For example, at the d-line wavelength of a sodium lamp, $\lambda = 0.5893$ μm, silver has [25] $\varepsilon/\varepsilon_0 = -11.793 + j0.688$ and $\mu \sim \mu_0 = 4\pi\,10^7$ H/m (with $\varepsilon_0 \sim (1/36\pi)10^{-9}$ F/m); in this case, $n_1^2 \sim \varepsilon_1 < 0$.

It is interesting to compare the expression for the propagation constant, β_z, with the more traditional expression in the literature. We note that nonmagnetic materials have $\mu_{1,3} \sim \mu_0$. This simplifies the expression for β_z a little because in that case $n_{1,3}^2 \sim \varepsilon_{1,3}$ and $\beta_z = (\omega/c)\sqrt{\varepsilon_1\varepsilon_3/(\varepsilon_1 + \varepsilon_3)}$.

Using Eqs. (11.4), the decay constants are given by (κ_1, $\kappa_3 > 0$)

$$\kappa_1^2 = \beta_z^2 - n_1^2\frac{\omega^2}{c^2} = \frac{\omega^2}{c^2}\frac{\varepsilon_3^2 n_1^2 - \varepsilon_1^2 n_3^2}{\varepsilon_3^2 - \varepsilon_1^2} - n_1^2\frac{\omega^2}{c^2} = \frac{\omega^2}{c^2}\varepsilon_1^2\frac{n_1^2 - n_3^2}{\varepsilon_3^2 - \varepsilon_1^2},$$

$$\kappa_3^2 = \beta_z^2 - n_3^2\frac{\omega^2}{c^2} = \frac{\omega^2}{c^2}\frac{\varepsilon_3^2 n_1^2 - \varepsilon_1^2 n_3^2}{\varepsilon_3^2 - \varepsilon_1^2} - n_3^2\frac{\omega^2}{c^2} = \frac{\omega^2}{c^2}\varepsilon_3^2\frac{n_1^2 - n_3^2}{\varepsilon_3^2 - \varepsilon_1^2}.$$

The solution for the propagating **E** field is (up to a constant):

$$E_x = -\frac{1}{j\omega\varepsilon_1}\frac{\partial H_y}{\partial z} = Const\left(\frac{\beta_z}{\omega\varepsilon_1}\right)\exp(+\kappa_1 x)\exp[-j(\omega t - \beta_z z)] \quad x < 0;$$

$$E_z = +\frac{1}{j\omega\varepsilon_1}\frac{\partial H_y}{\partial x} = Const\left(\frac{\kappa_1}{\omega\varepsilon_1}\right)\exp(+\kappa_1 x)\exp[-j(\omega t - \beta_z z)] \quad x < 0;$$

$$E_x = -\frac{1}{j\omega\varepsilon_3}\frac{\partial H_y}{\partial z} = Const\left(\frac{\beta_z}{\omega\varepsilon_1}\right)\exp(-\kappa_3 x)\exp[-j(\omega t - \beta_z z)] \quad x > 0;$$

$$E_z = +\frac{1}{j\omega\varepsilon_3}\frac{\partial H_y}{\partial x} = Const\left(\frac{-\kappa_3}{\omega\varepsilon_3}\right)\exp(-\kappa_3 x)\exp[-j(\omega t - \beta_z z)] \quad x > 0.$$

$$(11.8)$$

Because the electric field has components parallel and perpendicular to the direction of propagation, one may conclude that the field has an elliptical polarization in either medium. Similarly to a sea wave, the electric field is rolling in the direction of wave propagation (or away from it, as in the case of the dielectric).

One may repeat the entire calculation for the TE modes, which means that instead of H_y we are considering E_y. The relationships between the **E** and **H** fields are as follows: $H_x = +(1/j\omega\mu)\partial E_y/\partial z$ and $H_z = (-1/j\omega\mu)\partial E_y/\partial x$. Effectively, this means that we replace H_y by E_y and ε by $-\mu$ in the above equations.

The characteristic equation will now read

$$\tanh(2\kappa_2 d) = -\frac{\kappa_2}{\mu_2}\frac{\kappa_1/\mu_1 + \kappa_3/\mu_3}{(\kappa_2/\mu_2)^2 + (\kappa_1/\mu_1)(\kappa_3/\mu_3)}. \qquad (11.9)$$

Again, a positive dielectric guide layer will result in an imaginary value for κ_2.

When setting $\mu_1 > 0$ and $\varepsilon_1 < 0$ (say, a plasmonic metal), a TE mode does not exist because the numerator is always positive and cannot be set to zero; $\kappa_1/\mu_1 + \kappa_3/\mu_3 > 0$ (remember that layer 3 is made of a positive dielectric).

When setting $Re\,\mu < 0$ and $Re\,\varepsilon > 0$ (a spintronic medium) a surface TE wave may be sustained by a lossy medium for which $Im\,\mu > 0$. Again we assume $Im\,\mu \sim 0$ for the time being. As long as $n_1^2 < 0$ and $n_3^2 > 0$, such a surface mode exists.

For an interface between a dielectric, such as air ($\varepsilon_3 \sim 1$), and a plasmonic type metal (for example, silver) with ε_1 such that $|\varepsilon_1| \gg |\varepsilon_3|$, we may simplify the equations. Such a condition leads to $\kappa_1 \gg \kappa_3$, and the wave propagates mainly in the dielectric, confined to the interface. Attenuation is the result of the mode propagating in the metal. The attenuation coefficient is related to the imaginary part of the propagation constant. For $\mu_1 \sim \mu_3 \sim \mu_0$, we arrive at the following text book result:

$$\beta_z = \frac{\omega}{c} \left[\frac{\varepsilon_1 \varepsilon_3}{\varepsilon_1 + \varepsilon_3} \right]^{1/2},$$

where the dielectric constant for the cladding is ε_3, a real number, and, for the metal, $\varepsilon_1 = \varepsilon_{1R} + j\varepsilon_{1I}$, as identified for silver before.

Metamaterial and dielectric interfaces

Considering now the interface between a metamaterial region and a dielectric, the calculations may be repeated for two half-spaces: one is made of a dielectric as before ($Re\{\varepsilon_3, \mu_3\} > 0$) and the other, the substrate, is made of a metamaterial ($Re\{\varepsilon_1\,\mu_1\} < 0$) (see, for example, ref. [26]). In this case, the characteristic equation for the TM and TE modes remains the same. The condition that is deduced from these equations, $0 = \kappa_1/\varepsilon_1 + \kappa_3/\varepsilon_3$ for the TM, or p polarization, mode, is also valid. The propagation constant is positive, $\beta_z^2 = (\omega^2/c^2)[(\varepsilon_3^2 n_1^2 - \varepsilon_1^2 n_3^2)/(\varepsilon_3^2 - \varepsilon_1^2)]$; this sets the conditions for the relationship between the various dielectric constants. We note that $n_{1,3}^2 > 0$ for both regions. The same holds true for the TE, or s polarization, mode: the condition for a surface mode is $0 = \kappa_1/\mu_1 + \kappa_3/\mu_3$, and the propagation constant is given by $\beta_z^2 = (\omega^2/c^2)[(\mu_3^2 n_1^2 - \mu_1^2 n_3^2)/(\mu_3^2 - \mu_1^2)]$.

Metamaterial and metal interfaces

We consider surface waves between a metal and a metamaterial. We retain region 1 as a metal ($Re\,\varepsilon_1 < 0$, $Re\,\mu_1 > 0$) and make region 3 a metamaterial ($Re\{\varepsilon_3, \mu_3\} < 0$). The condition $\beta_z^2 > 0$ remains valid. For the TM (p polarization) mode $n_1^2 < 0$,

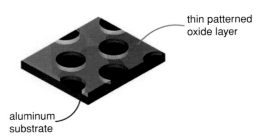

Figure 11.4. Surface guide made of an aluminum substrate with a very thin ($\sim$50 nm) oxide layer, which is patterned with a hexagonal array of holes.

which makes the numerator negative, and therefore we need to set $(\varepsilon_3^2 - \varepsilon_1^2) < 0$. For the TE ($s$ polarization) mode we set $(\mu_3^2 - \mu_1^2) < 0$. We note that, for plasmonic metals, the real part of the refractive index is indeed $Re\, n_1^2 \ll 0$, yet the refractive index of the metal is in general a complex number, as we saw for silver. As we mentioned before, a large Goos–Hänchen effect may occur if we place a thin metamaterial layer between a positive dielectric and a metal [24].

11.2.4 Periodically patterned interfaces

Consider now a very thin (of the order of $\lambda/10$) interfacial layer between the dielectric and metal. This interfacial layer is also patterned, as shown in Fig. 11.4, and has a dielectric constant $\varepsilon_2 \sim 1.77$, compared with $\varepsilon_1 = 1$ and $\varepsilon_3 = -38.74 + j10.42$ at $\lambda = 515$ nm. The pattern comprises holes in the dielectric layer: the hole diameter is 20 nm and the array pitch is 90 nm. The array has hexagonal symmetry.

Since the pattern is periodic, one may expand it in a Fourier series: $\varepsilon_2(y, z) = \sum_{q_y, q_z} \varepsilon_2(q_y, q_z) \exp(jq_y y + jq_z z)$. One may use, then, a perturbation approach. The magnetic field distribution is taken as a linear combination of the solutions for a system without the perturbation: $H_y(x, y, z) = \sum_m A_m(z) H_{my}(x, y) \exp(j\beta_{mz} z)$. Here $\mathbf{q} = (q_y, q_z)$ is the wavevector of the pattern; if the pattern is periodic, it may be viewed as a lattice. In our case, the intermediate layer is very thin and the lattice is two-dimensional (2-D). Therefore, one may set $\mathbf{q} = \mathbf{G}$, a vector of the reciprocal lattice.

This perturbation approach may be questionable in the presence of strong scattering, namely when higher orders of diffraction participate in the scattering process. The reason is that we have ignored the gradient terms of both ε and μ in Eqs. (11.2) when treating homogeneous layers (each layer was considered separately, and

(a)   (b)

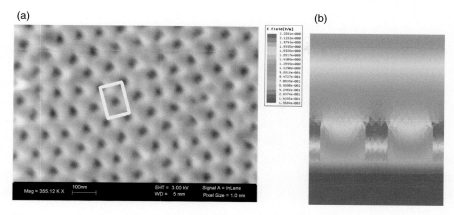

Figure 11.5. (a) Hole array fabricated in 50 nm oxide layer on top of aluminum (anodized aluminum oxide, or AAO). The rectangle indicates the simulation cell. (b) Cross sectional view of electric field maps when the incident beam is launched at normal incidence. The refractive index for the oxide was 1.77. The electric field intensity values range from 2.25 V/m to 10^{-3} V/m. The incident beam was taken as 1 V/m [27]. Simulations were obtained with an Ansoft tool. See color plates section.

we tailored the final solution by demanding continuity of the solution and its derivatives).

One immediate observation, however, is that the solution is no longer uniform along the y direction because $\varepsilon_2 = \varepsilon_2(y)$, and gradients of the dielectric constant would appear in Eq. (11.2b). The solution will be periodic along the y direction through the use of the Floquet theorem [12]. Also, the propagation constant along the z direction will be modified ($\beta_z \rightarrow \beta_z + q_z$) because it experiences scatterings from a well defined structure. Under certain conditions, the wave may be reflected back and forth, and a standing wave may be formed. At this point, it appears that the problem becomes too complex for an analytical approach, and we seek a numerical solution through one of the available computer aided design (CAD) tools. The solution is presented as an electric field distribution and is shown in Fig. 11.5. An incident plane wave is launched at normal incidence on the surface of Fig. 11.4. One obtains the intensity field plots (thermal plots) for a cell defined in Fig. 11.5(a). We assume that the pattern repeats itself. Similar plots may be obtained as the sample is tilted and rotated. What we take out of this and other simulations is that the field is distributed at the top of the interfacial thin dielectric layer and is mainly concentrated at the hole area. Such observations are also confirmed by analytical solutions to a one-dimensional problem of a loaded microwave guide; loaded waveguides are guides with a thin dielectric coating [12].

11.2.5 Suspended periodic metallic structures

Free-standing thick metal screens are metallic structures with a hole array. The array may have a 2-D crystallographic symmetry, such as square or hexagonal symmetry (Fig. 11.5). The complementary structure (metal patches where the openings are and dielectric material where the metal area is) requires a dielectric backing. Here, the metallic layer is surrounded by a dielectric cladding, typically air. The thickness of the screen, t, is considered thin when $t/\lambda \ll 1$ or thick when $t/\lambda > 0.1$. Such screens have been studied for a long time [3]. In general, periodic metallo-dielectric structures are able to discriminate desired infrared (IR) signals from more energetic short wavelength radiation, allow color temperature measurements, provide order sorting for grating spectrometers, and improve the signal-to-noise ratio of Fourier transform spectrometers. Free-standing metal screens are commercially available and have been used as band pass filters [28], reflectors for long IR wavelengths, Fabry–Perot etalons [29], antennas [3], and platforms for bio-chemical characterizations [30, 31].

In principle, such periodic structures enable transmission (or reflection) of specific frequency bands by invoking standing wave surface modes; the incident beam is coupled to the surface waves by the periodic structure, which in turn supports standing waves [32]. The reason these structures are mentioned here is because they too utilize surface plasmonic modes as described before. Namely, layers 1 and 3 are dielectrics and layer 2 is the perforated plasmonic layer with $Re\, \varepsilon_2 \ll 0$. Propagation of electromagnetic radiation is enabled only at the metal–dielectric surfaces. These surface modes are coupled by the holes in the metal, therefore enabling transmission of energy flux from one side of the screen to the other. The screens exhibit transmission of a certain frequency band, which is related to the periodicity of the array. Although there was a great deal of excitement in the literature about the so-called extraordinary transmission through periodic metallic structures [33], this excitement was built upon the erroneous premise that the transmission ought to be related to the relative area of the screen's openings. This is definitely not the case here, and, for that matter, is not true for a large class of resonating structures (for example, a Fabry–Perot etalon, comprising two mirrors facing each other, transmits 100% of a certain wavelength even if the reflectivity of the mirrors approaches 100%).

The reason why these screens are able to achieve large transmission at certain wavelengths (in some cases, transmission as high as 100%) is because of an efficient coupling between the two surface modes on either side of the screen. This occurs when the surface modes become standing waves, thus enabling coupling via the waveguide modes in the screen openings. The transmission changes as a function of tilt and rotation of the screen and depends on the local field polarization [34].

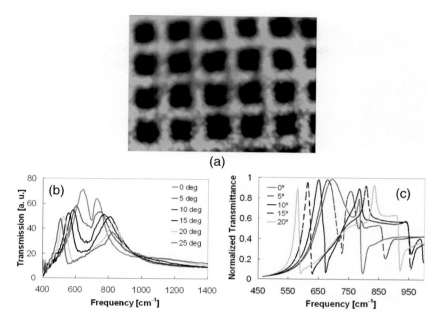

(a)

Figure 11.6. (a) Flat metal screens: $7.6 \times 7.6\ \mu m^2$ openings, arranged in a square lattice with lattice constant $a = 12.7\ \mu m$. (b) Transmitted p polarization through a screen tilted at various angles. (c) Simulations of transmission of a p-polarized incident beam as a function of tilt angle [30]. The peaks become farther apart as the tilt angle between the incident beam and the screen increases.

The coupling to the surface plasmons is achieved by coherent scatterings from the periodic structure of openings.

As previously described, the wavenumber of the propagating surface mode, β, is modified by the periodic structure such that $\beta \to \beta + \mathbf{G}$, where $\mathbf{G}$ is a translation vector of the lattice reciprocal to the structure. A reciprocal lattice vector $\mathbf{G}$ is a linear combination of the primitive translation vectors $\mathbf{G}_1$ and $\mathbf{G}_2$ of the reciprocal lattice. For holes arranged in a square lattice, $\mathbf{G} = q_1 \mathbf{G}_1 + q_2 \mathbf{G}_2$, where $\mathbf{G}_1 = (2\pi/a, 0), \mathbf{G}_2 = (0, 2\pi/a)$, a is the lattice constant, and q_1 and q_2 are integers. For a given pitch the free-standing screen will transmit a certain frequency band. The peak frequency (wavelength) will split and the frequency gap between the peaks will vary as a function of tilt angle, as shown in Fig. 11.6. The frequency split is due to the existence of two standing waves, each centered at a slightly different frequency. The transmission for s polarization (the oscillation mode which is parallel to the screen) exhibits a pass-band with a single transmission peak; the peak is down shifted as a function of the tilt angle. It is clear from the preceding discussion that, in order to transmit a *given* incident wavelength through a given

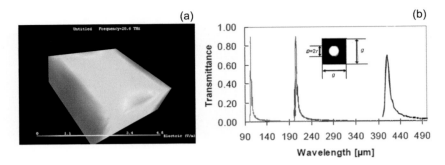

Figure 11.7. (a) Simulations of the electric field intensity distribution at the surface of a copper screen; opening $= 7.6 \times 7.6 \ \mu m^2$ and pitch $= 12 \ \mu m$. The electric field is peaked at the edges of the screen opening when the wavelength is at resonance with the structure. The relatively small opening allows for only the lowest order waveguide mode to propagate and couple to the surface plasmon on the other side of the screen. (b) Simulations of transmission for three screens with *periodicities* of 100, 200, and 400 μm, respectively. The screen was made of round holes in a square lattice. The screens' thicknesses were, respectively, 2, 5, and 5 μm. The hole's radius was one-fifth of the screen pitch. Note how narrow and how large the transmission may be when the screen is relatively thick and the opening is relatively narrow.

periodic structure, we need to tilt and rotate the screen with respect to the incident beam in order to obtain optimal coupling to the surface plasmon modes. Unlike Fig. 11.4, the electric field under the resonance condition (maximum transmission) is concentrated at the edges of the screen opening (Fig. 11.7).

The next question would be how does one form a standing surface plasmon wave? The answer is rather simple: we invoke second order diffractions. We clarify it by example; suppose that we employ a metal screen with a square symmetry of holes such that $a = a_0$. At normal incidence, the optimal wavelength coupled to the screen's surface mode would be $\lambda = a_0$. The propagation constant will be $\beta = 2\pi/\lambda = 2\pi/a_0 = G$. Efficient backscattering along the surface occurs at the Bragg condition [35], which takes the form $2\beta = mG$ when the surface waves are scattered in the direction normal to the periodic interfaces. Therefore it is clear that $m = 2$ and the back reflection of the surface wave occurs *via* the second order of diffraction or twice the spatial pitch frequency. The surface mode will be scattered back and forth, and a standing mode will be formed. We will return to this point later on.

11.2.6 Energy considerations, dispersion, and loss

The energy carried by an electromagnetic wave may be divided into two parts: the internal energy and the energy flux. The energy flux, **S**, also known as the Poynting vector, is the electromagnetic energy that is transfered through a known surface. The internal energy, or the energy density W (if defined per unit volume), is the energy contained at a given time in a given volume. Conservation of energy dictates that

$$\text{div } \mathbf{S} + \frac{\partial W}{\partial t} = 0, \qquad (11.10)$$

similar to the relationship between the charge density ρ and the current $\mathbf{J}$: div $\mathbf{J}$ + $\partial\rho/\partial t = 0$. The energy flux averaged over one period of the wave is given by $\langle \mathbf{S} \rangle = (1/2)Re\{\mathbf{E} \times \mathbf{H}^*\}$. For a TM ($p$ polarization) waveguide, or for surface modes propagating along the guide axis in the z direction, $\langle S_z \rangle = (1/2)(E_x H_y)$. The triad $\mathbf{E}$, $\mathbf{H}$, and $\mathbf{S}$ always forms a right-handed system. This expression may be written differently, i.e. $\langle \mathbf{S}(x) \rangle = (\boldsymbol{\beta}/2\omega\mu)|E(x)|^2$, where β is the propagation constant of the wave in the medium. For a homogeneous medium, the propagation constant may be written as $\beta \approx k_0 n$, and, therefore, if $\mu < 0$, then $n < 0$ also in order to maintain a positive energy flux. (Note that in the case $\varepsilon > 0$, $n = \sqrt{\varepsilon\mu}$ is imaginary and the energy flux is decaying over a short propagation distance.) This means that negative refractive index materials have phase propagation (the direction of the $\mathbf{k}$ vector) opposite to the direction of the energy flux. In our case, $\langle S_z(x) \rangle = (\beta_z/2\omega\mu)|E_x(x)|^2$, and we arrive at similar conclusions.

Considering surface plasmons and a given direction of propagation β_z along the surface, the energy flux in the dielectric and in the metal are in opposite directions: this necessitates the formation of an energy flux vortex. Therefore, in a waveguide made of a metamaterial core with a dielectric cladding, there ought to be two energy flux vortices to sustain propagation in the waveguide [22]. The propagation of phase, however, is not a physical quantity, and therefore may not be important for metamaterials. When both ε and μ are negative and real, the triad $\mathbf{E}$, $\mathbf{H}$, and $\boldsymbol{\beta}$ forms a left-handed system ($\mathbf{E} \times \mathbf{H}$ is opposite to $\boldsymbol{\beta}$), hence the name "left-handed materials."

The problem for homogeneous metamaterials lies with the internal energy expression. This expression equals the photon energy $\hbar\omega$. The internal energy is written as $\langle W \rangle = (1/4)Re\,[\mathbf{E} \cdot \mathbf{D}^* + \mathbf{H} \cdot \mathbf{B}^*] = \hbar\omega$. Using the constitutive relations $\mathbf{D} = \varepsilon\mathbf{E}$ and $\mathbf{B} = \mu\mathbf{H}$, we may write the internal energy as $\langle W \rangle = (\varepsilon/4)|E|^2 + (\mu/4)|H|^2$. If both ε and μ are negative and real, then the internal energy is negative, which means that the photon energy is negative. Thermodynamically, the electromagnetic energy density is the difference between the free energies of the system with and without the wave, respectively. If ε, μ are negative, the thermodynamic system becomes unstable.

The practical solution is to avoid homogeneous structures altogether. We recognize that the above energy density term is correct only for homogeneous materials, and that the constitutive relations are valid only for nondispersive media. In general, the entities ε and μ are 3×3 matrices (namely $[\varepsilon]$ and $[\mu]$ have different values along different directions) and depend on the frequency of the wave. The constitutive relations for the component i of the field are written as follows [16] (summation over repeating indices is assumed):

$$D_i(\mathbf{r}, t) = E_i(\mathbf{r}, t) + \int d^3r' \int dt' \, f_{ik}(t', \mathbf{r}, \mathbf{r}')E_k(\mathbf{r}', t - t'), \quad (11.11a)$$

$$B_i(\mathbf{r}, t) = H_i(\mathbf{r}, t) + \int d^3r' \int dt' \, f_{ik}(t', \mathbf{r}, \mathbf{r}')H_k(\mathbf{r}', t - t'). \quad (11.11b)$$

The integral over time is taken in the range $0 < t' < \infty$ (although one might argue that it should be extended to $-\infty < t < +\infty$ for finite resonating structures to include advanced and retarded potentials). Here, the function $f_{ik}(t', \mathbf{r}, \mathbf{r}')$ is a correlation function, which relates the electric field $\mathbf{E}$ to the $\mathbf{D}$ field through events that occurred over time and space. In essence one may identify f_{ik} with the susceptibility χ.

In the simple case where the electric and magnetic fields are taken to be slowly varying plane waves, $\mathbf{E} = \mathbf{E}_0(\mathbf{r}, t) \exp[-j(\omega t - \boldsymbol{\beta} \cdot \mathbf{r})]$, we may approximate the f_{ik} function by an averaged function and identify $\langle f_{ik} \rangle = \partial(\varepsilon_{ik}E_k)/\partial t$ from the general constitutive relations $D_i = \varepsilon_{ik}E_k$. Taking the Fourier transform of the preceding equation (and emphasizing again that the wave is monochromatic), we get $\langle f_{ik} \rangle E_k = -j\omega\varepsilon_{ik}(\omega, \beta)E_k$. The latter is achieved by writing a spatial Fourier transform and interchanging the order of integration. For a monochromatic wave propagating in an anisotropic medium, the energy density and the energy flux may be written as follows:

$$\langle W \rangle = \frac{1}{4} \frac{\partial(\omega\varepsilon_{ik})}{\partial\omega} E_i E_k^* + \frac{\partial(\omega\mu_{ik})}{\partial\omega} H_i H_k^*, \quad (11.12a)$$

$$\langle S \rangle = \frac{1}{2} Re[\mathbf{E} \times \mathbf{H}^*] - \frac{\omega}{4} \frac{\partial\varepsilon_{ik}}{\partial\boldsymbol{\beta}} E_i E_k^*. \quad (11.12b)$$

The purpose for writing these expressions is this: one may resolve the issue of negative energy densities associated with homogeneous metamaterials by instead turning to anisotropic and dispersive media. Since the derivatives in front of the amplitude square terms may become positive, even if the local real values of the permittivity and permeability constants are negative, the overall energy expression remains positive as desired. The energy flux is also modified by the spatial–temporal dispersion, as can be seen from Eqs. (11.12). Typical dispersion relations for plasmonics and spintronics are

$$\varepsilon(\omega) = \varepsilon_0 \left(1 - \omega_p^2/\omega^2\right) \quad \text{and} \quad \mu(\omega) = \mu_0 \left[1 - C_1\omega_2/(\omega^2 - \omega_0^2)\right], \quad (11.13)$$

respectively, where ε_0, μ_0 are the permittivity and permeability constants of the vacuum, respectively; ω_p is the plasma frequency; C_1 is a constant; and ω_0 is the resonance frequency of the magnetic elements. Therefore, the way to handle metamaterials is to return to Eqs. (11.5), (11.6), and (11.4b) with the now dispersive and complex values for the permittivity and permeability operators. As was pointed out in ref. [36], dispersion and loss are inherent to metamaterial structures, and therefore cannot be ignored under any circumstances.

On the sign of the temporal phase factor and the imaginary part of the permittivity and permeability constants

The choice of the phase $\pm(\omega t - \mathbf{k} \cdot \mathbf{r})$ is a matter of definition, and the bewildered reader may find it written in publications either way. In general, the electric or magnetic fields may be written as: $\mathbf{A}(\mathbf{r}) \exp(-j\omega t) + \mathbf{A}^*(\mathbf{r}) \exp(+j\omega t)$, with the second term being the complex conjugate of the first one. Most physics and engineering books define the fields as we do (with a negative sign in front of the temporal phase term):

$$(\mathbf{E}(x, y, z, t), \mathbf{H}(x, y, z, t)) = Re\{(\mathbf{E}(x, y, z), \mathbf{H}(x, y, z)) \exp(-j\omega t)\}.$$

This means that the constitutive relations for a monochromatic wave, e.g. $\mathbf{D} = \varepsilon\mathbf{E}$, will determine the properties of the permittivity, because the $\mathbf{D}$ field may be written as

$$\mathbf{D} = \varepsilon(\omega)\mathbf{E}_0(\mathbf{r}) \exp(-j\omega t) + \varepsilon^*(-\omega)\mathbf{E}_0(\mathbf{r}) \exp(+j\omega t)$$

for real frequencies. This also means that when $\varepsilon(\omega)$ is complex, namely when it is written as $\varepsilon(\omega) = \varepsilon_R(\omega) + j\varepsilon_I(\omega)$, the imaginary part $\varepsilon_I(\omega)$ is always positive (and an odd function of frequency) and the real part may be either positive or negative (but an even function of frequency). This is also consistent with Beer–Lambert's law that absorption of the propagation intensity along the z direction $I(z)$ behaves as $I(z) = I_0 \exp(-\alpha z)$, with $\alpha = 2(\omega/c)Im\{n(\omega)\}$. With $n = (\varepsilon\mu)^{1/2}$ complex, one may write $n = (\varepsilon\mu)^{1/2} = \{[\varepsilon_R(\omega) + j\varepsilon_I(\omega)][\mu_R(\omega) + j\mu_I(\omega)]\}^{1/2}$ and extract the real and imaginary parts of the refractive index n. More on the constitutive relations in metamaterials may be found in ref. [37].

Dispersion: another look

At a given frequency ω_0, the dispersion relation may be approximated as $\boldsymbol{\beta}(\omega_0) = \boldsymbol{\beta}_0 + \delta\omega(\partial\boldsymbol{\beta}/\partial\omega)_{\beta_0}$ plus higher order terms. We deliberately keep the wavenumber (which is the absolute value of the wavevector) as $\boldsymbol{\beta} \cdot \boldsymbol{\beta} = \beta^2$. In this case, $\beta^2/\omega^2 = (\boldsymbol{\beta}_0 + \delta\omega(\partial\boldsymbol{\beta}/\partial\omega)_{\beta_0})^2/\omega^2 = (1/c_n)^2 + (\delta\omega/\omega^2)\boldsymbol{\beta}_0 \cdot (\partial\boldsymbol{\beta}/\partial\omega)_{\beta_0}$. The term in the second set of parentheses of this equation is the definition of the group velocity,

$(\mathbf{v}_g)^{-1} = (\partial\boldsymbol{\beta}/\partial\omega)$. This is the velocity of a band of frequencies (a pulse) that is $\delta\omega$ wide. Therefore, $\beta^2/\omega^2 = (1/c_n)^2 + 2(\delta\omega/\omega^2)(\boldsymbol{\beta}_0 \cdot \mathbf{v}_g)/(v_g)^2$. On the other hand, the ratio $\delta\omega/\omega = Q^{-1}$, where Q is the quality factor, i.e. the energy stored divided by the energy dissipated. If the material is constructed such that it is at resonance with the wave along one direction, yet the wave is free to propagate along other directions (e.g. a Bragg grating), the propagation is skewed. A wave oscillating along the dispersionless directions has $(v_g)^{-1} = (\partial\beta/\partial\omega) = \beta/\omega = c_n^{-1}$; in general, for a wave oscillating along the resonating direction, $(v_g)^{-1} = (\partial\beta_z/\partial\omega) \neq \beta_z/\omega = c_n^{-1}$. When coupling to such a Bragg reflector at an angle, the wave may be made to scatter back and forth within the structure, thus interfering with itself and resulting in a partial standing wave mode (resonance conditions). There will be a band of frequencies, $\delta\omega$ wide, satisfying such a condition. In such periodic structures there will be one mode per frequency band $\Delta\omega$ reflected, whereas the other remaining, say $N - 1$, modes within the entire frequency range are transmitted. The ratio $\omega/\delta\omega = Q$ defines the number of participating frequencies. Therefore, for high-Q resonators, the group velocity in the direction of the resonating structure would be very close to the phase velocity, whereas it will be very different along the other directions. Such spatial–temporal dispersion may skew the beam direction of propagation, and higher orders of diffraction need to be considered. It will also alter the expression for the energy flux, as we have already seen. The question in this case is, is there a physical meaning to the phase or even the group velocities in contrast to the physical entity of energy velocity? As for molecules coupled to a resonator, where the emission from the molecule is modified by the presence of a resonator, one could argue that the propagation will be modified as well [38].

To summarize, surface waves rely on large negative permittivity or permeability operators which exhibit dispersion and loss. The remaining question is whether one can compensate for these dispersions and losses with structured films exhibiting gain. As we have shown earlier, the largest interaction with a patterned film will be when we form standing surface waves; we will take advantage of these when assessing signal enhancement of Raman spectroscopy and surface plasmon lasers.

11.3 Raman spectroscopy with metamaterials

11.3.1 Fields and resonance effects (colloids and structured surfaces)

Raman spectroscopy is a widely used spectroscopic tool for characterizing vibration modes of molecules. This nonlinear process couples the pump light intensity to the vibrating molecule. The frequency of the scattered light is shifted by a vibration frequency, thereby providing direct information about the molecule. The coupling process is very weak, however. Surface enhanced Raman spectroscopy (SERS) is a

Figure 11.8. A large local field is predicted inside a dimer consisting of metal spheres.

modified version of Raman spectroscopy; the usually weak signal of a nonresonant, spontaneous Raman line is amplified via coupling of the Raman-active, optical phonons to localized electric fields. In a typical arrangement, the probed molecule is adsorbed on a rough metallic surface, such as silver, gold, or copper colloids [39]. Periodic structures are often utilized to couple the pumping laser light to surface charge waves (surface plasmons). In these structures, the lattice constants are typically one-half of either the exciting or the excited (scattered) optical wavelength [40]. With surface plasmons in mind, one turns to silver, gold, aluminum, and copper. Most metals are not compatible with bio-species, which is particularly true for silver, a widely used SERS metal. An aluminum surface is rapidly oxidized forming a nanoscale-thin but durable alumina film, which is bio-compatible [41]. An aluminum surface may be anodized to produce macroscopically homogeneous and hexagonally packed nanopores with tunable diameter and pitch. However, the oxide layer on aluminum was an impediment in the minds of many because it was thought that SERS requires a direct contact with the molecule under test.

The gain of the Raman process is proportional to the electric field intensity (namely, $\gamma \sim |E|^2$). This triggered an approach to create artificially "hot spots," namely, spots with large local fields. An example of a seemingly large field inside a dimer is shown in Fig. 11.8. In a nutshell, the external electric field induces charge separation in each metal sphere. The accumulated charges on the spheres' interfaces which face each other are responsible for the increased local field [42]. The problem is that the field is minimal directly between the spheres, and the molecule needs to be precisely positioned on either of the metal surfaces.

Other points of concern are as follows.

(a) Most spontaneous SERS data do not exhibit the line-shift which might be expected in such a high field environment.

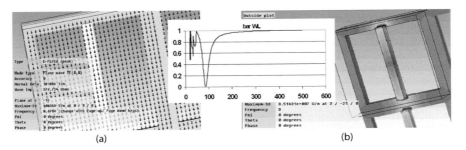

Figure 11.9. (a) Dipole structure where the incident beam (in the infra red in this case) is polarized parallel to the dipole. Note that the field is "hot" only at the very vicinity of the dipole edge (b). The resonance is relatively narrow.[1] Simulations were made with the Micro Suite program by CST. The transmission is plotted as a function of wavelength in micrometers.

(b) While some indications are that luminescence is also enhanced, it is not enhanced by the same factor as the Raman signal.

(c) No correlation has been found between Mie scattering and SERS.

(d) There is a substantial efficiency difference between gold and silver gratings, yet their optical conductivity values are not that far apart.

(e) While the Raman spectrum of a solution such as methanol is well known, there is no indication of its corresponding SERS effect. *This is particularly puzzling since, if correct, large local fields should always impact the molecules of the solution.*

The hot spot concept could be found in arrays of nano-holes [40, 43]. Using the Babinet theorem, we may investigate the complementary structure in which the holes are replaced by metal patches and the metallic regions are replaced by dielectric material. The latter structures may be viewed as patched antennas (one example would be a dipole antenna) or, more precisely, resonating structures. Such is the periodic dipole structure shown in Fig. 11.9. The exciting incident beam is parallel to the dipole direction. While these simulations are conducted at low-THz frequencies, they are applicable up to the region of surface plasmons' frequencies [31].

With the idea that one needs to amplify *both* the pump *and* the scattered frequencies, we extend this concept to metamaterials in the hope that these structures possess a wider resonance range. More precisely, one would like to analyze structures such as the horse shoe shown in Figs. 11.10 and 11.11. The resonance is indeed fairly wide, and some hot spots appear near the antenna's edges.

---

[1] Internal memo from J. Budd, R. Cevallos, D. Moeller, and H. Grebel.

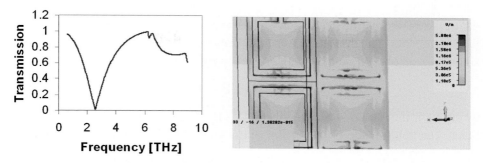

Figure 11.10. Transmission (left) showing the large absorption/reflection at resonance, and the field intensity plots (right). The polarization was set along the *y* direction. See color plates section.

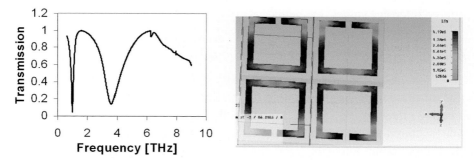

Figure 11.11. Transmission (left) and electric field thermal plots (right) when the polarization is along the *x* direction. The frequency of the peak has changed, and so has the field distribution. See color plates section.

From Figs. 11.10 and 11.11 one may conclude that the *y* polarization is slightly better than the *x* polarization in terms of generating extended hot spots near the relatively wide antenna sides. Thus, in principle, one may attempt to place molecules near, or at the gap between, the antenna loops. The connection to Raman spectroscopy will be apparent in the following. Another view on Raman amplification will be provided later on, which will help us understand the idea behind surface lasers.

Raman spectroscopy is a nonlinear process as is always the case when the frequency of the incident beam is shifted. The description involves the calculation of the material response (polarization) as a result of the incident pump beam. As shown below, the nonlinear polarization depends on the intensity of the pump (incident) beam. The constitutive relations may be written as $\mathbf{D} = \varepsilon_0 \mathbf{E} + \mathbf{P}$, where $\mathbf{P}$ is the material polarization or its response to the external field. The polarization may be assumed to be linearly divided between the linear part and the nonlinear part such that $\mathbf{P} = \mathbf{P}_L + \mathbf{P}_{NL}$. The linear part of the polarization is absorbed in the definition of the permittivity such that $\mathbf{D} = \varepsilon\mathbf{E} + \mathbf{P}_{NL}$. We are interested in the final term. The scattered nonlinear polarization, $\mathbf{P}_S$, that gives rise to the Raman signal as a result of a pump field $\mathbf{E}_P$ is given by [44]:

$$\mathbf{P}_S^{(3)} = 3\varepsilon_0\chi_R|\mathbf{E}_P|^2\mathbf{E}_S , \qquad (11.14)$$

where χ_R is the Raman susceptibility and $\mathbf{E}_S$ is the scattered (Stokes or anti-Stokes) field. The field scattered from a film of thickness t may be written directly as

$$\mathbf{E}_S(t) = \mathbf{E}_S(0)\exp[3j\omega_S\chi_R I_P(0)t/2\varepsilon_0 c^2 n_S n_P] , \qquad (11.15)$$

where $I_P(0)$ is the incident pump intensity, ω_S is the scattered radial frequency, and n_P and n_S are the refractive indices of the sample (structure and molecule) at the pump and scattered wavelengths, respectively. Linearizing Eq. (11.15) for small scattered intensity values, the signal-to-noise ratio (SNR) may be written as

$$\text{SNR} \approx \frac{I_S}{I_N} \approx 1 - \frac{Im\{\chi_R\}I_P(0)t}{\varepsilon_0 c^2 n_S n_P} . \qquad (11.16)$$

If we assume that the optical noise intensity I_N (originating mainly from coherent linear substrate scattering) is the seed for the Raman process, then $I_N \approx I_S(0)$. Note the strong effect on the scattered mode as the refractive index at either the pump or the scattered wavelengths (n_S and/or n_P) approaches zero. Also note that, for a given input pump intensity $I_P(0)$ and molecular susceptibility χ_R, the change in the SNR value, and thus in the enhancement factor, depends only on the effective refractive index.

Such were the results of a model for two plasmonic spheres [45]. By solving a non-linear dipole equation for the spheres with a vibrating molecule, one can show that the polarization of the complex behaves as $(1/n_{eff})^4$, with n_{eff} the real value of the effective refractive index of the complex. Such a refractive index will be affected mostly by the refractive index of the metallic spheres. We note that this value is rather small when we consider the real part of the refractive index of bulk silver or gold, as these are 0.067 and 0.16 at red laser light, respectively. Therefore, with two metallic spheres, the amplification factors, which are related to the impact of the refractive index, are given by $(1/n_S^2)^2 \sim 50\,000$ and 1500 for silver and gold, respectively. Other effects add to an overall amplification of nine orders of magnitude. The effective refractive index may be due to the material itself (such as in the case of two spheres made of silver or gold), or may be due to a resonating structure, such as a cavity or a periodic surface structure. Therefore, the above explains the superiority of silver over gold and postulates the advantage of metamaterials at the point where either the permittivity or the permeability approaches zero.

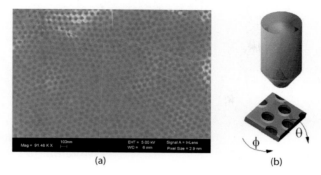

Figure 11.12. (a) Graphene-coated platform, made of anodized aluminum oxide (AAO). Hole diameter = 20 nm; pitch = 90 nm. (b) The experimental arrangement. A laser beam is focused on the substrate via a lens (objective). The Raman signal is measured in a confocal arrangement using a beam splitter, a spectrometer, and a cooled detector array.

To summarize, we are looking at a plasmonic periodic surface structure, preferably made of silver or gold, which, under certain conditions, will yield a standing wave pattern. The choice of a surface wave is because of the large interaction between the electric field and the molecule, as was shown in Fig. 11.4. The periodic structure provides for coupling and maintaining the standing wave, as explained in Section 11.2.5. The enhancement of the Raman signal with such platforms exhibiting standing waves may be experimentally validated by tilting the samples towards and away from the resonance conditions. Such procedures should produce variations in the Raman peak intensities [27]. We will expand on these momentum conservation concepts later on.

11.3.2 Examples (sensors, etc.)

The experimental set up is depicted in Fig. 11.12. Anodized aluminum oxide is our platform of choice: it has a hexagonal array of nano-holes and is bio-compatible. The structure may be coated with graphene – a monolayer of graphite – in order to sustain the molecule under test at the surface level of the platform, preferably above a hole. As mentioned earlier, the area above the hole exhibits a large density of the electric field intensity (Fig. 11.5). The platform is tilted (along the θ direction) and rotated (along the ϕ direction) such that the surface plasmon resonance condition is obtained.

Measured Raman signals of stilbene, a known dye molecule, as a function of tilt angle are shown in Fig. 11.13. We have chosen stilbene because its fluorescence wavelength is well in the blue region (420 nm), and interference to the Raman signals is thereby minimized. Peaks of signal amplification over the noise are

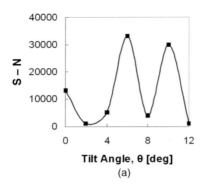
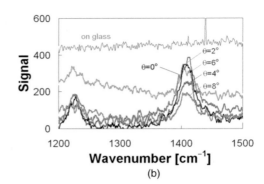

(a)   (b)

Figure 11.13. (a) Signal minus the noise as a function of tilt angle of the 1230 cm^{-1} Raman line of stilbene. The molecule was placed on a graphenated AAO substrate and the sample was rotated in-plane in order to obtain optimal conditions [46]. In this case, the first resonance occurs at 2°. (b) Raman spectra of 10^{-5} M streptavidin in water. The protein was placed in a microfluidic channel, which was resting on either glass or AAO substrates. The Raman signal is plotted as a function of tilt angle. The first resonance occurs for a tilt angle of 0° (normal incidence); it was affected by the presence of water as the top dielectric layer [47].

clearly demonstrated. The minimum at 8° is identified with the SP resonance for the pump wavelength (at 514.5 nm). The minimum at 12° is identified with the SP resonance for the scattered wavelength (at 548 nm). Details on the identification of angles are provided in Section 11.4, where, for example, Eq. (11.18) relates the angle of incidence to a particular wavelength.

11.4 Gain and feedback with structured metallo-dielectric surfaces

One could argue that the loss in metallic structures and, for that matter, the loss associated with metamaterials could be compensated for by gain [48]. Although Kramers–Kronig relations associate dispersion with loss and there would, therefore, be a limit to such compensation [49], one may nevertheless try to resolve the issue for a relatively narrow band of frequencies. With the concept of standing waves in mind, we are set to explore surface plasmon lasers. Lasers require a gain medium and a feedback mechanism, which forces light to scatter back and forth or repeat its path (as in ring lasers), thus maintaining a steady state solution. Such a process extends the interaction between the wave and matter, and oscillation occurs when the gain overcomes the overall loss [50]. Feedback may be provided by a Fabry–Perot etalon (two mirrors facing each other at a given distance), or by a periodic structure; the periodic structure is either external (distributed Bragg reflectors, DBR lasers), or exists as part of the gain medium itself (distributed feedback, DFB, lasers). In an external grating, the zero or first order of diffraction at almost grazing angles was efficient enough to obtain laser oscillations at a single frequency [51]. This concept has also been employed in semiconductor lasers, including quantum well lasers [52]. In a semiconductor planar construction the grating was integrated with the laser waveguide itself, often fabricated on the top surface. The feedback typically invokes high order harmonics of Bragg diffraction because of fabrication constraints. In an attempt to avoid large scattering losses, only the tail of the waveguide mode is made to experience Bragg scattering. This concept has been extended to the far-IR regime where the otherwise dielectric grating was replaced with a two-metal grating [53]. It is clear by now that, for a SP laser, the surface mode needs to interact intimately with a gain medium. At optical frequencies this creates a problem because of the large losses induced by the metallic free carriers. The solution was to provide high gain in a semiconductor structure encapsulated in a metallic cavity [54], or to separate a nano-semiconductor wire (the gain) by a thin insulating film from the metallic surface [55]. Such separation of gain from the metal surface was a key to the realization of the first SP laser in the visible frequency range [56], when it was recognized that the SP mode has its highest intensity at the separating layer interface (as we saw earlier in Fig. 11.4). The construction of that laser deviated substantially from the original proposal [57]. A demonstration of gain and line narrowing was achieved with metal spheres coated with a dielectric layer; the layer was impregnated with a dye material which provided the gain [58]. Finally, one may realize a sub-wavelength laser (in this case at far-IR wavelengths) by using strictly circuit concepts [59].

11.4.1 From local to extended feedback mechanisms

Feedback is provided by the highly scattering array of holes. Coupling to surface waves or surface plasmons (SPs) is most conveniently made using a periodic perturbation, obeying the momentum conservation form of

$$\mathbf{k}_{xy} + q\mathbf{G} = \boldsymbol{\beta}. \tag{11.17}$$

Here, $\boldsymbol{\beta}$ is the wavevector of the SP mode propagating along the surface; $\mathbf{k}_{xy}$ is the projection of the wavevector of the incident light on the surface, launched at an angle of incidence θ with respect to the surface normal (note that $|\mathbf{k}_{xy}| = |\mathbf{k}_0| \sin \theta$); and $\mathbf{G}$ is a reciprocal lattice vector of holes. The parameter q takes both negative and positive values. We note that, when $G \geq k$, only the TM mode (p polarization), with its oscillating electric field within the plane of incidence, may propagate along the surface. If a molecule is placed on the surface, signal enhancement occurs whenever the pump or the scattered wavelengths is in resonance with the periodic structure. This is true for Raman signals, as we have seen in the preceding section, or for photoluminescence signals, as we will see in the following. Unlike

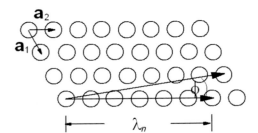

Figure 11.14. Symmetry of the hole-array. The hexagonal crystallography dictates 60° azimuthal symmetry (the cycle of ϕ upon in-plane rotations should be 60°). The effect of standing waves halves that value to 30°.

typical photonic crystal structures [60, 61], where the structure pitch is of the order of a propagating wavelength in the material, we are extending the resonance condition to sub-wavelength structures. This was also shown to be relevant for self-imaging properties of periodic structures [17]. The feedback to the SP waves may be provided by a periodic hole array with the caveat that the pitch is much smaller than the incident wavelength. This type of resonance extends over several hole planes, and hence the parameter q is sub-integer (fractional): for example, for an SP wavelength of 540 nm and hole-array pitch of $a = 90$ nm, resonance at normal incidence dictates $q = \pm 1/6$. Coupling to and from these surface modes is provided by the same periodic array of holes.

Let us consider an ideal 2-D hexagonal case (Fig. 11.14), and start by identifying two lattice vectors, $\mathbf{a}_1 = \hat{x}a$ and $\mathbf{a}_2 = \hat{x}a/2 + \hat{y}a\sqrt{3}/2$, where a is the distance between the nearest neighbor holes (Fig. 11.13). The primitive translation vectors of the reciprocal lattice are $\mathbf{G}_1 = \hat{x}2\pi/a - \hat{y}2\pi/a\sqrt{3}$ and $\mathbf{G}_2 = \hat{y}4\pi/a\sqrt{3}$. Light is coupled to the periodic array of holes by momentum conservation, as described by Eq. (11.17). Suppose that the surface component of the wave along the AAO sample is scattered from the same periodic structure. This may happen because the AAO is made of densely packed holes with an inter-hole spacing that is much smaller than the propagating wavelength. This also happens because of the abrupt nature of the hole in the AAO slab, invoking higher orders and sub-orders of diffraction. One can envision that such Bragg diffraction creates a standing wave within the hole-array. These considerations are true for the pump as well as for the fluorescing wavelengths.

The scattering of laser light from such a periodic structure is shown in Fig. 11.15. The scattering cycle upon in-plane rotation follows the 30° symmetry. The sample was tilted at a tilt angle $\theta = 8°$, the angle at which surface plasmon waves are launched.

The optimal launching conditions for a surface mode are achieved by a small tilt and in-plane rotation of the sample with respect to the incident beam due to

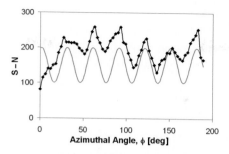

Figure 11.15. Laser line scattering (the 514.5 nm peak minus the noise floor) from an AAO substrate at a tilt angle $\theta = 8°$. The shifted sinusoidal curve, $A + B \sin^2(6\phi + \phi_0)$, accentuates the 30° symmetry.

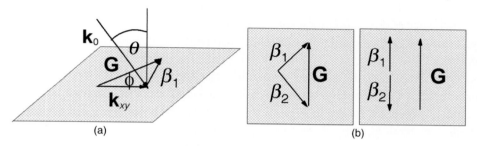

(a)   (b)

Figure 11.16. (a) Coupling into an SP wave: the projection of the incident wavevector $\mathbf{k}_{xy}$ is coupled to the surface plasmon wave by a reciprocal lattice vector $\mathbf{G}$. (b) Standing wave scenarios: $\boldsymbol{\beta}_1$ and $\boldsymbol{\beta}_2$ are two surface plasmon wavevectors that are coupled by another reciprocal lattice vector $\mathbf{G}$ such that $\boldsymbol{\beta}_1 - \boldsymbol{\beta}_2 = \mathbf{G}$.

the incomplete gap throughout the Brillouin zone for the hole-array, and may be computed through Bragg scattering. Following Fig. 11.16, the projection of $\mathbf{k}_0$ onto the array of holes is given by $\mathbf{k}_{xy}$. The coupling to an SP mode is made via momentum conservation: $\boldsymbol{\beta} = \mathbf{k}_{xy} \pm \mathbf{G}$. Therefore, $\beta^2 = |\boldsymbol{\beta}|^2 = |\mathbf{k}_{xy} \pm \mathbf{G}|^2$. The angle between the reciprocal lattice vector for the hole array $\mathbf{G}$ and $\mathbf{k}_{xy}$ is ϕ. We define $\beta/k_0 = n_{eff}$, $|\mathbf{k}_{xy}|/k_0 = \sin\theta$, and $G' = G/k_0$, and write an equation for θ as follows:

$$\sin^2\theta - 2G'\cos\phi\,\sin\theta + (G')^2 - n_{eff}^2 = 0.$$

The solution to this equation is given by $\sin\theta = G'\cos\phi \pm [n_{eff}^2 - (G')^2(1 - \cos^2\phi)]^{1/2}$. When the coupling is co-linear, $\phi = 0$, the tilt angle is given by

$$\sin\theta = \frac{\lambda_0}{a}\sqrt{\frac{4}{3}(q_1^2 - q_1 q_2 + q_2^2)} - n_{eff}. \qquad (11.18)$$

Here we have identified $\mathbf{G} = (q_1\mathbf{G}_1, q_2\mathbf{G}_2)$, where q_1, q_2 are the inverses of the numbers of planes used in the coupling process. Therefore, q is a sub-integer (namely, 1/integer). For example, Eq. (11.18) predicts $\theta \sim 5°$ and $\theta \sim 0°$ with $q_2 = 0$, $q_1 = 1/6$ for the pump wavelength at 514.5 nm. In the case of Raman scattering, the G Raman line of graphene (or a carbon nanotube) is 558 nm. Therefore, the 1600 cm^{-1} Stokes line of graphene corresponds to a +43 nm shift from the laser wavelength, or $\lambda = 558$ nm. The equation predicts $\theta \sim 4°$ for the 597 nm Stokes shift of ~ 2700 cm^{-1}, which is the Raman D' line used to identify the number of graphene layers. This is at resonance with $q_2 = 0$, $q_1 = 1/7$. In fact, one may find numerous resonances in the range between 5 and 12 degrees because of the densely packed hole-array structure. One of the strongest resonances is for a tilt angle $\theta = 8°$, as we have seen before in Fig. 11.12(a) with $q_2 = -1/10$, $q_1 = 1/10$ for the 514.5 nm laser line. For the SP laser, the idea is to find an angle of incidence θ that is common (or pretty close) to both the incident and scattered wavelengths, either incident and Stokes wavelengths, or pump and luminescence wavelengths.

After launching the surface wave, the standing wave conditions need to be established. The standing wave condition $|\boldsymbol{\beta} - q\mathbf{G}| = \beta$ has several scenarios, some of which are shown in Fig. 11.16(b). Due to the holes being closely packed there are many opportunities to couple into standing wave modes in such an environment. As stated before, at normal incidence we utilize the vector $\mathbf{G}$ for coupling to an SP mode and the vector $2\mathbf{G}$ for establishing a standing wave.

As we have seen earlier for suspended screens, there are two frequencies associated with a given launching angle θ: these are ω^- and ω^+, corresponding to SP propagating wave vectors $\boldsymbol{\beta}^-$ and $\boldsymbol{\beta}^+$, respectively. The difference frequency, $\delta\Omega = \omega^+ - \omega^-$, defines a frequency gap, as was shown in Fig. 11.6. The scattering process and the standing wave formation for each wavevector is known as band-edge coupling. Again note that these wavevectors, although differing by a reciprocal lattice vector, belong to two different frequencies. Each one will be associated with a standing surface wave. At normal incidence, $\theta = 0°$, the incident beam generates two counter-propagating surface waves, with $\beta = \pm qG$. The propagation constant of the surface mode is given by $\beta = k_0 n_{eff}$, with n_{eff} the effective index of the surface guide including the patterned interfacial layer. Waveguides that are interfaced with a thin dielectric layer (patterned or not) are known as loaded guides. Since the interfacial layer is very thin, $n_{eff} \sim 1$ due to the large negative permittivity of the aluminum and the small thickness of this oxide layer. Such a frequency–wavenumber plot is shown schematically in Fig. 11.17. The plot is folded at the Brillouin zone boundary, $G/2m$.

Momentum conservation for the surface plasmon polariton (SPP) ought to include all scattering processes. As mentioned earlier, the SP plasmon is

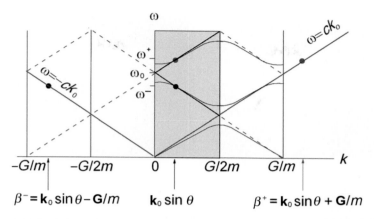

Figure 11.17. Schematic of the dispersion relations used. The folded Brillouin zone (shaded area) is scaled by m across the light lines, $\omega = \pm ck_0$. Each angle of incidence θ is associated with two frequencies ω^+ and ω^- and SP wave vectors β^+ and β^-, respectively. The same frequencies marked by dots belong to wave vectors separated by a reciprocal lattice vector G/m [56]. One could argue for a frequency gap at $\beta = 0$ because of the higher harmonics involved in the scatterings.

concentrated at the structure's holes at the interface away from the metal. We have used graphene to "hold" the molecules at these points. However, graphene has phonon modes which may propagate along the surface as well. Therefore, any scattering process needs to include all modes that affect the coupling to and from the sample. Consider the following momentum conservation condition with incident, scattered, and optical surface phonon modes all propagating along the nano-hole array:

$$(\beta_i + q_i \mathbf{G}) + (\beta_{sc} + q_{sc}\mathbf{G}) + (\mathbf{K} + q_K \mathbf{G}) = 0. \tag{11.19}$$

Here, β_i is the wavevector of the surface plasmon (SP) mode at the frequency of the incident beam; β_{sc} is the wavevector of the SP mode at the frequency of the scattered beam (Stokes or anti-Stokes processes as appropriate); $\mathbf{K}$ is the wavevector of a phonon propagating in the graphene layer; $\mathbf{G}$ is a reciprocal lattice vector; the various q values are the reciprocals of the number of sub-lattices involved (namely, if scattering occurs every six lattice constants apart, then, $q = 1/6$). Equation (11.19) sets the conditions for coupling between all scattered components involved.

Standing waves or resonance conditions for one or several components occur when $|\beta_i - q_i \mathbf{G}| = \beta_i$ or $|\beta_{sc} - q_{sc}\mathbf{G}| = \beta_{sc}$ or $|\mathbf{K} - q_K \mathbf{G}| = K$. The latter relationship requires a very small q_K factor; it may be observed, however, in graphenated *micron* size screens. In general, optimal coupling to a surface mode ensures the resonance condition.

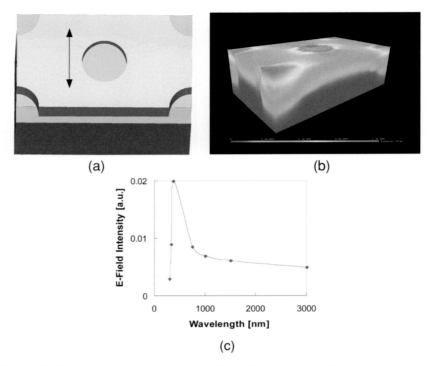

(a)   (b)

(c)

Figure 11.18. (a) The simulation cell: a 50 nm thick hexagonally perforated oxide is lying on top of aluminum. The hole radius is 40 nm and the refractive index of the oxide layer is 3.3. Graphene is deposited on top of the oxide layer. The arrow points to the direction of the incident TEM polarization state. (b) Linear electric field intensities ranging from 1.3×10^{-6} V/m to 0 V/m. (c) Field intensity at the hole center as a function of incident optical wavelength. See color plates section.

To summarize, we aim at coupling to a surface mode which forms a standing wave. By tilting and rotating the sample, one taps into two resonant conditions: one for the incident wavelength and one for the scattered wavelength. One aims at a condition that includes both resonances (with different hole-planes involved) having the same tilt angle. Alternatively, we may design a substrate such that both the incident frequency ω^{+} and the scattered frequency ω^{-} are coupled through the same periodic structure. The confocal arrangement dictates that the angle of incidence for one frequency coincides (as much as possible) with the angle of scattering for the other frequency.

11.4.2 Electric field distribution

We have used a commercial code (MicroStripes) to calculate the field distribution at the surface of the perforated substrates. The computational hexagonal structure is shown in Fig. 11.18(a). A perforated oxide is lying on top of aluminum. The

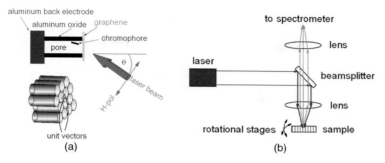

Figure 11.19. (a) Platform set up. (b) Experimental configuration.

oxide layer is coated with graphene, which for calculation purposes is treated as a lossy dielectric (a dielectric with some conduction). The incident TEM mode is at resonance with the structure.

From Fig. 11.18(b) we observe that the electric field is concentrated at the holes, just above the hole/air interface. This is similar to what was shown in Fig. 11.5 although with two distinctions: (1) the refractive index of the oxide was taken as 3.3 to examine its effect on the electric field distribution and (2) graphene (the gray layer in Fig. 11.18(a)) was added on top of the interfacial layer. A larger refractive index further concentrates the field into the hole. Figure 11.18(c) implies that the maximum field intensity is obtained for a wavelength of 370 nm, approximately four times the structure pitch. While the wavelength is somewhat affected by the presence of the oxide layer, the majority of the wave tail propagates in the cladding (air) above the oxide layer. This translates to $q_1 = q_2 = 1/4$ for $n_{eff} = 1.02$. This also means that the optimal "hot spot" conditions are achieved for this 90 nm pitched AAO substrate and may be useful for Raman spectroscopy with UV radiation.

11.4.3 Examples (enhanced fluorescence and SP lasers)

One could contemplate that fluorescence would also be enhanced (and the corresponding time constant shortened) in periodic or simply resonating structures [62–64] (with the caveat of how such amplification is calculated). Our own contribution is detailed below. The experimental system is shown in Fig. 11.19. This is a confocal arrangement similar to the one used in the Raman experiments. The sample was tilted and azimuthally rotated (in-plane rotations) with respect to the linearly polarized incident beam until optimal conditions for launching the SP waves were reached. Experimentally, we found that the best tilt angle for a substrate with a 90 nm pitch for fluorescence purposes was $\theta \sim 8°$, fairly close to the $\theta = 9°$ predicted by Eq. (11.18). This also corroborated previous results with Raman spectroscopy.

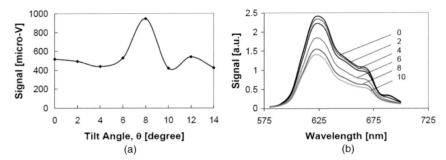

Figure 11.20. (a) The 560 nm fluorescence peak of fluorescein imbedded AAO (no graphene) as a function of tilt angle. The substrate pitch was 90 nm. The 560 nm peak is enhanced at $\theta = 8°$ (resonance for the pump wavelength at 514.5 nm) and $12°$ (resonance for the scattered wavelength at 560 nm) [66]. (b) Quantum dots on graphenated AAO. The substrate pitch was 120 nm. The 630 nm signal is maximized at a tilt angle of $\theta = 2°$, in agreement with the prediction of $\theta \sim 0° - 2°$; the related scattering sub-planes involved are $(q_1, q_2) = (-1/8, 1/8)$ and $(1/6, 0)$ for the pump and scattered waves, respectively.[2]

Examples with dye material and quantum dots as gain media are shown in Fig. 11.20 for varying platform pitch. The dye on AAO (without graphene) portrays two fluorescence peaks, $\theta = 8°$ (corresponding to the pump wavelength of 514.5 nm in p polarization mode) and $12°$ (corresponding to the scattered wavelength at 560 nm). The 630 nm signal from quantum dots on graphenated AAO with a 120 nm pitch peaks at $\theta = 2°$. In general, comparing intensity values (fluorescence, Raman, etc.) from various substrates suffers from uncertainty in the radiation coupling to and from the samples. By tilting the sample, we achieve more direct data at on- and off-resonance conditions, thus minimizing such uncertainty.

The fluorescence of a dye (fluorescein) as a function of wavelength is shown in Fig. 11.21 for samples pumped with an Ar ion laser at 514.5 nm. Upon increase of the intensity, either graphene bound or unbound samples exhibited a fluorescence peak shift from 560 nm towards 550 nm. This is due to the diminishing effect of longer fluorescing wavelengths by the confining periodic matrix. A careful examination of the fluorescence peaks as functions of the pump intensity reveals a 3 nm spectral line width, narrowing for graphene bound samples, compared with a 3 nm spectral line width, broadening for the unbound waveguide samples. The intensity curves exhibited some undulations, which are due to the relatively long lived excited mode of the fluorescein dye (up to ~20 s).

While encouraged by these results obtained with a continuous wave source, it seems that a pulsed optical source will be better suited for the task. Several

---

[2] Internal memo from S. Trivedi and H. Grebel.

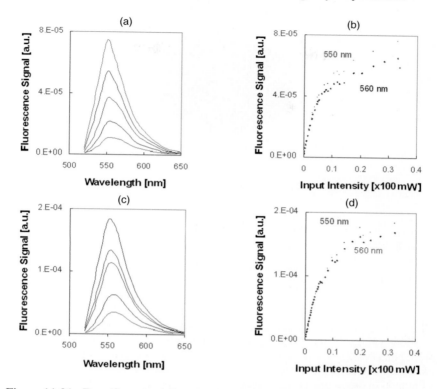

Figure 11.21. Dye (fluorescein) on AAO substrate, pumped with Ar laser: (a), (b) AAO with graphene and (c), (d) without it. (b) and (d) Fluorescence as a function of input intensity. The spectral line width narrowed by 3 nm for the graphene bound SP; the spectral line width broadened by 3 nm for the unbounded SP mode as the intensity of the pump laser increased.

mechanisms interfere with the gain process: long lived excitation modes, bleaching, and, for a gain medium composed of quantum dots, an Auger process as well (nonlinear induced transitions that are related to the free carriers in the dots). While we have demonstrated a dye SP laser [56], better results have been achieved with quantum dots as a gain medium [65]. The dots were dropped on graphenated AAO substrates, as shown in Fig. 11.22. The experimental set up is similar to the one shown in Fig. 11.18(b).

While excited by these results, one has to be careful not to confuse the threshold and line narrowing with ultimate lasing. Amplified stimulated emission exhibits line narrowing as well [50]. This is more severe for large volume, or long length, gain media such as suspended colloids coated with dye material, or fiber amplifiers doped with erbium ions. As the volume of the gain medium increases, each segment of the volume (length) acts as a source of radiation for the next one, and, while

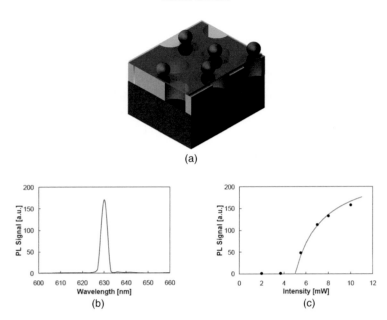

(a)

(b)   (c)

Figure 11.22. (a) Quantum dots on graphenated AAO substrates. (b) Photolumi-
nescence (PL) as a function of wavelength at pump intensity of 10 mW, 10 ns
pulse Nd:YAG laser. The curve is perfectly gaussian with a line width of 2.7 nm,
compared to a 12 nm line width at much lower power. (c) The 630 nm PL peak as
a function of pump intensity at the optimal tilt angle of $\theta = 16°$. The curve is a fit
to the data.

the process is not coherent (as in a laser), nevertheless larger amplification may
occur. We calculated the possibility of amplified spontaneous emission (ASE) in
our system and ruled it out: our lasers are simply too small to account for such a
process.

11.5 Concluding remarks

We have described surface polariton modes on various substrates and their possible
usefulness in detecting molecules as well as sources for coherent radiation (lasers).
In the course of their description, we had to re-affirm assumptions typically made
in connection with wave propagation. For example, isotropic materials are such
that the material features are considered much smaller than the propagating optical
wavelength within the material, $\lambda_n = \lambda/n \gg \zeta$, where ζ is some characteristic
length. One would also assume that the wave propagating in a medium is far
from any molecular resonance. For good metals, such as silver, the permittivity
constant is a large negative number, $\varepsilon \ll 0$, because of the fast collective response
of the electronic cloud. The same electronic cloud is responsible for the masking of

magnetic effects (diamagnetic properties). Therefore, one would imagine placing very small resonators within an otherwise isotropic material (e.g. artificial dielectric with small spheres or rods). In principle, resonances may be tailored such that the effective permittivity and permeability constants become negative, $\varepsilon, \mu < 0$. Despite the fact that periodic structures typically resonate at dimensions comparable to λ_n/n, those structures with a large extinction ratio (e.g. photonic crystals) may exhibit nontraditional refraction at a much smaller scale. As we have shown, coherent scattering from sub-wavelength periodic structures may be maintained even if the periodicity is $1/5$ or even $1/7$ of the propagating wavelength. In fact, the optimal scattering for such patterns is a pitch-to-wavelength ratio of $1:4$. These spatial–temporal dispersive structures may not be considered as homogeneous in the first place.

The quadratic wave equation may not distinguish between propagation in homogeneous negative and positive refractive media. It is when we analyze the energy flux and the energy density (which is also the change in the overall free-energy of the system) that we realize that homogenous negative refractive index media are thermodynamically unstable. One may argue, however, that, by turning to surface modes, a stable propagation system may be maintained.

Finally, while the future of negative index materials is unknown, one thing is for sure: such materials have triggered a re-examination of wave propagation concepts that we have taken for granted.

References

[1] J. Zenneck, *Elektromagnetische Schwingungen und drahtlose Telegraphie* (Stuttgart: F. Enke, 1905); trans. *Electromagnetic Oscillations and Wireless Telegraphy.*

[2] J. C. Maxwell Garnett, "Colours in metal glasses, in metallic films, and in metallic solutions. II," *Phil. Trans. Roy. Soc. Lond. A* **205**, 237–288 (1906).

[3] B. A. Munk, *Frequency Selective Surfaces* (New York: John Wiley & Sons, Inc., 2000).

[4] J. Takahara, S. Yamagishi, H. Taki, A. Morimoto, and T. Kobayashi, "Guiding of a one-dimensional optical beam with nanometer diameter," *Opt. Lett.* **22**, 475–477 (1997).

[5] G. Veronis and S. Fan, "Guided subwavelength plasmonic mode supported by a slot in a thin metal film," *Opt. Lett.* **30**, 3359–3361 (2005).

[6] A. W. Sanders, D. A. Routenberg, B. J. Wiley, Y. Xia, E. R. Dufresne, and M. A. Reed, "Observation of plasmon propagation, redirection, and fan-out in silver nanowires," *Nano Lett.* **6**, 1822–1826 (2006).

[7] W. Wang, T. Lee, and M. A. Reed, "Mechanism of electron conduction in self-assembled alkanethiol monolayer devices," *Phys. Rev. B* **68**, 035416(1-7) (2003).

[8] Z. Jacob, L. V. Alekseyev, and E. Narimanov, "Optical hyperlens: far-field imaging beyond the diffraction limit," *Opt. Express* **14**, 8247–8256 (2006).

[9] B. A. Munk, *Metamaterials: Critique and Alternatives* (Hoboken, NJ: John Wiley, 2009).

[10] S. C. Wu and H. Grebel, "Phase shifts in coplanar waveguides with patterned conductive top cover," *J. Phys. D: Appl. Phys.* **28**, 437–439, (1995).

[11] W. E. Kock, "Metallic delay lenses," *Bell Syst. Tech. J.* **27**, 58–82, (1948).

[12] R. E. Collin, *Field Theory of Guided Waves*, 2nd edn, IEEE Press Series on Electromagnetic Wave Theory, ed. D. G. Dudley (New York: John Wiley & Sons, Inc., 1990).

[13] J. C. Bose, "On the rotation of plane of polarization of electric waves by a twisted structure," *Proc. Roy. Soc. A* **63**, 146–152 (1898).

[14] A. Lakhtakia, V. V. Varadan, and V. K. Varadan, "A parametric study of microwave reflection characteristics of planar achiral-chiral interface," *IEEE Trans. Electromag. Compatibility* **EMC-28**, 90–95 (1986).

[15] M. Chien, Y. Kim, and H. Grebel, "Mode conversion in optically active and isotropic waveguides," Opt. Lett. **14**, 826–828 (1989).

[16] L. D. Landau, E. M. Lifshitz, and L. P. Pitaevskii, *Electrodynamics of Continuous Media*, 2nd edn (Oxford: Pergamon Press, 1984).

[17] J. M. Tobias and H. Grebel, "Self-imaging in photonic crystals in a sub-wavelength range," Opt. Lett. **24**, 1660-1662 (1999).

[18] H. F. Talbot, "Facts relating to optical science, No. IV," *Phil. Mag.* **9**, 401–407 (1836).

[19] Lord Rayleigh, "On copying diffraction gratings, and on some phenomena connected therewith," *Phil. Mag.* **11**, 196–205 (1881).

[20] V. G. Veselago, "The electrodynamics of substances with simultaneously negative values of permittivity and permeability," *Sov. Phys. Usp.* **10**, 509–514 (1968).

[21] R. Ruppin, "Surface polaritons of a left-handed material slab," *J. Phys. Condens. Mat.* **13**, 1811–1819 (2001).

[22] I. V. Shadrivov, A. A. Sukhorukov, and Y. S. Kivshar, "Guided modes in negative-refractive-index waveguides," *Phys. Rev. E* **67**, 057602(1-4) (2003).

[23] R. Benard, "Total reflection: a new evolution of the Goos–Hänchen shift," *J. Opt. Soc. Am.* **54**, 1190–1197 (1964).

[24] M. Cheng, Y. Zhou, Y. Li, and X. Li, "Large positive and negative generalized Goos–Hänchen shifts from a double negative metamaterial slab backed by a metal," *J. Opt. Soc. Am. B* **25**, 773–776 (2008).

[25] M. Born and E. Wolf, *Principles of Optics,* 6th edn (New York: Pergamon Press, 1984).

[26] R. Ruppin, "Surface polaritons of a left-handed medium," *Phys. Lett. A* **277**, 61–64 (2000).

[27] C. Zhang, K. Abdijalilov, and H. Grebel, "Surface enhanced Raman with anodized aluminum oxide films," *J. Chem. Phys.* **127**, 044701(1-5) (2007).

[28] G. M. Ressler and K. D. Möller, "Far infrared bandpass filters and measurements on a reciprocal grid," *Appl. Opt.* **6**, 893–896 (1967).

[29] R. Ulrich, K. F. Renk, and L. Genzel, "Tunable sub-millimeter interferometers of the Fabry-Perot type," *IEEE Trans. Micro. Th. Tech.* **MTT-11**, 363–371 (1963).

[30] M. E. Stewart, N. Mack, V. Malyarchuk, J. A. N. T. Soares, T.-W. Lee, S. K. Gray, R. G. Nuzzo, and J. A. Rogers, "Quantitative multispectral biosensing and 1D imaging using quasi-3D plasmonic crystals," *Proc. Natl Acad. Sci.* **103**, 17143–17148 (2006).

[31] A. Banerjee, D. Moeller, A. I. Smirnov, and H. Grebel, "Graphenated IR screens," *IEEE Sensors J.* **10**, 419–422 (2010).

[32] O. Sternberg, K. P. Stewart, Y. Hor *et al.* "Square-shaped metal screens in the IR to THz spectral region: resonance frequency, band gap and bandpass filter characteristics," *J. Appl. Phys.* **104**, 023103(1-7) (2008).

[33] T. W. Ebbesen, H. J. Lezec, H. F. Ghaemi, T. Thio, and P. A. Wolff, "Extraordinary optical transmission through sub-wavelength hole arrays," *Nature* **391**, 667–669 (1998).

[34] A. Banerjee, D. Sliwinski, K. P. Stewart, K. D. Möller, and H. Grebel, "Curved infrared screens," *Opt. Lett.* **35**, 1635–1637 (2010).

[35] W. L. Bragg, "The diffraction of short electromagnetic waves by a crystal," *Proc. Camb. Phil. Soc.* **17**, 43–57 (1913).

[36] A. Reza, M. M. Dignam, and S. Hughes, "Can light be stopped in realistic metamaterials?" *Nature* **455**, E10–E11 (2008).

[37] A. Ishimaru, S.-W. Lee, Y. Kuga, and V. Jandhyala, "Generalized constitutive relations for metamaterials based on the quasi-static Lorentz theory," *IEEE Trans. Antenn. Propag.* **AP-51**, 2550–2557 (2003).

[38] D. Kleppner, "Inhibited spontaneous emission," *Phys. Rev. Lett.* **47**, 233–235 (1981).

[39] M. Fleischmann, P. J. Hendra, and A. J. McQuillan, "Raman spectra of pyridine adsorbed at a silver electrode," *Chem. Phys. Lett.* **26**, 163–166 (1974).

[40] H. Grebel, Z. Iqbal, and A. Lan, "Detecting single wall nanotubes with surface enhanced Raman scattering from metal coated periodic structures," *Chem. Phys. Lett.* **348**, 203–208 (2001).

[41] R. Langer and J. P. Vacanti, "Tissue engineering," *Science* **260**, 920-926 (1993).

[42] H. Xu, J. Aizourua, M. Käll, and P. Apell, "Electromagnetic contributions to single-molecule sensitivity in surface-enhanced Raman scattering," *Phys. Rev. E* **62**, 4318-4324 (2000).

[43] A. G. Brolo, E. Arctander, R. Gordon, B. Leathem, and K. L. Kavanagh, "Nanohole-enhanced Raman scattering," *Nano Lett.* **4**, 2015–2018 (2004).

[44] P. N. Butcher and D. Cotter, *The Elements of Nonlinear Optics* (New York: Cambridge University Press, 1991).

[45] H. Grebel, "Surface enhanced Raman scattering – phenomenological approach," *J. Opt. Soc. Am. B* **21**, 429–435 (2004).

[46] A. Banerjee, R. Li, and H. Grebel, "Raman spectroscopy with graphenated anodized aluminum oxide substrates," *Nanotechnol.* **20**, 295502(1-7) (2009).

[47] A. Banerjee, R. Perez-Castillejos, D. Hahn, A. Smirnov, and H. Grebel, "Microfluidic channels on nanopatterned substrates: monitoring protein binding to lipid bilayers with surface-enhanced Raman spectroscopy," *Chem. Phys. Lett.* **489**, 121–126 (2010).

[48] M. A. Noginov, V. A. Podolskiy, G. Zhu, M. Mayy, M. Bahoura, J. A. Adegoke, B. A. Ritzo, and K. Reynolds, "Compensation of loss in propagating surface plasmon polariton by gain in adjacent dielectric medium," *Opt. Express* **16** 1385–1392 (2008).

[49] M. I. Stockman, "Criterion for negative refraction with low losses from a fundamental principle of causality," *Phys. Rev. Lett.* **98**, 177404(1-4) (2007).

[50] J. T. Verdeyen, *Laser Electronics*, 3rd edn (Englewood Cliffs, NJ: Prentice Hall, 1995).

[51] M. G. Littman and H. J. Metcalf, "Spectrally narrow pulsed dye laser without beam expander," *Appl. Opt.* **17**, 2224–2227 (1978).

[52] H. Kogelnik and C. V. Shank, "Stimulated emission in a periodic structure," *Appl. Phys. Lett.* **18**, 152–154 (1971).

[53] A. Tredicucci, C. Gmachl, F. Capasso, A. L. Hutchinson, D. L. Sivco, and A. Y. Cho, "Single-mode surface-plasmon laser," *Appl. Phys. Lett.* **76**, 2164–2166 (2000).

[54] M. T. Hill, Y.-S. Oei, B. Smalbrugge *et al.*, "Lasing in metallic-coated nanocavities," *Nature Photon.* **1**, 589–594 (2007).

[55] R. F. Oulton, V. J. Sorger, T. Zentgraf, R.-M. Ma, C. Gladden, L. Dai, G. Bartal, and X. Zhang, "Plasmon lasers at deep subwavelength scale," *Nature* **461**, 629–632 (2009).

[56] R. Li, A. Banerjee, and H. Grebel, "The possibility for surface plasmon lasers," *Opt. Express* **17**, 1622–1627 (2009).

[57] D. J. Bergman and M. I. Stockman, "Surface plasmon amplification by stimulated emission of radiation: quantum generation of coherent surface plasmons in nanosystems," *Phys. Rev. Lett.* **90**, 027402(1-4), (2003).

[58] M. A. Noginov, G. Zhu, A. M. Belgrave *et al.*, "Demonstration of a spaser-based nanolaser," *Nature* **460** 1110–1112 (2009).

[59] C. Walther, G. Scalari, M. I. Amanti, M. Beck, and J. Faist, "Microcavity laser oscillating in a circuit-based resonator," *Science* **327**, 1495–1497 (2010).

[60] K. Sakoda, *Optical Properties of Photonic Crystals* (New York: Springer, 2005).

[61] M. Notomi, "Theory of light propagation in strongly modulated photonic crystals: refractionlike behavior in the vicinity of the photonic band gap," *Phys. Rev. B* **62**, 10696–10705 (2000).

[62] Y. Liu and S. Blair, "Fluorescence enhancement from an array of subwavelength metal apertures," *Opt. Lett.* **28**, 507–509 (2003).

[63] S. Vukovic, S. Corni, and B. Mennucci, "Fluorescence enhancement of chromophores close to metal nanoparticles. Optimal setup revealed by the polarizable continuum model, " *J. Phys. Chem. C* **113**, 121–133 (2009).

[64] S.-H. Guo, J. J. Heetderks, H.-C. Kan, and R. J. Phaneuf, "Enhanced fluorescence and near-field intensity for Ag nanowire/nanocolumn arrays: evidence for the role of surface plasmon standing waves," *Opt. Express* **16**, 18417–18425 (2008).

[65] A. Banerjee, R. Li, and H. Grebel, "Surface plasmon lasers with quantum dots as gain mediaâ" Appl. Phys. Lett. **95**, 251106(1-3) (2009).

[66] R. Li and H. Grebel, "Surface enhanced fluorescence," *IEEE Sensors J.* **10**, 465–468 (2010).

Index

Printed in the United States
by Baker & Taylor Publisher Services